Bíos

CARY WOLFE, SERIES EDITOR

5 *Dorsality: Thinking Back through Technology and Politics*
David Wills

4 *Bíos: Biopolitics and Philosophy*
Roberto Esposito

3 *When Species Meet*
Donna J. Haraway

2 *The Poetics of DNA*
Judith Roof

1 *The Parasite*
Michel Serres

Bíos

Biopolitics and Philosophy

Roberto Esposito

Translated and with an Introduction by Timothy Campbell

posthumanities 4

University of Minnesota Press
Minneapolis
London

The University of Minnesota Press gratefully acknowledges the assistance provided for the publication of this book by the McKnight Foundation.

Published by the University of Minnesota Press
111 Third Avenue South, Suite 290
Minneapolis, MN 55401-2520
http://www.upress.umn.edu

Library of Congress Cataloging-in-Publication Data

Esposito, Roberto, 1950–
[Bíos. English.]
Bíos : biopolitics and philosophy / Roberto Esposito ; translated and with an introduction by Timothy Campbell.
p. cm. — (Posthumanities series ; v. 4)
Originally published: Bios: Biopolitica e filosofia.
Includes bibliographical references and index.
ISBN-13: 978-0-8166-4989-1 (hc : alk. paper)
ISBN-10: 0-8166-4989-8 (hc : alk. paper)
ISBN-13: 978-0-8166-4990-7 (pb : alk. paper)
ISBN-10: 0-8166-4990-1 (pb : alk. paper)
1. Biopolitics. 2. Political science—Philosophy. I. Title.
JA80.E7713 2008
320.01—dc22 2007045837

Printed in the United States of America on acid-free paper

15 14 13 12 10 9 8 7 6 5 4

Contents

TRANSLATOR'S INTRODUCTION

Bíos, Immunity, Life

The Thought of Roberto Esposito

Timothy Campbell

The name of Roberto Esposito is largely unknown in the United States. Outside of a few Romance Studies departments who know him primarily for *Communitas: The Origin and Destiny of the Community*, the work of this Italian philosopher over the past twenty-five years remains completely untranslated into English.[1] That his introduction to an American audience should occur now and concern his most recent study, *Bíos: Biopolitics and Philosophy*, is owing in no small part to the particular (bio)political situation in which we find ourselves today: the ever-increasing concern of power with the life biology of its subjects, be it American businesses urging, indeed forcing, workers to be more active physically so as to save on health care costs, or the American government's attempts in the "war on terror" to expose the lives of foreign nationals to death, "fighting them there" so as to "protect" American lives here.[2] Yet this politicization of biology, the biopolitics that forms the object of Esposito's study, has a long and terrible history in the twentieth century. Indeed, *Bíos* may be profitably read as nothing short of a modern genealogy of biopolitics that begins and ends in philosophy.

In the following pages, I will sketch the parameters of this genealogy and Esposito's contribution to our current understanding of biopolitics, particularly as it relates to the conceptual centerpiece of *Bíos*, what Esposito calls the "paradigm of immunization." Immunity has a long and well-known history in recent critical thought. Niklas Luhmann placed immunity at the heart of his systems theory in his 1984 opus *Soziale Systeme;* Donna Haraway deployed "an immune system discourse" in her seminal reading of postmodern bodies from 1988; Jean Baudrillard in the early 1990s spoke of artificial sterilization compensating for "faltering internal immunological defenses."[3] For them and for many writing today on immunity, the term quickly folds into autoimmunity, becoming the ultimate horizon in which contemporary politics inscribes itself. Others continued to discuss immunity throughout the 1990s—Agnes Heller most prominently—as well as Mark C. Taylor, but no one placed it more forcefully at the center of contemporary politics than did Jacques Derrida in a series of interviews and writings after the "events" of September 11.[4] Speaking of autoimmunity aggression and suicidal autoimmunity, Derrida affiliates the figure of immunity with trauma and a repetition compulsion.[5] As the reader will soon discover, much sets apart Esposito's use of immunity from Derrida's, as well as the others just mentioned, especially as it relates to Esposito's radical inversion of immunity in its communal antinomy and the subsequent effects on our understanding of biopolitics. In the first section, therefore, I attempt to trace where Esposito's use of the immunity paradigm converges and diverges with Derrida and others.

In the second part, I situate *Bíos* more broadly within current American and European thinking on biopolitics. Here obviously the work of Michel Foucault in his seminars from 1975 and 1976 on biopolitics and racism merits considerable attention for it is precisely on these discourses that Esposito will draw his own reflections in *Bíos*.[6] But as anyone who has followed the recent fortunes of the term "biopolitics" knows, two other figures dominate contemporary discussions of life in all its forms and they both originate in Italy: Giorgio Agamben and Antonio Negri. In *Homo Sacer, Remnants of Auschwitz,* and *The Open,* Giorgio Agamben declines biopolitics negatively, anchoring it to the sovereign state of exception that separates bare life *(zōē)* from political forms of life *(bíos)*.[7] For Antonio Negri, writing with Michael Hardt, biopolitics takes on a distinctly positive tonality when thought together with the multitude.[8] It is between these two contradictory poles that Esposito's focus on *bíos* must be understood. Indeed, as I argue here, *Bíos* comes to resemble something like a synthesis of both Agamben's and Negri's positions, with Esposito co-opting Agamben's negative analysis of biopolitics early on, only to criticize later the antihistorical moves that characterize Agamben's association of biopolitics to the state of exception.[9] In some of *Bíos*'s most compelling pages, Esposito argues instead for the modern origin of biopolitics in the immunizing features of sovereignty, property, and liberty as they emerge in the writings of Hobbes and Locke. It is at this point that the differences with Hardt and Negri become clear;

they concern not only what Esposito argues is their misguided appropriation of the term "biopolitics" from Foucault, but also their failure to register the thanatopolitical declension of twentieth-century biopolitics. Essentially, Esposito argues that Hardt and Negri aren't wrong in pushing for an affirmative biopolitics—a project that Esposito himself shares—but that it can emerge only after a thoroughgoing deconstruction of the intersection of biology and politics that originates in immunity.

Clearly, understanding Italian contributions to biopolitical discourse is crucial if we are to register the originality of Esposito's argument. Equally, though, other critical texts will also help us in situating *Bíos* within contemporary work on biopolitics—Judith Butler's reflections on mourning and community in *Precarious Life* and *Giving an Account Oneself* come to mind, as do Keith Ansell Pearson's Deleuzian musings on symbiosis and viroid life, as well as Jürgen Habermas's recent *The Future of Human Nature* and Ronald Dworkin's essays on euthanasia and abortion.[10] Here too Esposito's work shares a number of areas of contact with them, ranging from the notion of community to the genetic engineering that promises to prevent "lives unworthy of life" in Binding and Hoche's phrase.[11] But other texts figure as well, especially as they relate to Esposito's reading of community/immunity. I will introduce them at appropriate moments and then in my conclusion tie up some of the loose ends that inevitably result when broad introductions of the sort I am attempting are made. Most important will be asking after the use value of *bíos* for imagining a public culture no longer inscribed in a negative horizon of biopolitics.

Community/Immunity

In order to appreciate the originality of Esposito's understanding of biopolitics, I first want to rehearse the relation of community to immunity as Esposito sketches it, not only in *Bíos* but in his two earlier works, *Communitas: Origin and Destiny of the Community* and *Immunitas: The Protection and Negation of Life.*[12] Reading the terms dialectically, Esposito asks if the relation between community and immunity is ultimately one of contrast and juxtaposition, or rather if the relation isn't part of a larger move in which each term is inscribed reciprocally in the logic of the other. The launching pad for his reflections concerns the principles on which communities are founded. Typically, of course, when we think of community, we immediately think of the common, of that which is shared among the members of a group. So too for Esposito: community is inhabited by the communal, by that which is not my own, indeed that begins where "my own" ends. It is what belongs to all or most and is therefore "public in juxtaposition to 'private,' or 'general' (but also 'collective') in contrast to particular."[13] Yet Esposito notes three further meanings of *communitas,* all associated with the term from which it originates: the Latin *munus.* The first two meanings of *munus*—*onus* and *officium*—concern obligation and office, while the third centers paradoxically on the term *donum,* which Esposito glosses as a form of gift that combines the features of the previous two. Drawing on the classic linguistic studies of Benveniste and Mauss, Esposito marks the specific tonality of this communal *donum,* to signify not simply any gift but a category of gift that requires, even demands, an exchange in return.[14] "Once one has accepted the *munus,*" Esposito writes, then "one is obliged to return the *onus,* in the form of either goods or services *[officium].*"[15] *Munus* is, therefore, a much more intense form of *donum* because it requires a subsequent response from the receiver.

At this point, Esposito can distill the political connotations of *munus.* Unlike *donum, munus* subsequently marks "the gift that one gives, not the gift that one receives," "the contractual obligation one has vis-à-vis the other," and finally "the gratitude that *demands* new donations" on the part of the recipient (emphasis in original).[16] Here Esposito's particular declension of community becomes clear: thinking community through *communitas* will name the gift that keeps on giving, a reciprocity in the giving of a gift that doesn't, indeed cannot, belong to oneself. At its (missing) origin, *communitas* is constructed around an absent gift, one that members of community cannot keep for themselves. According to Esposito, this debt or obligation of gift giving operates as a kind of originary defect for all those belonging to a community. The defect revolves around the pernicious effects of reciprocal donation on individual identity. Accepting the *munus* directly undermines the capacity of the individual to identify himself or herself as such and not as part of the community.

I want to hold the defective features of *communitas* in reserve for the moment and reintroduce the question of immunity because it is precisely the immunitary mechanism that will link community to biopolitics.[17] For Esposito, immunity is coterminus with community. It does not simply negate *communitas* by protecting it from what is external, but rather is inscribed in the horizon of the communal *munus.* Immune is he—and immunity is clearly gendered as masculine in the examples from classical Rome that Esposito cites—who is exonerated or has received a *dispensatio* from reciprocal gift giving. He who has been freed from communal obligations or who enjoys an originary autonomy or successive freeing from a previously contracted debt enjoys the condition of *immunitas*. The relationship immunity maintains with individual identity emerges clearly here. Immunity connotes the means by which the individual is defended from the "expropriative effects" of the community, protecting the one who carries it from the risk of contact with those who do not (the risk being precisely the loss of individual identity).[18] As a result, the borders separating what is one's own from the communal are reinstituted when the "substitution of private or individualistic models for communitarian forms of organization" takes place.[19] It follows therefore that the condition of immunity signifies both not to be and not to have in common.[20] Seen from this perspective, immunity presupposes community but also negates it, so that rather than centered simply on reciprocity, community doubles upon itself, protecting itself from a presupposed excess of communal gift giving. For Esposito, the conclusion can only be that "to survive, the community, every community, is forced to introject the negativity of its own opposite, even if the opposite remains precisely a lacking and contrastive mode of being of the community itself."[21] It is this introjection of negativity or immunity that will form the basis of Esposito's reading of modern biopolitics. Esposito will argue that the modern subject who enjoys civil and political rights is itself an attempt to attain immunity from the contagion of the possibility of community. Such an attempt to immunize the individual from what is common ends up putting at risk the community as immunity turns upon itself and its constituent element.

Immunity and Modernity

Those familiar with Jean-Luc Nancy's writings on the inoperative community or Alphonso Lingis's reflections on the shared nothingness of community will surely hear echoes of both in much of the preceding synopsis.[22] What sets Esposito's analysis apart from them is the degree to which he reads immunity as a historical category inextricably linked to modernity:

> That politics has always in some way been preoccupied with defending life doesn't detract from the fact that beginning from a certain moment that coincides exactly with the origins of modernity, such a self-defensive requirement was identified not only and simply as a given, but as both a problem and a strategic option. By this it is understood that all civilizations past and present faced (and in some way solved) the needs of their own

> immunization, but that it is only in the modern ones that immunization constitutes its most intimate essence. One might come to affirm that it wasn't modernity that raised the question of the self-preservation of life, but that self-preservation is itself raised in modernity's being *[essere]*, which is to say it invents modernity as a historical and categorical apparatus able to cope with it.[23]

For Esposito, modernity doesn't begin simply in the institution of sovereign power and its theorization in Hobbes, as Foucault argues. Rather, modernity appears precisely when it becomes possible to theorize a relation between the communitarian *munus*, which Esposito associates with a Hobbesian state of generalized conflict, and the institution of sovereign power that acts to protect, or better to immunize, the community from a threatened return to conflict.

If we were to push Esposito's argument, it might be more appropriate to speak of the sovereign who immunizes the community from the community's own implicit excesses: the desire to acquire the goods of another, and the violence implicated in such a relation. When its individual members become subject to sovereign power, that is, when it is no longer possible to accept the numerous threats the community poses to itself and to its individual members, the community immunizes itself by instituting sovereign power. With the risk of conflict inscribed at the very heart of community, consisting as it does in interaction, or perhaps better, in the equality between its members, immunization doesn't precede or follow the moment of community but appears simultaneously as its "intimate essence." The moment when the immunitary aporia of community is recognized as the strategic problem for nascent European nation-states signals the advent of modernity because it is then that sovereign power is linked theoretically to communal self-preservation and self-negation.[24]

Two further reflections ought to be made at this point. First, by focusing on the immunizing features of sovereignty as it emerges in modernity, Esposito takes issue with a distinction Foucault makes between the paradigm of sovereignty and that of governmentality. For Foucault, governmentality marks the "tactics of government which make possible the continual definition and redefinition of what is within the competence of the State and what is not, the public versus the private, and so on." These tactics are linked to the emergence of the population as an objective of power that culminates at the end of the eighteenth century, particularly regarding campaigns to reduce mortality.[25] A full-fledged regime of governmentality for Foucault cannot be thought separately from the emergence of biopower that takes control of "life in general—with the body as one pole and the population as the other" in the nineteenth century.[26] Esposito, however, shows how Foucault oscillates between sovereignty and governmentality precisely because of his failure to theorize the immunitary declension of both terms. Both are inscribed in a modern biopolitical horizon thanks to a modernity that strengthens exponentially its own immunitary characteristics.

Second, Esposito's focus on immunity ought to be compared to recent attempts, most notably by Judith Butler, to construct a conceptual language for describing gender and sexuality as modes of relation, one that would "provide a way of thinking about how we are not only constituted by our relations but also dispossessed by them as well."[27] Esposito's language of an always already immunized and immunizing *munus* suggests that Butler is clearly right in affirming the importance of relationality for imagining community, but at the same time that any hoped-for future community constructed on "the social vulnerability of bodies" will founder on the implicit threat contained in any relation among the same socially constituted bodies.[28] In other words, an ecology of socially interdependent bodies doesn't necessarily ensure vulnerability, but might actually augment calls for protection. Thus the frequent suggestion of immunity in Butler whenever the body appears in all its vulnerability or the threat of contagion symbolically produced by the presumed enemy.[29] For his part, Esposito is attempting something different: the articulation of a political semantics that can lead to a nonimmunized (or radically communitized) life.[30]

Autoimmunity after September 11

Yet Esposito's diagnosis of the present biopolitical scene doesn't rest exclusively on reading the antinomies of community in immunity or, for that matter, on the modern roots of immunization in the institution of sovereignty. In *Bíos* and *Immunitas,* Esposito sketches the outlines of a global autoimmunity crisis that grows more dangerous and lethal by the day. The reason, Esposito argues, has primarily to do with our continuing failure to appreciate how much of our current political crisis is the result of a collective failure to interrogate the immunitary logic associated with modern political thought. In somewhat similar fashion, Jacques Derrida also urged forward an autoimmunity diagnosis of the current political moment, beginning in his writings on religion with Gianni Vattimo, then in *The Politics of Friendship,* and most famously in his interviews in the aftermath of September 11.

I want to summarize briefly how Derrida conjoins politics to autoimmunity so as to distinguish Esposito's own use of the term from Derrida's. Setting out their differences is a necessary step to understanding more fully the contemporary formation of power and what strategies are available to resolve the current moment of political autoimmunity crisis.

In "Faith and Knowledge," his contribution to Gianni Vattimo's volume titled *On Religion,* Derrida utilizes the optic of immunity to describe a situation in which religion returns to the forefront of political discourse. Interestingly, the change will be found in religion's relation to immunity. For Derrida, (auto)immunity names the mode by which religion and science are reciprocally inscribed in each other. And so any contemporary analysis of religion must begin with the recognition that religion at the end of the millennium "accompanies and precedes" what he calls "the critical and tele-technoscience reason," or better those technologies that decrease the distance and increase the speed of communications globally, which he links to capitalism and the Anglo-American idiom.[31] The same movement that makes religion and the tele-technoscience coextensive results in a countermove of immunity. Drawing upon the etymological roots of religion in *religio,* which he associates with repetition and then with performance, Derrida shows how religion's iterability presupposes the automatic and the machinelike—in other words, presupposes a technique that marks the possibility of faith. Delivering technique (technology) over to a faith in iterability shared with religion allows him to identify the autoimmunitary logic underpinning the current moment of religious revival and crisis. He writes: "It [the movement that renders religion and tele-technoscientific reason] secretes its own antidote but also its own power of auto-immunity. We are here in a space where all self-protection of the unscathed, of the safe and sound, of the sacred (*heilig,* holy) must protect itself against its own protection, its own police, its own power of rejection, in short against its own, which is to say, against its own immunity."[32]

In the context of the overlapping fields of religion and tele-technoscientific reason, immunity is always autoimmunity for Derrida and hence always destructive. It is immunal because, on the one hand, religion—he will substitute the term "faith" repeatedly for it—cannot allow itself to share performativity with tele-reason as the effects of that same reason inevitably lead to an undermining of the basis for religion in tradition, that is, in maintaining a holy space apart from its iterable features. Furthermore, it is autoimmunal to the degree that the protection of the sacred space, the "unscathed" of the preceding quote, is created precisely thanks to the same iterability, the same features of performance that it shares with tele-technoscientific reason. The result is a protective attack against protection itself, or a crisis in autoimmunity.

Not surprisingly, religious (auto)immunity also has a biopolitical declension for Derrida, though he never refers to it as such. Thus, in the mechanical principle by which religions say they value life, they do so only by privileging a transcendental form of life. "Life" for many religions, Derrida writes, "is sacred, holy, infinitely respectable only in the name of what is worth more than it and what is not restricted to the naturalness of the bio-zoological (sacrificeable)."[33] In this, biological life is repeatedly transcended or made the supplement religion provides to life. So doing, transcendence opens up the community, constitutively formed around the living, to the "space of death that is linked to the automaton . . . to technics, the machine, prosthesis: in a word, to the dimensions of the auto-immune and self-sacrificial supplementarity, to this death drive that is silently at work in every community, every *auto-co-immunity*."[34] For Derrida (as for Esposito) the aporia of immunity operates in every community, based on "a principle of sacrificial self-destruction ruining the principle of self-protection."[35] At the origin of religious immunity lies the distinction between bio-zoological or anthropo-theological life and transcendental, sacred life that calls forth sacrifices in almost parasitical form so as to protect its own dignity. If there is a biopolitical moment to be found in Derrida's analysis of religion and autoimmunity, it will be found here in the difference between biological life and transcendental life that will continually require the difference between the two to be maintained. It is, needless to say, despite the contemporary context that informs Derrida's analysis, a conceptual aporia that precedes the discussion of capitalism, life, and late-twentieth-century technology. Writing in 1994, Derrida gestures to these changes, but in his analysis of the resurgence of religion within a certain kind of political discourse, autoimmunity co-originates with religion in the West.

Whether the same holds true in the political dimension, Derrida doesn't actually answer, at least not in his important work from 1997, *The Politics of Friendship*. There instead, after the requisite footnote marking the debt he owes Blanchot, Bataille, and Nancy, Derrida emphasizes a different political declension of (political) community, one based on a certain form of friendship of separation undergirding philosophical attempts to think a future community of solitary friends:

> Thus is announced the anchoritic community of those who love in separation ... The invitation comes to you from those who can *love only at a distance, in separation* ... Those who love only in cutting ties are the uncompromising friends of solitary singularity. They invite you to enter into this community of social disaggregation *[déliaison]*, which is not necessarily a secret society, a conjuration, the occult sharing of esoteric or crypto-poetic knowledge. The classical concept of the secret belongs to a thought of the community, solidarity, or the sect—initiation or private space which represents the very thing the friend who speaks to you as a friend of solitude has rebelled against.[36]

Here a different form of political relationship emerges, one linked to Bataille's "community of those without community," and one at least initially distinct from the autoimmunizing features of religion. Derrida suggests as much with his gesture here to the Deleuzian singularity, those separate entities whose very separateness functions as the invitation to the common.[37] At the same time, Derrida does preface the remarks with the adjective *anchoritic*, thereby associating the form of distant love afforded those who have withdrawn for religious reasons from the world with a political dimension. Derrida suggests that in the separateness of singularity it may be possible to avoid some of the immunizing features of community that emerged with his discussion on faith.

If I have focused initially on these two pieces in an introduction to Esposito's thought, it is because they inform much of Derrida's important reflections on global autoimmunity in the wake of September 11. Without rehearsing here all of the intricacies of his analysis, the reintroduction of the notion of autoimmunity into a more properly political discourse, both in his interviews with Giovanna Borradori after September 11 and in his later reflections on democracy in *Rogues*, shows Derrida extending the autoimmune process to two related fronts: first, to a constituent "pervertibility of democracy" at the heart of defining democracy, and second the suicidal, autoimmune crisis that has marked American foreign policy since the 1980s. As for the first, democracy for Derrida appears to have at its heart a paradoxical meaning, one in which it continually postpones both the moment when it can be fully realized as the political government in which the many rule and simultaneously the possibility that when such an event comes, the many may precisely vote to suspend democracy. Writing with the recent experience of 1990s Algeria in mind, Derrida argues that "democracy has always been suicidal" because there are always some who do not form part of the many and who must be excluded or sent off.[38] The result, and it is one that we ought to keep in mind when attempting to think Esposito's thought on community/immunity, is that "the autoimmune topology always dictates that democracy be *sent off [renvoyer]* elsewhere, that it be excluded or rejected, expelled under the pretext of protecting it on the inside by expelling, rejecting, or sending off to the outside the domestic enemies of democracy."[39] For Derrida, autoimmunity is inscribed "right onto the concept of democracy" so that "democracy is never properly what it is, never *itself.* For what is lacking in democracy is proper meaning, the very *[même]* meaning of the selfsame *[même]*... the it-self *[soi-même]*, the selfsame, the properly selfsame of the itself."[40] A fundamental, constitutive lack of the proper marks democracy.

Esposito's analysis of the immunity aporia of community does, much like Derrida's analysis of democracy, implicitly evoke in community something like democracy, but we ought to be careful in linking the two discussions on autoimmunity too closely—first, because Esposito clearly refuses to collapse the process of immunization into a full-blown autoimmune suicidal tendency at the heart of community. That he doesn't has to do primarily with the larger project of which *Bíos* and *Immunitas* are a part, namely, how to think an affirmative biopolitics through the lens of immunity. Esposito's stunning elaboration of a positive immunity evidenced by mother and fetus in *Immunitas* is the proof that immunity doesn't necessarily degenerate—and that sense is hardly unavoidable in Derrida's discussion—into a suicidal autoimmunity crisis. In this, Esposito sketches the outlines of an affirmative model of biopolitical immunity, whereas rarely if ever does Derrida make explicit the conceptual language of biopolitics that undergirds his analysis.

But, as I mentioned, Derrida speaks of autoimmunity in a different context, one that characterizes American foreign policy after September 11 as essentially an autoimmune reaction to previous cold-war policy that armed and trained former freedom fighters during the cold war's hot phase in Afghanistan in the early 1980s. He says:

> Immigrated, trained, prepared for their act in the United States by the United States, these *hijackers* incorporate so to speak, two suicides in one; their own (and one will remain forever defenseless in the face of a suicidal, autoimmunitary aggression—and that is what terrorizes most) but also the suicide of those who welcomed, armed and trained them.[41]

The soul-searching among the British in response to the bombings in London in the summer of 2005 is clearly proof of the correctness of Derrida's analysis; in the United States, a similar analogy might be found with the Oklahoma City bombings (though there was clearly less reflection on the elements that contributed to that instance of suicidal immunity than in the United Kingdom). In any case, by linking American foreign policy to suicide via autoimmunity, Derrida not only acknowledges an important historical context for understanding September 11, but implicitly links "these *hijackers*" to technical proficiency and high-tech knowledge and, so it would seem, to his earlier analysis of tele-reason and technology as reciprocally implicated in religious iterability. Although space doesn't allow me more than a mere mention, it might be useful to probe further the overdetermined connection of the "religious" in radical Islamic fundamentalism with just such a technological prowess. In any case, for the present discussion what matters most is that Derrida believes that September 11 cannot be thought independently of the figure of immunity; indeed, that as long as the United States continues to play the role of "guarantor or guardian of the entire world order," autoimmunitary aggression will continue, provoked in turn by future traumatizing events that may be far worse than September 11.

How, then, does Esposito's reading of an immunological lexicon in biopolitics differ from Derrida's? Where Derrida's emphasis falls repeatedly on autoimmunity as the privileged outcome of American geopolitics in the period preceding September 11, Esposito carefully avoids conflating immunity with autoimmunity; instead, he repeatedly returns to the question of *munus* and modernity's attempts to immunize itself against the ever-present threat, from its perspective, of immunity's reversal into the communal, from immunization to communization.[42] Writing at length in *Immunitas* on the imperative of security that assails all contemporary social systems and the process by which risk and protection strengthen each other reciprocally, he describes the autoimmunity crisis of biopolitics and with it the possibility of a dialectical reversal into community. "Evidently, we are dealing," Esposito writes, "with a limit point beyond which the entire biopolitical horizon risks entering in a lethal contradiction with itself." He continues:

> This doesn't mean that we can turn back the clock, perhaps reactivating the ancient figures of sovereign power. It isn't possible today to imagine a politics that doesn't turn to life as such, that doesn't look at the citizen from the point of view of his living body. But this can happen reciprocally in

> opposite forms that put into play the different meanings of biopolitics: on the one hand the self-destructive revolt of immunity against itself or the opening to its reversal in community.[43]

Looking back today at the series of attempts after September 11 in the United States to immunize the "homeland" from future attack—the term itself a powerful immunizing operator—it isn't hard to imagine that we are in the midst of a full-scale autoimmunity crisis whose symptomology Derrida and Esposito diagnose.

Yet a political autoimmunity crisis isn't the only possible biopolitical outcome of the present moment. Esposito suggests that another possibility exists, one to which his own affirmative biopolitics is directed, namely, creating the conditions in which it becomes possible to identify and deconstruct the principal twentieth-century biopolitical, or better, thanatopolitical, *dispositifs* that have historically characterized the modern immunitary paradigm. Only after we have sufficiently understood the extent to which our political categories operate to immunize the collective political body from a different set of categories associated with community can we reorient ourselves to the affirmative biopolitical opening presented by the current crisis in immunity. This opening to community as the site in which an affirmative biopolitics can emerge is the result of a dialectical reversal at the heart of the immunitary paradigm: once we recognize that immunization is the mode by which biopolitics has been declined since the dawn of modernity, the question becomes how to rupture the juncture between biology and politics, between *bíos* and *politikos.* The necessary first step is moving away from a rationality of bodies when attempting to locate the object of politics, and so shifting the conceptual ground on which immunization depends. An affirmative biopolitics thought through the *munus* of community proceeds with the recognition that a new logic is required to conceptualize and represent a new community, a coming "virtual" community, Esposito will say with Deleuze, characterized by its impersonal singularity or its singular impersonality, whose confines will run from men to plants, to animals independent of the material of their individuation.[44]

Biopolitics and Contemporary Italian Thought

The reference to a virtual, future community immediately recalls two other contemporary thinkers from Italy who are deeply engaged with the notion of biopolitics in its contemporary configuration. Of course, I am speaking of Antonio Negri and Giorgio Agamben. That modern Italian political philosophy has emerged as perhaps the primary locus for research related to biopolitics is not happenstance. Few places have been as fertile for Foucault's teachings; few places so well primed historically and politically to reflect on and extend his work. The reasons, it seems to me, have to do principally with a rich tradition of political philosophy in Italy—we need only remember Machiavelli, Vico, de Sanctis, Croce, and Gramsci, for instance—associated with the specificity of the Italian history and a political scene characterized by the immunizing city-state.[45] Many other reasons may account for it, but what they together spell is an ongoing engagement in Italy with politics thought in a biopolitical key.[46]

With that said, the more one reads of recent Italian contributions to biopolitics, the more two diverging lines appear to characterize them: one associated with the figure of Agamben and the negative tonality he awards biopolitics; the other a radically affirmative biopolitics given in the writings of Michael Hardt and Toni Negri. As the originality of Esposito's reading of modern biopolitics cannot be appreciated apart from the implicit dialogue that runs through *Bíos* with both Agamben, and Hardt and Negri, I want to summarize these two often competing notions of biopolitics. What emerges in Esposito's analysis is a thorough critique of both Agamben and Negri; his pinpointing of their failures to think through the immunity aporia that characterizes their respective configurations of biopolitics leads to his own attempt to design a future, affirmative biopolitics. That all three launch their reflections from essentially the same series of texts, namely, Foucault's series of lectures collected in English in *Society Must Be Defended* and the fifth chapter of *The History of Sexuality*, suggests that we ought to begin there for an initial definition of biopolitics before turning to their respective appropriations of Foucault.

For Foucault, biopolitics is another name for a technology of power, a biopower, which needs to be distinguished from the mechanisms of discipline that emerge at the end of the eighteenth century. This new configuration of power aims to take "control of life and the biological processes of man as species and of ensuring that they are not disciplined but regularized."[47] The biopolitical apparatus includes "forecasts, statistical estimates, and overall measures," in a word "security mechanisms [that] have to be installed around the random element inherent in a population of living beings so as to optimize a state of life."[48] As such, biopolitics is juxtaposed in Foucault's analysis to the power of sovereignty leading to the important distinction between them: "It [biopower] is the power to make live. Sovereignty took life and let live. And now we have the emergence of a power that I would call the power of regularization, and it, in contrast, consists in making live and letting die."[49] Biopower thus is that which guarantees the continuous living of the human species. What turns out to be of almost greater importance, however, for Agamben, Negri, and Esposito, is the relation Foucault will draw between an emerging biopower at the end of the eighteenth century, often in opposition to individual disciplinary mechanisms and its culmination in Nazism. For Foucault, what links eighteenth-century biopower to Nazi biopower is precisely their shared mission in limiting the aleatory element of life and death. Thus, "[C]ontrolling the random element inherent in biological processes was one of the regime's immediate objectives."[50] This is not to say that the Nazis simply operated one-dimensionally on the body politic; as Foucault notes repeatedly, the Nazis had recourse again and again to disciplinary power; in fact "no State could have more disciplinary power than the Nazi regime," presumably because the attempts to amplify biopower depended on certain concurrent disciplinary tools.[51] For Foucault, the specificity of the Nazis' lethal biopower resides in its ability to combine and thereby intensify the power directed both to the individual and to the collective body.

Certainly, other vectors crisscross biopolitics in Foucault's analysis, and a number of scholars have done remarkable jobs in locating them, but the outline above is sufficient for describing the basis on which Agamben, Hardt and Negri, and Esposito frame their respective analyses.[52] Thus Agamben's notion of biopolitics is certainly indebted to the one sketched above—the impression that modernity produces a certain form of biopolitical body is inescapable reading Agamben as it is one implicit in Foucault. But Agamben's principal insight for thinking biopolitics concerns precisely the distinction between *bíos* and *zōē* and the process by which he links the sovereign exception to the production of a biopolitical, or better a zoo-political, body. Indeed, *Homo Sacer* opens with precisely this distinction:

> The Greeks had no single term to express what we mean by the word "life." They used two terms that, although traceable to a common etymological root, are semantically and morphologically distinct: zoē, which expressed the simple fact of living common to all living beings (animals, men, or gods) and bios, which indicated the form or way of life proper to an individual or group.[53]

Leaving aside for the moment whether in fact these terms exhaust the Greek lexicon for life, Agamben attempts to demonstrate the preponderance of *zōē* for the production of the biopolitical body.[54] The reason will be found in what Agamben, following Carl Schmitt, calls the sovereign exception, that is, the process by which sovereign power is premised on the exclusion of those who are simply alive when seen from the perspective of the *polis*.[55] Thus Agamben speaks of an inclusive exclusion of *zōē* from political life, "almost as if politics were the place in which life had to transform itself into good and in which what had to be politicized were always already bare life."[56] A number of factors come together to condition politics as the site of exclusion, but chief among them is the role of language, by which man "separates and opposes himself to his own bare life and, at the same time, maintains himself in relation to that bare life in an inclusive exclusion."[57] The *homo sacer* is precisely the political figure that embodies what is for Agamben the originary political relation: it is the name of the life excluded from the political life (*bíos*) that sovereignty institutes, not so much an ontology of the one excluded (and therefore featuring an unconditional capacity to be killed), but more the product of the relation in which *bíos* is premised not upon another form of life but rather on *zōē* (because *zōē* is not by definition such a form), and its principal characteristic of being merely alive and hence killable.

In such a scheme, the weight afforded the classical state of exception is great indeed, and so at least initially biopolitics for Agamben is always already inscribed in the sovereign exception. Thus Agamben will de-emphasize the Foucauldian analysis of the emergence of biopower in the late nineteenth century, for it represents less a radical rupture with sovereignty or for that matter a disciplinary society, and will instead foreground the means by which biopolitics intensifies to the point that in the twentieth century it will be transformed into thanatopolitics for both totalitarian and democratic states. Certainly, a number of differences remain between the classic and modern models of biopolitics—notably the dispersal of sovereign power to the physician and scientist so that the *homo sacer* no longer is simply an analogue to the sovereign—and of course Agamben will go out of his way to show how the political space of modernity is in fact a biopolitical space linked to "the birth of the camps."[58] But the overwhelming impression is of a kind of flattening of the specificity of a modern biopolitics in favor of a metaphysical reading of the originary and infinite state of exception that has since its inception eroded the political foundations of social life. For

Agamben, an authentically political *bíos* always withdraws in favor of the merely biological.[59] The result is a politics that is potentially forever in ruins in Marco Revelli's description, or a politics that is always already declined negatively as biopolitical.[60]

Where Agamben's negative characterization of contemporary biopolitics as thanatopolitics depends on the predominance of *zōē* over *bíos*, Hardt and Negri's radical affirmation of biopolitics centers instead on the productive features of *bíos*, and "identifying the materialist dimension of the concept beyond any conception that is purely naturalistic (life as '*zōē*') or simply anthropological (as Agamben in particular has a tendency to do, making the concept in effect indifferent)."[61] Leaving aside for the moment the descriptor "indifferent," which it seems to me fails to mark the radical negativity of Agamben's use of the term, what stands out in Hardt and Negri's reading of biopolitics is the mode by which they join contemporary forms of collective subjectivity to the transformations in the nature of labor to what a number of Italian Marxist thinkers have termed immaterial labor.[62] Thinking together these changes in forms of labor—ones characterized not by the factory but rather by "the intellectual, immaterial, and communicative labor power" affiliated with new communication technologies—through Foucault's category of biopower allows Hardt and Negri to see biopolitics as both the locus in which power exerts itself in empire and the site in which new subjectivities, what they call social singularities, subsequently emerge. Thus the term "biopolitical" characterizes not only the new social formation of singularities called the multitude but also the emergence of a new, democratic sovereignty, one joined to a radically different understanding of the common.

As Hardt and Negri themselves readily admit, reading the multitude ontologically as a biopolitical social formation represents a significant reversal if not outright break with Foucault's conception of biopolitics. Where Foucault often associates the negative features of biopower with its object, a biopolitical subject, Hardt and Negri deanchor biopolitics from its base in biopower in the current moment of empire to read it primarily and affirmatively as a social category. Thus: "Biopolitical production is a matter of ontology in that it constantly creates a new social being, a new human nature" linked to the "continuous encounters, communications, and concatenations of bodies."[63] They do the same in their reading of Agamben, forgoing his declension of a twentieth-century thanatopolitics by evoking instead a new form of sovereignty in which the state of exception presumably either no longer operates or is soon overwhelmed by the rhizomatic production of singular multitudes, unveiling the illusory nature of modern sovereignty.[64] In its place the multitude produces a concept of the common, which "breaks the continuity of modern state sovereignty and attacks biopower at its heart, demystifying its sacred core. All that is general or public must be reappropriated and managed by the multitude and thus become common."[65] Transposing into the biopolitical language we have used to this point, Hardt and Negri juxtapose the affirmative biopolitics associated with the multitude and the common to biopower and its privileging of modern sovereignty.

In *Bíos* Esposito takes up a position directly opposite both Agamben and Hardt and Negri and their conflicting uses of biopolitics. First Agamben. Certainly, Esposito's genealogy of biopolitics shares many features with Agamben's reading of modern biopolitics through the figure of the *homo sacer.* Indeed, the chapter on thanatopolitics and the cycle of *genos* is nothing short of an explicit dialogue with Agamben and his biopolitical interpretation of Nazism, as well as an implicit critique of Agamben's biopolitics. To see why, we need to rehearse briefly the chief lines of argument Esposito develops for working through the coordinates of Nazi biopolitics. Significantly, Esposito first pinpoints an oscillation in Foucault's reading of Nazism. On the one hand, Nazism for Foucault shares the same biopolitical valence with a number of modern regimes, specifically socialist, which Foucault links to a racist matrix. On the other hand, the mode by which Foucault frames his interpretations of Nazism privileges the singular nature of the "Nazi event," as Esposito calls it. The result is an underlying inconsistency in Foucault's reading: either Nazi biopolitics is inscribed along with socialism as racism, and hence is no longer a singular event, or it maintains its singularity when the focus turns to its relation to modernity.[66]

The second line will be found in Esposito's principal question concerning the position of life in Nazi biopolitics. "Unlike all the other forms past and present," he asks, "why did Nazism propel the homicidal temptation of biopolitics to its most complete realization?"[67] That his answer will move through the category of immunization suggests that Esposito refuses to superimpose Nazi thanatopolitics too directly over contemporary biopolitics.[68] Rather, he attempts to inscribe the most significant elements of the Nazi biopolitical apparatus in the larger project of immunizing life through the production of death. In so doing, death becomes both the object and the therapeutic instrument for curing the German body politic, simultaneously the cause and the remedy of "illness." Esposito dedicates much of the final third of *Bíos* to elaborating the immunizing features of Nazi biopolitics in order to reconstruct the move from a modern biopolitics to a Nazi thanatopolitics. The Nazi immunitary apparatus, he theorizes, is characterized by the absolute normativization of life, the double enclosure of the body, and the anticipatory suppression of life. Space doesn't allow me to analyze each, though the reader will certainly find some of the most compelling pages of *Bíos* here. More useful is to ask where Esposito's overall portrayal of Nazi biopolitics diverges from that of Agamben in immunization. By focusing on the ways in which *bíos* becomes a juridical category and *nomos* (law) a biologized one, Esposito doesn't directly challenge Agamben's reading of the state of exception as an aporia of Western politics, one the Nazis intensified enormously so that the state of exception becomes the norm. Rather, he privileges the figure of immunization as the ultimate horizon within which to understand Nazi political, social, juridical, and medical policies. In a sense he folds the state of exception in the more global reading of modern immunity *dispositifs*.

Implicit in the optic of immunity is a critique of the categories by which Nazism has been understood, two of which are primarily sovereignty and the state of exception.[69] By privileging the immunitary paradigm for an understanding of Nazi biopolitics, Esposito forgoes Agamben's folding of sovereignty into biopolitics (and so bypasses the *Musulmann* as the embodiment of the twentieth-century *homo sacer*), focusing instead on the biocratic elements of the Nazi dictatorship. He notes, for instance, the requirement that doctors had to legitimate Nazi political decisions, which previously had been translated into the Reich's new legal codes, as well as the required presence of a physician in all aspects of the workings of the concentration camp from selection to the crematoria. Esposito's analysis not only draws upon Robert Lifton's classic description of the Nazi state as a "biocracy," but more importantly urges forward the overarching role that immunization plays in the Nazi understanding of its own political goals; indeed, the Nazi politicization of medicine cannot be fully understood apart from the attempt to immunize the Aryan race.[70] Central therefore to Esposito's reading of the biopolitical tonality of the Nazi dictatorship is the recognition of the therapeutic goal the Nazis assigned the concentration camp: only by exterminating the Jews did the Nazis believe that the German *genos* could be strengthened and protected. And so for Esposito the specificity of the Nazi experience for modernity resides in the actualization of biology, when the transcendental of Nazism becomes life, its subject race, and its lexicon biological.[71]

An Affirmative Biopolitics?

The same reasons underlying Esposito's critique of Agamben's biopolitics also spell out his differences with Hardt and Negri. Not only does Esposito explicitly distance himself from their reading of the multitude as an affirmative biopolitical actor who resists biopower—he notes how their line of interpretation pushes well beyond Foucault's manifest intentions when delimiting biopolitics, beyond the resistance of life to power—but he asks a decisive question for their use of biopolitics as an organizing principle around which they posit their critique of empire. "If life is stronger than the power that besieges it, if its resistance doesn't allow it to bow to the pressure of power, then how do we account for the outcome obtained in modernity of the mass production of death?"[72] In a number of interviews Esposito has continued to challenge Hardt and Negri's reading of biopolitics. What troubles Esposito principally is a categorical (or historical) amnesia vis-à-vis modernity's negative inflection of biopolitics.[73]

Essentially, Esposito charges that Hardt and Negri's reading of the multitude is riven by the same immunitary aporia that characterizes Agamben's negative biopolitics. In what way does the biopolitical multitude escape the immunitary aporia that resides at the heart of any creation of the common? Although he doesn't state so explicitly, Esposito's analysis suggests that folding biopower into the social in no way saves Hardt and Negri from the long and deadly genealogy of biopolitics in which life is protected and strengthened through death, in what Esposito calls the "enigma" of biopolitics. Esposito laid some of the groundwork for such a critique in the early 1990s when, in a series of reflections on the impolitical, he urged forward a thorough deconstruction of many of the same political categories that undergird Hardt and Negri's analysis, most particularly sovereignty. It certainly is plausible (and productive) to read *Bíos* through an impolitical lens, in which Esposito offers biopolitics as the latest and ultimate of all the modern politics categories that require deconstruction. Indeed, it's not by chance that the first chapter of *Bíos* aggressively positions biopolitics not only as one of the most significant ways of organizing contemporary political discourse, but also as the principal challenger to the classic political category of sovereignty. For Esposito, sovereignty, be it a new global sovereignty called empire or the long-lived national variety, doesn't transcend biopolitics but rather is immanent to the workings of the immunitary mechanism that he sees driving all forms of modern (bio)politics. The multitude remains inscribed in modern sovereignty, whose final horizon, following Esposito's reading of Foucault, is the immunitary paradigm itself. In other words, the multitude remains anchored to a genealogy of biopolitics. Thus Esposito not only deeply questions the hermeneutic value of sovereignty for understanding the contemporary political scene or for imagining a progressive politics oriented to the future, but also points to a sovereign remainder in the figure of multitude.

Bíos also offers another less explicit objection to Esposito's analysis of Hardt and Negri's use of the term "biopolitics." We recall that for Hardt and Negri the multitude produces a new concept of the common, which corresponds to their belief that the multitude represents a rupture with all forms of state sovereignty. This occurs thanks to the economic and biopolitical activity of the multitude, which coincides with a "commonality created by the positive externalities or by the new informational networks, and more generally by all the cooperative and communicative forms of labor."[74] The multitude mobilizes the common in the move from a *res-publica* to a *res-communis,* in which the multitude comes to embody ever more the expansive logic of singularity-commonality. However, Esposito's reading of *communitas/immunitas* sketched above suggests that there is no common obligation joining members of a community *in potentia* that can be thought apart from attempts to immunize the community, or in this case the multitude. As Esposito notes, "without this immunity apparatus individual and common life would die away."[75] The impolitical question Esposito raises for Hardt and Negri is precisely whether the new biopolitical multitude somehow transcends the political aporia of immunity that undergirds every conception of community. Perhaps in the new configuration of the common that they describe and the fundamental changes in the nature of immaterial production, the global *munus* changes as well, so that, unlike every previous form of community, the multitude no longer has any need of immunizing itself from the perils of *communitas.* Just such a reading is suggested by Hardt and Negri's repeated troping of the multitude as a network of rhizomatic singularities, who presumably would have less need of immunizing themselves because the network itself provides the proper threshold of virtual contact. Esposito in *Bíos* implicitly raises the question of whether these singularities acting in common and so forming "a new race or, rather, a new humanity" don't also produce new forms of immunity.[76]

Immunity, we recall, emerges as a constituent element of community for Esposito, when the common threatens personal identity. Thus it isn't difficult to read those pages in *Bíos* dedicated, for instance, to the immunitary mechanism in Locke as aimed as well at Hardt and Negri. Writing apropos of the potential risk of a world that is given in common (and therefore exposed to an unlimited indistinction) is neutralized by an element presupposed in the originary manifestation . . . namely, that of the relationship one has with oneself in the form of personal identity, Esposito once again situates personal identity as the subject and object of immunitary protection.[77] The *res-communis* that Hardt and Negri see as one of the most important productions of the multitude is in Esposito's reading of Locke always seen as a threat to a *res propria*. Following this line of inquiry, *Bíos* asks us, what becomes of personal identity when the multitude produces the new sense of the common? Is it now less a threat given new forms of communication and labor, or rather does the threat to individual identity increase given the sheer power of extension Hardt and Negri award the multitude? What is at stake isn't only a question of identity or difference here, but the prevalence of one or the other in the multitude. Seen in this optic, their emphasis on the singularity and commonality of the multitude may in fact be an attempt to ward off any suggestion of an underlying antinomy between the multitude as a radically new social formation and personal identity.

A Communal *Bíos*

Given these differences, the obvious question will be what form Esposito awards his own conception of biopolitics such that it avoids the kinds of difficulties raised in these other contributions. After two illuminating readings of *bíos* in Arendt and Heidegger—which may be read as dialoging with Agamben's discussion of *homo sacer* and his appropriation of "the open" via Heidegger—Esposito sets out to construct just such an affirmative vision by "opening the black box of biopolitics," returning to the three *dispositifs* that he had previously used to characterize the Nazi bio-thanatological project and then reversing them. These are the normativization of life, the double enclosure of the body, and the anticipatory suppression of life that I noted earlier. The effect of appropriating them so as to reverse Nazi immunitary procedures will surprise and certainly challenge many readers. Esposito clearly is aware of such a possible reaction and his response merits a longer citation:

> Yet what does it mean exactly to overturn them and then to turn them *inside out?* The attempt we want to make is that of assuming the same categories of "life," "body," and "birth," and then of converting their immunitary (which is to say their self-negating) declension in a direction that is open to a more originary and intense sense of *communitas.* Only in this way—at the point of intersection and tension between contemporary reflections that have moved in such a direction—will it be possible to trace the initial features of a biopolitics that is finally affirmative. No longer over life but of life.[78]

Esposito recontextualizes his earlier work on *communitas* as the basis for an affirmative biopolitics: following his terminology, the term becomes the operator whereby a long-standing immunitary declension of *bíos* as a form of life can be reversed.[79] He premises such a reading on the belief that contemporary philosophy has fundamentally failed to grasp the relation between Nazi bio-thanatological practices and biopolitics today. "The truth," he writes, "is that many simply believed that the collapse of Nazism would also drag the categories that had characterized it into the inferno from which it had emerged."[80] Only by identifying the immunitary apparatus of the Nazi biopolitical machine and then overturning it—the word Esposito uses is *rovesciare,* which connotes the act of turning inside out—can contemporary philosophy come to terms with the fundamental immunitary features of today's global biopolitics and so devise a new lexicon able to confront and alter it.

It's precisely here that Esposito synthesizes Agamben's negative vision of biopolitics with Hardt and Negri's notion of the common as signaling a new affirmative biopolitics. Esposito doesn't offer a simple choice between immunity and community that will once and for all announce the arrival of a new human nature and with it an affirmative biopolitics. The continuum between Nazi and contemporary biopolitics that characterizes Agamben's approach is less significant from this point of view than the continuum of immunity and community. At the risk of reducing Esposito's line of argument, he suggests that if Nazi thanatopolitics is the most radically negative expression of immunization, then inverting the terms, or changing the negative to a positive, might offer contemporary thought a series of possibilities for thinking *bíos,* a qualified form of life, as the communal form of life. Such a positive conception of biopolitics can only emerge, however, if one simultaneously develops a conception of life that is aporetically exposed to others in such a way that the individual escapes an immunization of the self (and hence is no longer an individual proper).[81] For Esposito, it is less a matter of exposure than of openness to what is held in common with others.[82] The reader will find much of interest in the way Esposito draws on the work of Merleau-Ponty, Levinas, and Deleuze when elaborating such a conception.[83]

The reference to the singular and the common also echoes those pages of Agamben's *The Coming Community,* especially the sections in which Agamben anchors a nude, exposed life to incommunicability. We recall that the coming community for Agamben begins when a meaningful context for life emerges in which death has meaning, that is, when it can be communicated. Only when the previously meaningless and unfelt death of the individual takes on meaning can one speak more properly of singularities without identity who enjoy the possibility of communication. Such a community will consequently be "without presuppositions and without subjects" and move "into a communication without the incommunicable."[84] So too for Esposito, though *Bíos* doesn't offer many details on the communicative aspects of an affirmative biopolitical community. To find them we need to turn to *Communitas,* where Esposito links forms of communication to singular lives open to each other in a community. There the differences with Agamben can be reassumed around their respective readings of Heidegger and Bataille. Thus, when Agamben emphasizes death as the means by which a life may uncover (or recover) an authentic opening into *Dasein,* he rehearses those moments of Heidegger's thought that celebrate death as the final horizon of our existence. For Esposito, such a perspective is too limiting for thinking future forms of community. "Death," he writes, glossing Bataille, "is our communal impossibility of being that which we endeavor to remain—isolated individuals."[85]

In that sense, Agamben and Esposito certainly agree on the antinomy between individuals or subjects and community. But for Bataille as for Esposito, the crucial thought for a future community concerns precisely what puts members of the community outside themselves; not their own death, "since that is inaccessible," but rather "the death of the other."[86] In such a reading, communication occurs when beings lose a part of themselves, the Bataillian rent or a wound, that unites them in communication while separating them from their identity.[87] It is in Bataille's notion of "strong communication" linked to sacrifice that Esposito locates the key for unlocking a contemporary *communitas,* one in which communication will name "a contagion provoked by the breaking of individual boundaries and by the reciprocal infection of wounds" in a sort of arch-event of contagion and communication.[88] The implicit question for Esposito appears to be how to create conditions in which such a contagion can be contained without involving the entire immunitary machinery. To do so we need to develop a new vocabulary for thinking the boundaries of life and its other, in biojuridical forms that recognize the one in the other such that any living being is thought in "the unity of life," in a co-belonging of what is different.[89] Essentially, then, Esposito's emphasis on difference is linked to his larger defense of personal identity throughout *Bíos,* which is deeply inflected, as the reader will discover, in chapter 3 by Esposito's encounter with a hyperindividualistic Nietzsche. This may explain in part his defense of *bíos* as individuated life as opposed to *zōē.*

Birth and Autoimmunity

Esposito's emphasis on man and his relation to his living being (as opposed to Heidegger's distinction between life and existence) calls to mind other attempts to think nonontologically the difference between living beings through other perspectives on life. Keith Ansell-Pearson's privileging of symbiosis and of inherited bacterial symbionts is perhaps the most sophisticated, in his attempts to show how "amid cell gorgings and aborted invasions" a reciprocal infection arises such that the bacteria "are reinvigorated by the incorporation of their permanent disease." The human becomes nothing more than a viroid life, "an integrated colony of ameboid beings," not distinct from a larger history of symbiosis that sees germs "not simply as 'disease-causing,' but as 'life-giving' entities."[90] Consequently, anthropocentric readings of human nature will give way to perspectives that no longer focus on one particular species, such as humankind, but rather on those that allow us to think life together across its different forms (biological, social, economic). The reference to disease as life-giving certainly recalls Esposito's own reading of Nietzsche and the category of *compensatio* in *Immunitas,* as well as Machiavelli's category of productive social conflict, suggesting that some forms of immunity do not necessarily close off access to an authentically political form of life. Indeed, reading the immunitary system as only self-destructive fails to see other interpretive perspectives in which immunity doesn't protect by attacking an authentic *bíos* grounded in a common *munus,* but rather augments its members' capacity to interact with their environment, so that community can actually be fortified by immunity.

The primary example Esposito offers for such an immunitary opening to community will be found in birth. In *Immunitas,* Esposito introduces pregnancy as a model for an immunity that augments the ability of the fetus and mother to remain healthy as the pregnancy runs its course. Their interaction takes place, however, in an immunitary framework in which the mother's system of self-defense is reined in so that the fetus does not become the object of the mother's own immunization. The immunity system of the mother "immunizes itself against an excess of immunization" thanks to the extraneousness of the fetus to the mother.[91] It isn't that the mother's body fails to attack the fetus—it does—but the immunological reaction winds up protecting the fetus and not destroying it. In the example of pregnancy with its productive immunitary features, Esposito finds a suggestive metaphor for an immunity in which the greater the diversity of the other, which would in traditional immunitary terms lead to an all-out immunitary struggle against it, is only one possibility. Another is an immunization that, rather than attacking its communal antinomy, fortifies it. *Bíos* as a political form of life, a community, emerges out of an immunization that successfully immunizes itself against attacking what is other, with the result that a more general defense of the system itself, the community, occurs.

This may account for the distance Esposito is willing to travel in awarding birth a political valence. In some of *Bíos*'s most rewarding pages, Esposito suggests that immunization isn't the only category capable of preserving or protecting life from death, but rather that birth, or the continual rebirth of all life in different guises, can function similarly. Drawing on Spinoza's theory of life and Gilbert Simondon's reflections on individuation, Esposito extends the category of birth to those moments in which the subject, "moving past one threshold," experiences a new form of individuation. He assumes a stratum of life that all living beings share, a common *bíos* that is always already political as it is the basis on which the continued birth of individuation occurs. So doing, he elaborates *bíos* in such a way that *zōē* will in turn be inscribed within it: there is no life without individuation through birth. Although Esposito doesn't say so explicitly, the suggestion is that a new affirmative biopolitics might begin by shuffling the terms by which we think of the preservation of life. Life is no longer linked exclusively to those deemed worthy of it along with those who are not, but now comes to mark every form of life that appears thanks to individuation. He writes:

> If one thinks about it, life and birth are both the contrary of death: the first synchronically and the second diachronically. The only way for life to defer death isn't to preserve it as such (perhaps in the immunitary form of negative protection), but rather to be reborn continually in different guises.[92]

An ontology of the individual or the subject becomes less a concern than the process of individuation associated with the appearance of life, be it individual or collective. Attempts to immunize life against death give way to strategies that seek to promote new forms of individuation. The emphasis on individuation (and not the individual) allows Esposito to argue that the individual is the subject that produces itself through individuation, which is to say that the individual "is not definable outside of the political relationship with those that share the vital experience." So too the collective, which is no longer seen as the "neutralization of individuality" but rather as a more elaborated form of individuation.[93] Rather than limiting *bíos* to the immunization of life, Esposito imagines an affirmative *bíos* that privileges those conditions in which life as manifested across different forms is equipped for individuation. There will be no life that isn't born anew and hence that isn't inscribed in the horizon of *bíos.* Thus Esposito repositions *bíos* as the living common to all beings that allows for individuation to take place, not through the notion of a common body—for that too assumes an immunizing function—but rather through a *bíos* that is inscribed in the flesh of the world. Those pages dedicated to Francis Bacon are significant here for Esposito sees in Bacon's paintings not only a reversal of the Nazi biopolitical practice of animalizing man, but also an opening to flesh as describing the condition of the majority of humanity. Or more than an opening to the category of flesh, we might well speak of a nonbelonging or an interbelonging among bodies that makes certain that what is different isn't closed hermetically within itself but remains in contact with the outside. Essentially, Esposito is describing not an exteriorization of the body but rather an internal, even Bataillian rending, that impedes the body's own absolute immanence. It is on this basis that an affirmative biopolitics can begin to be imagined.

The Biopolitics of Biotechnology

What does the opening to *bíos* as a political category that humanity shares tell us about that other development that so decidedly marks the current biopolitical moment, namely, biotechnology? The question isn't posed in the reflections and exchanges with regard to biotechnology between Jürgen Habermas and Ronald Dworkin; indeed, missing is precisely a reflection on the role biotechnology plays for contemporary biopolitics.[94] The uncovering of the immunitary paradigm in *Bíos,* however, allows us to see just where biopolitics and the ethical uncertainty surrounding biotechnology might intersect. Consider first Habermas's objection that genetic programming, which allows individuals to enhance what they believe to be the desirable features of future offspring, places the future of human nature at risk. Describing a new type of interpersonal relationship "that arises when a person makes an irreversible decision about the natural traits of another person," Habermas argues that our self-understanding as members of the species will be altered when a person or persons can manipulate the genetic basis of life of another; the basis of free societies that are premised on relations "between free and equal human beings" will be undermined. He adds: "This new type of relationship offends our moral sensibility because it constitutes a foreign body in the legally institutionalized relations of recognition in modern societies."[95] The reference to foreign bodies in new recognition protocols makes it clear that Habermas's language is one largely indebted to the language of immunity. What's more, the impression is that for Habermas symmetrical relations among the members of a group are homologous to the foundation of a moral and ethical community; he assumes something like an unproblematic origin of community that is both the cause and the effect of "human nature." With the genetic manipulation of the human, the development of certain individuals becomes unhinged from their free and unhindered growth. Knowing that others are responsible for who and what they are not only alters how they see themselves and the kinds of narratives they construct about their individual lives, but also jeopardizes how others will see them (as privileged, as escaping somehow from the natural development of characteristics that occur in interactions with others). These social foundations of society will be irreparably damaged when some members are allowed to intervene genetically in the development of others.

Certainly, Esposito's analysis in *Bíos* and elsewhere shares a number of features with Habermas's symptomology of a catastrophic neoliberal eugenic regime in which individual choice on future genetic programming operates, in not so different form, to immunize certain individuals from the community. But Esposito parts ways with Habermas in two areas. First, by disclosing the negative modality of community in immunity, Esposito deconstructs the transcendental conception of community that for Habermas is structured by "forms of communication through which we reach an understanding with one another."[96] For Esposito, there is no originary moment of individual self-understanding that brings together subjects to form a community, but rather an impolitical immunitary mechanism operating at the heart of the genesis of community: everyone is joined together in their subtraction from community to the degree the gift of the *munus* does not belong to the subject. There is "nothing in common," as he titles a chapter in *Communitas*, and hence no self-understanding that can bridge the irreducible difference between subjects. If there is to be a defense of community against the threat of future members whose genetically altered bodies undercut the shared life experiences of all, it cannot be premised on the effects of biotechnology to subtract certain members from the communal giving of the *munus*. A critique of the dangers of contemporary eugenics based on the threat it raises for the biological conformity of its members runs aground therefore on the impulse to create a transcending norm of biological life.

This by no means precludes a thoroughgoing critique on Esposito's part of the biopolitical lexicon in which neoliberal eugenic practices are inscribed. Although Esposito in *Bíos* doesn't discuss current neoliberal eugenics, certainly genetic programming cannot be thought apart from a history of twentieth-century immunizing biopolitics. Thus, in genetic enhancement one observes the domination of the private sphere in questions of public interest, which is captured in the blurring between therapeutic and enhancing interventions. As Esposito shows, such a blurring was already a part of early-twentieth-century eugenics beginning in the United States. The result is that in the realm of biotechnology and genetic engineering, politics continues to center on—Esposito will say to be crushed by—the purely biological. But there is more. Neoliberal eugenics often appears to combine within it the three immunitary procedures sketched above that Esposito locates in a Nazi thanatopolitics. The enormous influence that biologists enjoy today for how individual life may unfold later suggests that the absolute normativization of life has increased exponentially, witnessed in the example with which Esposito opens *Bíos* of the French child, born with serious genetic lesions, who sued his mother's doctor for a missed diagnosis. One can easily imagine other such cases in the near future in which a failure to intervene genetically might well lead to similar cases against parents or doctors. So too the second immunitary procedure in which the bodies of a future generation of genetically enhanced individuals can be said to belong no longer to themselves, but rather to the individuals who had earlier decided on their genetic makeup. A hereditary patrimony based on the elimination of weaker elements will occur no longer primarily through euthanasia or sterilization, but rather by selecting beforehand the desired characteristics. In this sense, where the bodies of the German people during Nazism were said to belong to the *Führer,* neoliberal eugenics disperses the choice to the marketplace and science that together will determine which genetic features are deemed of value. Thus, in ever more rapid fashion bioengineered bodies may be said to belong to the mechanisms of profit and science. So too the preemptory suppression of birth that now takes place routinely in those instances in which the risk of genetic defects surrounding a birth leads to early termination of the pregnancy. This is not to say, of course, that Nazi thanatopolitics and contemporary neoliberal eugenics are coterminus for Esposito. In his recent discussion of totalitarianism and biopolitics, Esposito anticipates objections to any kind of superimposition of Nazism and liberalism:

> If for Nazism man *is* his body and only his body, for liberalism, beginning with Locke, man *has* a body, which is to say he possesses his body—and therefore can use it, transform it, and sell it much like an internal slave. In this sense liberalism—naturally I'm speaking of the category that founds it—overturns the Nazi perspective, transferring the property of the body from the State to the individual, but within the same biopolitical lexicon.[97]

Here Esposito implicitly marks the shared vocabulary of liberalism that collaborates deeply with capitalism and twentieth-century thanatopolitics—not the double of Nazi biopolitics or its return, but their shared indebtedness to the terms of an immunizing modern biopolitics.

Dworkin and Life's Norm

The acuteness of Esposito's angle of vision on liberalism also allows us to situate his position with regard to Ronald Dworkin's discussion of abortion, euthanasia, and biotechnology. What we find is a thoroughgoing deconstruction of the biopolitical and immunizing features of many of the terms Dworkin employs. To review: in *Life's Dominion* from 1994, Dworkin speaks of the sacred and inviolable characteristics of "human life" in current debates on euthanasia and abortion in an attempt to undercut any arguments about the fetus as enjoying any intrinsic rights as a person. His argument hinges on a reading of the sacred as embedded in human and "artistic creation":

> Our special concern for art and culture reflects the respect in which we hold artistic creation, and our special concern for the survival of animal species reflects a parallel respect for what nature, understood as divine or as secular,

> has produced. These twin bases of the sacred come together in the case of survival of our own species, because we treat it as crucially important that we survive not only biologically but culturally, that our species not only lives but thrives.[98]

Naturally, the sacred life Dworkin defends is not *bíos* at all but what he calls subjective life, the "personal value we have in mind when we say that normally a person's life is the most important thing he or she has," which is to say bare life. Such a conflation of bare life and *bíos* accounts for his failure to think life across different forms; a sacred life is one limited almost entirely to bare life and hence to all the associations that it calls forth.

Not surprisingly, the emphasis he places on artistic and divine creation appears again in his most recent defense of biotechnology. There the inviolability of life is linked to a defense of biotechnology via the notion of creation. In an essay titled "Playing God," Dworkin strongly pushes for what appears to be a neoliberal eugenics program masked by the term "ethic individualism." "There is nothing in itself wrong," he writes, "with the detached ambition to make the lives of future generations of human beings longer and more full of talent and hence achievement." "On the contrary," he continues, "if playing God means struggling to improve our species, bringing into our conscious designs a resolution to improve what God deliberately or nature blindly has evolved over eons, then the first principle of ethical individualism commands that struggle, and its second principle forbids, in the absence of positive evidence of danger, hobbling the scientists and doctors who volunteer to lead it."[99] To the degree the weight we afford human lives is contingent on a notion of creation, the "playing God" of the title, biotechnology cannot be separated from the implicit sacred nature of created life in all its forms. The emphasis on creation (and not creationism, we should be clear) leads Dworkin down the path of a robust defense of biotechnology. Who, the argument runs, would disagree with the implicit desire of the not-yet-born individual to live a longer and more successful life?[100]

Here too Esposito offers a rejoinder. By focusing on the inviolability of individual human life, Dworkin fails to weigh properly the singularity of all life, which is to say that as long as the emphasis is placed on the individual and other traditional forms used to decline the subject, Dworkin's perspective on life is disastrous for any affirmative biopolitics. What's more, in such a scheme, ethic individualism quickly becomes the norm that transcends life; it is a norm of life that limits life to the confines of an individual subject and individual body; in this it operates, as it has traditionally done, to immunize the community and modernity itself, from the immanence of impersonal, singular life. Such an immanence Esposito anchors to the *bíos* of *communitas*—not one based, as Dworkin would have it, on a community of citizens who "recognize that the community has a communal life," but rather an ecumenical community that runs to all life-forms and one that is not always and everywhere transcended by notions of citizenship and individuality.[101] In other words, Dworkin's explicit linking of the "sacred" nature of biotechnology and bare life depends not simply on the function of creation but more importantly is riven through with a debt owed the notion of the individual. It isn't simply that the government and commerce ought to "fuel, restrain, or shape these developments [in biotechnology]," but rather that life understood as the opening to the impersonal singularity and to the trans- or preindividual cannot emerge as the immanent impulse of life so long as the norm of life is only thought in terms of the individual subject.[102] The open question is to what degree the marriage between biotechnology and the individual subject represents a radical jump in quality of the immunizing paradigm. How one answers that will determine the prospects for a coming, affirmative biopolitics.

A Fortified *Bíos*?

How, then, can we set about reversing the current thanatopolitical inflection of biotechnics and biopolitics? Esposito's final answer in *Bíos* will be found by rethinking precisely the relation between norm and life in opposition to Nazi semantics by developing another semantics in which no fundamental norm exists from which the others can be derived. This is because "every behavior carries with it the norm that places it in existence within a more general natural order. Considering that there are as many multiple individuals as there are infinite modes of substance means that the norms will be multiplied by a corresponding number."[103] Once the notion of individual no longer marks an individual subject but the process of individuation linked to the birth of all forms of life, our attention will then shift to producing a multiplicity of norms within the sphere of law. The individual will no longer be seen as simply the site in which previous genetic programming is executed, no mere hardware for a genetic software, but instead the space in which individuation takes place thanks to every living form's interdependence with other living forms. Norms for individuals will give way to individualizing norms that respect the fact that the human body "lives in an infinite series of relations with the bodies of others."[104] Here as elsewhere Esposito is drawing on Spinoza for his elaboration of a new, nonimmunitary semantics of a multiplicity of norms, in which norms cannot be thought outside the "movement of life," one in which the value of every norm is linked to its traducibility from one system to another. The result is the continual deconstruction of any absolute normative system, be it Nazi thanatopolitics or contemporary capitalist bioengineering of the human. The result is both a defense of difference among life-forms and their associated norms and an explicit critique of otherness, which for Esposito inevitably calls forth immunization from the implicit threat of contagion and death.[105] The emphasis on difference (and not otherness) among life-forms in the closing pages of *Bíos* is linked to change, which Esposito sees not only as a prerogative of the living, but as the basis for elaborating a radical tolerance toward a world understood as a multiplicity of different living forms.

The question, finally, is how to fortify a life's opening to other lives without at the same time inscribing it in an immunitary paradigm. For Esposito, the answer, as I suggested when addressing Dworkin's neoliberal perspective on biotechnology, lies in destabilizing the absolute immanence of the individual life by forgoing an emphasis on the individual life in favor of an "indefinite life." The reference to Deleuze's last essay, "Pure Immanence," allows Esposito to counterpose the absolute immanence of individual life to the absolute singularity of a "life." The relevant quote from Deleuze merits citation:

> The life of the individual gives way to an impersonal and yet singular life that releases a pure event freed from the accidents of internal and external life, that is, from subjectivity and objectivity of what happens: a "Homo tantum" with whom everyone empathizes and who attains a sort of beatitude. It is haecceity no longer of individuation but of singularization: a life of pure immanence, neutral beyond good and evil, for it was only the subject that incarnated it in the midst of things that made it good or bad. The life of such an individuality fades away in favor of the singular life immanent to a man who no longer has a name, though he can be mistaken for no other. A singular essence, a life.[106]

Esposito's excursus on flesh and individuating birth attempts to articulate the necessary conditions in which the characteristics of just such a singular *homo tantum* can be actualized; implicit in the figure of the *homo tantum* is a "norm of life that doesn't subject life to the transcendence of a norm, but makes the norm the immanent impulse of life."[107] If we were to express such a figure biopolitically, the category of *bíos* will name the biopolitical thought that is able to think life across all its manifestations or forms as a unity. There is no *zōē* that can be separated from *bíos* because "every life is a form of life and every form refers to life."[108] Esposito here translates Deleuze's singular life as the reversal of the thanatopolitics he sees underpinning the Nazi normative project in which some lives were not considered forms and hence closed off from *bíos.* The opening to an affirmative biopolitics takes place precisely when we recognize that harming one part of life or one life harms all lives. The radical toleration of life-forms that epitomizes Esposito's reading of contemporary biopolitics is therefore based on the conviction that every life is inscribed in *bíos.*

No greater obstacle to fortifying *bíos* exists today than those biopolitical practices that separate out *zōē* from *bíos,* practices that go hand in hand with the workings of the immunitary paradigm. Esposito seems to be suggesting that our opening to an affirmative biopolitics becomes thinkable only when a certain moment has been reached when a philosophy of life appears possible in the folds of an ontology of death, when the immunitary mechanisms of the twenty-first century reach the point of no return. In such an event, when the immunitary apparatus attacks *bíos* by producing *zōē,* a space opens in which it becomes possible to posit *bíos* as not in opposition to *zōē* but as its ultimate horizon. Thus the subject of *Bíos* is life at the beginning of the twenty-first century, its fortunes inextricably joined to a ductile immunitary mechanism five hundred years or so in operation. Five hundred years is a long time, but the conditions, Esposito argues, may be right for a fundamental and long overdue rearticulation or reinscription of *bíos* in a still to be completed political lexicon that is radically humanistic to the degree that there can be no *zōē* that isn't already *bíos.* One of the shorthands Esposito offers in *Bíos* for thinking the difference will be found in the juxtaposition between a "politics of mastery and the negation of life" and another future, affirmative politics of life.[109]

Life as *Bíos*

These are, it seems to me, the most significant elements of Esposito's genealogy and ontology of contemporary biopolitics. What I would like to do in the remaining pages is to suggest possible areas of contact between *Bíos* and contemporary public culture.

Esposito's uncovering of the reciprocity between community and immunity captures brilliantly the stalemate that continues to characterize debates about the choice between security and freedom. One need only recall the Patriot Act and the justification for its attacks on civil liberties in the name of "homeland security" to see where the disastrous effects of excessive immunization on a community will be registered: precisely in immunity's closing to community. Once we see immunity/community as a continuum we can understand the precise meaning of "the war on terror begins at home" as directed against the radical opening to social relations that are implicit in the gift and obligation of the *munus*, both globally and locally. We are living, Esposito suggests, in one of the most lethal immunitary mechanisms of the modern period, lethal for both global relations, which now are principally based on war, and the concurrent repression sanctioned by security concerns. As I have noted repeatedly, recognizing the dangers of immunization for meaningful and productive relations between individual members and among communities doesn't in any way lead Esposito, however, to argue for a return to some privileged origin of community. Attempts to locate such an origin are doomed to a melancholic search for community that can never be met. At the same time, recognizing the futility of such a search creates an opportunity, thanks to the contemporary immunity crisis, to think again what the basis for community might be. What needs to take place therefore is thinking through a dialectic of how to singularize "we." Esposito's itinerary that moves through immunities that fortify singular "we's" thanks to the articulation of individuation can help make us not only more attentive to our encounters with others and the other, but also to examine more deeply the kinds of motivations that undergird these kinds of encounters.

Obviously, the opportunity for thinking anew the assumptions on which communities come together will have a profound impact on the kind of public culture we wish for ourselves. What kind of public culture, for instance, makes possible and nourishes an opening to the common flesh of all, one that is capable of vitalizing all forms of life? Is there already implicit in the notion of public culture a private space that can have no truck with the kinds of retooled relations Esposito is describing? These kinds of questions are not easily asked in the current war on terror, a war founded precisely on excluding "terrorists" from the horizon of *bíos*, that is, as forms of life (now enemy combatants) who do not merit any political qualification. Thus, when President Bush speaks of terrorism as representing "a mortal danger to all humanity" or when he describes "tense borders" under assault, the implicit connection to an immunitary paradigm becomes obvious.[110]

It is because terrorism represents a war on humanity that it is a war against life itself, that borders must be defended and strengthened. Not simply geographic borders but, more significantly, the borders of the kind of life that can and cannot be inscribed in *bíos.* The result is once again the politicization of life and with it the demarcation of those lives outside *bíos.* The effect of limiting *bíos* to only those on one side of the border isn't simply to mark, however, those who can be sacrificed as *homo sacer,* as Agamben would have it, but rather to attack with violence the *munus* immunity shares with community. Interestingly, in some of his speeches President George W. Bush also speaks of liberty as the vital catalyst for improving "the lives of all"; leaving aside just what he intends for liberty, clearly today liberty is disclosed ever more readily as an effect of the immunity modality, much as Esposito describes it in those pages dedicated to Locke.[111] In perhaps more obvious fashion than in recent memory, liberty is spectacularly reduced to the security of the subject; a subject who possesses liberty is the secure(d) citizen. Although Esposito doesn't elaborate on the relation of the modern subject to the citizen—as the closing pages of *Bíos* make clear, his research is moving necessarily toward a genealogy of "the person"—he does explicitly suggest that a semantics of the individual or the citizen has always functioned within an immunitary paradigm.[112] As tempting as it might be to read liberty as a vital multiplier of community in opposition to immunity, such a strategy is doomed to failure as well, given liberty's historical failure to maintain any autonomy with regard to the protection of life.

If we read Esposito carefully, the first step to a public culture made vital by *communitas* begins with the recognition that the lives of "terrorists" can in no way be detached from a political qualification that is originary to life. Rather than merely agreeing to their exteriorization to *bíos,* which appears as both an ethical and a philosophical failure of enormous magnitude, what we need to do is to understand and practice differently the unity of *bíos* and politics in such a way that we no longer reinforce the politicization of life (which is precisely what the war on terror is intended to do), but instead create the conditions for what he calls a "vitalization of politics."[113] No greater task confronts us today than imagining the form such a vitalized politics might take, as that is precisely the direction in which an originary and intense sense of *communitas* resides.

Bíos

Introduction

France, November 2000. A decision of the French Appeals Court opens a lacerating conflict in French jurisprudence. Two appeals are overturned, which had in turn reversed the previous sentences. The court recognized that a baby by the name of Nicolas Perruche, who was born with serious genetic lesions, had the right to sue the doctor who had misdiagnosed a case of German measles in the pregnant mother. Against her expressed wishes, she was prevented from aborting. What appears to be the legally irresolvable object of controversy in the entire incident is attributing to small Nicolas the right *not* to be born. At issue is not the proven error of the medical laboratory, but rather the status of the subject who contests it. How can an individual have legal recourse against the only circumstance that furnishes him with juridical subjectivity, namely, that of his own birth? The difficulty is both of a logical and an ontological order. If it is already problematic that a being can invoke his or her right not to be, it is even more difficult to think of a nonbeing (which is precisely who has not yet been born) that claims the right to remain as such, and therefore not to enter into the sphere of being. What appears undecidable in terms of the law is the relation between biological realty and the juridical person, that is, between natural life and a form of life. It is true that being born into such conditions, the baby incurred harm. But who if not he himself could have decided to avoid it, eliminating beforehand his own being as the subject of life, the life proper of a subject? Not only. Because every subjective right corresponds to the obligation of not obstructing those who are in a condition to do so signifies that the mother would have been forced to abort irrespective of her choice. The right of the fetus not to be born would be configured therefore as a preventive duty on the part of the person who conceived to eliminate him *[sopprimerlo]*, instituting in such a way a eugenic caesura, one that is legally recognized, between a juridical life that is judged as valid and another "life unworthy of life," to use the Nazi phrase.

Afghanistan, November 2001. Two months after the terrorist attacks of September 11, a new kind of "humanitarian" war takes shape in the skies above Afghanistan. The adjective *humanitarian* no longer concerns the reasons behind the conflict—as had occurred in Bosnia and Kosovo, namely, to defend entire populations from the threat of ethnic genocide—but its privileged instrument, which is to say air bombardments. And so we find that both highly destructive bombs were released along with provisions and medicine on the same territory at the same time. We must not lose sight of the threshold that is crossed here. The problem doesn't lie only in the dubious juridical legitimacy of wars fought in the name of universal rights on the basis of arbitrary or biased decisions on the part of those who had the force to impose and execute them, and not even in the lack of uniformity often established between proposed ends and the results that are obtained. The most acute oxymoron of humanitarian bombardment lies rather in the superimposition that is manifested in it between the declared intention to defend life and to produce actual death. The wars of the twentieth century have made us accustomed to the reversal of the proportion between military deaths (which was largely the case before) and civilian victims (which are today far superior to the former). From time immemorial racial persecutions have been based on the presupposition that the death of some strengthens the life of others, but it is precisely for this reason that the demarcation of a clear division between lives to destroy and lives to save endures and indeed grows. It is precisely such a distinction that is tendentiously erased in the logic of bombardments that are destined to kill and protect the same people. The root of such an indistinction is not to be sought, as is often done, in a structural mutation of war, but rather in the much more radical transformation of the idea of *humanitas* that subtends it. Presumed for centuries as what places human beings *[gli uomini]* above the simple common life of other living species (and therefore charged with a political value), *humanitas* increasingly comes to adhere to its own biological material. But once it is reduced to its pure vital substance and for that reason removed from every juridical-political form, the humanity of man remains necessarily exposed to what both saves and annihilates it.

Russia, October 2002. Special groups of the Russian state police raid the Dubrovska Theater in Moscow, where a Chechen commando unit is holding almost a thousand people hostage. The incursion results in the death of 128 hostages as well as almost all of the terrorists thanks to an incapacitating and lethal gas. The episode, justified and indeed praised by other governments as a model of firmness, marks another step with respect to the others I've already described. Even if in this case the term "humanitarian" was not used, the underlying logic is no different: the deaths here emerge out of the same desire to save as many lives as possible. Without lingering over other troubling circumstances (such as the use of a gas that was prohibited by international treaties or the impossibility of making available adequate antidotes while keeping secret their very nature), let's consider the point that interests us most. The death of the hostages wasn't an indirect and accidental effect of the raid by law enforcement, which can happen in cases such as these. It wasn't the Chechens, who, surprised by the police assault, killed the hostages, but the police who killed them directly. Frequently one speaks of the specularity of the methods between terrorists and those that face off against them. This is understandable and under certain limits inevitable. But never before does one see governmental agents, charged with saving prisoners from a possible death, carry out the massacre themselves, which the terrorists had themselves only threatened. Various factors weighed in the Russian president's decision: the desire to discourage other attempts of the sort; the message to the Chechens that their fight had no hope of succeeding; and a display of sovereign power in a time of its apparent crisis. But, fundamentally, something else constitutes its tacit assumption. The blitz on the Dubrovska Theater not only marks, as I said, the withdrawal of politics in the face of brute force, nor is it irreducible to the unveiling of an originary connection between politics and evil *[male]*. It is the extreme expression that politics can assume when it faces, without any mediation, the question of the survival of human beings suspended between life and death. To keep them alive at all cost, one can even decide to hasten their death.

China, February 2003. The Western media circulates the news (strongly censored by the Chinese government) that in the sole province of Henan there are a million and a half Chinese who are seropositive, with some villages such as Donghu having a percentage that reaches upwards of 80 percent of the population. Unlike other Third World countries, the contagion does not have a natural or a sociocultural cause, but an immediate economic and political one. At its origin is not unprotected sexual relations nor dirty drug needles, but rather the sale en masse of blood, which the central government encouraged and organized. The blood, which the government had extracted from peasants who were in need of money, was centrifuged in large containers that separated the plasma from the red globules. While the former was sent to rich buyers, the latter was again injected into the donors so as to avoid anemia and to force them into repeating the operation. But it only took one of them to be infected to contaminate the entire stock of blood contained in the huge cauldrons. Thus, entire villages were filled with those who were seropositive, which, given the lack of medicine, became a death sentence. It is true that China has recently sold cheap anti-AIDS medicines produced locally on the market, but it did not make them available to the peasants of Henan, whom it not only ignored, but whom it obliged to keep quiet at the risk of imprisonment. The affair was revealed by someone who, left alone after the deaths of his relatives, preferred dying in prison rather than in his own hut alone. It's enough to move our gaze onto another, larger phenomenon to see that biological selection in a country that continues to define itself as communist isn't only of class, but also of sex. This happens at the moment when the state policy of "a single child" (which was intended to halt a growing demographic) is joined to the technology of ecography, causing the abortion of a large number of those who would have become future women. This made the former traditional practice in the countryside, of drowning female infants upon birth, unnecessary, but it was bound to augment the numerical disproportion between males and females. It has been calculated that in less than twenty years it will be difficult for Chinese men to find a wife, if they don't tear her away from her family as an adolescent. Perhaps it's for this reason that in China the relation between female and male suicides is five to one.

Rwanda, April 2004. A United Nations report tells us that around ten thousand babies of the same age are the biological result of mass ethnic rapes that occurred ten years ago during the genocide that the Hutu committed on the Tutsi. As occurred later in Bosnia and other parts of the world, such a practice modified in original ways the relation between life and death that had until then been recognized in traditional wars and even in those so-called asymmetrical wars against terrorists. While in these wars death always comes from life—and even comes *through* life as in kamikaze suicide attacks—in the act of ethnic rape it is also life that emerges from death, from violence, and from the terror of women who were made pregnant while unconscious from the blows they had received or immobilized with a knife to their throat. It is an example of "positive" eugenics that is not juxtaposed to the negative one practiced in China or elsewhere, but rather constitutes its counterfactual result. Whereas the Nazis and all their imitators carried out genocide by preemptively destroying birth, those of today do so through forced birth and therefore in the most drastic perversion of the event that brings essence to self *[in sé l'essenza]*, other than the promise of life. Contrary to those who saw in the newness of birth the symbolic and real presupposition for renewed political action, ethnic rape makes it the most acute point of connection between life and death, but which occurs in the tragic paradox of a new generation of life. That all Rwandan mothers of the war, when asked about their own experiences, declared their love for their children born from hate signifies that the force of life prevails once again over that of death. Furthermore, the most extreme immunitary practice, which is to say affirming the superiority of one's own blood to the point of imposing it on those with whom one does not share it, is destined to be turned against itself, producing exactly what it wanted to avoid. The Hutu children of Tutsi women, or the Tutsi children of Hutu men, are the objective communitarian, which is to say multiethnic outcome of the most violent racial immunization. We are faced here too with a sort of undecidability, or a double-faced phenomenon in which life and politics are joined in a relation whose interpretation demands a new conceptual language.

At the center of such a language is the notion of biopolitics. It is by starting with biopolitics that events such as those I've just described, which escape a more traditional interpretation, find a complex of meaning that moves beyond their simple manifestation. It is true that they provide an extreme image (though certainly not unfaithful) of a dynamic that already involves all the most important political phenomena of our time. From the war of and against terrorism to mass migrations; from the politics of public health to those of demography; from measures of security to the unlimited extension of emergency legislation—there is no phenomenon of international importance that is extraneous to the double tendency that situates the episodes I've just described within a single of line of meaning. On the one hand, a growing superimposition between the domain of power or of law *[diritto]* and that of life; on the other, an equally close implication that seems to have been derived with regard to death. It is exactly the tragic paradox that Michel Foucault, in a series of writings dating back to the middle of the 1970s, examined. Why does a politics of life always risk being reversed into a work of death?

I think I can say, without failing to acknowledge the extraordinary analytic power of his work, that Foucault never fully answered the question; or better, that he always hesitated choosing from among different responses, responses that were for their part tributaries of different modes of approaching the question that he himself had raised. The opposite interpretations of biopolitics, the one radically negative and the other absolutely euphoric that today lead the field, do nothing except make absolute (by spreading them apart) the two hermeneutic options between which Foucault never decided. Without anticipating here a more detailed reconstruction of the affair, my impression is that this situation of philosophical and political stalemate originates with a question that is either missing or has been insufficiently posed concerning the presuppositions of the theme in question: not just what biopolitics signifies but how it was born. How is it configured over time and which aporias does it continue to carry? It's enough to extend research on the diachronic axis as well the horizontal level to recognize that Foucault's decisive theorizations are nothing but the final segment (as well as the most accomplished) of a line of discourse that goes rather further back in time, to the beginning of the last century. To bring to light this lexical tradition (for the first time I would add), revealing its contiguity and semantic intervals, obviously doesn't only have a philological emphasis, because only a similar kind of operation of excavation promotes the force and originality of Foucault's thesis through differences with it; but above all because it allows us to peer into the black box of biopolitics from a variety of angles and with a greater breadth of gaze. It becomes possible to construct a critical perspective on the interpretive path that Foucault himself created; for example, with reference to the complex relationship, which he instituted, between the biopolitical regime and sovereign power. We will return in more detail to this specific point further on, but what ought to draw our attention—because it involves the very same meaning of the category in question—is the relation between the politics of life and the ensemble of modern political categories. Does biopolitics precede, follow, or coincide temporally with modernity? Does it have a historical, epochal, or originary dimension? Foucault's response to such a question is not completely clear, a question that is decisive because it is logically connected to the interpretation of contemporary experience. He oscillates between a continuist attitude and another that is more inclined to mark differential thresholds.

My thesis is that this kind of an epistemological uncertainty is attributable to the failure to use a more ductile paradigm, one that is capable of articulating in a more intrinsic manner the two lemmas that are enclosed in the concept in question, which I have for some time now referred to in terms of immunization. Without expanding here on its overall meaning (which I've had occasion to define elsewhere in all its projections of sense), the element that quickly needs to be established is the peculiar knot that immunization posits between biopolitics and modernity.[1] I say quickly because it restores the missing link of Foucault's argumentation. What I want to say is that only when biopolitics is linked conceptually to the immunitary dynamic of the negative protection of life does biopolitics reveal its specifically modern genesis. This is not because its roots are missing in other preceding epochs (they aren't), but because only modernity makes of individual self-preservation the presupposition of all other political categories, from sovereignty to liberty. Naturally, the fact that modern biopolitics is also embodied through the mediation of categories that are still ascribable to the idea of order (understood as the transcendental of the relation between power and subjects) means that the politicity of *bíos* is still not affirmed absolutely. So that it might be, which is to say so that life is *immediately* translatable into politics or so that politics might assume an *intrinsically* biological characterization, we have to wait for the totalitarian turning point of the 1930s, in particular for Nazism. There, not only the negative (which is to say the work of death) will be functionalized to stabilize order (as certainly was still the case in the modern period), but it will be produced in growing quantities according to a thanatopolitical dialectic that is bound to condition the strengthening of life vis-à-vis the ever more extensive realization of death.

In the point of passage from the first to the second form of immunization will be found the works of Nietzsche, to whom I've dedicated an entire chapter of this book. I have done so not only for his underlying biopolitical relevance, but because he constitutes an extraordinary seismograph of the exhaustion of modern political categories when mediating between politics and life. To assume the will of power as the fundamental vital impulse means affirming at the same time that life has a constitutively political dimension and that politics has no other object than the maintenance and expansion of life. It is precisely in the relationship between these two ultimate modes of referring to *bíos* that the innovative or conservative, or active or reactive character of forces facing each other is established. Nietzsche himself and the meaning of his works is part of this comparison and struggle, in the sense that together they express the most explicit criticism of the modern immunitary loss of meaning and an element of acceleration from within. From here a categorical as well as stylistic splitting occurs between two tonalities of thought juxtaposed and interwoven that constitutes the most typical cipher of the Nietzschean text: destined on the one side to anticipate, at least on the theoretical level, the destructive and self-destructive slippage of twentieth-century biocracy, and on the other the prefiguration of the lines of an affirmative biopolitics that has yet to come.

The final section of the book is dedicated to the relation between philosophy and biopolitics *after Nazism.* Why do I insist on referring philosophy to what wanted to be the most explicit negation of philosophy as ever appeared? Well, first because it is precisely a similar negation that demands to be understood philosophically in its darkest corners. And then because Nazism negated philosophy not only generically, but in favor of biology, of which it considered itself to be the most accomplished realization. I examine in detail this thesis in an extensive chapter here, corroborating its truthfulness, at least in the literal sense that the Nazi regime brought the biologization of politics to a point that had never been reached previously. Nazism treated the German people as an organic body that needed a radical cure, which consisted in the violent removal of a part that was already considered spiritually dead. From this perspective and in contrast to communism (which is still joined in posthumous homage to the category of totalitarianism), Nazism is no longer inscribable in the self-preserving dynamic of both the early and later modernities; and certainly not because it is extraneous to immunitary logic. On the contrary, Nazism works within that logic in such a paroxysmal manner as to turn the protective apparatus against its own body, which is precisely what happens in autoimmune diseases. The final orders of self-destruction put forward by Hitler barricaded in his Berlin bunker offer overwhelming proof. From this point of view, one can say that the Nazi experience represents the culmination of biopolitics, at least in that qualified expression of being absolutely indistinct from its reversal into thanatopolitics. But precisely for this reason the catastrophe in which it is immersed constitutes the occasion for an epochal rethinking of a category that, far from disappearing, every day acquires more meaning, not only in the events I noted above, but also in the overall configuration of contemporary experience, and above all from the moment when the implosion of Soviet communism cleared the field of the last philosophy of modern history, delivering us over to a world that is completely globalized.

It is at this level that discourse today is to be conducted: the body that experiences ever more intensely the indistinction between power and life is no longer that of the individual, nor is it that sovereign body of nations, but that body of the world that is both torn and unified. Never before as today do the conflicts, wounds, and fears that tear the body to pieces seem to put into play nothing less than life itself in a singular reversal between the classic philosophical theme of the "world of life" and that theme heard so often today of the "life of the world." This is the reason that contemporary thought cannot fool itself (as still happens today) in belatedly defending modern political categories that have been shaken and overturned. Contemporary thought cannot and must not do anything of the sort, because biopolitics originates precisely in these political categories, before it rebels against them; and then because the heart of the problem that we are facing, which is to say the modification of *bíos* by a part of politics identified with technology *[tecnica]*, was posed for the first time (in a manner that would be insufficient to define as apocalyptic), precisely in the antiphilosophical and biological philosophy of Hitlerism. I do realize how delicate this kind of statement may seem in its contents and still more in its resonance, but it isn't possible to place questions of expediency before the truth of the matters at hand. From another perspective, twentieth-century thought has from the beginning implicitly understood this, accepting the comparison and the struggle with radical evil on its own terrain. It was so for Heidegger, along an itinerary that brought him so close to that vortex that he risked letting himself be swallowed by it. But the same was also true for Arendt and Foucault, both of whom were conscious, albeit in different ways, that one could rise above Nazism only by knowing its drifts and its precipices. It is the path that I myself have tried to follow here, working back to front within three Nazi *dispositifs:* the *absolute normativization of life,* the *double enclosure of the body,* and the *anticipatory suppression of birth.* I have traced them with the intention of profiling the admittedly approximate and provisional contours of an affirmative biopolitics that is capable of overturning the Nazi politics of death in a politics that is no longer over life but *of* life.

Here there is a final point that seems to me useful to clarify before proceeding. Without denying the legitimacy of other interpretions or other normative projects, I do not believe the task of philosophy—even when biopolitics challenges it—is that of proposing models of political action that make biopolitics the flag of a revolutionary manifesto or merely something reformist. This isn't because it is too radical a concept but because it isn't radical enough. This would, moreover, contradict the initial presupposition according to which it is no longer possible to disarticulate politics and life in a form in which the former can provide orientation to the latter. This is not to say, of course, that politics is incapable of acting on what is both its object and subject; loosening the grip of new sovereign powers is possible and necessary. Perhaps what we need today, at least for those who practice philosophy, is the converse: not so much to think life as a function of politics, but to think politics within the same form of life. It is a step that is anything but easy because it would be concerned with bringing life into relation with biopolitics not from the outside—in the modality of accepting or refusing—but from within; to open life to the point at which something emerges which had until today remained out of view because it is held tightly in the grip of its opposite. I have attempted to offer more than one example of such a possibility and of such a demand with regard to the figure of *flesh, norm,* and *birth* thought inversely with respect to body, law, and nation. But the most general and intense dimension of this constructive deconstruction has to do precisely with that immunitary paradigm that constitutes the distinctive mode in which biopolitics has until now been put forward. Never more than in this case does its semantics, that of the negative protection of life, reveal a fundamental relation with its communitarian opposite. If *immunitas* is not even thinkable outside of the common *munus* that also negates it, perhaps biopolitics, which until now has been folded tightly into it, can also turn its negative sign into a different, positive sense.

CHAPTER ONE

The Enigma of Biopolitics

Bio/politics

Recently, not only has the notion of "biopolitics" moved to the center of international debate, but the term has opened a completely new phase in contemporary thought. From the moment that Michel Foucault reproposed and redefined the concept (when not coining it), the entire frame of political philosophy emerged as profoundly modified. It wasn't that classical categories such as those of "law" *[diritto]*, "sovereignty," and "democracy" suddenly left the scene—they continue to organize current political discourse—but that their effective meaning always appears weaker and lacking any real interpretive capacity. Rather than explaining a reality that everywhere slips through their analytic grip, these categories themselves demand to be subjected to the scrutiny of a more penetrating gaze that both deconstructs and explains them. Let's consider, for instance, law *[legge]*. Differently from what many have argued, there is nothing that suggests that such a domain has somehow been reduced. On the contrary, the impression is that the domain of law is gaining terrain both domestically and internationally; that the process of normativization is investing increasingly wider spaces. Nevertheless, this doesn't mean that juridical language per se reveals itself to be incapable of illuminating the profound logic of such a change. When one speaks of "human rights," for example, rather than referring to established juridical subjects, one refers to individuals defined by nothing other than the simple fact of being alive. Something analogous can be said about the political *dispositif* of sovereignty. Anything but destined to weaken as some had rashly forecast (at least with regard to the world's greatest power), sovereignty seems to have extended and intensified its range of action—beyond a repertoire that for centuries had characterized its relation to both citizens and other state structures. With the clear distinction between inside and outside weakened (and therefore also the distinction between war and peace that had characterized sovereign power for so long), sovereignty finds itself directly engaged with questions of life and death that no longer have to do with single areas, but with the world in all of its extensions. Therefore, if we take up any perspective, we see that something that goes beyond the customary language appears to involve directly law and politics, dragging them into a dimension that is outside their conceptual apparatuses. This "something"—this element and this substance, this substrate and this upheaval—is precisely the object of biopolitics.

Yet there doesn't appear to be an adequate categorical exactitude that corresponds to the epochal relevance of biopolitics. Far from having acquired a definitive order, the concept of biopolitics appears to be traversed by an uncertainty, by an uneasiness that impedes every stable connotation. Indeed, I would go further. Biopolitics is exposed to a growing hermeneutic pressure that seems to make it not only the instrument but also the object of a bitter philosophical and political fight over the configuration and destiny of the current age. From here its oscillation (though one could well say its disruption) between interpretations, and before that even its different, indeed conflicting tonalities. What is at stake of course is the nature of the relation that forces together the two terms that make up the category of biopolitics. But even before that its definition: what do we understand by *bíos* and how do we want to think a politics that directly addresses it? The reference to the classic figure of *bíos politikos* doesn't help, since the semantics in question become meaningful precisely when the meaning of the term withdraws. If we want to remain with the Greek (and in particular with the Aristotelian) lexicon, biopolitics refers, if anything, to the dimension of *zōē*, which is to say to life in its simple biological capacity *[tenuta]*, more than it does to *bíos*, understood as "qualified life" or "form of life," or at least to the line of conjugation along which *bíos* is exposed to *zōē*, naturalizing *bíos* as well. But precisely with regard to this terminological exchange, the idea of biopolitics appears to be situated in a zone of double indiscernibility, first because it is inhabited by a term that does not belong to it and indeed risks distorting it. And then because it is fixed by a concept, precisely that of *zōē*, which is stripped of every formal connotation. *Zōē* itself can only be defined problematically: what, assuming it is even conceivable, is an absolutely natural life? It's even more the case today, when the human body appears to be increasingly challenged and also literally traversed by technology *[tecnica]*.[1] Politics penetrates directly in life and life becomes other from itself. Thus, if a natural life doesn't exist that isn't at the same time technological as well; if the relation between *bíos* and *zōē* needs by now (or has always needed) to include in it a third correlated term, *technē*—then how do we hypothesize an exclusive relation between politics and life?

Here too the concept of biopolitics seems to withdraw or be emptied of content in the same moment in which it is formulated. What remains clear is its negative value, what it is *not* or the horizon of sense that marks its closing. Biopolitics has to do with that complex of mediations, oppositions, and dialectical operations that in an extended phase made possible the modern political order, at least according to current interpretation. With respect to these and the questions and problems to which they correspond relative to the definition of power, to the measure of its exercise and to the delineation of its limits, it's indisputable that a general shift of field, logic, and the object of politics has taken place. At the moment in which on one side the modern distinctions between public and private, state and society, local and global collapse, and on the other that all other sources of legitimacy dry up, life becomes encamped in the center of every political procedure. No other politics is conceivable other than a politics of life, in the objective and subjective sense of the term. But it is precisely with reference to the relation between the subject and object of politics that the interpretive divergence to which I alluded earlier appears again: How are we to comprehend a political government of life? In what sense does life govern politics or in what sense does politics govern life? Does it concern a governing *of* or *over* life? It is the same conceptual alternative that one can express through the lexical bifurcation between the terms, used indifferently sometimes, of "biopolitics" and "biopower." By the first is meant a politics in the name of life and by the second a life subjected to the command of politics. But here too in this mode the paradigm that seeks a conceptual linking between the terms emerges as split, as if it had been cut in two by the very same movement. Compressed (and at the same time destabilized) by competing readings and subject to continuous rotations of meaning around its own axis, the concept of biopolitics risks losing its identity and becoming an enigma.

To understand why, it isn't enough to limit our perspective simply to Foucault's observations. Rather, we need to return to those texts and to authors (often not cited) that Foucault's discussion derives from, and against which he repositions himself, while critically deconstructing them. These can be cataloged in three distinct and successive blocks in time (at least those that explicitly refer to the concept of biopolitics). They are characterized, respectively, by an approach that is organistic, anthropological, and naturalistic. In the first instance, they refer to a substantial series of essays, primarily German, that are joined by a vitalistic conception of the state, such as Karl Binding's *Zum Werden und Leben der Staaten* (1920), of which we will have occasion to speak later; Eberhard Dennert's *Der Staat als lebendiger Organismus* (1920); and Edward Hahn's *Der Staat, ein Lebenwesen* (1926).[2] Our attention will be focused, however, most intently on the Swede Rudolph Kjellén, probably because he was the first to employ the term "biopolitics" (we also owe him the expression "geopolitics" that Friedrich Ratzel and Karl Haushofer will later elaborate in a decidedly racist key). With respect to such a racist propensity, which will shortly thereafter culminate in the Nazi theorization of a "vital space" *(Lebensraum)* we should note that Kjellén's position remains less conspicuous, despite his proclaimed sympathy for Wilhelminian German as well as a certain propensity for an aggressive foreign policy. As he had previously argued in his book of 1905 on the great powers, vigorous states, endowed with a limited territory, discover the need for extending their borders through the conquest, fusion, and colonialization of other lands.[3] But it's in the volume from 1916 titled *The State as Form of Life* that Kjellén sees this geopolitical demand as existing in close relation to an organistic conception that is irreducible to constitutional theories of a liberal framework.[4] While these latter represent the state as the artificial product of a free choice of individuals that have created it, he understands it to be a "living form" (*som livsform* in Swedish or *als Lebensform* in German), to the extent that it is furnished with instincts and natural drives. Already here in this transformation of the idea of the state, according to which the state is no longer a subject of law born from a voluntary contract but a whole that is integrated by men and which behaves as a single individual both spiritual and corporeal, we can trace the originary nucleus of biopolitical semantics. In *Outline for a Political System,* Kjellén brings together a compendium of the preceding theses:

> This tension that is characteristic of life itself... pushed me to denominate such a discipline *biopolitics*, which is analogous with the science of life, namely, biology. In so doing we gain much, considering that the Greek word *bíos* designates not only natural and physical life, but perhaps just as significantly cultural life. Naming it in this way also expresses that dependence of the laws of life that society manifests and that promote, more than anything else, the state itself to that role of arbiter or at a minimum of mediator.[5]

These are expressions that take us beyond the ancient metaphor of the body-state with all its multiple metamorphoses of post-Romantic inspiration. What begins to be glimpsed here is the reference to a natural substrate, to a substantial principle that is resistant and that underlies any abstraction or construction of institutional character. The idea of the impossibility of a true overcoming of the natural state in that of the political emerges in opposition to the modern conception derived from Hobbes that one can preserve life only by instituting an artificial barrier with regard to nature, which is itself incapable of neutralizing the conflict (and indeed is bound to strengthen it). Anything but the negation of nature, the political is nothing else but the continuation of nature at another level and therefore destined to incorporate and reproduce nature's original characteristics.

If this process of the naturalization of politics in Kjellén remains inscribed within a historical-cultural apparatus, it experiences a decisive acceleration in the essay that is destined to become famous precisely in the field of comparative biology. I am referring to *Staatsbiologie*, which was also published in 1920 by Baron Jakob von Uexküll with the symptomatic subtitle *Anatomy, Physiology, and Pathology of the State.*[6] Here, as with Kjellén, the discourse revolves around the biological configuration of a state-body that is unified by harmonic relations of its own organs, representative of different professions and competencies, but with a dual (and anything but irrelevant) lexical shift with respect to the preceding model. Here what is spoken about is not any state but the German state with its peculiar characteristics and vital demands. What makes the difference, however, is chiefly the emphasis that pathology assumes with respect to what is subordinated to it, namely, anatomy and physiology. Here we can already spot the harbinger of a theoretical weaving—that of the degenerative syndrome and the consequent regenerative program—fated to reach its macabre splendors in the following decades. Threatening the public health of the German body is a series of diseases, which obviously, referring to the revolutionary traumas of the time, are located in subversive trade unionism, electoral democracy, and the right to strike: tumors that grow in the tissues of the state, causing anarchy and finally the state's dissolution. It would be "as if the majority of the cells in our body (rather than those in our brain) decided which impulses to communicate to the nerves."[7] But even more relevant, if we consider the direction of future totalitarian developments, is the biopolitical reference to those "parasites" which, having penetrated the political body, organize themselves to the disadvantage of other citizens. These are divided between "symbionts" from different races who under certain circumstances can be useful to the state and true parasites, which install themselves as an extraneous living body within the state, and which feed off of the same vital substance. Uexküll's threateningly prophetic conclusion is that one needs to create a class of state doctors to fight the parasites, or to confer on the state a medical competency that is capable of bringing it back to health by removing the causes of the disease and by expelling the carriers of germs. He writes: "What we are still lacking is an academy with a forward-looking vision not only for creating a class of state doctors, but also for instituting a state system of medicine. We possess no organ to which we can trust the hygiene of the state."[8]

The third text that should hold our attention—because it is expressly dedicated to the category in question—is *Bio-politics.* Written by the Englishman Morley Roberts, it was published in London in 1938 with the subtitle *An Essay in the Physiology, Pathology and Politics of the Social and Somatic Organism.*[9] Here too the underlying assumption, which Roberts sets forth immediately in the book's introduction, is the connection, not only analogical, but real, between politics and biology, and particularly medicine. His perspective is not so distant fundamentally from that of Uexküll. If physiology is indivisible from the pathology from which it derives its meaning and emphasis, the state organism cannot be truly known or guided except by evaluating its actual and potential diseases. More than a simple risk, these diseases represent the ultimate truth because it is principally a living entity that in fact can die. For this reason, biopolitics has the assignment on the one hand of recognizing the organic risks that jeopardize the body politic and on the other of locating and predisposing mechanisms of defense against them; these too are rooted in the same biological terrain. The most innovative part of Roberts's book is connected precisely to this ultimate demand and is constituted by an extraordinary comparison between the defensive apparatus of the state and the immunitary system that anticipates an interpretive paradigm to which we will return:

> The simplest way to think of immunity is to look on the human body as a complex social organism, and the national organism as a simpler functional individual, or "person," both of which are exposed to dangers of innumerable kinds for which they must continually provide. This provision is immunity in action.[10]

Beginning with this first formulation, Roberts develops a parallel between the state and the human body involving the entire immunological repertoire—from antigens to antibodies, from the function of tolerance to the reticuloendothelial system—and finds in each biological element its political equivalent. The most significant step, however, one that moves in the direction previously taken by Uexküll, is perhaps constituted by the reference to mechanisms of immunitary repulsion and expulsion of the racial sort:

> The student of political biology should study national mass attitudes and their results as if they were actual secretions or excretion. National or international repulsions may rest on little. To put the matter at once on the lowest physiological level, it is well known that the smell of one race may offend as much or even more than different habits and customs.[11]

That Roberts's text closes with a comparison between an immunitary rejection of the Jews by the English and an anaphylactic shock of the political body in the year in which the Second World War begins is indicative of the increasingly slippery slope that the first biopolitical elaboration takes on: a politics constructed directly on *bíos* always risks violently subjecting *bíos* to politics.

The second wave of interest in the thematic of biopolitics is registered in France in the 1960s. The difference from the first wave is all too obvious and it couldn't be otherwise in a historical frame that was profoundly modified by the epochal defeat of Nazi biocracy. The new biopolitical theory appeared to be conscious of the necessity of a semantic reformulation even at the cost of weakening the specificity of the category in favor of a more domesticated neohumanistic declension, with respect not only to Nazi biocracy, but also to organistic theories that had in some way anticipated their themes and accents. The volume that in 1960 virtually opened this new stage of study was programmatically titled *La biopolitique: Essai d'interprétation de l'histoire de l'humanité et des civilisations* [Biopolitics: An essay on the interpretation and history of humanity and civilization], and it takes exactly this step.[12] Already the double reference to history and humanity as the coordinates of a discourse intentionally oriented toward *bíos* expresses the central direction and conciliatory path of Aroon Starobinski's essay. When he writes that "biopolitics is an attempt to explain the history of civilization on the basis of the laws of cellular life as well as the most elementary biological life," he does not in fact intend to push his treatment toward a sort of naturalistic outcome.[13] On the contrary, the author argues (sometimes even acknowledging the negative connotations that the natural powers *[potenze]* of life enjoy), for the possibility as well as the necessity that politics incorporates spiritual elements that are capable of governing these natural powers in function of metapolitical values:

> Biopolitics doesn't negate in any way the blind forces of violence and the will to power, nor the forces of self-destruction that exist in man and in human civilization. On the contrary, biopolitics affirms their existence in a way that is completely particular because these forces are the elementary forces of life. But biopolitics denies that these forces are fatal and that they cannot be opposed and directed by spiritual forces: the forces of justice, charity, and truth.[14]

That the concept of biopolitics thus risks being whittled down to the point of losing its meaning, that is, of being overturned into a sort of traditional humanism, is also made clear in a second text published four years later by an author destined for greater fortune. I am referring to Edgar Morin's *Introduction à une politique de l'homme.*[15] Here the "fields" that are truly "biopolitical of life and of survival" are included in a more sweeping aggregate of the "anthropolitical" type, which in turn refers to the project of a "multidimensional politics of man."[16] Rather than tightening the biological-political nexus, Morin situates his perspective on the problematic connection in which the infrapolitical themes of minimal survival are productively crossed with those that are suprapolitical or philosophical, relative to the sense of life itself. The result, more than a biopolitics in the strict sense of the expression, is a sort of "onto-politics," which is given the task of circumscribing the development of the human species, limiting the tendency to see it as economic and productive. "And so all the paths of life and all the paths of politics begin to intersect and then to penetrate one another. They announce an onto-politics that is becoming ever more intimately and globally man's being."[17] Although Morin, in the following book dedicated to the paradigm of human nature, contests in a partially self-critical key the humanistic mythology that defines man in opposition to the animal, culture in opposition to nature, and order in opposition to disorder, there doesn't seem to emerge from of all this an idea of biopolitics endowed with a convincing physiognomy.[18]

Here we are dealing with a theoretical weakness as well as a semantic uncertainty to which the two volumes of *Cahiers de la biopolitique*, published in Paris at the end of the 1960s by the Organisation au Service de la Vie, certainly do not put an end. It is true that with respect to the preceding essay we can recognize in them a more concrete attention to the real conditions of life of the world's population, exposed to a double checkmate of neocapitalism and socialist realism—both incapable of guiding productive development in a direction that is compatible with a significant increase in the quality of life. And it is also true that in several of these texts criticism of the current economic and political model is substantiated in references concerning technology, city planning, and medicine (or better the spaces and the material forms of living beings). Still, not even here can we say that the definition of biopolitics avoids a categorical genericness that will wind up reducing its hermeneutic scope: "Biopolitics was defined as a science by the conduct of states and human collectives, determined by laws, the natural environment, and ontological givens that support life and determine man's activities."[19] There is, however, no suggestion in such a definition of what the specific statute of its object or a critical analysis of its effects might be. Much like the Days of Biopolitical Research held in Bordeaux in December 1966, so too these works have difficulty freeing the concept of biopolitics from a mannerist formulation into a meaningful conceptual elaboration.[20]

The third resumption of biopolitical studies took place in the Anglo-Saxon world and it is one that is still ongoing. We can locate its formal introduction in 1973, when the International Political Science Association officially opened a research site on biology and politics. After that various international conventions were organized, the first of which took place in Paris in 1975 at the École des Hautes Études en Sciences Humaines and another at Bellagio, in Warsaw, Chicago, and New York. In 1983, the Association for Politics and the Life Sciences was founded, as was the journal *Politics and Life Sciences* two years later, as well as the series *Research in Biopolitics* (of which a number of volumes were published).[21] But to locate the beginning of this sort of research we need to return to the middle of the 1960s when two texts appeared that elaborated the biopolitical lexicon. If Lynton K.

Caldwell was the first to adopt the term in question in his 1964 article "Biopolitics: Science, Ethics, and Public Policy," the two polarities within which is inscribed the general sense of this new biopolitical thematization can be traced to the previous year's *Human Nature in Politics* by James C. Davies.[22] It is no coincidence that when Roger D. Masters attempts to systematize the thesis in a volume (dedicated, however, to Leo Strauss) twenty years later, he will eventually give it a similar title, *The Nature of Politics*.[23] These are precisely the two terms that constitute both the object and the perspective of a biopolitical discourse, which after its organistic declension in the 1920s and 1930s and its neohumanistic one of the 1960s in France, now acquires a marked naturalistic character. Leaving aside the quality of this production, which in general is admittedly mediocre, its symptomatic value resides precisely in the direct and insistent reference made to the sphere of nature as a privileged parameter of political determination. What emerges—not always with full theoretical knowledge on the part of the authors—is a considerable categorical shift with respect to the principal line of modern political philosophy. While political philosophy presupposes nature as the problem to resolve (or the obstacle to overcome) through the constitution of the political order, American biopolitics sees in nature its same condition of existence: not only the genetic origin and the first material, but also the sole controlling reference. Politics is anything but able to dominate nature or "conform" *[formare]* to its ends and so itself emerges "informed" in such a way that it leaves no space for other constructive possibilities.

At the origin of such an approach can be distinguished two matrices: on the one side, Darwinian evolution (or more precisely social Darwinism), and, on the other, the ethological research, developed principally in Germany at the end of the 1930s. With regard to the first, the most important point of departure is to be sought in *Physics and Politics* by Walter Bagehot within a horizon that includes authors as diverse as Spencer and Sumner, Ratzel and Gumplowitz.[24] The clear warning, however, is that the emphasis of the biopolitical perspective resides in the passage from a physical paradigm to one that is exactly biological, something that Thomas Thorson underscores forcefully in his book from 1970 with the programmatic title *Biopolitics*.[25] What matters, therefore, is not so much conferring the label of an exact science on politics as referring it back to its natural domain, by which is understood the vital terrain from which it emerges and to which it inevitably returns.[26] Above all, we are dealing with the contingent condition of our body, which keeps human action within the limits of a determinate anatomical and physical possibility, but also the biological or indeed genetic baggage of the subject in question (to use the lexicon of a nascent sociobiology). Against the thesis that social events require complex historic explanations, they refer here finally to dynamics that are tied to evolutive demands of a species such as ours, different quantitatively but not qualitatively from the animal that precedes and comprises our species. In this way, not only does the predominantly aggressive behavior of man (as well as the cooperative) refer to an instinctive modality of the animal sort, but insofar as it inheres in our feral nature, war ends up taking on a characteristic of inevitability.[27] All political behavior that repeats itself with a certain frequency in history—from the control of territory to social hierarchy to the domination of women—is deeply rooted in a prehuman layer not only to which we remain tied, but which is usually bound to resurface. In this interpretive framework, democratic societies are not impossible in themselves, but appear in the form of parentheses that are destined to be quickly closed (or that at least allow one to see the dark depths out of which they contradictorily emerge). The implicit and often explicit conclusion of the reasoning is that any institution or subjective option that doesn't conform, or at least adapt, to such a given is destined to fail.

The biopolitical notion that emerges at this point is sufficiently clear, as Somit and Peterson, the most credentialed theoreticians of this interpretive line express it.[28] What remains problematic, however, is the final point, which is to say the relation between the analytic-descriptive relation and that of the propositional-normative (all because it is one thing to study, explain, and forecast and another to prescribe). Yet it is precisely in this postponement from the first to the second meaning, that is, from the level of being to that of requirement, that the densest ideological valence is concentrated in the entire discourse.[29] The semantic passage is conducted through the double versant of fact and value in the concept of nature. It is used as both a given and a task, as the presupposition and the result, and as the origin and the end. If political behavior is inextricably embedded in the dimension of *bíos* and if *bíos* is what connects human beings *[l'uomo]* to the sphere of nature, it follows that the only politics possible will be the one that is already inscribed in our natural code. Of course, we cannot miss the rhetorical short-circuit on which the entire argument rests: no longer does the theory interpret reality, but reality determines a theory that in turn is destined to corroborate it. The response is announced even before the analysis is begun: human beings cannot be other than what they have always been. Brought back to its natural, innermost part, politics remains in the grip of biology without being able to reply. Human history is nothing but our nature repeated, sometimes misshapen, but never really different. The role of science (but especially of politics) is that of impeding the opening of too broad a gap between nature and history; making our nature, in the final analysis, our only history. The enigma of biopolitics appears resolved, but in a form that assumes exactly what needs to be "researched."

Politics, Nature, History

From a certain point of view it's understandable that Foucault never gestured to the different biopolitical interpretations that preceded his own—from the moment in which his extraordinary survey is born precisely from the distance he takes up with regard to his predecessors. This doesn't mean that no points of contact exist, if not with their positive contents, then with the critical demand that follows from them, which refers more broadly to a general dissatisfaction with how modernity has constructed the relation among politics, nature, and history. It is only here that the work begun by Foucault in the middle of the 1970s manifests a complexity and a radicality that are utterly incomparable with the preceding theorizations. It isn't irrelevant that Foucault's specific biopolitical perspective is indebted in the first place to Nietzschean genealogy. This is because it is precisely from genealogy that Foucault derives that oblique capacity for disassembly and conceptual reelaboration that gives his work the originality that everyone has recognized. When Foucault, returning to the Kantian question surrounding the meaning of the Enlightenment, establishes a contemporary point of view, he doesn't simply allude to a different mode of seeing things that the past receives from the present, but also to the interval that such a point of view of the present opens between the past and its self-interpretation. From this perspective, Foucault doesn't think of the end of the modern epoch—or at least the analytic block of its categories highlighted by the first biopolitical theorizations—as a point or a line that interrupts an epochal journey, but rather as the disruption of its trajectory produced by a different sort of gaze: if the present isn't what (or only what) we have assumed it to be until now; if its meanings begin to cluster around a different semantic epicenter; if something novel or ancient emerges from within that contests the mannerist image; this means, then, that the past, which nonetheless the present derives from, is no longer necessarily the same. This can reveal a face, an aspect, or a profile that before was obscured or perhaps hidden by a superimposed (and at times imposed) narrative; not necessarily a false narrative, but instead functional to its prevailing logic, and for this reason partial, when not tendentious.

Foucault identifies this narrative, which compresses or represses with increasing difficulty something that is heterogeneous to its own language, with the discourse on sovereignty. Despite the infinite variations and transformations to which it has been subjected in the course of modernity on the part of those who have made use of it, sovereignty has always been based on the same figural schema: that of the existence of two distinct entities, namely, the totality of individuals and power that at a certain point enters into relation between individuals in the modalities defined by a third element, which is constituted by the law. We can say that all modern philosophies, despite their heterogeneity or apparent opposition, are arranged within this triangular grid, now one, now the other, of its poles. That these affirm the absolute character of sovereign power according to the Hobbesian model or that, on the contrary, they insist on its limits in line with the liberal tradition; that they subtract or subject the monarch with respect to the laws that he himself has promulgated; that they subject or distinguish the principles of legality and of legitimacy—what remains common to all these conceptions is the *ratio* that subtends them, which is precisely the one characterized by the preexistence of subjects to sovereign power that these conceptions introduce and therefore by the rights *[diritto]* that in this mode they maintain in relation to subjects. Even apart from the breadth of such rights—one that moves from the minimum of the preservation of life and the maximum of participation in political government—the role of counterweight that is assigned to subjects in relation to sovereign decision is clear. The result is a sort of a zero-sum relation: the more rights one has, the less power there is and vice versa. The entire modern philosophical-juridical debate is inscribed to varying degrees within this topological alternative that sees politics and law *[legge]*, decision and the norm as situated on opposite poles of a dialectic that has as its object the relation between subjects *[sudditi]* and the sovereign.[30] Their respective weight depends on the prevalence that is periodically assigned to the two terms being compared. When, at the end of this tradition, Hans Kelsen and Carl Schmitt will argue (the one, normativism, armed against the other, decisionism), they do nothing but replicate the same topological contrast that from Bodin on, indeed in Bodin, seemed to oppose the versant of law to that of power.

It is in the breaking of this categorical frame that Foucault consciously works.[31] Resisting what he himself will define as a new form of knowledge (or better, a different order of discourse with that of all modern philosophical-political theories) doesn't mean, of course, erasing the figure or reducing the decisively objective role of the sovereign paradigm, but rather recognizing the real mechanism by which it functions. It isn't that of regulating relations between subjects or between them and power, but rather their subjugation *at the same time* to a specific juridical and political order. On the one side, rights will emerge as nothing other the instrument that the sovereign uses for imposing his own domination. Correspondingly, the sovereign can dominate only on the basis of the right that legitimates the whole operation. In this way, what appeared as split in an alternative bipolarity between law and power, legality and legitimacy, and norm and exception finds its unity in a same regime of sense. Yet this is nothing but the first effect of the reversal of perspective that Foucault undertakes, one that intersects with another effect relative to the line of division no longer internal to the categorical apparatus of the sovereign *dispositif*, but now immanent to the social body. This perspective claimed to unify it through the rhetorical procedure of polar oppositions. It is as if Foucault undertook the dual work of deconstructing or outflanking the modern narration, which, while suturing an apparent divergence, located a real distinction. It is precisely the recomposition of the duality between power and right, excavated by the sovereign paradigm that makes visible a conflict just as real that separates and opposes groups of diverse ethnicity in the predominance over a given territory. The presumed conflict between sovereignty and law is displaced by the far more real conflict between potential rivals who fight over the use of resources and their control because of their different racial makeup. This doesn't mean in any way that the mechanism of juridical legitimation fails, but rather than preceding and regulating the struggle under way, it constitutes the result and instrument used by those who now and again emerge as victorious. It isn't that the discourse of rights *[diritto]* determines war, but rather that war adopts the discourse of rights in order to consecrate the relation of forces that war itself defines.

Already this unearthing of the constituitive character of war—not its background or its limit, but instead its origin and form of politics—inaugurates an analytic horizon whose historical import we can only begin to see today. But the reference to the conflict between races, a topic to which Foucault dedicated his course in 1976 at the Collège de France, indicates something else, which brings us directly to our underlying theme. That such a conflict concerns so-called populations from an ethnic point of view refers to an element that is destined to disrupt in a much more radical way the modern political and philosophical apparatus. I am referring to *bíos*, a life presupposed simultaneously in its general and specific dimension of biological fact. This is both the object and the subject of the conflict and therefore of the politics that it forms:

> It seems to me that one of the basic phenomena of the nineteenth century was what might be called power's hold over life. What I mean is the acquisition of power over man insofar as man is a living being, that the biological came under State control, that there was at least a certain tendency that leads to what might be termed State control of the biological.[32]

This phrase that opens the lecture of March 17, 1976, and appears to be a new formulation, is in fact already the point of arrival of a trajectory of thought that was inaugurated at least a biennial before. That the first utilization of the term in Foucault's lexicon can be traced directly back to the conference in Rio in 1974, in which Foucault said that "for capitalist society it is the biopolitical that is important before everything else; the biological, the somatic, the corporeal. The body is a biopolitical reality; medicine is a biopolitical strategy" doesn't have much importance.[33] What counts is that all his texts from those years seem to converge in a theoretical step within which every discursive segment comes to assume a meaning that isn't completely perceptible if it is analyzed separately or outside of a biopolitical semantics.

Already in *Discipline and Punish,* the crisis of the classical model of sovereignty, which was represented by the decline of its deadly rituals, is marked by the emergence of a new disciplinary power, which is addressed rather to the life of the subjects that it invests.[34] Although capital punishment through the dismemberment of the convicted responds well to the individual's breaking of the contract (making him guilty of injuring the Majesty), from a certain moment every individual death now is assumed and interpreted in relation to a vital requirement of society in its totality. Yet it is in the course Foucault offered simultaneously titled *Abnormal* that the process of deconstruction of the sovereign paradigm in both its state-power declination and its juridical identity of subject culminates: the entrance and then the subtle colonization of medical knowledge in what was first the competence of law *[diritto]* establishes a true shift in regime, one that pivots no longer on the abstraction of juridical relations but on the taking on of life in the same body of those who are its carriers.[35] In the moment in which the criminal act is no longer to be charged to the will of the subject, but rather to a psychopathological configuration, we enter into a zone of indistinction between law and medicine in whose depths we can make out a new rationality centered on the question of life—of its preservation, its development, and its management. Of course, we must not confuse levels of discourse: such a problematic was always at the center of sociopolitical dynamics, but it is only at a certain point that its centrality reaches a threshold of awareness. Modernity is the place more than the time of this transition and turning *[svolta]*. By this I mean that while, for a long period of time, the relation between politics and life is posed indirectly—which is to say mediated by a series of categories that are capable of distilling or facilitating it as a sort of clearinghouse—beginning at a certain point these partitions are broken and life enters directly into the mechanisms and *dispositifs* of governing human beings.

Without retracing the steps that articulate this process of the governmentalization of life in Foucauldian genealogy—from "pastoral power" to the reason of state to the expertise of the "police"—let's keep our attention on the outcome: on the one side, all political practices that governments put into action (or even those practices that oppose them) turn to life, to its process, to its needs, and to its fractures. On the other side, life enters into power relations not only on the side of its critical thresholds or its pathological exceptions, but in all its extension, articulation, and duration. From this perspective, life everywhere exceeds the juridical constraints used to trap it. This doesn't imply, as I already suggested, some kind of withdrawal or contraction of the field that is subjected to the law. Rather, it is the latter that is progressively transferred from the transcendental level of codes and sanctions that essentially have to do with subjects of will to the immanent level of rules and norms that are addressed instead to bodies: "these power mechanisms are, at least in part, those that, beginning in the eighteenth century, took charge of men's existence, men as living bodies."[36] It is the same premise of the biopolitical regime. More than a removal of life from the pressure that is exercised upon it by law, it is presented rather as delivering their relation to a dimension that both determines and exceeds them both. It is with regard to this meaning that the apparently contradictory expression needs to be understood according to which "it was life more than the law that became the issue of political struggles, even if the latter were formulated through affirmations concerning rights."[37] What is in question is no longer the distribution of power or its subordination to the law, nor the kind of regime nor the consensus that is obtained, but something that precedes it because it pertains to its "primary material." Behind the declarations and the silences, the mediations and the conflicts that have characterized the dynamics of modernity—the dialectic that up until a certain stage we have named with the terms of liberty, equality, democracy (or, on the contrary, tyranny, force, and domination)—Foucault's analysis uncovers in *bíos* the concrete power from which these terms originate and toward which they are directed.

Regarding such a conclusion, Foucault's perspective would seem to be close to that of American biopolitics. Certainly, he too places life at the center of the frame and he too, as we have seen, does so polemically vis-à-vis the juridical subjectivism and humanistic historicism of modern political philosophy. But the *bíos* that he opposes to the discourse of rights and its effects on domination is also configured in terms of a historical semantics that is also symmetrically reversed with respect to the legitimating one of sovereign power. Nothing more than life—in the lines of development in which it is inscribed or in the vortexes in which it contracts—is touched, crossed, and modified in its innermost being by history. This was the lesson that Foucault drew from the Nietzschean genealogy, when he places it within a theoretical frame that substituted a search for the origin (or the prefiguration of the end) with that of a force field freed from the succession of events and conflict between bodies. Yet he also was influenced by Darwinian evolution, whose enduring actuality doesn't reside in having substituted "the grand old biological metaphor of life and evolution" for history, but, on the contrary, in having recognized in life the marks, the intervals, and the risks of history.[38] It is precisely from Darwin, in fact, that the knowledge comes that "life evolved, that the evolution of the species is determined, by a certain degree, by accidents of a historical nature."[39] And so it makes little sense to oppose a natural paradigm to a historical one within the frame of life, or locate in nature the hardened shell in which life is immobilized or loses its historical content. This is because, contrary to the underlying presupposition of Anglo-Saxon *biopolitics*, something like a definable and identifiable human nature doesn't exist as such, independent from the meanings that culture and therefore history have, over the course of time, imprinted on it. And then because the same knowledges that have thematized it contain within them a precise historical connotation outside of which their theoretical direction risks remaining indeterminate. Biology itself is born around the end of the eighteenth century, thanks to the appearance of new scientific categories that gave way to a concept of life that is radically different from what was in use before. "I would say," Foucault will say in this regard, "that the notion of life is not a *scientific concept;* it has been an *epistemological indicator* of which the classifying, delimiting, and other functions had an effect on scientific discussions, and not on what they were talking about."[40]

It is almost too obvious the shift (though one could also rightly say the reversal) that such an epistemological deconstruction impresses on the category of biopolitics. That it is always historically qualified according to a modality that Foucault defines with the term "biohistory" as anything but limited to its simple, natural casting implies a further step that to this point has been excluded from all the preceding interpretations. Biopolitics doesn't refer only or most prevalently to the way in which politics is captured—limited, compressed, and determined—by life, but also and above all by the way in which politics grasps, challenges, and penetrates life:

> If one can apply the term *bio-history* to the pressures through which the movements of life and processes of history interfere with one another, one would have to speak of *bio-power* to designate what brought life and its mechanisms into the realm of explicit calculations and made knowledge-power an agent of transformation of human life.[41]

We can already glimpse in this formulation the radical novelty of the Foucauldian approach. What in the preceding declensions of biopolitics was presented as an unalterable given—nature or life, insofar as it is human—now becomes a problem; not a presupposition but a "site," the product of a series of causes, forces, and tensions that themselves emerge as modified in an incessant game of action and reaction, of pushing and resisting. History and nature, life and politics cross, propel, and violate each other according to a rhythm that makes one simultaneously the matrix and the provisional outcome of the other. But it is also a sagittal gaze that deprives it of its presumed fullness, as well as of every presumption of mastery of the entire field of knowledge. Just as Foucault adopts the category of life so as to break apart the modern discourse of sovereignty and its laws from within, so too in turn does that of history remove from life the naturalistic flattening to which the American biopolitical exposes it:

> It is history that designs these complexes [the genetic variations from which the various populations arise] before erasing them; there is no need to search for brute and definitive biological facts that from the depths of "nature" would impose themselves on history.[42]

It is as if the philosopher makes use of a conceptual instrument that is necessary for taking apart a given order of discourse in order to give it other meanings, at the moment in which it tends to assume a similarly pervasive behavior. Or additionally that it is separated from itself, having been placed in the interval in such a way as to be subject to the same effect of knowledge that it allows externally. From here we can see the continual movement, the rotation of perspective, along a margin that, rather than distinguishing concepts, dismantles and reassembles them in topologies that are irreducible to a monolinear logic. Life as such doesn't belong either to the order of nature or to that of history. It cannot be simply ontologized, nor completely historicized, but is inscribed in the moving margin of their intersection and their tension. The meaning of biopolitics is sought "in this dual position of life that placed it at the same time outside history, in its biological environment, and inside human historicity, penetrated by the latter's techniques of knowledge and power."[43]

The complexity of Foucault's perspective, that is, of his biopolitical *cantiere*, doesn't end here. It doesn't only concern his own position, which is situated precisely between what he calls "the threshold of modernity," on the limit in which modern knowledge folds upon itself, carried in this way outside itself.[44] Rather, it is also the effect of meaning that from an undecidable threshold communicates with the notion defined thusly: once the dialectic between politics and life is reconstructed in a form that is irreducible to every monocausal synthesis, what is the consequence that derives for each of the two terms and for their combination? And so we return to the question with which I opened this chapter on the ultimate meaning of biopolitics. What does biopolitics mean, what outcomes does it produce, and how is a world continually more governed by biopolitics configured? Certainly, we are concerned with a mechanism or a productive *dispositif*, from the moment that the reality that invests and encompasses it is not left unaltered. But productive of what? What is the *effect* of biopolitics? At this point Foucault's response seems to diverge in directions that involve two other notions that are implicated from the outset in the concept of *bíos*, but which are situated on the extremes of its semantic extension: these are *subjectivization* and *death*. With respect to life, both constitute more than two possibilities. They are at the same time life's form and its background, origin, and destination; in each case, however, according to a divergence that seems not to admit any mediation: it is either one or the other. Either biopolitics produces subjectivity or it produces death. Either it makes the subject its own object or it decisively objectifies it. Either it is a politics of life or a politics over life. Once again the category of biopolitics folds in upon itself without disclosing the solution to its own enigma.

Politics of Life

In this interpretive divergence there is something that moves beyond the simple difficulty of definition, which touches the profound structure of the concept of biopolitics. It is as if it were traversed initially and indeed constituted by an interval of difference or a semantic layer that cuts and opens it into two elements that are not constituted reciprocally. Or that the elements are constituted only at the price of a certain violence that subjects one to the domination of the other, conditioning their superimposition to an obligatory positioning-under *[sotto-posizione]*. It is as if the two terms from which biopolitics is formed (life and politics) cannot be articulated except through a modality that simultaneously juxtaposes them. More than combining them or even arranging them along the same line of signification, they appear to be opposed in a long-lasting struggle, the stakes of which are for each the appropriation and the domination of the other. From here the never-released tension, that lacerating effect from which the notion of biopolitics never seems to be able to liberate itself because biopolitics produces the effect in the form of an alternative between the two that cannot be bypassed. Either life holds politics back, pinning it to its impassable natural limit, or, on the contrary, it is life that is captured and prey to a politics that strains to imprison its innovative potential. Between the two possibilities there is a breach in signification, a blind spot that risks dragging the entire category into vacuum of sense. It is as if biopolitics is missing something (an intermediary segment or a logical juncture) that is capable of unbinding the absoluteness of irreconcilable perspectives in the elaboration of a more complex paradigm that, without losing the specificity of its elements, seizes hold of the internal connection or indicates a common horizon.

Before attempting a definition, it is to be noted that not even Foucault is able to escape completely from such a deadlock, and this despite working in a profoundly new framework with respect to the preceding formulations. Foucault too ends up reproducing the stalemate in the form of a further "indecisiveness"—no longer relative to the already acquired impact of power on life, but relative to its effects, measured along a moving line that, as was said, has at one head the production of new subjectivity and at the other its radical destruction. That these contrastive possibilities cohabit within the same analytic axis, the logical extremes of which they constitute, doesn't detract from the fact that their different accentuations determine an oscillation in the entire discourse in opposite directions both from the interpretive and the stylistic point of view. Such a dyscrasia is recognizable in a series of logical gaps and of small lexical incongruences or of sudden changes in tonality, on which it is not possible to linger in detail here. When taken together, however, they mark a difficulty that is never overcome—or, more precisely, an underlying hesitation between two orientations that tempt Foucault equally. Yet he never decisively opts for one over the other. The most symptomatic indication of such an uncertainty is constituted by the definitions of the category, which he from time to time puts into play. Notwithstanding the significant distortions (owing to the different contexts in which they appear), the definitions are mostly expressed indirectly. This was already the case for perhaps Foucault's most celebrated formulation, according to which "for millennia, man remained what he was for Aristotle: a living animal with the additional capacity for a political existence; modern man is an animal whose politics places his existence as a living being in question."[45] This is even more the case where the notion of biopolitics is derived from the contrast with the sovereign paradigm. In this case too a negative modality prevails: biopolitics is primarily that which is *not* sovereignty. More than having its own source of light, biopolitics is illuminated by the twilight of something that precedes it, by sovereignty's advance into the shadows.

Nevertheless, it is precisely here in the articulation of the relation between the two regimes that the prospective splitting to which I gestured previously reappears, a split that is destined in this case to invest both the level of historical reconstruction and that of conceptual determination. How are sovereignty and biopolitics to be related? Chronologically or by a differing superimposition? It is said that one emerges out of the background of the other, but what are we to make of such a background? Is it the definitive withdrawal of a preceding presence, or rather is it the horizon that embraces and holds what newly emerges within it? And is such an emergence really new or is it already inadvertently installed in the categorical framework that it will also modify? On this point too Foucault refuses to respond definitively. He continues to oscillate between the two opposing hypotheses without opting conclusively for either one or the other. Or better: he adopts both with that characteristic, optical effect of splitting or doubling that confers on his text the slight dizziness that simultaneously seduces and disorients the reader.

The steps in which discontinuity seems to prevail are at first sight univocal. Not only is biopolitics other than sovereignty, but between the two a clear and irreversible caesura passes. Foucault writes of that disciplinary power that constitutes the first segment of the *dispositif* that is truly biopolitical: "An important phenomenon occurred in the seventeenth and eighteenth centuries: the appearance—one should say the invention—of a *new* mechanism of power which had very specific procedures, completely *new* instruments, and *very different* equipment. It was, I believe, *absolutely incompatible* with relations of sovereignty."[46] It is new because it turns most of all on the control of bodies and of that which they do, rather than on the appropriation of the earth and its products. From this side, the contrast appears frontally and without any nuances: "It seems to me that this type of power is the exact, point-for-point opposite of the mechanics of power that the theory of sovereignty described or tried to transcribe."[47] For this reason, it "can therefore *no longer* be transcribed in terms of sovereignty."[48]

What is it that makes biopolitics completely unassimilable to the sovereign? Foucault telescopes such a difference in a formula, justifiably famous for its synthetic efficacy, which appears at the end of *The History of Sexuality:* "One might say that the ancient right to *take* life or *let* live was replaced by a power to *foster* life or *disallow* it to the point of death."[49] The opposition couldn't be any plainer: whereas in the sovereign regime life is nothing but the residue or the remainder left over, saved from the right of taking life, in biopolitics life encamps at the center of a scenario of which death constitutes the external limit or the necessary contour. Moreover, whereas in the first instance life is seen from the perspective opened by death, in the second death acquires importance only in the light radiated by life. But what precisely does affirming life mean? *To make* live, rather than limiting oneself to allowing to live? The internal articulations of the Foucauldian discourse are well known: the distinction—here too defined in terms of succession and a totality of copresence—between the disciplinary apparatus and *dispositifs* of control; the techniques put into action by power with regard first to individual bodies and then of populations as a whole; the sectors—school, barracks, hospital, factory—in which they drill and the domains—birth, disease, mortality—that they affect. But to grasp in its complexity the affirmative semantics that—at least in this first declension of the Foucauldian lexicon—the new regime of power connotes, we need to turn again to the three categories of *subjectivization, making immanent,* and *production* that characterize it. Linked between them by the same orientation of sense, they are distinctly recognizable in three genealogical branches in which the biopolitical code is born and then develops, which is to say those that Foucault defines as the pastoral power, the art of government, and the police sciences.

The first alludes to that modality of government of men that in the Jewish-Christian tradition especially moves through a strict and one-to-one relation between shepherd and flock. Unlike the Greek or the Roman models, what counts is not so much the legitimacy of power fixed by law or the maintenance of the harmony between citizens, but the concern that the shepherd devotes to protecting his own flock. The relation between them is perfectly unique: as the sheep follow the will of him who leads them without hesitation, in the same way the shepherd takes care of the life of each of them, to the point, when necessary, of being able to risk his own life. But what connotes the pastoral practice even more is the mode in which such a result is realized: that of a capillary direction, that is both collective and individualized, of the bodies and souls of subjects. At the center of such a process is that durable *dispositif* constituted by the practice of confession on which Foucault confers a peculiar emphasis, precisely because it is the channel through which the process of subjectivization is produced of what remains the object of power.[50] Here for the first time the fundamental meaning of the complex figure of subjection is disclosed. Far from being reduced to a simple objectivization, confession refers rather to a movement that conditions the domination over the object to its subjective participation in the act of domination. Confessing—and in this way placing oneself in the hands of the authority of him who will apprehend and judge its truth—the object of pastoral power is subjugated to its own objectivization and is objectivized in the constitution of its subjectivity. The medium of this crisscrossing effect is the construction of the individual. Forcing him into exposing his subjective truth, controlling the most intimate sounds of his conscience, power singles out the one that it subjects as its own object, and so doing recognizes him as an individual awarded with a specific subjectivity:

> It is a form of a power that makes individuals subjects. There are two meanings of the word "subject": subject to someone else by control and dependence; and tied to his own identity by a conscience or self-knowledge. Both meanings suggest a form of power which subjugates and makes subject to.[51]

If the direction of the conscience by the pastors of souls opens the movement of the subjectivization of the object, the conduct of government, which was theorized and practiced in the form of the reason of state, translates and determines the progressive shift of power from the outside to within the confines of that on which it is exercised. Although the Machiavellian principle still preserves a relation of singularity and of transcendence with regard to its own principality, the art of governing induces a double movement of making immanent and pluralization. On the one side, power is no longer in circular relation with itself, which is to say to the preservation or the amplification of its own order, but in relation to the life of those that it governs, in the sense that its ultimate end is not simply that of obedience but also the welfare of the governed. Power, more than dominating men and territories from on high, adheres to their demands, inscribes its own operation in the processes that the governed establish, and draws forth its own force from that of the subjects *[sudditi]*. But to do so, that is, to collect and satisfy all the requests that arrive from the body of the population, power is forced into multiplying its own services for the areas that relate to subjects—from that of defense, to the economy, to that of public health. From here there is a double move that intersects: the first is a vertical sort that moves from the top toward the bottom, placing in continuous communication the sphere of the state with that of the population and families, reaching finally that of single individuals; the other the horizontal, which places in productive relation the practices and the languages of life in a form that amplifies the horizons, improves the services, and intensifies the performance. With respect to the inflection of sovereign power that is primarily negative, the difference is obvious. If sovereign power was exercised in terms of subtraction and extraction of goods, services, and blood from its own subjects, governmental power, on the contrary, is addressed to the subjects' lives, not only in the sense of their defense, but also with regard to how to deploy, strengthen, and maximize life. Sovereign power removed, extracted, and finally destroyed. Governmental power reinforces, augments, and stimulates. With respect to the salvific tendency of the pastoral power, governmental power shifts decisively its attention onto the secular level of health, longevity, and wealth.

Yet in order that the genealogy of biopolitics can be manifested in all its breadth, a final step is missing. This is represented by the science of the police. Police science is not to be understood in any way as a specific technology within the apparatus of the state as we understand it today. It is rather the productive modality that its government assumes in all sectors of individual and collective experience—from justice, to finance, to work, to health care, to pleasure. More than avoiding harm *[mali]*, the police need to produce goods *[beni]*. Here the process of the positive reconversion of the ancient sovereign right of death reaches its zenith. If the meaning of the term *Politik* remains the negative one of the defense from internal and external enemies, the semantics of *Polizei* is absolutely positive. It is ordered to favor life in all its magnitude, along its entire extension, through all its articulations. And, as Nicolas De Lamare wrote in his compendium, there is even more to be reckoned with. The police are given the task of doing what is necessary as well as what is opportune and pleasurable: "In short, life is the object of the police: the indispensable, the useful, and the superfluous. That people survive, live, and even do better than just that: this is what the police have to ensure."[52] In his *Elements of Police,* Johann Heinrich Gottlob von Justi aims the lens even further ahead: if the object of the police is defined here too as "live individuals living in society," a more ambitious understanding is that of creating a virtuous circle between the vital development of individuals and the strengthening of the forces of the state:[53]

> [T]he police has to keep the citizens happy—happiness being understood as survival, life, and improved living . . . to develop those elements constituitive of individuals' lives in such a way that their development also fosters the strength of the state.[54]

The affirmative character is already fully delineated above, those features (at least from this perspective) that Foucault seems to assign to biopolitics in contrast to the commanding tendency of the sovereign regime. In opposition to it, biopolitics does not limit or coerce *[violenta]* life, but expands it in a manner proportional to its development. More than two parallel flows, we ought to speak of a singular expansive process in which power and life constitute the two opposing and complementary faces. To strengthen itself, power is forced at the same time into strengthening the object on which it discharges itself; not only, but, as we saw, it is also forced to render it subject to its own subjugation *[assoggettamento]*. Moreover, if it wants to stimulate the action of subjects, power must not only presuppose but also produce the conditions of freedom of the subjects to whom it addresses itself. But—and here Foucault's discourse tends toward the maximum point of its own semantic extension—if we are free *for* power, we are also free *against* power. We are able not only to support power and increase it, but also to resist and oppose power. In fact, Foucault concludes that "where there is power, there is resistance, and yet, or rather consequently, this resistance is never in a position of exteriority in relation to power."[55] This doesn't mean, as Foucault quickly points out, that resistance is always already subjected to power against which it seems to be opposed, but rather that power needs a point of contrast against which it can measure itself in a dialectic that doesn't have any definitive outcome. It is as if power, in order to reinforce itself, needs continually to divide itself and fight against itself, or to create a projection that pulls it where it wasn't before. This line of fracture or protrusion is life itself. It is the place that is both the object and the subject of resistance. At the moment in which it is directly invested by power, life recoils against power, against the same striking force that gave rise to it:

> Moreover, against this power that was still new in the nineteenth century, the forces that resisted relied for support on the very thing it invested, that is, on life and man as a living being . . . life as a political object was in a sense taken at face value and turned back against the system that was bent on controlling it.[56]

Simultaneously within and outside of power, life appears to dominate the entire scenario of existence; even when it is exposed to the pressure of power—and indeed, never more than in such a case—life seems capable of taking back what had deprived it before and of incorporating it into its infinite folds.

Politics over Life

This, however, isn't Foucault's entire response, nor is it his only. Certainly, there is an internal coherence therein, as is testified by an entire interpretive line, which not only has made itself the standard-bearer of Foucault's position, but which has pushed Foucault's response well beyond his own manifest intentions.[57] Be that as it may, this doesn't eliminate an impression of insufficiency, or indeed of an underlying reservation concerning a definitive outcome. It is as if Foucault himself wasn't completely satisfied by his own historical-conceptual reconstruction or that he believed it to be only partial and incapable of exhausting the problem; indeed, it is bound to leave unanswered a decisive question: if life is stronger than the power that besieges it, if its resistance doesn't allow it to bow to the pressure of power, then how do we account for the outcome obtained in modernity of the mass production of death?[58] How do we explain that the culmination of a politics of life generated a lethal power that contradicts the productive impulse? This is the paradox, the impassable stumbling block that not only twentieth-century totalitarianism, but also nuclear power asks philosophy with regard to a resolutely affirmative declension of biopolitics. How is it possible that a power of life is exercised against life itself? Why are we not dealing with two parallel processes or simply two simultaneous processes? Foucault accents the direct and proportional relation that runs between the development of biopower and the incremental growth in homicidal capacity. There have never been so many bloody and genocidal wars as have occurred in the last two centuries, which is to say in a completely biopolitical period. It is enough to recall that the maximum international effort for organizing health, the so-called Beveridge Plan, was elaborated in the middle of a war that produced 50 million dead: "One could symbolize such a coincidence by a slogan: Go get slaughtered and we promise you a long and pleasant life. Life insurance is connected with a death command."[59] Why? Why does a power that functions by insuring, protecting, and augmenting life express such a potential for death? It is true that wars and mass destruction are no longer perpetrated in the name of a politics of power *[potenza]*—at least according to the declared intentions of those who conduct these wars—but in the name of the survival itself of populations that are involved. But it is precisely what reinforces the tragic aporia of a death that is necessary to preserve life, of a life nourished by the deaths of others, and finally, as in the case of Nazism, by its own death.[60]

Once again we are faced with that enigma, that terrible unsaid, that the "bio" placed before politics holds for the term's meaning. Why does biopolitics continually threaten to be reversed into thanatopolitics? Here too the response to such an interrogative seems to reside in the problematic point of intersection between sovereignty and biopolitics. But seen now from an angle of refraction that bars an interpretation linearly in opposition to the two types of regime. The Foucauldian text marks a passage to a different representation of their relation by the slight but meaningful semantic slip between the verb "to substitute" (which still connotes discontinuity) and the verb "to complement," which alludes differently to a process of progressive and continuous mutation:

> And I think that one of the greatest transformations that the political right underwent in the nineteenth century was precisely that, I wouldn't say exactly that sovereignty's old right—to take life or let live—was *replaced,* but it came to be *complemented* by a new right which does not erase the old right but which does penetrate it, permeate it.[61]

It isn't that Foucault softens the typological distinction as well as the opposition between the two kinds of power: these are defined as they were previously. It is only that, rather than deploying the distinction along a single sliding line, he returns it to a logic of copresence. From this point of view, the same steps that were read before in a discontinuous key now appear to be articulated according to a different argumentative strategy:

> This power cannot be described or justified in terms of the theory of sovereignty. It is radically heterogeneous and should logically have led to the complete disappearance of the great juridical edifice of the theory of sovereignty. In fact, the theory of sovereignty not only continued to exist as, if you like, an ideology of right; it also continued to organize the juridical codes that nineteenth-century Europe adopted after the Napoleonic codes.[62]

Foucault furnishes an initial explanation of the ideological-functional kind vis-à-vis such a persistence, in the sense that the use of the theory of the sovereign, once it has been transferred from the monarch to the people, would have allowed both a concealment and a juridicization of the *dispositifs* of control put into action by biopower. From here the institution of a double level that is intertwined between an effective practice of the biological kind and a formal representation of juridical character. Contractualist philosophies would have constituted from this point of view the natural terrain of contact between the old sovereign order and the new governmental apparatus, applied this time not only to the individual sphere, but also to the area of population in its totality. And yet, this reconstruction, insofar as it is plausible on the historical level, doesn't completely answer the question on the theoretical level. It is as if between the two models, sovereignty and biopolitics, there passes a relation at once more secret and essential, one that is irreducible both to the category of analogy and to that of contiguity. What Foucault seems to refer to is rather a copresence of opposing vectors superimposed in a threshold of originary indistinction that makes one both the ground and the projection, the truth and the surplus of the other. It is this antinomic crossing, this aporetic knot, that prevents us from interpreting the association of sovereignty and biopolitics in a monolinear form or in the sense of contemporaneity or succession. Neither the one nor the other restores the complexity of an association that is much more antithetical. In their mutual relation, different times are compressed within a singular epochal segment constituted and simultaneously altered by their reciprocal tension. Just as the sovereign model incorporates the ancient pastoral power—the first genealogical incunabulum of biopower—so too biopolitics carries within it the sharp blade of a sovereign power that both crosses and surpasses it. If we consider the Nazi state, we can say indifferently, as Foucault himself does, that it was the old sovereign power that adopts biological racism for itself, a racism born in opposition to it. Or, on the contrary, that it is the new biopolitical power that made use of the sovereign right of death in order to give life to state racism. If we have recourse to the first interpretive model, biopolitics becomes an internal articulation of sovereignty; if we privilege the second, sovereignty is reduced to a formal schema of biopolitics. The antinomy emerges more strongly with regard to nuclear equilibrium. Do we need to look at it from the perspective of life that, notwithstanding everything, has been able to ensure it or from the perspective of total and mass death that continues to threaten us?

> So the power that is being exercised in this atomic power is exercised in such a way that it is capable of suppressing life itself. And, therefore, to suppress itself insofar as it is the power that guarantees life. Either it is sovereign and uses the atomic bomb, and therefore cannot be power, biopower, or the power to guarantee life, as it has been ever since the nineteenth century. Or, at the opposite extreme, you no longer have a sovereign right that is in excess of biopower, but a biopower that is in excess of sovereign right.[63]

Once again, after having defined the terms of an alternating hermeneutic between two opposing theses, Foucault never opts decisively for one or the other. On the one hand, he hypothesizes something like a return to the sovereign paradigm within a biopolitical horizon. In that case, we would be dealing with a literally phantasmal event, in the technical sense of a reappearance of death—of the destitute sovereign decapitated by the grand revolution—on the scene of life; as if a tear suddenly opened in the reign of immunization (which is precisely that of biopolitics), from which the blade of transcendence once again vibrates, the ancient sovereign power of taking life. On the other hand, Foucault introduces the opposing hypothesis, which says that it was precisely the final disappearance of the sovereign paradigm that liberates a vital force so dense as to overflow and be turned against itself. With the balancing constituted by sovereign power diminished in its double orientation of absolute power and individual rights, life would become the sole field in which power that was otherwise defeated is exercised:

> The excess of biopower appears when it becomes technologically and politically possible for man not only to manage life but to make it proliferate, to create living matter, to build the monster, and ultimately, to build viruses that cannot be controlled and that are universally destructive. This formidable extension of biopower, unlike what I was just saying about atomic power, will put it beyond all human sovereignty.[64]

Perhaps we have arrived at the point of maximum tension, as well as at the point of potential internal fracture of the Foucauldian discourse. At the center remains the relation (not only historical, but conceptual and theoretical) between sovereignty and politics, or more generally between modernity and what precedes it, between present and past. Is that past truly past or does it extend as a shadow that reaches up to the present until it covers it entirely? In this irresolution there is something more than a simple exchange between a topological approach of the horizontal sort and another, more epochal, of the vertical kind; or we are dealing with both a retrospective and a prospective gaze.[65] There is indecision concerning the underlying meaning of secularization. Is it nothing other than the channel, the secret passage through which death has returned to capture "life" again? Or, on the contrary, was it precisely the absolute disappearance of death, its conclusive death without remainder that sparks in the living a lethal battle against itself? Once again, how do we wish to think the sovereign paradigm within the biopolitical order, and then what does it represent? Is it a residue that is delayed in consuming itself, a spark that doesn't go out, a compensatory ideology or the ultimate truth, because it is prior to and originary of its own installation, its own profound subsurface, its own underlying structure? And when it pushes with greater force so as to resurface (or, on the contrary, when it ultimately collapses), does death rise again in the heart of life until it makes it burst open?

What remains suspended here isn't only the question of the relation of modernity with its "pre," but also that of the relation with its "post." What was twentieth-century totalitarianism with respect to the society that preceded it? Was it a limit point, a tear, a surplus in which the mechanism of biopower broke free, got out of hand, or, on the contrary, was it society's sole and natural outcome? Did it interrupt or did it fulfill it? Once again the problem concerns the relation with the sovereign paradigm: does

Nazism (but also true *[reale]* communism) stand on the outside or inside vis-à-vis it? Do they mark the end or the return? Do they reveal the most intimate linking or the ultimate disjunction between sovereignty and biopolitics? It isn't surprising that Foucault's response is split into lines of argument that are substantially at odds with each other. Totalitarianism and modernity are at the same time continuous and discontinuous, not assimilable and indistinguishable:

> One of the numerous reasons why [fascism and Stalinism] are, for us, so puzzling is that in spite of their historical weakness they are not quite original. They used and extended mechanisms already present in most other societies. More than that: in spite of their internal madness, they used to a large extent the ideas and the devices of our political rationality.[66]

The reason Foucault is prevented from responding less paradoxically is clear: if the thesis of indistinction between sovereignty, biopolitics, and totalitarianism were to prevail—the continuist hypothesis—he would be forced to assume genocide as the constituitive paradigm (or at least as the inevitable outcome) of the entire parabola of modernity.[67] Doing so would contrast with his sense of historical distinctions, which is always keen. If instead the hypothesis of difference were to prevail—the discontinuist hypothesis—his conception of biopower would be invalidated every time that death is projected inside the circle of life, not only during the first half of the 1900s, but also after. If totalitarianism were the result of what came before it, power would always have to enclose and keep watch over life relentlessly. If it were the temporary and contingent displacement, it would mean that life over time is capable of beating back every power that wants to violate it. In the first case, biopolitics would be an absolute power over life; in the second, an absolute power of life. Held between these two opposing possibilities and blocked in the aporia that is established when they intersect, Foucault continues to run simultaneously in both directions. He doesn't cut the knot, and the result is to keep his ingenious intuitions unfinished on the link between politics and life.

Evidently, Foucault's difficulty and his indecision move well beyond a simple question of historical periodization or genealogical articulation between the paradigms of sovereignty and biopolitics to invest the same logical and semantic configuration of the latter. My impression is that such a hermeneutic impasse is connected to the fact that, notwithstanding the theorization of their reciprocal implication, or perhaps because of this, the two terms of life and politics are to be thought as originally distinct and only later joined in a manner that is still extraneous to them. It is precisely for this reason that politics and life remain indefinite in profile and in qualification. What, precisely, are "politics" and "life" for Foucault? How are they to be understood and in what way does their definition reflect on their relationship? Or, on the contrary, how does their relation impact on their respective definitions? If one begins to think them separately in their absoluteness, it becomes difficult and even contradictory to condense them in a single concept. Not only, but one risks blocking a more profound understanding, relating precisely to the originary and elemental character of that association. It has sometimes been said that Foucault, absorbed for the most part in the question of power, never sufficiently articulated the concept of politics—to the point of substantially superimposing the expressions of "biopower" and "biopolitics." But an analogous observation—a conceptual elaboration that is lacking or insufficient—could be raised as well in relation to the other term of the relation, which is to say that of life; that despite describing the term analytically in its historical-institutional, economic, social, and productive nervature, life remains, nevertheless, little problematized with regard to its epistemological constitution. What is life in its essence and even before that, does life have an essence—a recognizable and describable designation outside of the relation with other lives and with what is not life? Does there exist a simple life—a bare life—or does it emerge from the beginning as formed, as put into form by something that pushes it beyond itself? From this perspective as well, the category of biopolitics seems to demand a new horizon of meaning, a different interpretive key that is capable of linking the two polarities together in a way that is at the same time more limited and more complex.

CHAPTER TWO
The Paradigm of Immunization

Immunity

For my part, I believe I've traced the interpretive key in the paradigm of "immunization" that seems to have eluded Foucault. How and in what sense can immunization fill that semantic void, that interval of meaning which remains open in Foucault's text between the constitutive poles of the concept of biopolitics, namely, biology and politics? Let's begin by observing that the category of "immunity," even in its current meaning, is inscribed precisely in their intersection, that is, on the tangential line that links the sphere of life with that of law. Where the term "immunity" for the biomedical sphere refers to a condition of natural or induced refractoriness on the part of a living organism when faced with a given disease, immunity in political-juridical language alludes to a temporary or definitive exemption on the part of subject with regard to concrete obligations or responsibilities that under normal circumstances would bind one to others. At this point, however, we still remain only at the outermost side of the question: many political terms of biological derivation (or at least of assonance) such as those of "body," "nation," and "constitution" come to mind. Yet in the notion of immunization something more determines its specificity when compared with the Foucauldian notion of biopolitics. It concerns the intrinsic character that forces together the two elements that compose biopolitics. Rather than being superimposed or juxtaposed in an external form that subjects one to the domination of the other, in the immunitary paradigm, *bíos* and *nomos*, life and politics, emerge as the two constituent elements of a single, indivisible whole that assumes meaning from their interrelation.

Not simply the relation that joins life to power, immunity is the power to preserve life. Contrary to what is presupposed in the concept of biopolitics—understood as the result of an encounter that arises at a certain moment between the two components—in this perspective no power exists external to life, just as life is never given outside of relations of power. From this angle, politics is nothing other than the possibility or the instrument for keeping life alive *[in vita la vita]*.

Yet the category of immunization enables us to take another step forward (or, perhaps better, laterally) to the bifurcation that runs between the two principal declinations of the biopolitical paradigm: one affirmative and productive and the other negative and lethal. We have seen how the two terms tend to be constituted in an alternating and reciprocal form that doesn't take into account points of contact. Thus, either power negates life or enhances its development; or violates life and excludes it or protects and reproduces it; objectivizes life or subjectifies it—without any terms that might mediate between them. Now the hermeneutic advantage of the immunitary model lies precisely in the circumstance that these two modalities, these two effects of sense—positive and negative, preservative and destructive—finally find an internal articulation, a semantic juncture that organizes them into a causal relation (albeit of a negative kind). This means that the negation doesn't take the form of the violent subordination that power imposes on life from the outside, but rather is the intrinsically antinomic mode by which life preserves itself through power. From this perspective, we can say that immunization is a negative [form] of the protection of life. It saves, insures, and preserves the organism, either individual or collective, to which it pertains, but it does not do so directly, immediately, or frontally; on the contrary, it subjects the organism to a condition that simultaneously negates or reduces its power to expand. Just as in the medical practice of vaccinating the individual body, so the immunization of the political body functions similarly, introducing within it a fragment of the same pathogen from which it wants to protect itself, by blocking and contradicting natural development. In this sense we can certainly trace back a prototype to Hobbesian political philosophy: when Hobbes not only places the problem of the *conservatio vitae* at the center of his own thought, but conditions it to the subordination of a constitutive power that is external to it, namely, to sovereign power, the immunitary principle has virutally already been founded.

Naturally, we must not confound the objective genesis of a theory with that of its self-interpretation, which obviously occurs later. Hobbes, and with him a large part of modern political philosophy, is not fully cognizant of the specificity (and therefore also of the contrafactual consequences) of the conceptual paradigm that he in point of fact also inaugurates. In order for the power of the contradiction that is implicit in an immunitary logic to come to light, we need to turn away from the level of irreflexive elaboration to that of conscious reflection. In other words, we need to introduce Hegel into the discussion. It has been noted that Hegel was the first to assume the negative not just as the price—an unwanted residue, a necessary penalty—paid for the positive to be realized, but rather as the motor of the positive, the fuel that allows it to function. Of course, Hegel doesn't adopt the term or the concept of immunization as such. The life to which the Hegelian dialectic refers concerns that of reality and of thought in their constitutive indistinctness, rather than that of animal-man assumed as individual and as species (even if the constitution of subjectivity in some of his fundamental texts occurs thanks to a challenge with a death that is also biological).[1] The first knowingly to use such a transition is Nietzsche. When Nietzsche transfers the center of the analysis from the soul to the body—or better, when he assumes the soul as the immunitary form that protects and imprisons the body at the same time—the paradigm acquires its specific critical weight. Here we are dealing not only with the metaphor of a virulent vaccination that Nietzsche imparts to the common man, contaminating him with man's own madness, but also with the interpretation of an entire civilization in terms of self-protection and immunity. All of knowledge and power's *dispositifs* play the role of protective containment in the face of a vital power *[potenza]* that is led to expand without limits. What Nietzsche's judgment might be about such an epochal occurrence—double, ambivalent—we will see shortly. The fact remains, however, that with Nietzsche, the category of immunization has already been completely elaborated.

From that moment on, the most innovative part of twentieth-century culture begins to make implicit use of the paradigm. The negative—that which contradicts order, norms, values—is taken on not only as an indispensable element of human history in all its singular or social configurations that it assumes periodically, but indeed as history's productive impulse. Without that obstacle or lack represented by the negative, the life of the individual and of the species would never find enough energy to develop on its own. Instead it would remain dominated by the jumble of natural impulses from which it needs to free itself in order to be able to open itself to the sphere of greater performance *[prestazioni]*. Thus Émile Durkheim refers precisely to immunology when considering an ineliminable and functional polarity of human behavior that appeared as pathological in a social environment:

> Smallpox, a vaccine of which we use to inoculate ourselves, is a true disease that we give ourselves voluntarily, yet it increases our chance of survival. There may be many other cases where the damage caused by the sickness is insignificant compared with the immunities that it confers upon us.[2]

But it is perhaps with the philosophical anthropology developed in Germany in the middle of the last century that the lexical horizon in which the dialectical notion of *compensatio* acquires its most explicit immunitarian valence. From Max Scheler to Helmuth Plessner, ending with Arnold Gehlen, the *conditio humana* is literally constituted by the negativity that separates it from itself.[3] It is precisely for this reason that the human is placed above other species that surpass the human on the level of those natural elements required to live. In ways different from Marx, not only can the alienation of man not be reintegrated, but indeed it represents the indispensable condition of our own identity. And so the man whom Herder had already defined as an "invalid of his superior forces" can be transformed into the "armed combatant of his inferior forces," into a "Proteus of surrogates" who is able to turn his own initial lack into a gain.[4] It is precisely these "transcendences in the here and now"—what Gehlen defines as institutions—that are destined to immunize us from the excess of subjectivity through an objective mechanism that simultaneously liberates and deprives *[destituisce]* us.[5]

Yet if we are to recognize the immunitary semantics at the center of modern self-representation, we need to move to the point of intersection between two rather different (albeit converging) hermeneutic lines. The first is that which extends from Freud to Norbert Elias along a theoretical line marked by the knowledge of civilization's necessarily inhibiting character. When Elias speaks of the transformation of hetero-constrictions into self-constrictions that characterize the move from the late-classical period to the modern one, he doesn't simply allude to a progressive marginalization of violence, but rather to its enclosure within the confines of the individual psyche. Thus, while physical conflict is subjected to a social regulation that becomes always more severe, "at the same time the battlefield, is, in a sense, moved within. Part of the tensions and passions that were earlier directly released in the struggle of man and man, must now be worked out within the human being."[6] This means that on one side the negative, in this case conflict, is neutralized with respect to its most disruptive effects; on the other that the equilibrium arrived at in such a way is for its part marked by a negative that undermines it from within. The life of the ego, divided between the driving power of the unconscious and the inhibiting one of the superego, is the site in which such an immunitary dialectic is expressed in its most concentrated form.

The scene doesn't change if we shift our attention to the outside. As was already noted, this is what results when other lines intersect with the first (albeit less critically). I am referring to the critical route that leads us to Parson's functionalism and Luhmann's systems theory. That Parsons himself linked his own research to the "Hobbesian problem of order" is in this sense doubly indicative of its immunitary declension: first because it directly joins up with the philosopher with whom our genealogy began, namely, Hobbes; and second for the semantic and conceptual slippage that occurs vis-à-vis Hobbes, relative to the overcoming of the acute alternative between order and conflict and the regulated assumption of conflict within order. Just as society needs to integrate into itself that individual who negates its essence, so too is order the result of a conflict that is both preserved and dominated.[7]

Niklas Luhmann is the one who has derived the most radical consequences from immunization, particularly regarding terminology. To affirm, precisely as he does, that "the system does not immunize itself against the no but with the help of the no" or, "to put this in terms of an older distinction, it protects through negation against annihilation," means getting right to the heart of the question, leaving aside the apologetic or at least the neutral connotations with which the author frames it.[8] His thesis that systems function not by rejecting conflicts and contradictions, but by producing them as necessary antigens for reactivating their own antibodies, places the entire Luhmannian discourse within the semantic orbit of immunity.[9] Not only does Luhmann affirm that a series of historical tendencies point to a growing concern to realize a social immunology from the onset of modernity, particularly from the eighteenth century onwards, but he pinpoints "society's specific immunitary system" in the legal system.[10] When the internal development of a true immunological science—beginning at least with the work of Burnet—doesn't just offer an analogical border to this complex of argumentations but something more, then the immunitary paradigm comes to constitute the neuralgic epicenter between intellectual experiences and traditions of thinking that are rather different.[11] While cognitive scientists such as Dan Sperber theorize that cultural dynamics can be treated as biological phenomena and therefore become subject to the same epidemiological laws that regulate living organisms, Donna Haraway, in critical dialogue with Foucault, comes to argue that "the immune system is a plan for meaningful action to construct and maintain the boundaries for what may count as self and other in the dialectics of Western biopolitics."[12] Similarly, whereas Odo Marquard interprets the aestheticization of postmodern reality as a form of preventive anesthetization, incipient globalization furnishes another area of research, or rather the definitive background to our paradigm.[13] Just as communicative hypertrophy caused by telematics is the reverse sign of a generalized immunization, so too the calls for immunized identities of small states are nothing but the countereffect or the crisis of an allergic rejection to global contamination.[14]

The new element that I have proposed in this debate concerns what appears to me to be the first systematic elaboration of the immunitary paradigm held on one side by the contrastive symmetry with the concept of community—itself reread in the light of its original meaning—and on the other by its specifically modern characterization.[15] The two questions quickly show themselves to be intertwined. Tracing it back to its etymological roots, *immunitas* is revealed as the negative or lacking *[privativa]* form of *communitas.* If *communitas* is that relation, which in binding its members to an obligation of reciprocal donation, jeopardizes individual identity, *immunitas* is the condition of dispensation from such an obligation and therefore the defense against the expropriating features of *communitas. Dispensatio* is precisely that which relieves the *pensum* of a weighty obligation, just as it frees the exemption *[l'esonero]* of that onus, which from its origin is traceable to the semantics of a reciprocal *munus.*[16] Now the point of impact becomes clear between this etymological and theoretical vector and the historical or more properly genealogical one. One can say that generally *immunitas,* to the degree it protects the one who bears it from risky contact with those who lack it, restores its own borders that were jeopardized by the common. But if immunization implies a substitution or an opposition of private or individualistic models with a form of communi-tary organization—whatever meaning we may wish to attribute to such an expression—the structural connection with the processes of modernization is clear.

Of course, by instituting a structural connection between modernity and immunization, I do not intend to argue that modernity might be interpretable only through an immunitary paradigm, nor that it is reducible only to the modern. In other words, I do not deny the heuristic productivity of more consolidated exegetical models of use such as "rationalization" (Weber), "secularization" (Löwith), or "legitimation" (Blumenberg). But it seems to me that all three can gain from a contamination with an explicative category, which is at the same time more complex and more profound, one that constitutes its underlying premise. This surplus of sense with respect to the above-mentioned models is attributable to two distinct and linked elements. The first has to do with the fact that while the modern epoch's self-interpretive constructions—the question of technology *[tecnica]* in the first case, that of the sacred in the second, and that of myth in the third—originate in a circumscribed thematic center, or rather are situated on a unique sliding axis, the immunization paradigm instead refers us to a semantic horizon that itself contains plural meanings—for instance, precisely that of *munus.* Investing a series of lexical areas of different provenance and destination, the *dispositif* of its neutralization will prove to be furnished by equal internal articulations, as is testified even today by the polyvalences that the term of immunity still maintains.

But this horizontal richness doesn't exhaust the hermeneutic potential of the category. It also needs to be investigated—and this is the second element noted above—by looking at the particular relation that the category, immunity, maintains with its antonym, community. We have already seen how the most incisive meaning of *immunitas* is inscribed in the reverse logic of *communitas:* immune is the "nonbeing" or the "not-having" anything in common. Yet it is precisely such a negative implication with its contrary that indicates that the concept of immunization presupposes that which it also negates. Not only does it appear to be derived logically, but it also appears to be internally inhabited by its opposite. Certainly, one can always observe that the paradigms of disillusion, secularization, and legitimation—to remain with those cited above—presupposed in a certain way their own alterity: illusion, the divine, and transcendence, respectively. But they also assume precisely that which at various times is consumed, which then lessens or at least changes into something different. For its part, the negative of *immunitas* (which is another way of saying *communitas*) doesn't only disappear from its area of relevance, but constitutes simultaneously its object and motor. What is immunized, in brief, is the same community in a form that both preserves and negates it, or better, preserves it through the negation of its original horizon of sense. From this point of view, one might say that more than the defensive apparatus superimposed on the community, immunization is its internal mechanism *[ingranaggio]:* the fold that in some way separates community from itself, sheltering it from an unbearable excess. The differential margin that prevents the community from coinciding with itself takes on the deep semantic intensity of its own concept. To survive, the community, every community, is forced to introject the negative modality of its opposite, even if the opposite remains precisely a lacking and contrastive mode of being of the community itself.[17]

But the structural connection between modernity and immunization allows us to take another step forward with reference to the "time" of biopolitics. I noted earlier how Foucault himself oscillates between two possible periodizations (and therefore interpretations) of the paradigm that he himself introduced.[18] If biopolitics is born with the end of sovereignty—supposing that it has really come to an end—this means that the history of biopolitics is largely modern and in a certain sense postmodern. If instead, as Foucault suggests on other occasions, biopolitics accompanies the sovereign regime, constituting a particular articulation or a specific tonality, then its genesis is more ancient, one that ultimately coincides with that of politics itself, which has always in one way or another been devoted to life. With regard to the second case, the question is, why did Foucault open up a new site of reflection? The semantics of immunity can provide us with an answer to this question to the degree in which immunity inserts biopolitics into a historically determined grid. Making use of the immunitary paradigm, one would then have to speak about biopolitics beginning with the ancient world. When does power penetrate most deeply into biological life if not in the long phase in which the bodies of slaves were fully available to the uncontrolled domination of their masters, and when prisoners of war could be legitimately run through with a victor's sword? And how can the power of life and death exercised by the Roman paterfamilias with respect to his own children be understood if not biopolitically?[19] What distinguishes the Egyptian agrarian politics or the politics of hygiene and health of Rome from protective procedures and the development of life set in motion by modern biopower? The only plausible response would, it seems to me, have to refer to the intrinsic immunitarian connotations of the latter, which were absent in the ancient world.

If one moves from the historical to the conceptual level, the difference appears even more evident. Consider the greatest philosopher of antiquity, Plato. In perhaps no one more than Plato can we identify a movement of thought that would seem to be oriented toward biopolitics. Not only does he take eugenic practices that Sparta adopted with respect to frail babies, and more generally with regard to those not seen as suitable for public life, as normal, indeed even as expedient, but—and this is what matters more—he enlarges the scope of political authority to include the reproductive process as well, going so far as to recommend that methods of breeding for dogs and other domestic animals be applied to the reproduction of offspring *(paidopoiia* or *teknopoiia)* of citizens or at least to the guardians *[guardiani]:*

> It follows from our conclusions so far that sex should preferably take place between men and women who are outstandingly good, and should occur as little as possible between men and women of a vastly inferior stamp. It also follows that the offspring of the first group shouldn't [reproduce]. This is how to maximize the potential of our flock. And the fact that all this is happening should be concealed from everyone except the rulers themselves, if the herd of guardians is to be as free as possible from conflict.[20]

Some have noted that passages of this sort—anything but rare if not always so explicit—may well have contributed to a biopolitical reading that Nazi propaganda took to an extreme.[21] Without wanting to introduce the rantings of Bannes or Gabler regarding the parallels between Plato and Hitler, it's enough merely to refer to the success of Hans F. K. Günther's *Platon als Hüter des Lebens* in order to identify the interesting outcome of a hermeneutical line that also includes authors such as Windelband.[22] When Günther interprets the Platonic *ekloge* in terms of *Auslese* or *Zucht* (from *züchten*), that is, as "selection," one cannot really speak of an out-and-out betrayal of the text, but rather of a kind of forcing in a biological sense that Plato himself in some way authorizes, or at a minimum allows (at least in *The Republic,* in *Politics,* and in *Laws,* unlike in the more avowedly dualistic dialogues). Undoubtedly, even if Plato doesn't directly state what happens to "defective" babies with an explicit reference to infanticide or to their abandonment, nevertheless, when seen in the context of his discourses, one can clearly infer Plato's disinterest toward them; the same holds true for the incurably ill, to whom it's not worthwhile devoting useless and expensive care.[23] Even if Aristotle tends to moderate the deeply eugenic and thanatopolitical sense of these texts, it remains the case that Plato revealed himself as sensitive to the demand for keeping pure the *genos* of the guardians and more generally of the governors of the polis according to rigid Spartan customs handed down by Critias and Senophone.[24]

Should we conclude from Plato's proximity to a biopolitical semantics that one can trace a Greek genesis for biopolitics? I would be careful in responding affirmatively, and not only because the Platonic "selection" does not have a specific ethnoracial inflection, nor more precisely a social one, but instead an aristocratic and aptitudinal one. Moreover, instead of moving in an immunitary direction, one that is oriented to the preservation of the individual, Plato's discourse is clearly directed to a communitarian sense, extended namely to the good of the *koinon.* It is this collective, public, communal, indeed immunitary demand that keeps Plato and the entire premodern culture more generally external to a completely biopolitical horizon. In his important studies on ancient medicine, Mario Vegetti has shown how Plato harshly criticizes the dietetics of Herodicus and Dione, precisely for this lacking, individualistic, and therefore necessarily impolitical tendency.[25] Contrary to the modern biocratic dream of medicalizing politics, Plato stops short of politicizing medicine.

Naturally, having said this, it's not my intention to argue that no one before modernity ever posed a question of immunity. On a typological level, the demand for self-preservation, strictly speaking, is far more ancient and long-lasting than the modern epoch. Indeed, one could plausibly claim that it is coextensive with the entire history of civilization from the moment that it constitutes the ultimate precondition, or better, the first condition, in the sense that no society can exist without a defensive apparatus, as primitive as it is, that is capable of protecting itself. What changes, however, is the moment one becomes aware of the question, and therefore of the kind of responses generated. That politics has always in some way been preoccupied with defending life doesn't detract from the fact that beginning from a certain moment that coincides exactly with the origins of modernity, such a self-defensive requirement was identified not only and simply as a given, but as both a problem and a strategic option. By this it is understood that all civilizations past and present faced (and in some way solved) the needs of their own immunization, but that it is only in the modern ones that immunization constitutes its most intimate essence. One might come to affirm that it wasn't modernity that raised the question of the self-preservation of life, but that self-preservation is itself raised in modernity's own being *[essere]*, which is to say it invents modernity as a historical and categorical apparatus able to cope with it. What we understand by modernity therefore in its complexity and its innermost being can be understood as that metalanguage that for a number of centuries has given expression to a request that originates in life's recesses through the elaboration of a series of narrations capable of responding to life in ways that become more effective and more sophisticated over time. This occurred when natural defenses were diminished; when defenses that had up to a certain point constituted the symbolic, protective shell of human experience were lessened, none more important than the transcendental order that was linked to the theological matrix. It is the tear that suddenly opens in the middle of the last millennium in that earlier immunitarian wrapping that determines the need for a different defensive apparatus of the artificial sort that can protect a world that is constitutively exposed to risk. Peter Sloterdijk sees the double and contradictory propensity of modern man originating here: on the one side, protected from an exteriority without ready-made shelter, on the other, precisely because of this, forced to make up for such a lack with the elaboration of new and ever stronger "immunitary baldachins," when faced with a life not only already exposed *[denudata]* but completely delivered over to itself.[26]

If that is true, then the most important political categories of modernity are not be interpreted in their absoluteness, that is, for what they declare themselves to be, and not exclusively on the basis of their historical configuration, but rather as the linguistic and institutional forms adopted by the immunitary logic in order to safeguard life from the risks that derive from its own collective configuration and conflagration. That such a logic expresses itself through historical-conceptual figures shows that the modern implication between politics and life is direct but not immediate. In order to be actualized effectively, life requires a series of mediations constituted precisely by these categories. So that life can be preserved and also develop, therefore, it needs to be ordered by artificial procedures that are capable of saving it from natural risks. Here passes the double line that distinguishes modern politics; on one side, from that which precedes it, and, on the other, from the condition that follows it. With regard to the first, modern politics already had a clear biopolitical tendency, in the precise sense that it is emphasized, beginning with the problem of *conservatio vitae.* Yet differently with respect to what will happen in a phase that we will call for now second modernity, the relationship between politics and life circulates through the problem of order and through historical-conceptual categories—sovereignty, property, liberty, power—in which it is innervated. It is this presupposition of order with respect to living subjectivity from which it objectively is generated that determines the aporetic structure of modern political philosophy; indeed, the fact that its response to the question of self-preservation from which it is born emerges not only as deviated but, as we will see soon enough, as also self-contradictory, is the consequence or the expression of a dialectic that is already in itself antinomic, as is the immunitary dialectic. If modern political philosophy is given the task of protecting life, which is always determined negatively, then the political categories organized to express it will end up rebounding against their own proper meanings, twisting against themselves. And that notwithstanding their specific contents: the pretense of responding to an immediacy—the question of *conservatio vitae*—is contradictory to the mediations, which are precisely the concepts of sovereignty, property, and liberty. That all of them at a certain point in their historical-semantic parabola are reduced to the security of the subject who appears to be the owner or beneficiary, is not to be understood either as a contingent derivation or as a destiny fixed beforehand, but rather as the consequence of the modality of immunity through which the Modern thinks the figure of the subject.[27] Heidegger more than anyone else understood the essence of the problem. To declare that modernity is the epoch of representation, that is, of the *subjectum* that positions itself as an *ens in se substantialiter completum* vis-à-vis its own object, entails bringing it back philosophically to the horizon of immunity:

> Representation is now, in keeping with the new freedom, a going forth—from out of itself—into the sphere, first to be made secure, of what is made secure . . . The *subjectum,* the fundamental certainty, is the being-represented-together-with—made secure at any time—of representing man together with the entity represented, whether something human or non-human, i.e. together with the objective.[28]

Yet to link the modern subject to such a horizon of immunitary guarantees also means recognizing the aporia in which the same experience remains captured: that of looking to shelter life in the same powers *[potenze]* that interdict its development.

Sovereignty

The conception of sovereignty constitutes the most acute expression of such a power. In relation to the analysis initiated by Foucault, sovereignty is understood not as a necessary compensatory ideology vis-à-vis the intrusiveness of control *dispositifs* nor as a phantasmal replica of the ancient power of death to the new biopolitical regime, but as the first and most influential that the biopolitical regime assumes. That accounts for its long persistence in a European juridical-political lexicon: sovereignty isn't before or after biopolitics, but cuts across the entire horizon, furnishing the most powerful response to the modern problem of the self-preservation of life. The importance of Hobbes's philosophy, even before his disruptive categorical innovations, resides in the absolute distinctness by which this transition is felt. Unlike the Greek conception—which generally thinks politics in the paradigmatic distinction with the biological dimension—in Hobbes not only does the question of *conservatio vitae* reenter fully in the political sphere, but it comes to constitute by far its most prevalent dimension. In order to qualify as such, to deploy in its forms, life must above all be maintained as such, be protected as such, and be protected from the dissipation that threatens it. Both the definition of natural right, that is, what man can do, and that of natural law, that is, what man must do, account for this original necessity:

> The Right of Nature, which Writers commonly call Jus Naturale, is the Liberty each man hath, to use his own power, as he will himselfe, for the preservation of his own Nature; that is to say, of his own Life; and consequently, of doing any thing, which in his own Judgement, and Reason, hee shall conceive to the aptest means thereunto.[29]

As for natural law, it is "a Precept, or generall Rule, found out by Reason, by which a man is forbidden to do that, which is destructive of his life, or taketh way the means of preserving the same, and to omit, that, by which he thinketh it may be best preserved.[30]

Already the setting up of the argumentation situates it in a clearly biopolitical frame. It's not by chance that the man to whom Hobbes turns his attention is one characterized essentially by the body, by its needs, by its impulses, and by its drives. And when one even adds the adjective "political," this doesn't qualitatively modify the subject to which it refers. With respect to the classic Aristotelian division, the body, considered politically, remains closer to the regions of *zōē* than to that of *bíos;* or better, it is situated precisely at the point in which such a distinction fades and loses meaning. What is at stake, or, more precisely, what is in constant danger of extinction, is life understood in its materiality, in its immediate physical intensity. It is for this reason that reason and law converge on the same point defined by the pressing demands of preserving life. But what sets in motion the argumentative Hobbesian machine is the circumstance that neither one nor the other is able by itself to achieve such an objective without a more complex apparatus in condition to guarantee it. The initial attempt at self-preservation *(conatus sese praeservandi)* is indeed destined to fail given the combined effects of the other natural impulses that accompany and precisely contradict the first, namely, the inexhaustible and acquisitive desire for everything, which condemns men to generalized conflict. Although it tends to self-perpetuation, the fact is that life isn't capable of doing so autonomously. On the contrary, it is subjected to a strong counterfactual movement such that the more life pushes in the direction of self-preservation, the more defensive and offensive means are mobilized to this end, given the fundamental equality among men, all of whom are capable of killing each other and thus, for the same reason, all capable of being killed:

> And therefore, as long as this naturall Right of every man to every thing endureth, there can be no security to any man, (how strong or wise soever he be), of living out the time, which Nature ordinarily alloweth men to live.[31]

It is here that the immunitary mechanism begins to operate. If life is abandoned to its internal powers, to its natural dynamics, human life is destined to self-destruct because it carries within itself something that ineluctably places it in contradiction with itself. Accordingly, in order to save itself, life needs to step out from itself and constitute a transcendental point from which it receives orders and shelter. It is in this interval or doubling of life with respect to itself that the move from nature to artifice is to be positioned. It has the same end of self-preservation as nature, but in order to actualize it, it needs to tear itself from nature, by following a strategy that is opposed to it. Only by negating itself can nature assert its own will to live. Preservation proceeds through the suspension or the alienation *[estraneazione]* of that which needs to be protected. Therefore the political state cannot be seen as the continuation or the reinforcement of nature, but rather as its negative converse. This doesn't mean that politics reduces life to its simple biological layer—that it denudes it of every qualitative form, as one might argue only by moving Hobbes to a lexicon in which he doesn't belong. It is no coincidence that he never speaks of "bare life," but on the contrary, in all his texts, implies it in terms that go well beyond simply maintaining life. If in *De Cive* he argues that "[B]ut by safety must be understood, not the sole preservation of life in what condition soever, but in order to its happiness," in *Elements* he stresses that with the judgment *(Salus populi suprema lex esto)* "must be understood, not the mere preservation of their lives, but generally their benefit and good," to conclude in *Leviathan* that "by safety here is not meant a bare preservation, but also all other contentments of life, which every man by lawful industry, without danger or hurt to the Commonwealth, shall acquire to himself."[32]

Nor does this mean that the category of life in the modern period replaces that of politics, with progressive depoliticization as its result. On the contrary, once the centrality of life is established, it is precisely politics that is awarded the responsibility for saving life, but—and here is the decisive point in the structure of the immunitary paradigm—it occurs through an antinomic *dispositif* that proceeds via the activation of its contrary. In order to be saved, life has to give up something that is integral to itself, what in fact constitutes it principal vector and its own power to expand; namely, the acquisitive desire for everything that places itself in the path of a deadly reprisal. Indeed, it is true that every living organism has within it a sort of natural immunitary system—reason—that defends it from the attack of external agents. But once its deficiencies, or rather its counterproductive effects, have been ascertained, it is substituted with an induced immunity, which is to say an artificial one that both realizes and negates the first. This occurs not only because it is situated outside the individual body, but also because it now is given the task of forcibly containing its primordial intensity.

This second immunitary (or better, meta-immunitary) *dispositif*, which is destined to protect life against an inefficient and essentially risky protection, is precisely sovereignty. So much has been said about its pactional inauguration and its prerogatives that it isn't the case to return to them here. What appears most relevant from our perspective is the constitutively aporetic relation that ties it to the subjects to whom it is directed. Nowhere more than in this case is the term to be understood in its double meaning: they are subjects of sovereignty to the extent to which they have voluntarily instituted it through a free contract. But they are subjects to sovereignty because, once it has been instituted, they cannot resist it for precisely the same reason: otherwise they would be resisting themselves. Because they are subjects of sovereignty, they are subjected to it. Their consensus is requested only once, after which they can no longer take it back.

Here we can begin to make out the constitutively negative character of sovereign immunization. It can be defined as an immanent transcendence situated outside the control of those that also produced it as the expression of their own will. This is precisely the contradictory structure that Hobbes assigns to the concept of representation: the one representing, that is, the sovereign, is simultaneously identical and different with respect to those that he represents. He is identical because he takes their place *[stare al loro posto]*, yet different from them because that "place" remains outside their range. The same spatial antinomy is seen temporally, that is, that which the instituting subjects declare to have put in place eludes them because it logically precedes them as their own same presupposition.[33] From this point of view, one could say that the immunization of the modern subject lies precisely in this exchange between cause and effect: he, the subject, can be presupposed, self-insured in Heidegger's terms, because he is already caught in a presupposition that precedes and determines him. It is the same relation that holds between sovereign power and individual rights. As Foucault explains it, these two elements must not be seen in an inversely proportional relationship that conditions the enlargement of the first to the shrinking of the second or vice versa. On the contrary, they mutually implicate themselves in a form that makes the first the complementary reverse of the other: only individuals who are considered equal with others can institute a sovereign that is capable of legitimately representing them. At the same time, only an absolute sovereign can free individuals from subjection to other despotic powers. As a more recent, discriminating historiography has made clear, absolutism and individualism, rather than excluding or contradicting each other, implicate each other in a relation that is ascribable to the same genetic process.[34] It is through absolutism that individuals realize themselves and at the same time negate themselves; presupposing their own presupposition, they are deprived insofar as they are constituted as subjects from the moment that the outcome of such a founding is nothing other than that which in turn constructs them.

Behind the self-legitimating account of modern immunization, the real biopolitical function that modern individualism performs is made clear. Presented as the discovery and the implementation of the subject's autonomy, individualism in reality functions as the immunitary ideologemme through which modern sovereignty implements the protection of life. We shouldn't lose sight of any intermediate passage in this dialectic. We know that in a natural state men also relate to each other according to a modality of the individual that leads to generalized conflict. But such a conflict is still always a horizontal relation that binds them to a communal dimension. Now, it is exactly this commonality—the danger that derives to each and every one—that is abolished through that artificial individualization constituted precisely by the sovereign *dispositif.* Moreover, the same echo is to be heard in the term "absolutism," not only in the independence of power from every external limit, but above all in the dissolution projected onto men: their transformation into individuals, equally absolute by subtracting from them the *munus* that keeps them bound communally. Sovereignty is the not being *[il non essere]* in common of individuals, the political form of their desocialization.

The negative of *immunitas* already fills our entire frame: in order to save itself unequivocally, life is made "private" in the two meanings of the expression. It is privatized and deprived of that relation that exposes it to its communal mark. Every external relationship to the vertical line that binds everyone to the sovereign command is cut at the root. Individual literally means this: to make indivisible, united in oneself, by the same line that divides one from everyone else. The individual appears protected from the negative border that makes him himself and not other (more than from the positive power of the sovereign). One might come to affirm that sovereignty, in the final analysis, is nothing other than the artificial vacuum created around every individual—the negative of the relation or the negative relation that exists between unrelated entities.

Yet it isn't only this. There is something else that Hobbes doesn't say explicitly, as he limits himself to letting it emerge from the creases or the internal shifts of the discourse itself. It concerns a remnant of violence that the immunitary apparatus cannot mediate because it has produced it itself. From this perspective, Foucault seizes on an important point that is not always underlined with the necessary emphasis in the Hobbesian literature: Hobbes is not the philosopher of conflict, as is often repeated in regard to "the war of every man against every man," but rather the philosopher of peace, or better of the neutralization of conflict, from the moment that the political state needs preemptively to insure against the possibility of internecine warfare.[35] Yet the neutralization of conflict doesn't completely provide for its elimination, but instead for its incorporation in the immunized organism as an antigen at once necessary to the continuous formation of antibodies. Not even the protection that the sovereign assures his subjects is exempt. Especially here is manifested the most strident form of antibody. Concurrently, in the order of instruments adopted to mitigate the fear of violent death that all feel toward the other, it remains a fear that is more acceptable because it is concentrated on one objective (though not for this reason essentially different from the one already overcome). In a certain sense, the asymmetric condition intensifies this fear, a condition in which the subject *[suddito]* finds himself vis-à-vis a sovereign who preserves that natural right deposited by all the other moments of the entrance into the civil state. What occurs from this, as a result, is the necessary linking of the preservation of life with the possibility—always present even if rarely utilized—of the taking away of life by the one who is also charged with insuring it. It is a right precisely of life and death, understood as the sovereign prerogative that cannot be contested precisely because it has been authorized by the same subject that endures it. The paradox that supports the entire logic lies in the circumstance that the sacrificial dynamic is unleashed not by the distance, but, on the contrary, by the assumed identification of individuals with the sovereign who represents them with their explicit will. Thus, "nothing the Soveraign Representative can doe to a subject, on what pretense soever, can properly be called an Injustice, or Injury: because every Subject is Author of every act the Soveraign doth."[36] It is exactly this superimposition between opposites that reintroduces the term of death in the discourse of life:

> And therefore it may and does often happen in Common-wealths, that a Subject may be put to death, by the command of the Soveraign Power, and yet neither doe the other wrong: As when Jeptha caused his daughter to be sacrificed: In which, and the like cases, he that so dieth, had Liberty to doe the action, for which he is neverthelesse, without Injury put to death. And the same holdeth also in a Soveraign Prince, that putteth to death an Innocent Subject.[37]

What emerges here with a severity that is only barely contained by the exceptional character in which the event appears circumscribed is the constitutive antinomy of the sovereign immunization, which is based not only on the always tense relationship between exception and norm, but on its normal character of exception (because anticipated by the same order that seems to exclude it). This exception—the liminal coincidence of preservation and capacity to be sacrificed of life—represents both a remainder that cannot be mediated and the structural antinomy on which the machine of immunitary mediation rests. At the same time, it is the residue of transcendence that immanence cannot reabsorb—the prominence of the "political" with respect to the juridical with which it is also identified—and the aporetic motor of their dialectic. It is as if the negative, keeping to its immunitary function of protecting life, suddenly moves outside the frame and on its reentry strikes life with uncontrollable violence.

Property

The same negative dialectic that unites individuals to sovereignty by separating them invests all the political-juridical categories of modernity as the inevitable result of their immunitary declension. This holds true in the first instance for that of "property." Indeed, one can say that property's constitutive relevance to the process of modern immunization is ever more accentuated with respect to the concept of sovereignty. And this for two reasons. First, thanks to the originary antithesis that juxtaposes "common" to "one's own" *[proprio]*, which by definition signifies "not common," "one's own" is as such always immune. And second, because the idea of property marks a qualitative intensification of the entire immunitary logic. As we just observed, while sovereign immunization emerges transcendent with respect to those who also create it, that of proprietary immunization adheres to them—or better, remains within the confines of their bodies. It concerns a process that conjoins making immanent *[immanentizzazione]* and specialization: it is as if the protective apparatus that is concentrated in the unitary figure of sovereignty is multiplied to the degree that sovereignty, once multiplied, is installed in biological organisms.

At the center of the conceptual transition will be found the work of John Locke. Here, just as in Hobbes, what is at stake is the preservation of life (*preservation of himself, desire of self-preservation* [trans: in English]), which Locke from the beginning declares to be "the first and strongest God Planted in Men,"[38] but in a form that conditions it to the presence of something, precisely the *res propria,* that contemporaneously arises from and reinforces it.

> For the desire, strong desire of Preserving his Life and Being having been Planted in him, as a Principle of Action by God himself, Reason, which was the Voice of God in him, could not but teach him and assure him, that pursuing that natural Inclination he had to preserve his Being, he followed the Will of his Maker, and therefore had the right to make use of those

> Creatures, which by his Reason or Senses he could discover would be serviceable thereunto. And thus Man's Property in the Creatures, was founded upon the right he had, to make use of those things, that were necessary or useful to his Being.[39]

The right of property is therefore the consequence as well as the factual precondition for the permanence in life. The two terms implicate each other in a constitutive connection that makes of one the necessary precondition of the other: without a life in which to inhere, property would not be given; but without something of one's own—indeed, without prolonging itself in property—life would not be able to satisfy its own primary demands and thus it would be extinguished. We mustn't lose sight of the essential steps in the argument. Locke doesn't always include life among the properties of the subject. It is true that in general he unifies *lives, liberties, and estates* [trans: in English] within the denomination of property, so that he can say that "civil goods are life, liberty, bodily health and freedom from pain, and the possession of outward things, such as lands, money, furniture, and the like."[40] But in other passages property assumes a more restricted sense, one that is limited to material goods to which life doesn't belong. How does one explain such an incongruence? I believe that to understand them less in obvious fashion, these two enunciative modalities should not be juxtaposed but integrated and superimposed in a singular effect of sense: life is contemporaneously inside and outside property. It is within from the point of view of having—as part of the goods with which everyone is endowed *[in dotazione]*. But beyond that, life is also the all of the subject if one looks at it from the point of view of being. Indeed, in this case it is property, any kind of property, that is part of life. One can say that the relationship and the exchange, which from time to time Locke sets up between these two optics, define his entire perspective. Life and property, being and having, person and thing are pressed up together in a mutual relation that makes of one both the content and the container of the other. When he declares that the natural state is a state of "Liberty to dispose, and order, as he lists, his Person, Actions, Possession, and his whole property, within the Allowance of those Laws under which he is; and therein not to be subject to the arbitrary Will of another, but freely to follow his own," on the one hand, he inscribes property in a form of life expressed in the personal action of an acting subject; on the other, he logically includes subject, action, and liberty in the figure of "one's own."[41] In this way it emerges as an "inside" that is inclusive of an "outside" that in turn subsumes it within.

The resulting antinomy will be found in the logical difficulty of placing property before the ordering regime that institutes it. Unlike in Hobbes (but also differently than Grozio and Pufendor), Locke's notion of property precedes sovereignty, which instead is ordered to defend it.[42] It is the presupposition and not the result of social organization. Yet—and here appears the question with which Locke himself explicitly begins—what if property is not rooted in a form of interhuman relation, in which property finds its own foundation within a world in which it is given in common? How can the common make itself "one's own" and "one's own" subdivide the common? What is the origin of "mine," of "yours," and of "his" in a universe of everyone? It is here that Locke impresses on his own discourse that biopolitical declension that folds it in an intensely immunitarian sense:

> Though the Earth, and all inferior Creatures be common to all Men, yet every Man has a Property in his own Person. This no Body has any Right to but himself. The Labour of his Body and the Work of his Hands, we may say, are properly his. Whatsoever then he removes out of the State that Nature hath provided, and left it in, he hath mixed his Labour with, and joyned to it something that is his own, and thereby makes it his Property.[43]

Locke's reasoning unravels through concentric circles whose center does not contain a political-juridical principle, but rather an immediately biological reference. The exclusion of someone else cannot be established except as part of the consequential chain that originated in the metaphysical proviso of bodily inclusion. Property is implicit in the work that modifies what is naturally given as work, which in turn is included in the body of the person who performs it. Just as work is an extension of the body, so is property an extension of work, a sort of prosthesis that through the operation of the arm connects it to the body in the same vital segment; not only because property is necessary for the material support of life, but because its prolongation is directed to corporeal formation. Here another transition is visible, indeed, even a shift in the trajectory with respect to the subjective self-insurance identified by Heidegger in the modern *repraesentatio:* the predominance over the object isn't established by the distance that separates it from the subject, but by the movement of its incorporation. The body is the primary site of property because it is the location of the first property, which is to say what each person holds over himself *[ha su se stesso]*. If the world was given to us by God in common, the body belongs solely to the individual who at the same time is constituted from it and who possesses it before any other appropriation, which is to say in originary form. It is in this exchange—together both a splitting and a doubling—between being (a body) and having one's own body that the Lockean individual finds its ontological and juridical, its onto-juridical foundation for every successive appropriation. Possessing one's own corporeal form *[persona]*, he owns all his performances, beginning with the transformation of the material object, which he appropriates as transitive property. From that moment every other individual loses the right over it, such that one can be legitimately killed in the case of theft. Seeing how through work the appropriate object is incorporated into the owner's body, it then becomes one with the same biological life, and is defended with the violent suppression of the one that threatens it as the object has now become an integral part of his life.

Already here the immunitary logic seizes and occupies the entire Lockean argumentative framework: the potential risk of a world given in common—and for this reason exposed to an unlimited indistinction—is neutralized by an element that is presupposed by its same originary manifestation because it is expressive of the relation that precedes and determines all the others: the relation of everyone with himself or herself in the form of personal identity. This is both the kernel and the shell, the content and the wrapping, the object and the subject of the immunitary protection. As property is protected by the subject that possesses it, a self-protecting capacity, preserved by the subject through his *proprium* and of that *proprium* through himself (through the same subjective substance), extends, strengthens, and reinforces it. Once the proprietary logic is wedded to a solid underpinning such as belonging to one's own body, it can now expand into communal space. This is not directly negated, but is incorporated and recut in a division that turns it inside out into its opposite, in a multiplicity of things that have in common only the fact of being all one's own to the degree they have been appropriated by their respective owners:

> From all which it is evident, that though the things of Nature are given in common, yet Man (by being Master of himself, and *Proprietor of his Person,* and the Actions or Labour of it), had still in himself the great foundation of Property; and that which made up the great part of what he apllyed to the Support or Comfort of his being, when Invention and Arts had improved the conveniences of Life, was perfectly his own, and did not belong in common to others.[44]

Earlier I noted that we are dealing with an immunitary procedure that is much more potent than that of Hobbes because it inheres in the same form—though one could say in the material—of the individual. The increment of functionality that derives from it is nonetheless paid with a corresponding intensification of the contradiction on which the entire system rests, which is no longer situated in the point of connection and tension between individuals and the sovereign, as in the Hobbesian model, but in the complex relation that moves between subjectivity and property. What is at stake isn't only a question of identity or of difference—the divergence that is opened in the presupposed convergence between the two poles—but also and above all in the displacement of their prevalent relation. It is defined generally according to the following formulation: if the appropriated thing depends on the subject who possesses it such that it becomes one with the body, the owner in turn is rendered as such only by the thing that belongs to him—and therefore he himself depends on it. On the one hand, the subject dominates the thing in the specific sense that he places it within his domain. But, on the other hand, the thing in turn dominates the subject to the degree in which it constitutes the necessary objective of his acquisitive desire *[tensione]*. Without an appropriating subject, no appropriated thing. But without any appropriated thing, no appropriating subject—from the moment it that doesn't subsist outside of the constitutive relation with it. In this way, if Locke can hold that property is the continuation of subjective identity—or the extension of subjective identity outside itself—one sooner or later can respond that "with private property being incorporated in man himself and with man himself being recognized as its essence . . . carries to its logical conclusion the denial of man, since man himself no longer stands in an external relation of tension to the external substance of private property, but has himself become the essence of private property": its simple appendage.[45] We must not lose track of the reversible features that unite both conditions in one movement. It is precisely the indistinction between the two terms—as is originally established by Locke—that makes the one the *dominus* of the other, and which therefore constitutes them in their reciprocal subjection.

The point of transition and inversion between the two perspectives—from the mastery of the subject to that of the thing—is situated in the private *[privato]* character of appropriation.[46] It is through it that the appropriating act becomes at the same time exclusive of every other act, thanks to the thing itself: the privacy *[privatezza]* of possession is one with the subtraction *[privazione]* that specifies in whom privacy is not shared with the legitimate owner, which means the entire community of nonowners. From this point of view—not an alternative to, but speculative of the first—the negative clearly begins to prevail over the positive, or better, to manifest itself as its internal truth. It is "one's own" that is not common, that does not belong to others. The passive sense of every appropriation subtracts from every other one the appropriative *jus* toward the thing that has already been appropriated in the form of private property. But then also in the active sense, such that the progressive increase in individual property causes a progressive decrease in the goods that are at the disposition of others. Internecine conflict, exorcized from within the proprietary universe, in this way is clearly moved outside its confines, in the formless space of non-property. It is true that in principle Locke institutes a double limit to the increase of property in the obligation to leave for others the things necessary for their maintenance *[conservazione]* and in the prohibition of appropriating for oneself what isn't possible to consume. But then he considers it inoperative at the moment when goods become commutable into money and therefore infinitely capable of being accumulated without fearing that they might be lost.[47] From that point on, private property conclusively breaks down the relation of proportionality that regulates the relation of one to another, but it also weakens that which unites the owner of property to himself. This occurs when property, both private and subtractive *[privativa]*, begins to be emancipated (from the body from which it seems to depend) to take on a configuration of purely juridical stamp. The intermediate point of this long process is constituted by the breaking of the link, introduced by Locke, between property and work. As we know, it was precisely this that joins *proprium* within the confines of the body. When such a connection begins to be considered as no longer necessary—according to a reasoning set in motion by Hume and perfected by modern political economy—one witnesses a true and particular desubstantialization of property, theorized in its most accomplished form in the Kantian distinction between *possessio phaenomenon* (empirical possession) and *possessio noumenon* (intelligible possession), or, as it is also defined, *detentio* (possession without possession). At this point, what will be considered truly, even definitively, one's own is only that which is distant from the body of him who juridically possesses it. It is not physical possession that testifies to complete juridical possession. Originally thought within an indissoluble link with the body that works, property is already defined by its extraneousness to its own sphere.

> I can only call a corporeal thing or an object in space mine, when even though in physical possession of it, I am able to assert that I am in possession of it in another real non-physical sense. Thus, I am not entitled to call an apple mine merely because I hold it in my hand or possess it physically; but only when I am entitled to say "I possess it, although I have laid it out of my hand, and wherever it may lie.[48]

Distance is the condition, the testimonial of the duration of possession for a temporality that goes well beyond the personal life to whose preservation it is also ordered. Here already the contradiction implicit in proprietary logic fully emerges. Separated from the thing that it also inalienably possesses, the individual proprietor remains exposed to a risk of emptying out that is far more serious than the threat that he had tried to immunize himself from by acquiring property, precisely because it is the product of acquiring property. The appropriative procedure, represented by Locke as a personification of the thing—its incorporation in the proprietor's body—lends itself to be interpreted as the reification of the person, disembodied of its subjective substance. It is as if the metaphysical distance of modern representation were restored through the theorization of the incorporation of the object, but this time to the detriment of a subject who is isolated and absorbed by the autonomous power of the thing. Ordered to produce an increment in the subject, the proprietary logic inaugurates a path of inevitable desubjectification. This is a wild oscillation logic in the movement of self-refutation that seizes all the biopolitical categories of modernity. Here too in this case, but in a different form, with a result that converges with that of sovereign immunization, the proprietary paradigm's immunitary procedure is able to preserve life only by enclosing it in an orbit that is destined to drain it of its vital element. Where before the individual was displaced *[destituito]* by sovereign power that he himself instituted, so now too does the individual proprietor appear expropriated by the same appropriative power.

Liberty

The third immunitary wrapping of modernity is constituted by the category of liberty *[libertà]*.[49] As was already the case for those of sovereignty and property, and perhaps in a more pronounced manner, its historical-conceptual sequence is expressed by the general process of modern immunization, in the double sense that it reproduces its deportment and amplifies its internal logic. This may sound strange for a term so obviously charged with accents so constitutively refractory for every defensive tonality, and if anything oriented in the sense of an opening without reserve to the mutability of events. But it is precisely in relation to such a breadth of horizon—still protected in its etymon—that is possible to measure the process of semantic tightening and also of loss of meaning *[prosciugamento]* that marks its successive history.[50] Both the root *leuth* or *leudh*—from which originates the Greek *eleutheria* and the Latin *libertas*—and the Sanskrit root *frya*, which refers instead to the English freedom and the German *Freiheit*, refer us to something that has to do with an increase, a non-closing *[dischiudimento]*, a flowering, also in the typically vegetative meaning of the expression. If then we consider the double semantic chain that descends from it—which is to say that of love (*Lieben*, *lief*, love, as well as, differently, *libet* and *libido*) and that of friendship (friend, *Freund*)—we can deduce not only a confirmation of this original affirmative connotation: the concept of liberty, in its germinal nucleus, alludes to a connective power that grows and develops according to its own internal law, and to an expansion or to a deployment that unites its members in a shared dimension.

It is with respect to such an originary inflection that we should interrogate the negative reconversion that the concept of liberty undergoes in its modern formulation. It's certainly the case that from the beginning the idea of "free" *[libero]* logically implicates the contrastive reference to an opposite condition, that of the slave, understood precisely as "non-free."[51] But such a negation constitutes, more than the presupposition or even the prevailing content of the notion of liberty, its external limit: even though it is tied to an inevitable contrary symmetry, it isn't the concept of slave that confers significance on that of the free man, but the reverse. As it both refers to the belonging to a distinct people and to humanity in general, what has prevailed in the qualification of *eleutheros* has always been the positive connotation with respect to which the negative constitutes a sort of background or contour lacking an autonomous semantic resonance. And, as has repeatedly been brought to light, this relation is inverted in the modern period, when it begins to assume increasingly the features of a so-called negative liberty, with respect to that defined instead as "positive," as in "freedom from." What nevertheless has remained obscured in the ample literature is the fact that both meanings understood in this way—

compared to their initial meaning—in fact emerge within a negative horizon of meaning. If we assume the canonical distinction as Isaiah Berlin elaborates it, indeed not only does the first liberty—understood negatively as an absence of interference—but also the second, which he reads positively, appear quite distant from the characterization, both affirmative and relational, fixed at the origin of the concept:

> The "positive" sense of the word "liberty" derives from the wish on the part of the individual to be his own master. I wish my life and decisions to depend upon myself, *not* on external forces of whatever kind. I wish to be the instrument of my own, not of other men's, acts of will. I wish to be a subject, *not* an object . . . I wish to be somebody *not nobody.*[52]

The least that one can say, in relation to such a definition, is that it is manifestly unable to think liberty affirmatively in the modern conceptual lexicon of the individual, in terms of will and subject. It is as if each of these terms—and still more when placed together—irresistibly pushes liberty close to its "not," to the point of dragging it inside itself. Qualifying liberty—understood as the mastery of the individual subject over himself—is his not being disposed to, or his not being at the disposition of others. This oscillation or inclination of modern liberty toward its negative gives added significance to an observation of Heidegger's, according to which "not only are the individual conceptions of positive freedom different and ambiguous, but the concept of positive freedom as such is indefinite, especially if by positive freedom we provisionally understand the not-negative *[nicht negative]* freedom."[53] The reason for such a lexical exchange, which makes the positive, rather than affirmative, simply a nonnegative, ought to be sought in the break, which is implicit in the individualistic paradigm, of the constitutive link between liberty and otherness (or alteration).[54] It is that which encloses liberty in the relation of the subject with himself: he is free when no obstacle is placed between him and his will—or also between his will and its realization. When Thomas Aquinas translated the Aristotelian *proairesis* with *electio* (and the *boulēsis* with *voluntas*), the paradigmatic move is largely in operation: liberty will rapidly become the capacity to realize that which is presupposed in the possibility of the subject to be himself—not to be other than himself. Free will as the self-establishment of a subjectivity that is absolutely master of its own will. From this perspective, the historical-conceptual relation comes fully into view, which joins such a conception of liberty with other political categories of modernity, from that of sovereignty to that of equality. On the one hand, only free subjects can be made equal by a sovereign who legitimately represents them. On the other hand, such subjects are themselves conceived as equally sovereign within their own individuality—obliged to obey the sovereign because they are free to command themselves and vice versa.

The immunitary outcome—but one might also say the presupposition—of such a move cannot be avoided. In the moment in which liberty is no longer understood as a mode of being, but rather as a right to have something of one's own—more precisely the full predominance of oneself in relation to others—the subtractive or simply the negative sense is already destined to characterize it ever more dominantly. When this entropic process is joined to the self-preserving strategies of modern society, the overturning and emptying of ancient communal liberty *[libertates]* into its immune opposite will be complete. If the invention of the individual constitutes the medial segment of this passage—and therefore the sovereign frame in which it is inscribed—its absolutely prevailing language is that of protection. From this point of view, we need to be careful in not distorting the real sense of the battle against individual or collective *immunitates* fought on the whole by modernity. It isn't that of reducing but of intensifying and generalizing the immunitary paradigm. Without losing its typically polyvalent lexicon, immunity progressively transfers its own semantic center of gravity from the sense of "privilege" to that of "security." Unlike the ancient *libertates,* conferred at the discretion of a series of particular entities—classes, cities, bodies, convents—modern liberty consists essentially in the right of every single subject to be defended from the arbiters that undermine autonomy and, even before that, life itself. In the most general terms, modern liberty is that which insures the individual against the interference of others through the voluntary subordination to a more powerful order that guarantees it. It is here that the antinomical relation with the sphere of necessity originates that ends by reversing the idea of liberty into its opposites of law, obligation, and causality. In this sense it is a mistake to interpret the assumption of constricting elements as an internal contradiction or a conceptual error of the modern theorization of liberty. Instead, it is a direct consequence: necessity is nothing other than the modality that the modern subject assumes in the contrapuntal dialectic of its own liberty, or better, of liberty as the free appropriation of "one's own." The famous expression according to which the subject in chains is free is to be interpreted in this way—not in spite of but in reason of: as the self-dissolving effect of a liberty that is ever more overcome by its purely self-preserving function.

If for Machiavelli "a small part of the people wish to be free in order to command, but all the others who are countless, desire liberty in order to live in safety," Hobbes remains the most consequential and radical theoretician of this move: liberty preserves itself or preserves the subject that possesses it, losing itself and as a consequence losing the subject to the extent the subject is a subject of liberty.[55] That in him liberty is defined as "the absence of all impediments to action, that are not contained in the nature and the intrinsic quality of the agent," means that it is the negative result of a mechanical game of force within which its movement is inscribed and which therefore in the final analysis coincides with its own necessity.[56] In this way—if he who puts liberty to the test can do nothing other than what he has done—his de-liberation *[de-liberazione]* has the literal sense of a renouncing indeterminate liberty and of enclosing liberty in the bonds of its own predetermination:

> Every Deliberation is then sayd to end when that whereof they Deliberate is either done, or thought impossible; because till then wee retain the liberty of doing, or omitting according to our Appetite, or Aversion.[57]

As for Locke, the immunitary knot becomes ever more restrictive and absolute: as was already seen, it doesn't move through the direct subordination of individuals to the sovereign—on the contrary, their relation now begins to include a right of resistance—but rather through the dialectic of a preserving self-appropriation. It is true that, with respect to Hobbes's surrender of liberty, liberty for Locke is inalienable, but exactly for the same reasons we find in Hobbes, which is to say because it is indispensable to the physical existence of he who possesses it.

Consequently, it emerges as joined in an indissoluble triptych formed with property and life. On more than one occasion, Hobbes connects liberty and life, making the first a guarantee for the permanence of the second. Locke pushes even more resolutely in this direction. Indeed, liberty is "so necessary to, and closely joyned with a Man's Preservation, that he cannot part with it, but by what forfeits his Preservation and Life together."[58] Certainly, liberty isn't only a defense against the infringements of others; it is also the subjective right that corresponds to the biological-natural obligation to preserve oneself in life under the best possible conditions. That it is enlarged to include all other individuals according to the precept that no one "ought to harm another in his Life, Health, Liberty, or Possessions" doesn't alter the strictly immunitary logic that underpins the entire argument, which is to say the reduction of liberty to preserving life is understood as the inalienable property that each one has of himself.[59]

Beginning with such a drastic semantic resizing, which makes of liberty the biopolitical coincidence between property and preservation, its meaning tends to be stabilized ever nearer the imperative of security, until it coincides with it. If for Montesquieu political liberty "consists in security, or, at least, in the opinion that we enjoy security," it is Jeremy Bentham who takes the definitive step: "What means liberty? . . . Security is the political blessing I have in view; security as against malefactors, on the one hand, security as against the instruments of government on the other."[60] Already here the immunization of liberty appears as definitively actualized according to the dual direction of defense by the state and toward [the state]. But what qualifies it better still in its antinomical effects is the relation that is installed with its logical opposite, namely, coercion. The point of suture between the expression of liberty and what negates it from within—one could say between exposition and imposition—is constituted exactly by the demand for insurance *[assicurativa]:* it is what calls forth that apparatus of laws which, though not directly producing liberty, constitute nonetheless the necessary reversal: "Where there is no coercion, neither is there security . . . That which lies under the name of Liberty, which is so magnificent, as the inestimable and unreachable work of the Law, is not Libertà but security."[61] From this point of view, Bentham's work marks a crucial moment in the immunitary reconversion to which modern political categories seem to entrust their own survival. The preliminary condition of liberty is to be singled out in a control mechanism that blocks every contingency in the *dispositif* that anticipates it beforehand. The design of the famous Panopticon expresses most spectacularly this oscillation in meaning excavated in the heart of liberal culture.

As we know, it was Foucault who furnished a biopolitical interpretation of liberalism that would bring to light the fundamental antinomy on which it rests and which reproduces its power. To the degree that it isn't limited to the simple enunciation of the imperative of liberty but implicates the organization of conditions that make this effectively possible, liberalism contradicts its own premises. Needing to construct and channel liberty in a nondestructive direction for all of society, liberalism continually risks destroying what it says it wants to create.

> Liberalism, as I understand it, this liberalism that can be characterized as the new art of governing that is formed in the eighteenth century, implies an intrinsic relation of production/destruction with regard to liberty... With one hand it has to produce liberty, but this same gesture implies that with the other hand it must establish limitations, checks, coercions, obligations based on threats, etc.[62]

This explains, within the liberal governmental framework, the tendency to intervene legislatively, which has a contrafactual result with respect to the original intentions: it isn't possible to determine or define liberty except by contradicting it. The reason for such an aporia is obviously to be found in liberty's logical profile. But it is also revealed more tellingly when we consider the biopolitical frame in which Foucault from the beginning had placed it. Earlier Hannah Arendt gathered together the fundamental terms: "For politics, according to the same philosophy [of liberalism], must be concerned almost exclusively with the maintenance of life and the safeguarding of its interests. Now, where life is at stake all action is by definition under the sway of necessity, and the proper relation to take care of life's necessities."[63] Why? Why does the privileged reference to life force liberty into the jaws of necessity? Why does the rebellion of liberty against itself move through the emergence of life? Arendt's response, which in singular fashion adheres to the Foucauldian interpretive scenario, follows the passage, within the biopolitical paradigm, from the domain of individual preservation to that of the species:

> The rise of the political and social sciences in the nineteenth and twentieth centuries has even widened the breach between freedom and politics: for government, which since the beginning of the modern age had been identified with the total domain of the political, was now considered to be the appointed protector not so much of freedom as of the life process, the interests of society and its individuals. Security remained the decisive criterion, but not the individual's security against "violent death," as in Hobbes (where the condition of all liberty is freedom from fear), but a security which should permit an undisturbed development of the life process of society as a whole.[64]

The stipulation is of particular interest: it is the same culture of the individual—once immersed in the new horizon of self-preservation—that produces something that moves beyond it in terms of vital complex process. But Arendt doesn't make the decisive move that Foucault does, which consists in understanding the relation between individual and totality in terms of a tragic antinomy. When Foucault notes that the failure of modern political theories is owed neither to theory nor to politics but to a rationality that forces itself to integrate individuals within the totality of the state, he touches on the heart of the question.[65] If we superimpose his discourse on that elaborated by the anthropologist Luis Dumont regarding the nature and the destiny of individual modernism, we have a confirmation that takes us even further in the direction we are moving here. Asking after the reason first for the nationalistic and then the totalitarian opening *[sbocco]* of liberal individualism (which represents a further jump in quality), Dumont concludes that the political categories of modernity "function," which is to say they discharge the self-preserving function of life to which they are subordinated, including their own opposite or vice versa, or incorporating themselves in it. At a certain point, the culture of the individual also incorporates that which in principle is opposed to it, which is to say the primacy of all on the parts which it gives the name of "olism." The pathogenic effect that ever more derives from it is, according to Dumont, due to the fact that, when placed against its opposite, extraneous paradigms, such as those of individualism and "olism," these intensify the ideological force of their own representations so much that they give rise to an explosive mix.[66]

Tocqueville is the author who seems to have penetrated most deeply into this self-dissolving process. All of his analyses of American democracy are traversed by a modality that recognizes both the inevitability and the epochal risk of such a process. When he delineates the figure of the *homo democraticus* in the point of intersection and friction between atomism and massification, solitude and conformity, and autonomy and heteronomy, he does nothing other than recognize the entropic result of a parabola that has at its uppermost point precisely that self-immunization of liberty in which the new equality of conditions reflects itself in a distorted mirror.[67] To hold—as he does with the unparalleled intensity of a restrained pathos—that democracy separates man "from his contemporaries . . . it throws him back forever upon himself alone, and threatens in the end to confine him entirely within the solitude of his own heart," or that "equality places men side by side, unconnected by any common tie," means to have understood deeply (and with reference to its origin), the immunitary loss of meaning that afflicts modern politics.[68] At the moment when the democratic individual, afraid not to know how to defend the particular interests that move him, ends up surrendering "to the first master who appears," the itinerary will already be set in motion, one not so different from another which will push biopolitics nearer its own opposite, that of thanatopolitics: the herd, opportunistically domesticated, is already ready to recognize its willing shepherd.[69] At the end of the same century, it is Nietzsche who will be the most sensitive witness to such a process. As for freedom—a concept that seemed to Nietzsche to be "yet more proof of instinctual degeneration," he no longer has any doubt: "There is no one more inveterate or thorough in damaging freedom than liberal institutions.[70]

CHAPTER THREE

Biopower and Biopotentiality

Grand Politics

It's no coincidence that the preceding chapter closed with the name of Nietzsche. He, more than anyone else, registers the exhaustion of modern political categories and the consequent disclosing of a new horizon of sense. We already gestured to him in the brief genealogy first sketched of the immunitary paradigm, but that reference isn't enough to restore the strategic relevance that his thought has for my own analysis generally. Nietzsche isn't simply the one who brings the immunitary lexicon to its full development, but is also the one who makes evident its negative power, the uncontrollable nihilistic dissipation in meaning that pushes it in a self-dissolving direction. This is not to say that he is able to escape it, to withdraw himself completely from its growing shadow. Indeed, we will see that for an important part of his perspective, it will result in reproducing and making it more powerful than before.[1] Yet this doesn't erase the deconstructive force his work exercises on other texts with regard to modern immunization, which prefigures the lines of a different conceptual language.

The reasons why such a language, irrespective of its presumed affiliations, has never been elaborated, nor even fully deciphered, are many, not the least of which is the enigmatic character that increasingly comes to characterize Nietzsche's writing. My impression, nevertheless, is that these reasons refer on the whole to the missing or mistaken characterization of its internal logic or, better perhaps, its basic tonality of logic, that only today, precisely from the categorical scenario utilized by Foucault, can be seen in all its import. I am alluding not only to the two interventions that Foucault dedicated to Nietzsche—even if the second, "Nietzsche, Genealogy, History," more than any other (because it centered on the genealogical method), brings us directly to the question at hand: precisely how far does the Foucauldian analysis move within the biopolitical orbit? It is precisely the point of gravitation or the paradigmatic axis from which Nietzsche's entire production, with its internal twists and fractures, which begins to reveal a semantic nucleus that is inaccessible in the interpretive frames in which it has been placed until now. Otherwise, how would it be possible that something, let's call it a decisive stitch in the conceptual material, escaped our attention: that Nietzsche has been read not only in heterogeneous but in mutually opposing terms (even before he was *totus politicus* for some on the "right" or the left" and radically impolitical for others?)[2] Without even arriving at his more recent interpreters, if we simply compare Löwith's thesis that "this political perspective stands not at the margins of Nietzsche's philosophy but rather at its middle" with that of Georges Bataille, according to which "the movement of Nietzsche's thought implicates a defeat of the diverse possible foundations for contemporary politics," we can understand the *impasse* from which Nietzschean literature still seems unable to extricate itself.[3] Probably it is because both the "hyperpolitical" and the "impolitical" readings clash with mirror-like results within the notion of "politics"; Nietzsche's text is explicitly extraneous to such a notion, favoring instead another and different conceptual lexicon that today we can best describe as "biopolitical."

It is with respect to such a conclusion that Foucault's essay "Nietzsche, Genealogy, History" opens a significant tear in perspective.[4] In it Foucault essentially thematizes the opacity of the origin, the interval that separates the origin from itself, or better, from that which is presupposed in it as perfectly conforming to its intimate essence. Thus, what is put up for discussion isn't only the linearity of a history destined to substantiate the conformity of the origin to the end—the finality of the origin and the originality of the end—but also the entire conceptual foundation on which such a conception is based. The entire Nietzschean polemic vis-à-vis a history that is incapable of coming to terms with its own nonhistorical layer—and therefore to extend to itself that thorough historicization that it demands be applied to everything but itself—takes aim at the presumptive airs of universality on behalf of conceptual figures born as a result of specific demands to which it is tied in both their logic and development. When Nietzsche sees in the origin of things not the identity, unity, or purity of an uncontaminated essence, but rather the laceration, the multiplicity, and the alteration of something that never corresponds to that which it declares to be; when he discerns the tumult of bodies and the proliferation of errors as well as the usurpation of sense and vertigo of violence behind the ordered succession of events and the network of meanings in which they seem to consist; when, in short, he traces the dissociation and the contrast in the heart itself of their apparent conciliation, he profoundly questions the entire regulating form that European society has for centuries given itself. Furthermore, he interrogates the exchange that has often been verified, between cause and effect, function and value, and reality and appearance. This is true not only for modern juridical-political categories, beginning with equality, which practically all of the Nietzschean corpus contests, to that of liberty, deprived of its presumed absoluteness and reduced to the constitutive aporia that reverses it into its opposite, to law *[diritto]* itself, identified in its original semblance of naked command. It is especially true for the entire *dispositif* that constitutes both the analytic paradigm and the normative scenario of these categories, namely, that self-legitimating narrative according to which the forms of political power appear to be the intentional result of the combined will of single subjects united in a founding pact. When Nietzsche describes the state—which is to say the most developed juridical and political construct of the modern epoch—as "some horde or other of blond predatory animals, a race of conquerors and masters which, itself organized for war and with the strength to organize others, unhesitatingly lays its fearful paws on a population which may be hugely superior in numerical terms but remains shapeless and nomadic," one can consider "that sentimentalism which would have it begin with a 'contract'" liquidated.[5]

From these first annotations the thread that links them to the proposed hermeneutic activated a century afterwards by Foucault is already clear. If an individual subject of desire and knowledge is withdrawn from and antecedent to the forms of power that structure it; if what we call "peace" is nothing but the rhetorical representation of relations of force that emerge periodically out of continuous conflict; if rules and laws are nothing other than rituals destined to sanction the domination of one over another—all the instruments laid out by modern political philosophy are destined to reveal themselves as simultaneously false and ineffective. False, or purely apologetic, because they are incapable of restoring the effective dynamics in operation behind their surface figures. Ineffective because, as we saw in the preceding chapter, they bump up more and more violently against their own internal contradictions until they break apart. What breaks apart, precisely, more than the single categorical seams, is the logic itself of the mediation on which they depend, no longer able to hold or to strengthen a content that is in itself elusive of any formal control. What that content might be for Nietzsche is well known: it concerns the *bíos* that gives it the intensely biopolitical connotation in Nietzsche's discussion, to which I've already referred. All of Nietzschean criticism has accented the vital element—life as the only possible representation of being.[6] Nevertheless, what has a clear ontological relevance is always interpreted politically; not in the sense of any form that is superimposed from the outside onto the material of life—it is precisely this demand, experienced in all its possible combinations by modern political philosophy, which has been shown to be lacking in foundation. But, as the constitutive character of life itself, life is always already political, if by "political" one intends not what modernity wants—which is to say a neutralizing mediation of immunitary nature—but rather an originary modality in which the living *is* or in which being *lives.* Far from all the contemporary philosophies of life to which his position is from time to time compared, this is the manner in which Nietzsche thinks the political dimension of *bíos:* not as character, law, or destination of something that lives previously, but as the power that informs life from the beginning in all its extension, constitution, and intensity. That life as well as the will to power—according to the well-known Nietzschean formulation—doesn't mean that life desires power nor that power captures, directs, or develops a purely biological life. On the contrary, they signify that life does not know modes of being apart from those of its continual strengthening.

To grasp the characteristic trait that Nietzsche alludes to in the expression "grand politics," we need to look precisely at the indissoluble web of life and power *[potenza]:* in the double sense that living as such is only strengthened internally and that the power is imaginable only in terms of a living organism. Here as well emerges the essential sense of the Nietzschean project for constructing a "new party of life," less tied to contextual contingencies. Leaving aside the prescriptive, troubling contents with which he from time to time thought to fill them, what matters here in relation to our argument is the distance such a reference constitutes with regard to every mediated, dialectical, and external modality that seeks to understand the relation between politics and life. In this sense, we begin to see how much Nietzsche himself will say about it in *Beyond Good and Evil,* though such an observation could also be extended to his entire body of work. It is "in all essentials a *critique of modernity,* not excluding the modern sciences, modern arts, and even modern politics, along with pointers to a contrary type that is as little modern as possible—a noble, Yes-saying type."[7] Apart from the problematic identity of the kind prefigured by Nietzsche, what remains beyond any doubt is its polemical objective: modernity as the formal negation, or negative form, of its own vital content. What unifies his logical, aesthetic, and political categories is precisely the constitutive antinomy that wants to assume, preserve, and develop an immediate, what Nietzsche will call "life" through a series of mediations objectively destined to contradict them (because in fact they are obligated to negate their character of immediacy). From here the rejection not of this or of that institution, but of the institution, insofar as it is an institution and thus separated from and therefore given to destroying that power of life that it has also been charged with safeguarding. In a paragraph titled appropriately enough "Critique of Modernity," Nietzsche states that "our institutions are no longer any good: this is universally accepted. But it is not their fault, it is *ours.* Once we have lost all the instincts from which institutions grow, we lose the institutions themselves because we are no longer good enough for them."[8] What produces such a self-dissolving effect is the incapacity of modern institutions—from party to parliament to the state—to relate directly to life and therefore their tendency to slip into the same vacuum that such an interval of difference creates. This is separate from the political position chosen beforehand: what matters, negatively, is its not being biopolitical—the scission that opens between the two terms of the expression in a form that wrings *bíos* from politics and an originary politicity from life, or better, from its constitutive power.

From here, in the affirmative reversal of such a negativity, the positive meaning of "grand politics" emerges:

> The *grand* politics places physiology above all other questions—it wants to *rear [züchten]* humanity as a whole, it measures the range of the races, of peoples, of individuals according to . . . the guarantee of life they carry within them. Inexorably it puts an end to everything that is degenerate and parasitical to life.[9]

Before confronting with the requisite attention the most problematic part of the passage, that one relative to parasitic and degenerative pathology, let's linger over the passage's overall meaning. We know the emphasis Nietzsche placed on physiological studies in opposition to every form of idealistic thought. From this point of view, the placement of psychological studies in a culture is clear, and more so given the language strongly influenced by Darwin (despite whatever the relevant distinctions that separate Nietzsche from Darwin in a form that we will have occasion to examine in detail).[10] But we are not concerned only with Darwin. What Nietzsche wants to assert is that, at least beginning from a certain moment that coincides with the irreversible crisis of the modern political lexicon, the only politics not reduced to the mere preservation of already existing institutions is the one that confronts the problem of life from the perspective of the human species and of the mobile thresholds that define it, by contiguity or difference, with respect to other living species. Contrary to the presuppositions of modern individualism, the individual—which Nietzsche vindicates and exalts in its character of exceptionality—cannot be thought except against the backdrop of large ethnosocial aggregates that always emerge by way of contrast.

Nevertheless, this first consideration of method doesn't completely answer the question that Nietzsche poses, one that calls into question something whose extraordinary scope and ambivalent effects we are only able to make out today. It concerns the idea that the human species is never given once for all time, but is susceptible, in good and evil, to being molded in forms for which we do not have an exact knowledge, but which nevertheless constitute for us both an absolute risk and an inalienable challenge. "Why," Nietzsche asks himself in a crucial passage, "shouldn't we realize in man what the Chinese are able to do with the tree, so thus it produces on one side roses and on another pears? These natural processes of the *selection of man*, for example, which until now have been exercised in an infinitely slow and awkward way, could be taken over by man himself."[11] Rather than being disconcerted by the irregular approach of linking man to plant (not to mention that of breeding), what we need to foreground is Nietzsche's precocious understanding that in the centuries to come the political terrain of comparison and battle will be the one relative to redefining the human species in a scenario of progressive displacement of its borders with respect to what is not human, which is to say, on the one hand to the animal and on the other to the inorganic.

So too the central emphasis attributed to the body against its "disparagers" has to be traced back to the specificity to the biopolitical lexicon in the sense of the species. Naturally, a comprehensive polemic emerges that takes aim against a philosophical, spiritualistic, or abstractly rational tradition. We recall that reason just as soul is an integral part of an organism that has its unique expression in the body, which in turn doesn't weigh indifferently in the deconstruction of the most influential metaphysical categories. However, to reread the entire history of Europe through "the underlying theme of the body" is an option that cannot be truly understood outside of an established biopolitical lexicon. Certainly, using a physiological terminology in politics is anything but original. Still, the absolute originality of the Nietzschean text resides in the transferral of the relation between state and body from the classical level of analogy or metaphor, in which the ancient and modern tradition positions it, to that of an effectual reality: no politics exists other than that *of* bodies, conducted *on* bodies, *through* bodies. In this sense, one can rightly say that physiology, which Nietzsche never detaches from psychology, is the very same material of politics. It is its pulsating body. But if we are to reveal all of the political pregnancy of the body, we must also examine it from another angle, not only that of the physiological declination of politics, but also that of the political characterization of physiology. If the body is the material of politics, politics—naturally, in the sense that Nietzsche confers on the expression—takes the form of the body. It is this "form"—there is no life that isn't in some way formed, thus a "form of life"—that keeps Nietzsche distant from any type of biological determinism, as Heidegger well understood.[12] Not only because every conception of the body presupposes a later philosophical orientation, but because the body is constituted according to the principle of politics—struggle as the first and final dimension of existence. Struggle outside oneself, toward other bodies, but also within as the unstoppable conflict among its organic components. Before being in itself *[in-sé]*, the body is always *against*, even with respect to itself. In this sense, Nietzsche can say that "every philosophy that ranks peace above war" is "a *misunderstanding of the body.*"[13] This is because in its continual instability the body is nothing but the always provisional result of the conflict of forces that constitute it.

We know how much the Nietzschean conception of the body has weighed on contemporary biological and medical theories in authors such as Roux, Mayer, Foster, and Ribot.[14] Our perspective emphasizes, however, that all of them derive from Nietzsche the dual principle that the body is produced by determinate forces and that such forces are always in potential conflict among them.[15] It is not a *res extensa*, substance or material, but the material site of such a conflict and of the conditions of domination and subjection, and hierarchy and resistance, that from time to time determine it. From here it is a short distance to the essentially political and hence biopolitical semantics that the same definition of life assumes.

> One could define life as a durable form of *process* of *determinations of force* in which different forces in conflict grow in unequal measure. In this sense there is an opposition in obeying: one's own force is in fact not lost. In the same way, in commanding, we have to admit that the absolute force of the adversary is not defeated, absorbed, or dissolved. "To command" and "to obey" are complementary forms of the struggle.[16]

It is precisely because the power of single opponents is never absolute; he that provisionally loses always has a way of exerting his own residual forces such that the battle never ends. The battle never ends with a definitive victory or unconditional surrender. In the body neither sovereignty—the utter domination of another—nor the equality among many exists as they are perennially engaged in mutually overtaking each other. The uninterrupted polemic that Nietzsche wages against modern political philosophy has precisely to do with such a presupposition: if the battle within the single body is in itself infinite; if bodies therefore cannot distance themselves from the principle of struggle because struggle is the same form as life: how then can the order that conditions the survival of subjects to the neutralization of the conflict be realized? What condemns modern political concepts to ineffectuality is exactly this split between life and conflict—the idea of preserving life through the abolition of conflict. One could say that the heart of Nietzsche's philosophy will be found in his rebuttal of such a conception, which is to say in the extreme attempt to bring again to the surface that harsh and profound relation that holds together politics and life in the unending form of struggle.

Counterforces

From these initial considerations it is already clear that Nietzsche, without formulating the term, anticipated the entire biopolitical course that Foucault then defined and developed: from the centrality of the body as the genesis and termination of sociopolitical dynamics, to the founding role of struggle and also of war, to the configuration of juridical-institutional orders, to finally the function of resistance as the necessary counterpoint to the deployment of power. One can say that all the Foucauldian categories are present in a nutshell in Nietzsche's conceptual language: "War is another matter"—so Nietzsche notes in the text that functions as the definitive balance sheet of his entire work. "Being *able* to be an enemy, *being* an enemy—perhaps that presupposes a strong nature; in any case, it belongs to every strong nature. It needs objects of resistance; hence it *looks for* what resists: the *aggressive pathos* belongs just as necessary to strength as vengefulness and rancor belong to weakness."[17] Nevertheless, this passage already leads to an analytic landscape not limited to foreshadowing the Foucauldian theorization of biopolitics, but which in some ways also moves beyond it, or better, enriches it with a conceptual structure that contributes to untangling the underlying antinomy to which I referred in the opening chapter: to that immunitary paradigm that represents the peculiar figure of Nietzschean biopolitics. According to Nietzsche, reality is constituted by a complex of forces counterposed in a conflict that never ends conclusively because those who lose always maintain a potential of energy, which is able not only to limit the power of those who dominate, but, at times, to reverse the predominance in their own favor.

In Nietzsche's text, this systemic description, so to speak, is characterized by a tonality that is anything but neutral, but which is indeed decidedly critical: in the sense that once the play of forces has been defined from the objective point of view of quantity, assessing their quality remains open. Such forces, in short, are not in the least equivalent, so that it matters a great deal in a given phase which of these expands and which, on the contrary, contracts. Indeed, it is precisely on this that the larger trend depends—the "health," to adopt Nietzsche's lexicon—of the totality constituted by their struggle. There are forces that create and others that destroy; forces that strengthen and others that diminish; forces that stimulate and others that debilitate. Yet the peculiar characteristic of the Nietzschean logic is that the most important distinction between these forces doesn't pass through their constructive or destructive effect, but rather involves a more profound distinction, relative to the more or less original character of the forces themselves. The question of immunization bears upon this aspect, not only the objective emphasis that it comes to assume, but also the explicitly negative connotation that Nietzsche gives immunity, in an opposite trend to the positive connotation that modern philosophy has conferred upon it. Such a hermeneutic difference or even deviation doesn't relate to the preserving, salvific role that it exercises toward life—Nietzsche acknowledges it in the same way as does Hobbes—but instead to its logical-temporal arrangement in relation to the origin. To say this in the most concise way possible: while for Hobbes the immunitary demand comes first—it is the initial passion that moves men dominated by fear—for Nietzsche such a demand for protection is second with respect to another more original impulse, constituted we know by the will to power. It isn't that life doesn't demand its own preservation—otherwise the subject of every possible expansion would vanish—but it is in a form that, in contrast to all the modern philosophies of *conservatio,* is subordinated to the primary imperative of development, with respect to which it is reduced to a simple consequence:

> Physiologists need to think twice before putting the instinct of "preservation" as the cardinal instinct of an organic being. Above all, what lives wants to *give vent to* its own force; "preservation" is only one of the consequences of that.[18]

Here we are concerned with an argument to which Nietzsche himself assigns such prominence that he situates it exactly at the point of rupture with the entire tradition that precedes him: not only, he essentially adopts it against the philosopher to whom he otherwise is closest (even from this perspective), namely, "consumptive Spinoza":[19]

> The wish to preserve oneself is the symptom of a condition of distress, of a limitation of the really fundamental instinct of life, which aims at *the expansion of power,* and wishing for that, frequently risks and even sacrifices self-preservation.[20]

The text cited above appears even more clear-cut than the preceding one: preservation isn't to be considered only incidental and derivative with respect to the will to power, but in latent contradiction to it. And this is because the strengthening of the vital organism doesn't suffer limits or reductions, but, on the contrary, because it tends continually to move beyond and transgress them. It moves as a vortex or a flame, disrupting or burning every defensive partition, every liminal diaphragm, every border of definition. It crosses what is diverse and joins what is separate until it absorbs, incorporates, and devours everything that it meets. Life isn't only bound to overcome every obstacle that it comes up against, but is, in its own essence, the overcoming of the other and finally of itself: "And this secret life itself told me: 'Behold,' it said, '*I am that which must always overcome itself.*'"[21] By now Nietzsche's discourse bends in an ever more extreme direction, which seems to include its own contrary in a powerful self-deconstructive movement. Identifying life with its own overcoming means that it is no longer "in itself"—it is always projected beyond itself. But if life always pushes outside itself, or admits its outside within it, which is to say, to affirm itself, life must continually be altered and therefore be negated insofar as it is life. Its full realization coincides with a process of extroversion or exteriorization that is destined to carry it into contact with its own "not"; to make of it something that isn't simply life—neither only life nor life only—but something that is both more than life and other than life: precisely *not* life, if for "life" we understand something that is stable, as what remains essentially identical to itself. Nietzsche translates this intentionally paradoxical passage into the thesis that "human existence is merely an uninterrupted past tense, a thing that lives by denying and consuming itself, by opposing itself."[22] It is the same reason for which in *Beyond Good and Evil* he can write both that "life is *essentially* a process of appropriating, injuring, overpowering the alien and the weaker, oppressing, being harsh, imposing your own form, incorporating, and at least, the very least, exploiting" and simultaneously that life brings to the foreground "the feeling of fullness, of power that wants to overflow, the happiness associated with a high state of tension, the consciousness of a wealth that wants to make gifts and give way."[23]

At the bottom of such a conceptual tension, or indeed bipolarity, which seems to push Nietzsche's discourse in diverging directions, is a presupposition that is to be made explicit. Once again Nietzsche—in contrast to the largely dominant paradigm of modern anthropology, but also differently from the Darwinian conception of "struggle for existence"—holds that "in nature it isn't extreme angst that dominates, but rather superabundance and profusion pushed to the absurd."[24] Life doesn't evolve from an initial deficit but from an excess, which provides its double-edged impulse. On the one hand, it is dedicated to imposing itself over and incorporating everything that it meets. On the other hand, once it has been filled to the brim with its own acquisitive capacity, it is prone to tip over, dissipating its own surplus of goods, but also itself, what Nietzsche will define as "the bestowing virtue."[25] Here one already begins to glimpse the most troubling aspects of Nietzschean discourse: entrusted to itself, freed from its restraints, life tends to destroy and to destroy itself. It tends to dig a crevasse on every side as well as within, one into which life continually threatens to slip. Such a self-dissolving tendency isn't to be understood as a defect of nature or as a breach that is bound to damage an initial perfection. Nor is it an accident or the beginning that suddenly rises up or penetrates into life's domain. Rather, it is the constitutive character of life. Life doesn't fall in an abyss; rather, *it is* the abyss in which life itself risks falling. Not in a given moment, but already at the origin, from the moment that that abyss is nothing other than the interval of difference that withdraws the origin from every identifying consistency: the in/origineity of the origin that the Nietzschean genealogy ultimately traced to the source of being-in-life. In order to find an image or a conceptual figure of such a *deficiency for excess*, it is enough simply to return to one of the primary and most recurrent categories for Nietzsche, namely, that of the Dionysian. The Dionysian is life itself in absolute (or dissolute) form, unbound from any presupposition, abandoned to its original flow. Pure presence and therefore unrepresentable as such because it is without form, in perennial transformation, in the continuous overcoming of its own internal limits, of every principle of individuation and of separation between beings, genus, and species, but simultaneously of its external limits, that is, of its own categorical definition. How do we determine what not only escapes determinacy, but is also the greatest power of indeterminacy? And then do we differentiate what overwhelms all identities—and therefore all differences—in a sort of infinite metonymical contagion, that doesn't withhold anything, in a continual expropriation of everything distinct and the exteriorization of everything within? We can see in the Dionysian—understood as the in/original dimension of life in its entirety—the trace or the prefiguration of the common *munus* in all of its semantic ambivalence; as the donative elision of individual limits, but also as the infective and therefore destructive power of itself and the other. It is delinquency both in the literal significance of a lack and in the figurative sense of violence. Pure relation and therefore absence or implosion of subjects in relation to each other: a relation without subjects.

Against this possible semantic declension, against the vacuum of sense that opens at the heart of a life that is ecstatically full of itself, the general process of immunization is triggered, which coincides in the final analysis with all of Western civilization, but which finds in modernity its most representative space: "The democratization of Europe is, it seems, a link in the chain of those tremendous *prophylactic measures* which are the conception of modern times."[26] Nietzsche is the first not only to have intuited the absolute importance of immunization, but to have reconstructed its entire history in its genesis and internal articulations. Certainly, other authors—

from Hobbes to Tocqueville—recognized the onset of immunization first in the fear of violent death and then in the demand for protection with respect to the danger of individual passions that are highly combustible. But the absolute specificity of the Nietzschean perspective with regard to antecedent and successive diagnoses lies, on the one hand, in the return of the immunitary paradigm to its originary biological matrix, and, on the other, in the capacity to reconstruct critically the negative dialectic of the paradigm. As to the first, we note that Nietzsche refers all of the *dispositifs* of knowledge, which are apparently directed to the search for truth, to the function of preservation. Truth he defines as a lie—today we would say ideology—more suitable for sheltering us from that originary fracture of sense that coincides with the potentially unlimited expansion of life.[27] The same is true for the logical categories, from that of identity, to cause, to non-contradiction—all understood as biological instruments necessary to facilitate survival. They serve to save our existence from what is most unbearable about it; to create the minimal conditions to orient ourselves in a world that has no origin or end. They construct barriers, limits, and embankments with respect to that common *munus* that both strengthens and devastates life, pushing it continuously beyond itself. The procedures of reason raise up an immunitary *dispositif* against that vortex that in essence we are; against the trans-individual explosion of the Dionysian and against the contagion that derives from it, one that aims at restabilizing meaning and at redrawing lost boundaries, filling up the empty spaces deepened by the power of "outside." That outside is brought inside, or at least faced and then neutralized in the same way that what is open is contained and delimited in its most terrifying effects of incalculability, incomprehension, and unpredictability. Initially the Apollonian principle of individuation works to do this. Then, beginning with the grand Socratic therapy, followed by the entire Christian-bourgeois civilization (with an increasingly intensive and exclusive restorative expression) the following is attempted: to block the fury of becoming, the flow of transformation, the risk of metamorphosis in the "framework" of prevision and prevention.[28]

If this is the anesthetic or prophylactic role of the forms of knowledge, the same holds true for power and for the juridical and political institutions that flank moral and religious codes, reinforcing them in a logic of mutual legitimation. Above all, these institutions are born from ancestral fear, but are always secondary with respect to the originary will to power that grips man in a way unknown to other animals: "If one considers that man was for many hundreds of thousands of years an animal in the highest degree accessible to fear," it seems clear that the only way of mastering it is to construct the great immunitary involucres intended to protect the human species from the explosive potential that is implicit in its instinct for unconditional affirmation.[29] From Greek civilization onwards, institutions constructed by men "grow out of precautionary measures designed to make them safe from one another and from their inner *explosivity*."[30] The state is organized above all to defuse such explosivity, as, after all, modern political philosophy had already argued in a line of reasoning that saw in it the only way to master an otherwise lethal interindividual conflict. Nevertheless, it is precisely with regard to this last passage that Nietzsche grafts the change of theoretical paradigm that places him not only outside of that interpretive lineage, but in direct contrast with it: "The state is a prudent institution for the protection of individuals against one another," he admits, but then soon after adds, "if it is completed and perfected too far it will in the end enfeeble the individual and, indeed, dissolve him—that is to say, thwart the original purpose of the state in the most thorough way possible."[31] Evidently, what is at stake is not only the ability of the state to protect but more generally the overall evaluation of the immunitary logic, which Nietzsche diametrically reverses with respect to the substantially positive one of modern anthropology.

The thesis he advances is that such a logic cures illness *[male]* in a self-contradictory form because it produces a greater illness than the one it wanted to prevent. This occurs when the decided-upon compensation, with respect to the preceding vital order, is so considerable as to create a new and more deadly disequilibrium. Just as the state homologizes through forced obedience the same individuals that it intended to free, so too do all the systems of truth, which are also necessary for correcting harmful errors and superstitions, create new and more oppressive semantic blocks that are destined to obstruct the energetic flow of existence. In both of these cases, therefore, the stability and the duration that immunitary programs assure wind up inhibiting that innovative development that they need to stimulate. Impeding the possible dissolution of the organism, they also stop its growth, condemning it to stasis and impoverishment. This is the reason why Nietzsche defines morality, religion, and metaphysics simultaneously as both medicine and disease. Not only, but as diseases stronger than the medicines that work against them because they are produced for the same use: "[T]he worst sickness of mankind originated in the way in which they have combated their sicknesses, and what seemed to cure has in the long run produced something worse than that which it was supposed to overcome."[32]

With Nietzsche we are already in a position to reconstruct the entire diagram of immunization. Immunity, because it is secondary and derivative with respect to the force that it is intent on fighting, always remains subaltern to it. Immunity negates the power of negation, at least what it considers as such. Yet it is precisely because of this that immunity continues to speak the language of the negative, which it would like to annul: in order to avoid a potential evil, it produces a real one; it substitutes an excess with a defect, a fullness with a emptiness, a plus with a minus, negating what it affirms and so doing affirming nothing other than its negation. It is what Nietzsche means by the key concept of "resentment," which he identifies with all forms of resistance or of vengeance, and which is contrasted with the originary affirmative forces of life:

> For millennia this instinct for revenge has dominated humanity to such an extent that metaphysics, psychology and historical representation, and above all morality are marked by it. Wherever man has thought, even there, he has also inoculated the bacillus of revenge into things.[33]

Perhaps nowhere more than here does Nietzsche penetrate so deeply into the countereffective logic of the immunitary paradigm. Furthermore, Nietzsche explicitly recognizes this as the force—weakness is also a force, albeit one that degenerates from the will to power—that characterizes the entire process of civilization. If, as often happens, we do have full knowledge of it, this is because knowledge, just like all cognitive apparatuses, is also its product. Yet what counts even more is the mode in which this force acts—or, more precisely, "reacts." Just as in every medical immunization, immunization here too injects an antigenic nucleus into the social body, which is designed to activate protective antibodies. Doing so, however, it infects the organism in preventive fashion, weakening its primogenital forces: it risks killing what it is meant to keep alive. Nevertheless, it is what the ascetic priest or the pastor of souls does with regard to the sick flock: "He brings salves and balsam, there is no doubt; but he needs to wound before he can cure; then, in relieving the pain he has inflicted, *he poisons the wound.*"[34] More than a force that defends itself from a weakness, it is a weakness that draws off the force, draining it from within, separating it from itself. As Deleuze observed, the reactive force acts via decomposition and deviation, subtracting its power from the active force in order to appropriate some and to divert it from its originary destination.[35] So doing, however, it incorporates a force that is already exhausted, thwarting its capacity to react. This force continues to react, but in a debilitated form that isn't an active response, but rather a response without action, an action that is purely imaginary. Establishing itself within the organism, be it individual or collective that it aspires to defend, the organism itself is brought to ruin. Having destroyed the active forces in order to assimilate their power, nothing remains except to direct the poison point within, until it has destroyed itself as well.

Double Negation

What has been delineated above is a paradigm of great internal complexity. Not only forces and weaknesses clash and become entangled in a knot that doesn't allow for a stable distinction to be made, but what was a force can be weakened to such an extent that it turns into its opposite, just as an initial weakness can, at a certain moment, assume the form of a force that takes possession of power. Furthermore, the same element can simultaneously constitute a force for some and a weakness for others. This happens in Christianity as well and in religion generally, which the few use instrumentally to impose their own domination over the many and which is therefore destined to reinforce the former to the detriment of the latter. In addition, it also furnishes the latter with the means to retaliate on another level against the former and to drag them down into the same vortex. Something similar can be said for art and in particular for music. They can serve as potent stimulants for our senses according to the originary meaning of the term "aesthetic"; but they can also become a sort of subtle "anesthetic" with respect to the traumas of existence. This is what happens to music of the Romantic period until Wagner. Not any different, finally, is the double *[doppia]*, or better divided *[sdoppiata]*, reading that Nietzsche proposes of juridical-political institutions, beginning with that of the state; from one perspective, the state is seen as the necessary bulwark against destructive conflicts, and from another it is a mechanism that inhibits vital energies that have been completely scattered. Moreover, the entire process of civilization implies consequences that are reciprocally antinomic—precisely those that concern facilitating and weakening life. And doesn't Nietzsche define history as something useful and yet harmful? In short, to live, man needs in different situations (but at times in the same situations) both one thing and its opposite. He needs the historian and the nonhistorian, truth and lies, memory and forgetfulness, and health and disease, not to mention the dialectic between the Apollonian and the Dionysian into which all the other bipolarities finally devolve.

Such an ambivalence, or even aporeticity of judgment, derives from the mutability of perspective with which one views a given phenomenon, not to mention the always variable contingency in which it is situated. But digging deeper, the ambivalence is rooted in a contradiction that is as it were structural, according to which immunization, on the one hand, is necessary to the survival of any organism, but, on the other, is harmful because, blocking the organism's transformation, it impedes biological expansion. This in turn derives from the fact to which Nietzsche repeatedly draws attention, namely, that preservation and development, to the degree they are implicated in an indissoluble connection—that is, if something doesn't keep itself alive, it cannot develop—are in latent opposition when placed on another terrain, namely, the one decisive for the will to power. Not only, Nietzsche argues. In fact, what "is useful in relation to the acceleration of the rhythm of development is a 'use' which is different from that which refers to the maximum establishment and possible durability of what is developed," but "what is useful to the *duration* of the individual can become a disadvantage for its strength and its splendor, which is to say that what preserves the individual can hold it and block its development."[36] Development presupposes duration, but duration can delay or impede development. Preservation implies expansion, but expansion compromises and places preservation at risk. Here already the indissolubly tragic character of the Nietzschean perspective comes into view, not only because the effects are not directly referred back to their apparent cause, but because the wrinkle of a real autonomy opens between the one and the others: the survival of a force opposes the project of strengthening it. Limiting itself to survival, it weakens itself, flows back, and, to use the key word in Nietzschean semantics, *degenerates,* which is to say moves in the direction opposite its own generation. On the other hand, however, must we necessarily draw the paradoxical conclusion that to expand vitally, an organism has to cease to survive? Or, at a minimum, that it must face death?

This is the most extreme point of our inquiry, the conceptual intersection before which Nietzsche finds himself. In the course of his work (and frequently in the same texts), Nietzsche furnishes two kinds of responses, which sometimes appear to be superimposed, while at other moments seem to be incompatible. A good part of the question plays out in Nietzsche's difficult relation with Darwinian evolution, or better with what he, not always correctly, considers as such. We already know that Nietzsche rejects the idea of an initial deficit that would push men to struggle for their survival according to a selection that is destined to favor the fittest. He overturns this "progressive" reading with a different approach that—interpreting the origin of life in terms of exuberance and prodigality—anticipates conversely a discontinuous series of increments and decrements that are governed not by a selective adaptation but rather by the struggle within the will to power: of the reduction of the will to power for some and of its increase for others. But rather than being to the advantage of the strong and best, as Darwin would have it (at least the Darwin reread by Nietzsche through Spencer), this redounds to the benefit of the weak and the worst:

> What surprises me more than anything else when contemplating the grand destinies of man is to have always before my eyes the opposite of what Darwin with his school sees or *wants* to see: natural selection in favor of the stronger, the more gifted, the progress of the species. One can touch with one's hand the exact opposite: the elimination of cases to the contrary, the uselessness of types that are highly successful, the inevitable victory of the average and even of those *below* average.[37]

The reason for such a qualitative decrease is found, on the one hand, in the preponderance of the number of those less endowed with respect to the superior few and, on the other hand, in the organized strategy put in motion by the former against the latter. While the weak, gripped by fear, tend to protect themselves against the traps surrounding them (and by this increase them), the strong continually put their life on the line, for example, in war, exposing life to the risk of an early dissolution. What results finally is a process of degeneration that continually accelerates given that the remedies utilized form part of the same process: medicines implicated in the same disease that they intend to cure, which are constituted ultimately by the same poison. This is the dialectic of immunization that Nietzsche continually linked to decadence and to which he gave the name nihilism, especially in his later works.[38] Nihilism includes within itself the instruments by which it overcomes itself, beginning precisely with the category of decadence. Thus nihilism conceptually appears to be insurmountable: modernity doesn't have different languages apart from immunization, which is constitutively negative.

Not even Nietzsche is able to escape from such a conceptual constraint (and from this point of view Heidegger wasn't wrong in keeping him on this side of nihilism, or at least on its meridian). Indeed, he remains utterly implicated in at least one conspicuous vector of immunization. It is true that Nietzsche intends to oppose that process of immunitary degeneration which, rather than strengthening the organism, has the perverse effects of dehabilitating it further. The substitution of the will to power for the struggle for survival as both the ontogenetic and philogenetic horizons of reference constitutes the clearest confirmation. And yet precisely such a negation of immunization situates Nietzsche (or at least this Nietzsche) within its recharging mechanism. Negating the immunitary negation, Nietzsche undoubtedly remains the prisoner of the same negative lexicon. Rather than affirming his own perspective, Nietzsche limits himself to negating the opposite, remaining, so to speak, subaltern to it. Just as happens in every logic of the reactive type, whose structurally negative modality Nietzsche so effectively deconstructs, his critique of modern immunization responds to something that logically precedes it. The same idea of degeneration *(Entartung)*, from which Nietzsche derives the means of developing the antidote, has an intrinsically negative configuration: it is the contrary of generation, a generation folded upon itself and perverted—not an affirmative, but the negative of a negative, typical after all of the antigenic procedure. It isn't by coincidence that the more Nietzsche is determined to fight the immunitary syndrome, the more he falls into the semantics of infection and contamination. All the themes of purity, integrity, or perfection that obsessively return (even autobiographically) have this unmistakably reactive tonality, which is to say doubly negative toward a rampant impurity that constitutes the discourse's true *primum:*

> As has always been my wont—extreme uncleanliness *[Lauterkeit]* in relation to me is the presupposition of my existence; I perish under unclean conditions—I constantly swim and bathe and splash, as it were, in water—in some perfectly transparent and resplendent element . . . My whole *Zarathustra* is a dithyramb on solitude or, if I have been understood, on *cleanliness [Reinheit]*.[39]

Not only, but Nietzsche presents the degeneration as both the cause and the effect of the progressive contagion of the uncontaminated by the contaminated. It is these latter who, in order to reject the positive force of their own power, contaminate the former, and so swiftly extend the infected areas to the point that the decadence against which Nietzsche exhorts us to fight—more than a disease that can be easily eliminated as such—is unquestionably the advancing line of the contagion:

> Decadence is not something one can combat: it is absolutely necessary and belongs to every epoch and every people. What needs to be fought against with all one's strength is the contagion of the healthy parts of the organism.[40]

We cannot avoid the hyperimmunitary direction that this critique of immunization adopts. To refrain from an excess of protection—from the weaker species' obsession with self-preservation—protection is needed from their contagion. A stronger and more impenetrable barrier must be constructed, stronger than the one already in place. In so doing, the separation between the healthy and sick parts will be rendered definitive, where the biological distinction, or better opposition, between the physiological and the pathological has a transparent social meaning: "Life itself doesn't recognize either solidarity or 'equality of rights' among the healthy and diseased parts of an organism: the latter need to be lopped off or the whole will perish."[41] It would be superfluous to indicate to the reader the numerous passages in which Nietzsche insists on the necessity of preservation. More useful would be to accentuate the rigid disjunction Nietzsche makes between different classes, and in particular between the race of masters and slaves. His exaltation of incommunicable castes in India speaks volumes on the subject. What is to be emphasized here is the categorical contrast that also emerges vis-à-vis modern political philosophy: Nietzsche opposes liberal individualism and democratic universalism's *homo aequalis* to the premodern *homo iearchicus*, which serves to confirm the regressive and restorative character of this axis in the Nietzschean discourse. Moreover, the favorable citations of de Boulainviller, which a biopolitical Foucault quotes on more than occasion, move in the same antimodern direction.[42] De Boulainviller is one of the first to have contested the lexicon of sovereignty and of the one and indivisible nation in favor of an irreducible separation between conflicting classes and races. That Nietzsche's racism is of the horizontal or diagonal kind, in which he discriminates between diverse populations or makes a break within the same national community, is an undecidable question in the sense that he moves from one level to another according to the texts in question and the circumstances in which he is writing. But what deserves our attention in the conceptual profile sketched here is the obvious contradiction with regard to the thesis of originary abundance, of a zero-sum game according to which the elevation of the one is directly proportional to the coercion, and indeed the elimination, of others:

> The crucial thing about a good and healthy aristocracy, however, is that it . . . has no misgivings in condoning the sacrifice of a vast number of people who must *for its sake* be oppressed and diminished into incomplete people, slaves, tools.[43]

Of course, Nietzsche's position, as some have observed, isn't an isolated one when seen against the background of his time.[44] Accents of the sort can be found not only in conservative thought, but even in the liberal tradition, where reference is made to the destiny of extra-European peoples subject to colonization and racial exploitation. But what makes it relevant for our analysis is its intense biopolitical tonality. What is undoubtedly in question in this sacrificial balance, in which one level must necessarily drop down so that another can rise up, isn't only power, prestige, or work, but life itself. In order for life's biological substance to be intensified, life must be marked with an unyielding distinction that sets it against itself: life against life, or, more severely, the life of one against the nonlife of others: "What is *life*?—Life—that is: continually shedding something that wants to die."[45] Not only is life to be protected from the contagion of death, but death is to be made the mechanism for life's contrastive reproduction. The reference to the elimination of parasitic and degenerative species comes up again in all its crudeness, contained in the text I cited earlier on grand politics. That it concerns refusing to practice medicine on the incurable, or indeed of eliminating them directly; of impeding the procreation of unsuccessful biological types; or of urging those suffering from irreversibly hereditary traits to commit suicide—all of this can be interpreted as an atrocious link in the gallery of horrors running from the eugenics of the nineteenth century to the extermination camps of the twentieth. Personally, I share the hermeneutic option of not softening (either metaphorically or literally) passages and expressions of the sort, which Nietzsche himself shares with authors such as Lombroso, Emerson, Lapogue, Gumplowicz, and still others: for an implacable border divides human life, one that conditions the pleasure, knowledge, and power of the few to the struggle as well as the death of the many. If anything, the open question remains how to reconstruct the internal logic that pushes Nietzschean biopolitics into the shelter of its thanatopolitical contrary.

My impression is that such logic is firmly associated with that immunitary semantics against which Nietzsche too, from another point of view, struggles with clearly contradictory results. The epicenter of such a contradiction can be singled out in the point of intersection between a tendency to biologize existence and another, contrary and speculative, one, which is based on the existentialization or the purification of what also refers to the dimension of life. Or better: functionalizing the former so as to fulfill the latter. It is as if Nietzsche simultaneously moves in two opposite but convergent directions toward one objective: as we have already seen, on the one hand, he associates the metaphysical construct, which the theo-philosophical tradition defines as a "soul," to the body's biology; on the other hand, he withdraws the body from its natural degradation through an artificial regeneration that is capable of restoring its original essence. Only when *bíos* is forcibly brought back into the circle of *zōē* can *bíos* overcome itself in something that pushes it beyond itself. It isn't surprising that Nietzsche seeks the key to such a paradoxical move in the same Plato around whom his deconstruction turns. This is possible to the degree that Nietzsche substitutes a metaphysical Plato, the one of the separation and opposition of body and soul, for a biopolitical Plato. In this sense, he can argue that the true Platonic republic is a "state of geniuses," which is actualized through the elimination of lives that do not meet the required standards. At the center of the Platonic project, therefore, are the demands to maintain the purity of the "race of guardians" and through them to save the entire "human herd" from degenerative contagion. Leaving aside the legitimacy of similar interpretations of Plato—whose thanatopolitical folds we have seen, or will have occasion to see shortly—what counts most here at the end of our discourse is the intensely immunitary attitude that subtends the question. Not only is the solution to the degenerative impulse sought in the blocking of becoming, in a restoration of the initial condition, or in a return to a perfection of what is integral, pure, and permanent. Rather, such a restoration, or physical and spiritual reintegration (spiritual because it is physical), is strictly conditioned by the incorporation of the negative, both in the lethal sense of the annihilation of those that do not deserve to live, and in the sense of the crushing of the originary dimension of animality of those who remain. When Nietzsche insists on the definitive zoological connotation of terms such as *Züchtung* [breeding] or *Zähmung* [domestication], he is determined to assert (against the entire humanist culture) that man's vital potential lies in that profound belonging to what is still not, or is no longer, human, to something that constitutes for the human both the primogenital force and the specific negation. Only when man undergoes the same selective treatment applied to animals or to greenhouse plants will he be able to cultivate the self-generating capacity that degeneration has progressively consumed.

When this Platonism, now reversed by a biopolitical key, comes into contact with the contemporary theories on degeneration of Morel and of Faré—of whom I'll speak at length in the next chapter—the results appear to be devastating. Thus it isn't entirely unfounded to see in *this* Nietzsche, on the one hand, the nihilistic apex of nineteenth-century social Darwinism, and, on the other, that conceptual passage toward the eugenic activism that will be tragically on display in the next century.[46] Its specific axis of ideological elaboration emerges in the confluence of Galton's criminal pathology and the animal sociology of authors such as Espinas and Schneider.[47] If the origin of the criminal act lies more deeply in the biological conformity (and therefore in the genetic patrimony of the one who commits the crime than in a free individual choice), it's clear that punishment cannot but be characterized by both prevention and finality, relative not to the single individual but to the entire hereditary line from which it comes. Such a line, when not broken, is destined to be transmitted to its descendants. But this first superimposition between the mentally ill and the criminal involves a second and more extreme superimposition between the human and the animal species. From the moment that man appears bound by an unbreakable system of biological determinism, he can be reclaimed by his animal matrix from which he wrongly believes to have been emancipated (precisely on the strength of that distortion or perversion, civilization, which is nothing other than continual degeneration). Seen from this angle, we are well beyond the metaphor of the animal that originated with Hobbes, the man who is a wolf toward his equals. Taken literally, the wolf-man isn't actually what remains of a superior type already under attack, or better, one inhabited by another kind of inferior animal destined to devour him from within: the parasite, the bacillus, or the tick that sucks his blood and transmits it, now poisoned, to the rest of the species. With regard to such a biological risk (which is *therefore* also political), there can only be a similarly biopolitical response in the lethal sense in which such a term is reversed in the nihilist completion of the immunitary dialectic. Once again in question is the generation of the negation of degeneration, the effectuation of life in death:

> A sick person is a parasite on society. Once one has reached a certain state it is indecent to live any longer... Create a new kind of responsibility, the physicians, to apply in all cases where the highest interest of life, of *ascending* life, demands that degenerating life be ruthlessly pushed down and aside—for example in the case of the right to procreate, the right to be born, the right to live.[48]

Posthuman

Nonetheless, this isn't Nietzsche's last or only word on the subject. Certainly, it is the origin of a discursive line that is unequivocal in its conclusions and its effects of sense, whose categorical extraneousness from the most destructive results of nineteenth-century eugenics it would be arduous to demonstrate. But this line ought not to be separated from another perspective that is irreducible to the first, and indeed whose underlying inspiration runs contrary to it. The internal point of distinction between these two different semantics is to be found in the perspective that Nietzsche assumes with regard to the process of biological decadence, which is defined in terms of degeneration or of passive nihilism. How does one behave toward it? By trying to stop it, to slow it down, to hold it in check through immunitary *dispositifs* that are the same and contrary to those that it itself activated (and ultimately responsible for the decline under way); or, on the contrary, to push it toward completion, and so doing provoke its self-destruction? By erecting new and ever denser protective barriers against the wide-ranging contagion, or rather encouraging it as the means to the dissolution of the old organic equilibrium and therefore the occasion for a new morphogenetic configuration? By tracing more markedly the lines of separation between social classes, groups, and races to the point of conditioning the biological development of the one to the violent reduction of the others? Or instead by trying to find in their difference the productive energy for common expansion?

In the preceding paragraphs, we became familiar with Nietzsche's first response to these questions, along with its ideological presuppositions and the thanatopolitical consequences. Without being able to establish any chronological sequence between the two, it's opportune at this stage to note that at a certain point (that contrasts with and is superimposed upon his response), he appears to follow another track. The supporting idea is that only by accelerating what will nevertheless take place can one liberate the field for new affirmative powers *[potenze]*. Every other option—restorative, compensative, resistant—creates a worse stalemate than before:

> Even today there are still parties which dream of the crab-like *retrogression* of all things as their goal. But no one is free to be a crab. It is no use: we *have* to go forwards, i.e. *step by step further* in décadence (—this being *my* definition of modern "progress"...). You can *check* this development and, by checking it, dam up, accumulate degeneration itself, making it more vehement and *sudden:* no more can be done.[49]

Implicit in such expressions is the perspective (not extraneous to what will take the name of "eternal return") that, if a parabolic incline is continually increased, it ends up meeting itself in circular fashion at the point from which it began to move, returning again toward the top. It is exactly here that Nietzsche begins to deconstruct the hyperimmunitary machine that he himself set in motion against the debilitating effects of modern immunization. Where before he emphasized a strategy of containment, now enters another of mobilization and the unleashing of energy. Force, even reactive force, is unstoppable in itself: it can only recoil against itself. When pushed to a point of excess, every negation is destined to negate itself. After having annihilated everything that it encounters, negation cannot but fight against its own negativity and reverse itself in the affirmative. As Deleuze rightly argues, at the origin of this conceptual passage isn't the masked propensity for the dialectic (a sort of reverse Hegelianism), but rather the definitive release from its machinery: affirmation is not the synthetic result of a double negation, but instead the freeing of positive forces, which is produced by the self-suppression of the negation itself. As soon as the immunitary rejection, what Nietzsche calls "reaction," becomes intense enough to attack the same antibodies that provoked the rejection, the break with the old form becomes inevitable.

Of course, this seems to contradict what was said about the irreversibility of degeneration. In part it does, but only if we lose sight of the subtle line of reasoning that implies the possibility of its own reversal. As is customary for an author who distrusts the objectivity of the real, the question is one of perspective. The self-deconstruction of the immunity paradigm that Nietzsche operates (that runs counter to his eugenic aim) doesn't rest on a weakening of the vitalistic project, nor on an outright abandonment of the degenerative hypothesis. At stake isn't the centrality of the biopolitical relation between health and illness, but a different conception of one and the other and therefore of their relation. What fails in this more complex inflection of Nietzschean thought is the dividing line that separates them in the metaphysically presupposed form of the absolute distinction between good and evil. In this sense, then, Nietzsche can declare that "there is no health as such, and all attempts to define a thing that way have been wretched failures . . . there are innumerable healths of the body . . . and the more we abjure the dogma of the 'equality of men,' the more must the concept of a *normal* health along with a normal diet and the normal course of an illness, be abandoned by medical men."[50] Yet, if it isn't possible to settle on a canon of perfect health; if it isn't the norm that determines health, but health that creates its own norms in a manner that is increasingly plural and reversible—then every person has a different idea of health and therefore it inevitably follows that even an all-engaging definition of illness isn't possible. And not only in the logical sense that, if one doesn't know what health is, a stable conception of illness cannot be determined *[profilare]*, but in the biopolitical sense as well because health and illness are in a relation that is more complex than their simple exclusion. Illness, in short, isn't only the contrary of health, but is its presupposition, its means, and its path; illness is the something from which health originates and that it carries within as its inalienable internal component. No true health is possible that doesn't take in *[comprenda]*—in the dual sense of the expression: to know and to incorporate—illness:

> Finally, the great question would still remain whether we can really *dispense with* illness—even for the sake of our virtue—and whether our thirst for knowledge and self-knowledge in particular does not require the sick soul as much as the healthy, and whether, in brief, the will to health alone, is not a prejudice, cowardice, and perhaps a bit of very subtle barbarism and backwardness.[51]

At stake in this polemic against a will to health, one incapable of confronting its own opposite (and therefore also itself), is the challenge the relation between life and death continually presents to health. There's no need to imagine such a challenge as the battle between two juxtaposed forces, as a besieged city defending itself from an enemy intent on penetrating and conquering it. Not that an image of the sort is extraneous to the profound logic of Nietzschean discourse, as clearly results from its explicitly eugenic side. But, as has been said, such an image doesn't exhaust the logic. Indeed, one can assert that the extraordinary force of Nietzsche's work resides exactly in its intersection and contradiction of another analytic trajectory, which is situated within itself (and not worlds apart from it). The figure that emerges here is of a superimposition by way of contrast, all of whose logical passages (both in their succession and in their copresence) need to be recognized. We have seen how Nietzsche contests modernity's immunitary *dispositifs* not through negation, but instead by moving immunization from the institutional level to that of actual *[effettiva]* life; needing to be protected from the excess or the dispersion of life, no longer in the sense of a formal political order, but in the survival of the species as a whole. In a philogenetic framework of growing degeneration, such a possibility is conditioned both by the isolation and by the fencing in of those areas of life that are still whole with respect to the advancing contamination on the part of the weak whose life is ending, as well as by the reduction of the sick (in Malthusian fashion) in favor of the healthy. Nonetheless, we have seen how this prescription constitutes nothing other than the first hyperimmunitary or thanatopolitical stratum of the Nietzschean lexicon.

A second categorical vector draws alongside and is joined with it, one that moves in a direction that diverges from the first, or perhaps better, one that allows for a different reading. More than a revision, this vector moves through a semantic deferral of the preceding categories, beginning with that of "health" and "illness," bursting their nominal identity and placing them in direct contact with their contrary logic.[52] From this perspective (and with respect to the metaphor of the besieged city), the danger is also biological; it is no longer the enemy that makes an attempt on life from the outside, but the enemy is now life's own propulsive force. For this reason "the Greeks were certainly not possessed of a square and solid healthiness;—their secret was to honour even sickness as a god if only it had *power*."[53] Being "dangerously healthy, ever again health" means that this kind of health must necessarily traverse the sickness which it seems to fight.[54] Health is not separate from the mortal risk that runs through it, pushing it beyond itself, continuously updating its norms, overthrowing and re-creating rules for life. The result is a reversal that occurs by an intensification of the defensive and offensive logic that governs the eugenic strategy: if health is no longer separable from sickness; if sickness is part of health—then it will no longer be possible to separate the individual and social body according to insurmountable lines of prophylaxis and hierarchy. The entire immunitary semantic now seems to be rebutted, or perhaps better, to be reinterpreted in a perspective that simultaneously strengthens and overturns it, that confirms it and deconstructs it.

A paragraph in *Human, All Too Human* titled "Ennoblement through Degeneration" condenses in brief turns of phrase the entire trajectory that I've reconstructed to this point. At its center will be found the community held together by the equality of conditions and participation based on a shared faith. More than possible risks from the outside, what undermines the community's vitality is its stability: the more the community is preserved intact, the more the level of innovation is reduced. The greatest danger that the community faces is therefore its own preventive withdrawal from danger. Once immunized, the community doesn't run any risk of wounding, but it is precisely for this reason that it seals itself off blocking from within any possibility of relation with the outside and therefore any possibility of growth. Avoiding degeneration (according to the eugenic prescriptions of perfect health), the result is that the community loses its own self-generating potential. No longer capable of creating conditions of growth, it folds in upon itself. Saving it from such a decline are individuals who, free from the syndrome of self-preservation, are more inclined to experiment, although for the same reason they are biologically weaker. Disposed as they are to increasing the good that they possess (as well as their own vital substance), sooner or later they are bound not only to risk their lives, but also to damage the entire community. It is precisely here in the clench of this extreme risk, that the point of productive conjunction between generation and innovation is produced:

> It is precisely at this injured and weakened spot that the whole body is as it were *inoculated* with something new; its strength must, however, be as a whole sufficient to receive this new thing into its blood and to assimilate it. Degenerate natures are of the highest significance wherever progress is to be effected.[55]

This might seem to be mere theater for someone who elsewhere harps on defending the health of races and of individuals from the contagion of those who have degenerated. In reality, as we've already had occasion to discuss, the step in question is understood less as a distancing from the immunitary paradigm, and more as immunity's opening to its own communal reverse, to that form of self-dissolving gift giving that *communitas* names. The vocabulary that Nietzsche adopts indicates a similar semantic overlapping, which is situated precisely in the point of confluence between the lexicons of an immunity and community. I'm not speaking only of the identification of the new with infection, but also of the nobilizing effects produced by inoculation. Just as in the body of the community, so too in that of the individual, "the educator has to inflict injuries upon him, or employ the injuries inflicted on him by fate, and when he has thus come to experience pain and distress something new and noble can be inoculated into the injured places. It will be taken up into the totality of his nature, and later the traces of its nobility will be perceptible in the fruits of his nature."[56]

Clearly, the language Nietzsche adopts is immunitarian, that of vaccination—a viral fragment is placed into the individual or collective organism, which it is intended to strengthen. But the logic that underpins it is not directed to preserving identity or to simple survival, but rather to innovation and alteration. The difference between the two levels of discourse (and the slippage of one into the other) lies in the mode of understanding the relation with the "negative," and even before that with its own definition. That for which Nietzsche recommends the inoculation isn't an antigen destined to activate the antibodies, nor is it a sort of supplemental antibody intent on fortifying the defensive apparatus of the immunitary system. In short, it isn't a lesser negative used preventively to block the path of a greater negative. All of this is part of that dialectical procedure that Nietzsche criticizes as reactive and to which he poses instead a different modality according to which what is considered evil *[male]* upon first view (suffering, the unexpected, danger) is considered positively as characterizing a more intense existence. From this perspective, the negative not only is in turn detained, repressed, or rejected, but it is affirmed as such: as what forms an essential part of life, even if, indeed precisely because, it continually endangers it, pushing it on to a problematic fault line *[faglia]* to which it is both reduced and strengthened. Nietzsche sees the same role of philosophy—at least of that philosophy capable of abandoning the system of illusions to which it itself has contributed and so doing setting itself adrift—as a sort of voluntary intoxication. No longer the protecting Mother, but the Medusa that one cannot look upon without experiencing the lacerating power of unbearable contradictions. In this sense, the real philosopher "puts himself at risk," because he singles out the truth of life in something that continually overtakes it, in an exteriority that can never be completely interiorized, dominated, or neutralized in the name of other more comforting or obliging truths.[57]

Can we give the name of community to this exteriority with regard to the immunitary systems within which we endlessly seek refuge, just as Georges Bataille dared to do in his own time against an interpretive tendency oriented in the opposite direction?[58] Without wanting in any way to twist a philosophy whose entire layers and internal levels of contradiction I have tried to reconstitute, we can say that a series of texts induce a cautious, affirmative response. I am not referring only to those grouped around the theme of donation—of the "bestowing virtue"—whose deconstructive character cannot be avoided with respect to every appropriative or cumulative conception of the will to power.[59] Nor am I referring to those visionary passages concerning the "stellar friendship," also extended especially to those who are far removed and remote from us, even our enemies.[60] Rather, it concerns splinters, flashes of thought that are capable of suddenly illuminating (if only for an instant) that profound and enigmatic nexus between *hospes* and *hostis* (one that is situated at the origin of the Western tradition in a knot that we have still not been able to unravel). Certainly, all of this carries us along to the semantic threshold of that common *munus* whose opposite pole we have glimpsed.

Yet, if we adopt a more complex perspective, it is also the center, the incandescent nucleus of *immunitas*. In order to see it more clearly, we need to understand donation and also the friendship with the enemy not in an ethical sense (which would be completely extraneous to the Nietzschean lexicon, constitutively immune *[refrattario]* from all altruistic rhetoric), nor in a properly anthropological sense, but in a radically ontological sense. In Nietzsche, donation is not an opening to another man, but if anything to the other *of* man or also *from* man. It is the alteration of the self-belonging that an anything but exhausted humanistic tradition has attributed to man as one of the most proper to him of his essential properties—against which the Nietzschean text reminds us that man is still not, nor will ever be, what he considers himself to be. His being resides beyond this or beyond that side of the identity with himself. And indeed, he is not even a being as such, but a becoming that carries together within itself the traces of a different past and the prefiguration of a new future. At the center of this conceptual passage lies the theme of metamorphosis. With regard to the "retarding elements" of every species that is intent on constructing ever new means of preservation (who are determined to last as long as possible), the *Übermensch* (or however we may want to translate the expression) is characterized by an inexhaustible power of transformation. He literally is situated outside of himself, in a space that is no longer (nor was it ever) that of man as such. It isn't so important to know where or what he will become, because what he connotes is precisely becoming, a breaking through, a moving beyond his proper topos. It isn't that his life doesn't have form; it isn't a "form of life." Rather, it bears upon a form that itself is in perpetual movement toward a new form, traversed by an alterity from which it emerges simultaneously divided and multiplied.

In this sense, Nietzsche, the hyperindividualist, can say that the individual, the one undivided *[l'indiviso],* doesn't exist—that it is contradicted from its coming into the world by the genetic principle according to which "two are born from one and one from two."[61] It is no coincidence that birth, procreation, and pregnancy constitute perhaps the most symbolically charged figure of Nietzschean philosophy, one Nietzsche characterizes as falling under the sign of a painful delivery. This occurs because no term more than childbirth refers the theme of donation to its concrete biological dimension, which otherwise is simply metaphorical or classically intersubjective. Childbirth isn't only an offer of life, but it is the effective site in which a life makes itself two, in which it opens itself to the difference with itself according to a movement that in essence contradicts the immunitary logic of self-preservation. Against every presupposed interiorization, it exposes the body to the split that always traverses it as an outside of its inside, the exterior of the interior, the common of the immune. This holds true for the individual body, but also for the collective body, which emerges as naturally challenged, infiltrated, and hybridized by a diversity that isn't only external, but also internal. It is so for the *ethnos* and for the *genos,* that is, for the race that, despite all the illusions of eugenics, is never pure in itself, as well as for the species.[62] It is precisely with respect to the species, to what Nietzsche defines as human in order to distinguish it essentially from all the others, that he pushes the deconstruction or conversion of the immunitary paradigm farther and deeper into its opposite. Certainly, its superimposition with the animal sphere can be and has been interpreted in the most varied of ways. Undoubtedly, the sinister reference to "the beast of prey" or to "the breeding animal" contains within it echoes and a tonality that are attributable to the more deterministic and aggressive tendencies of social Darwinism. But in the animalization of man something else is felt that appears to mark more the future of the human species and less the ancestral past. In Nietzsche, the animal is never interpreted as the obscure abyss or the face of stone from which man escapes. On the contrary, it is tied to the destiny of "after-man" (as we could hazard translating *Übermensch*). It is his future not less than his past, or perhaps better, the discontinuous lines along with which the relationship between past and future assumes an irreducible configuration vis-à-vis all those that have preceded him. It's not by accident that the destiny of the animal is enigmatically connected *through* man to him who can exceed him in power and wisdom—to a man who is capable of redefining the meaning of his own species no longer in humanistic or anthropological terms, but in anthropocentric or biotechnological terms:

> What are the profound transformations that must derive from the theories according to which one asserts that there is no God that cares for us and that there is no eternal moral law (humanity as atheistically immoral)? That we are animals? That our life is transitory? That we have no responsibility? The wise one and the animal will grow closer and produce a new type [of human].[63]

Who or *what* this new "type" is naturally remains indeterminate, and not just for Nietzsche. But certainly Nietzsche understands (indeed, he was the first to seize with an absolute purity of a gaze) that we are at the threshold beyond which what is called "man" enters into a different relationship with his own species—beyond which, indeed, the same species becomes the object and the subject of a biopolitics potentially different from what we know because it is in relation not only to human life, but to what is outside life, to its other, to its after. The animalization of man in Nietzsche contains these two signs, which are perilously juxtaposed and superimposed: taken together, they form the point where a biopolitics precipitates into death and where the horizon of a new politics of life, which I outline here, begins.

CHAPTER FOUR

Thanatopolitics (The Cycle of *Genos*)

Regeneration

Michel Foucault was the first to provide us with a biopolitical interpretation of Nazism.[1] The force of his reading with respect to other possible readings lies in the distance he takes up with respect to all modern political categories. Nazism constitutes an irreducible protrusion for the history that precedes it because it introduces an antinomy that went unrecognized until then in its figure and in its effects. It is summarized in the principle that life defends itself and develops only through the progressive enlargement of the circle of death. Thus the paradigms of sovereignty and biopolitics, which seemed at a certain point to diverge, now experience a singular form of indistinction that makes one both the reverse and the complement of the other. Foucault locates the instrument of this process of superimposition in racism. Once racism has been inscribed in the practices of biopolitics, it performs a double function: that of producing a separation within the biological *continuum* between those that need to remain alive and those, conversely, who are to be killed; and that more essential function of establishing a direct relation between the two conditions, in the sense that it is precisely the deaths of the latter that enable and authorize the survival of the former. But that isn't all. In order to get to the bottom of the constitutively lethal logic of the Nazi conception [of life], we need to take a final step. Contrary to much of what we have been led to believe, such a conception doesn't concentrate the supreme power of killing only in the hands of the leader *[capo]*—as happens in classical dictatorships—but rather distributes it in equal parts to the entire social body. Its absolute newness lies in the fact that everyone, directly or indirectly, can legitimately kill everyone else. But if death as such (and here is the unavoidable conclusion of this line of "reasoning") constitutes the motor of development of the entire mechanism—which is to say that it needs to produce it in ever greater dimensions, first with regard to the external enemy, then to the internal, and then lastly to the German people themselves (as Hitler's final orders make perfectly clear)—then the result is an absolute coincidence of homicide and suicide that places it outside of every traditional hermeneutics.

Nevertheless, Foucault's interpretation isn't completely satisfying. I spoke earlier of the discontinuity that the interpretation aims at instituting in the modern conceptual lexicon.[2] Yet, the category assigned to fix more precisely the point of caesura of Nazi experience for history preceding it (namely, that of biopolitics) winds up constituting the part of their union: "Nazism was in fact the paroxysmal development of the new power mechanisms that had been established since the eighteenth century."[3] Certainly, Nazism carries the biopolitical procedures of modernity to the extreme point of their coercive power, reversing them into thanatological terms. But the process remains within the same semantics that seemed to have lacerated it. It extends onto the same terrain from which it appeared to tear itself away. In the Foucauldian reading, it is as if the tear were subjected to a more profound continuity that reincorporates its precision: "Of course Nazism alone took the play between the sovereign right to kill and the mechanisms of biopower to this paroxysmal point. But this play is in fact inscribed in the workings of all States."[4] Even if Foucault ultimately doubts such an affirmation, the comparison is by now established: even with its unmistakably new features, Nazism has much in common biopolitically with other modern regimes. The assimilation of Nazism to communism is even stronger; that too is traced back to a racist matrix and therefore to the notion of biopower that the matrix presupposes. We are already quite far from the discontinuist approach that seems to motivate Foucault's interpretation. It is as if, despite its contiguous and progressive steps, the generality of the framework prevails over the singularity of the Nazi event: both vertically in relation to the modern era and horizontally with regard to the communist regime. If the latter has a biopolitical context and if both inherit it from recent history, the power of rupture that Foucault had conferred on his own analysis is diminished or indeed has gone missing.[5]

It is precisely the comparison with communism (activated by the unwieldy category of totalitarianism) that allows us to focus on the absolute specificity of Nazi biopolitics.[6] Although the communist regime, in spite of its peculiarity, originates nonetheless in the modern era—its logic, its dynamics, and its wild swings in meaning—the Nazi regime is radically different. It isn't born from an exasperated modernity but from a decomposed modernity. If we can assert that communism always "carries out" *[realizzi]* one of its philosophical traditions (even in an aggravated form), nothing of the sort can be said of Nazism. Yet this is nothing more than a half-truth, which ought to be completed as follows: Nazism does not, nor can it, carry out a philosophy because it is an actualized *[realizzata]* biology. While the transcendental of communism is history, its subject class, and its lexicon economic, Nazism's transcendental is life, its subject race, and its lexicon biological. Certainly, the communists also believed that they were acting on the basis of a precise scientific vision, but only the Nazis identified their vision with the comparative biology of human races and animals. It is from this perspective that Rudolph Hess's declaration needs to be understood in the most restricted sense, according to which "National Socialism is nothing but applied biology."[7] In reality, Fritz Lenz, along with Erwin Baur and Eugen Fischer, used the expression for the first time in the successful manual *Rassenhygiene,* in a context in which they refer to Hitler as "the great German doctor" able to take "the final step in the defeat of that historicism and in the recognition of values that are purely biological.[8] In another influential medical text, Rudolph Ramm expressed his views similarly, asserting that "unlike any other political philosophy or any other party program, National Socialism is in agreement with natural history and the biology of man."[9]

We need to be careful not to lose sight of the utterly specific quality of this explicit reference to biology as opposed to philosophy. It marks the true breaking point with regard not only to a generic past, but also with respect to modern biopolitics. It's true, of course, that the political lexicon has always adopted biological metaphors, beginning with the long-standing notion of the state as body. And it is also true, as Foucault showed, that beginning with the eighteenth century the question of life progressively intersects with the sphere of political action. Yet both occurred thanks to a series of linguistic, conceptual, and institutional mediations that are completely missing in Nazism: every division collapses between politics and biology. What before had always been a vitalistic metaphor becomes a reality in Nazism, not in the sense that political power passes directly into the hands of biologists, but in the sense that politicians use biological processes as criteria with which to guide their own actions. In such a perspective we cannot even speak of simple instrumentalization: it isn't that Nazi politics limited itself to adopting biomedical research of the time for legitimizing its ends. They demanded that politics be identified directly with biology in a completely new form of biocracy. When Hans Reiter, speaking in the name of the Reich in occupied Paris, proclaimed that "this mode of thinking biologically needs to become little by little that of all the people," because at stake was the "substance" of the same "biological body of the nation," he understood well that he was speaking in the name of something that had never been part of a modern categorical lexicon.[10] "We find ourselves at the beginning of a new epoch," writes another ideo-biologist of the regime, Hans Weinert. "Man himself," Weinert continues, "recognizes the laws of life that model it individually and collectively; and the National Socialist state was given the right, insofar as it is in its power, to influence human becoming as the welfare of the people and the state demand."[11]

As long as we speak of biology, however, we remain on a level of discourse that is far too general. In order to get to the heart of the question, we need to focus our attention on medicine. We know the role that Nazi doctors played in the extermination effected by the regime. Certainly, the availability of the medical class for undertaking forms of thanatopolitics also occurred elsewhere—think of the role of psychiatrists in the diagnosis of mental illness for dissidents in Stalin's Soviet Union or in the vivisection practiced by Japanese doctors on American prisoners after Pearl Harbor. But it isn't simply about that in Nazi Germany. I am not speaking solely about experiments on "human guinea pigs" or anatomical findings that the camps directly provided prestigious German doctors, but of the medical profession's direct participation of in all of the phases of mass homicide: from the singling out of babies and then of adults condemned to a "merciful" death in the T4 program, to the extension of what was called "euthanasia" to prisoners of war, to lastly the enormous *therapia magna auschwitzciense:* the selection on the ramp leading into the camp, the start of the process of gassing, the declaration of being deceased, the extraction of gold from the teeth of the cadavers, and supervision of the procedures of cremation. No step in the production of death escaped medical verification. According to the precise legal disposition of Victor Brack, head of the Second "Euthanasia" Department of the Reich Chancellery, only doctors had the right to inject phenol into the heart of victims or to open the gas valve. If ultimate power wore the boots of the SS, supreme *auctoritas* was dressed in the white gown of the doctor. Zyklon-B was transported to Birkenau in Red Cross cars and the inscription that stood out in sharp relief at Mauthausen was "cleanliness and health." After all, it was the personal doctor to the *Euthanasie Programm* who constructed the gas chambers at Belzec, Sobibór, and Treblinka.

All of this is already well known and documented in the acts of the legal proceedings against those doctors believed to have been directly guilty of murder. But the paltry sentences with respect to the enormity of their acts testify to the fact that the underlying problem isn't so much determining the individual responsibility of single doctors (as necessary as that is), but defining the overall role that medicine played in Nazi ideology and practices. Why was the medical profession the one that adhered unconditionally to the regime, far surpassing any other? And why was such an extensive power of life and death conferred on doctors? Why was the sovereign's scepter given just to them—and before that the book of the clergyman as well? When Gerhard Wagner, führer of German doctors *[Reichsärzteführer]* before Leonardo Conti, stated that the physician "should go back to his origins, he should again become a priest, he should become a priest and physician in one," he does nothing other than state that the judgment over who is to be kept alive and who is to be condemned to death is vested in the physician and solely in the physician, that it is he and only he who possesses the knowledge of what qualifies as a valid life endowed with value, and therefore is able to fix the limits beyond which life can be legitimately extinguished.[12] Introducing *Das ärztliche Ethos* [The physician's ethos], the work of the great nineteenth-century doctor Christoph Wilhelm Hufeland, the head of Zyklon-B distribution at Auschwitz, Joachim Mrugowsky spoke of "the doctor's divine mission," and "the priest of the sacred flame of life."[13] In the no-man's-land of this new theo-biopolitics, or better theo-zoo-politics, doctors really do return to be the great priests of Baal, who after several millennia found themselves facing their ancient Jewish enemies, whom they could now finally devour at will.

We know that the Reich knew well how to compensate its doctors, not only with university professorships and honors, but also with something more concrete. If Conti was promoted directly under Himmler, the surgeon Karl Brandt, who had already been commissioned in operation "Euthanasia," became one of the most powerful men of the regime, subordinate only to the supreme authority of the Führer in his subject area, which was the unlimited one of the life and death of everyone (without dwelling on

Irmfried Eberl, promoted at thirty-two to commandant of Treblinka). Does this mean that all German doctors (or only those who supported Nazism) were simple butchers in white gowns? Although it would be convenient to think so, in reality this wasn't the case at all. Not only was German medical research one of the most advanced in the world (Wilhelm Hueper, father of American oncology, asked the Nazi minister of culture Bernhard Rust if he might return to work in the "new Germany"), but what's more the Nazis had launched the most powerful campaign of the period against cancer, restricting the use of asbestos, tobacco, pesticides, and colorants, encouraging the diffusion of organic vegetables and vegetarian cuisine, and alerting everyone to the potentially carcinogenic effects of X-rays. At Dachau, while the chimney smoked, biological honey was produced. In addition, Hitler himself detested smoking, was a vegetarian and an animal lover, besides being scrupulously attentive to questions of hygiene.[14]

What does all of this suggest? The thesis that emerges is that between this therapeutic attitude and the thanatological frame in which it is inscribed isn't a simple contradiction, but rather a profound connection; to the degree the doctors were obsessively preoccupied with the health of the German body, they made *[operare]* a deadly incision, in the specifically surgical sense of the expression, in its body. In short, and although it may seem paradoxical, it was in order to perform their therapeutic mission that they turned themselves into the executioners of those they considered either nonessential or harmful to improving public health. From this point of view, one can justifiably maintain that genocide was the result not of an absence, but of a presence, of a medical ethics perverted into its opposite.[15] It is no coincidence that the doctor, even before the sovereign or the priest, was equated with the heroic figure of the "soldier of life."[16] In corresponding fashion, Slavic soldiers who arrived from the East were considered not only adversaries of the Reich, but "enemies of life." It isn't enough to conclude, however, that the limits between healing and killing have been eliminated in the biomedical vision of Nazism. Instead we need to conceptualize them as two sides of the same project that makes one the necessary condition of the other: it is only by killing as many people as possible that one could heal *[risanare]* those who represented the true Germany. From this perspective it even appears plausible that at least some Nazi doctors actually believed that they were respecting the substance, if not the form, of the Hippocratic oath that they had taken, namely, not to harm in any way the patient *[malato]*. It's only that they identified the patient as the German people as a whole, rather than as a single individual. Caring for that body was precisely what required the death of all of those whose existence threatened its health. It's in this sense that we are forced to defend the hypothesis put forward earlier that the transcendental of Nazism was life rather than death, even if, paradoxically, death was considered the only medicine able to safeguard life. In Telegram Number 71 sent from his bunker in Berlin, Hitler ordered the destruction of the conditions of subsistence for the German people who had proven themselves too weak. Here the limit point of the Nazi antinomy becomes suddenly clear: the life of some, and finally the life of the one, is sanctioned only by the death of everyone.

At this point the question that opened the chapter presents itself again. Unlike all the other forms past and present, why did Nazism propel the homicidal temptation of biopolitics to its most complete realization? Why does Nazism (and only Nazism) reverse the proportion between life and death in favor of the latter to the point of hypothesizing its own self-destruction? The answer I would put forward refers again to the category of immunization because it is only immunization that lays bare the lethal paradox that pushes the protection of life over into its potential negation. Not only, but it also represents in the figure of the autoimmune illness the ultimate condition in which the protective apparatus becomes so aggressive that it turns against its own body (which is what it should protect), leading to its death. That this interpretive key captures better the specificity of Nazism is demonstrated on the other side by the particularity of the disease against which it intended to defend the German people. We aren't dealing with any ordinary sort of disease, but with an infective one. What needed to be avoided at all cost was the contagion of superior beings by those who are inferior. The regime propagated the fight to the death against the Jews as the resistance put up by the body (and originally the healthy blood) of the German nation against the invading germs that had penetrated within and whose intent it was to undermine the unity and life of the German nation itself. We know the epidemiological repertoire that the ideologues of the Reich adopted when portraying their supposed enemies, but especially the Jews: they are in turn and simultaneously "bacilli," "bacteria," "parasites," "viruses," and "microbes."[17] It is also true, as Andrzej Kaminski remembers, that Soviet detainees were sometimes designated with the same terms. And certainly the characterization of the Jews as parasites is part of the secular history of anti-Semitism. Nonetheless, such a definition acquires a different valence in the Nazi vocabulary. Here too it is as if what to a certain point remained a weighty analogy now actually took form: the Jews didn't *resemble* parasites; they didn't behave *as* bacteria—they *were* bacteria who were to be treated as such. In this sense, Nazi politics wasn't even a proper biopolitics, but more literally a *zoopolitics*, one expressly directed to human animals. Consequently, the correct term for their massacre—anything but the sacred "holocaust"—is "extermination": exactly the term used for insects, rats, and lice. *Soziale Desinfektion* it was called. "*Ein Laus, Ein Tod*"—a louse is your death was written on a washroom wall at Auschwitz, next to the couplet "*Nach dem Abort, vor dem Essen, Hände waschen, nicht vergessen*" (After the latrine, before eating, wash your hands, do not forget).[18]

It is for this reason that we need to award an absolute literality to the words Himmler addressed to the SS stationed at Kharkov according to which "anti-Semitism is like disinfestations. Keeping lice away is not an ideological question—it is a question of cleanliness."[19] And after all, it was Hitler himself who used an immunological terminology that is even more precise: "The discovery of the Jewish virus is one of the greatest revolutions of this world. The battle that we fight every day is equal to those fought in the last century by Pasteur and Koch."[20] We shouldn't blur the difference between such an approach, which is specifically bacteriological, with another that is simply racial. The final solution waged against the Jews has just such a biological-immunitarian characterization. Indeed, the gas used in the camps passed through shower tubes that were allocated for disinfections, but only that disinfecting the Jews seemed impossible from the moment that they were considered the bacteria from which one needed to rid oneself. The identification between men and pathogens reached such a point that the Warsaw ghetto was intentionally constructed in a zone that was already contaminated. And so, according to the modalities of a prophecy realized, the Jews fell victim to the same disease that was used to justify their ghettoization: finally they had become *really* infected and therefore were now agents of infection.[21] Accordingly, doctors had the right to exterminate them.

Degeneration

In the autoimmunitarian paroxysm of the Nazi vision, generalized homicide is therefore understood as the instrument for regenerating the German people. But this in turn is made necessary by a degenerative tendency that appears to undermine vital forces. The titles of two widely read books in the middle of the 1930s are indicative of such a syllogism: they are *Volk in*

Gefahr [Nation in danger] by Otto Helmut and *Völker am Abgrund* [Peoples on the precipice] by Friedrich Burgdörfer.[22] The task of the new Germany is that of saving the West from the threat presented by a growing degeneration. The prominence of this category—which we have already come across in Nietzsche—in the Nazi ideological machine should in no way be downplayed. It constitutes the conceptual passageway through which the biopolitics of the regime could present itself as the prosecution, and indeed the completion, of a discourse that circulated widely in the philosophical, juridical, and even medical culture of the period. Originally relative to the elimination of a thing with respect to the genus to which it belongs, the concept of degeneration progressively takes on an increasingly negative valence that assimilates it to terms such as "decadence," "degradation," and "deterioration," though with a specific biological characterization.[23] Thus, if in Buffon it still connotes the simple environmental variation of a organism with respect to the general features of his race—what Lamarck considered nothing other than a successful adaptation—Benedict-Augustin Morel's *Traité des dégénérescences* moves it decisively in a psychopathological direction.[24] The element that signals the change with respect to its original meaning isn't to be found only in the shift from anatomy to bioanthropology, but rather in the move from a static to a dynamic semantic: more than something given, the degenerative phenomenon is a *process* of dissolution. Produced by the intake of toxic agents, it can lead in a few generations to sterility and therefore to the extinction of a specific line. All of the multiple tests that were conducted on the subject between the end of the nineteenth century and the beginning of the next do nothing but reintroduce (in more or less the same arguments) the same schema: having only with difficulty survived the struggle for existence, the degenerate is he who carries imprinted within him the physical and psychological wounds in a form that is forced to become exponentially aggravated in the move from father to son. When in the 1880s Magnan and Legrain will transpose them to a clinical environment, the definition has already established its constitutive elements:

> Degeneration *[dégénérescence]* is the pathological state of being that, in comparison with generations closer to it in time, is constitutively weakened in its psychophysical resistance and only realizes in an incomplete manner the biological conditions of the hereditary struggle for life. This weakening that is translated into permanent stigma is essentially progressive, except for possible regeneration. When this life doesn't survive, it more or less rapidly leads to the annihilation of the species.[25]

Naturally, in order for the category to pass over into Nazi biopolitics, a series of cultural mediations will be needed—from Italian criminal anthropology to French hereditary theory, to a clear-cut racist reconversion of Mendelian genetics. But the most salient features are present in it, beginning with the enfolding of pathology into abnormality. What characterizes the degenerate above all is his distance from the norm: if the degenerate in Morel already expresses his deviation from the normal type, for Italian Giuseppe Sergi "it is impossible to find an invariable norm for his behavior in him."[26] What is intended here by "norm"? In the first instance it would seem a quality of the biological sort—the potentiality of a given organism for vital development understood both from a physical and a psychological point of view. Regarding precisely that, as the Englishman Edwin Ray Lankester makes clear, "degeneration can be defined as a gradual mutation in the structure in which the organism is able to adapt itself to less various and more complex conditions of life."[27] This doesn't mean that soon after a slippage in the definition of norm occurs from the morphogenetic level to that of the anthropological. The biological abnormality is nothing but the sign of a more general abnormality that links the degenerate subject to a condition that is steadily differentiated with regard to other individuals of the same species. But a second categorical move follows the first, which is destined to move abnormality from the intraspecies dimension to the limits of the human itself. To say that the degenerate is abnormal means pushing him toward a zone of indistinction that isn't completely included in the category of the human. Or perhaps better, it means enlarging the latter category so as to include its own negation: the non-man in man and therefore the man-animal *[uomo-bestia]*.[28] It is the Lombrosian conception of "atavism," in which all the possible degenerations are accounted for, that performs the function of the excluding inclusion. It is configured as a sort of biohistorical anachronism that reverses the line of human evolution until it has brought it back in contact with that of the animal. Degeneration is the animal element that reemerges in man in the form of an existence that isn't properly animal or human, but exactly their point of intersection: the contradictory copresence between two genera, two times, two organisms that are incapable of producing a unity of the person and consequently for the same reason incapable of forming a juridical subjectivity. The ascription of the degenerate type to an ever vaster number of social categories—alcoholics, syphilitics, homosexuals, prostitutes, the obese, even to the urban proletariat itself—reinstates the sign of this uncontrollable exchange between biological norm and juridical-political norm. What appears as the social result of a determinate biological configuration is in reality the biological representation of a prior political decision.

More than any other, the theory of heredity makes clear the improper exchange between biology and law *[diritto]*. At the same time that Morel's essay was published, Prosper Lucas's *Traité* appeared from the same Parisian editor, Ballière, on "natural heredity in the state of health and disease of the nervous system," followed at a distance of twenty years by Théodule Ribot's *L'hérédité: Étude psychologique sur ses phénomènes, ses lois, ses causes, ses conséquences.*[29] At the center of these texts, and of many others that followed, is a clear shift in perspective from that of the individual (understood in a modern sense as the subject of law and of judgment *[decisione]*), to the line of descent in which he constitutes only the final segment. A vertical relation linking fathers and sons and through them with their ancestors is substituted for the solidarity or the horizontal competition between brothers that is typical of liberal-democratic societies. Contrary to what pedagogical and social theories (inspired by the notion of equality) put forward, the difference that separates individuals appears insurmountable. Both somatic and psychological features are predetermined at birth according to a biological chain that neither individual will nor education can break. Just as for virtue and fortune, so too hereditary malformations take on the aspect of an inevitable destiny: no one can escape from oneself; no one can break the chain that inexorably ties one to one's past; no one can choose the direction of one's own life. It is as if death grabs life and holds it tightly: "Heredity governs the world," concludes Doctor Apert. "The living act, but the dead speak in them and make them what they are. Our ancestors live in us."[30] Life is nothing but the result of something that precedes it and defines it in all its movements. The Lombrosian figure of the "born delinquent" constitutes the most celebrated expression: as the ancient wisdom of the myth teaches us, the faults of the father always devolve upon their sons. Law *[diritto]*, which precisely originates in myth, can do nothing but model its procedures on this first law, which is stronger than any other because it is rooted in the most profound reasons of biology and blood. In Lucas's definition, heredity is "a *law*, a *force*, and a *fact*."[31] More precisely, it is a law that has the irresistible force of fact; it coincides with its own facticity.

Here emerges the reversal of the relation between *nomos* and *bíos* to which I referred earlier: what in reality is the effect is represented as the cause and vice versa. André Pichot has drawn our attention to the fact that the economic-juridical notion of heredity (which is apparently calculated using biological heredity) constitutes instead its foundation.[32] After all, the Latin term *hereditas* doesn't designate what is left to one's descendants at the moment of death. It is only from 1820 on that the word begins to be applied by analogy to the area of the transmission of biological characteristics. Proof of this will be found in the fact that the classic hereditary monarchy, which also refers to descent based on blood ("blue blood"), doesn't depend on a genetic type of conception, but rather on a juridical protocol that responds to a determinate social order. Motivated less by biology, the obligation of dynastic succession was also justified by arguments of a theological nature—the divine right of kings. In order for such a process to be secularized, however, we need to wait first for the birth of natural law and positive law; not, however, without a different tradition inserting itself between the two, namely, that originating in Calvinism (which reintroduces the idea of divine predestination that is applied to every individual). What needs to be highlighted is that post-Darwinian hereditary theory is situated exactly at the point of antinomic confluence between these two trajectories; on one side, it completely secularizes the dynastic tradition of the aristocratic sort; on the other, it reproduces the dogma of predestination in biopolitical terms. When the embryologist August Weismann defines germinative plasma, he will arrive at a singular form of "biological Calvinism" according to which the destiny of the living being is completely preformed—naturally, with the variant that the soul is not immortal, but rather blood, which is transmitted immutably through the bodies of successive generations.

This line of reasoning is grafted onto the theory of degeneration until it becomes its own presupposition: On the one hand, the degenerative process spreads via the transmission of hereditary characteristics. If blood that is inherited cannot be modified genetically (according to the theo-biological principle of germinative plasma), why then does the organic deficiency increase exponentially in the passage from father to son, until one arrives at sterility and the extinction of the hereditary line? On the other hand: if in the space of a few generations dissolution is inevitable, why then should one fear the phenomenon spreading? The answer has to do with the idea of contagion: degenerative pathology doesn't only multiply metonymically within the same body in a series of interrelated diseases, but spreads irresistibly from one body to the next. We can say that degeneration is always degenerative. It reproduces itself intensely and extends from inside to outside and vice versa. This contaminating power of an internal transmutation and of an external transposition is in fact its most characteristic feature. For this to be so, it must follow that it is *both* hereditary and contagious, which is to say contagious on the vertical level of lineage as well as on the horizontal level of social communication. What creates the difficulty is precisely this copresence: according to Weismann's law, if the germinative plasma cannot be modified, then it isn't susceptible to contagion. If instead it is a potential vehicle for contamination (as the theory of expanded degeneration would have it), this shows that the genetic structure is not unalterable. This logical difficulty, which has produced some confusion between contagious diseases (tuberculosis and syphilis, for example) and hereditary diseases, has been met by the intermediate thesis that the same tendency to contract the disease *[contagio]* can be hereditary. Thus, the external infection occurs thanks to internal predisposition and the internal predisposition thanks to an external infection. That degeneration is spread through hereditary transmission or through contagion matters less. In any case, what counts is the construction of the immunitary apparatus intent on blocking its advance. Some decades later, the illustrious German professors Fischer and Verschuer will split the research area in two: the first will study the blood of different ethnic groups, the second the hereditary lines of monozygote twins. Josef Mengele will produce the operative synthesis in his laboratory at Auschwitz.

Was such an outcome inevitable? Was it implicit in the logic of the category of degeneration? The answer isn't a simple yes. But that it has an immunitarian timbre is made evident by its explicitly reactive valence. Reactive, however, doesn't necessarily mean reactionary. I am referring not only to the important fact that many, who were not exponents of the Catholic right as well as progressive and socialist authors, make reference to such a category. What joins them all together fundamentally is the idea that degenerative pathology isn't simply the negative result of progress, but that one derives from the other. Not by chance the genesis of degenerative pathology is located in the years immediately following the French Revolution, when natural selection begins to be weakened by a protective stance with regard to the weakest parts of society. The classist connotation of such a line of argument (when not racist) is clear. But that doesn't cancel out a series of other vectors that seem to push the concept in the opposite direction, especially the conviction that a return to the past isn't possible (to simple, natural selection), but rather that one needs to have recourse to a series of artificial interventions (in particular the hypothesis of an unavoidable spread of the degenerative process in all social sectors and environments). Born in a part, degeneration winds up involving the whole. It is a global sickness that continually expands not only among inferior races, but also among superior ones. It is precisely the alleged connection with the dynamics of modernization—from industrialization to urbanization—that seems to tie it to the destiny of the bourgeois and intellectual classes.

As I noted, Lombroso had insisted earlier on the mysterious and worrying connection that exists between genius and madness: genius, insofar as it is a deviation from the norm, is a sophisticated form of degenerative neurosis. But it is the Hungarian doctor of Jewish origin Maximilian Südfeld, known to the larger public as Max Nordau, who more than any other localizes degeneration in the intellectual sphere. In his book dedicated to *Entartung*, Pre-Raphaelites, Parnassians, Nietzscheans, Zolians, Ibsenians, and so on are all included in this category—all assimilated on the typological level to those who "satisfy their insane instincts with the assassin's knife or with the dynamite's fuse rather than with pen or paintbrush."[33] It is impossible not to see the thread that ties similar evaluations with future Nazi lucubrations with regard to degenerate art. The point I want to emphasize will be found in the fact that if all of modern art is declared to be degenerate, then in corresponding fashion this indicates that degeneration has the same aesthetic nervature as is presupposed in the same category of "decadentism."

That degeneration, on the other hand, isn't only negative—or better, that it is a minus sign that can, from another point of view, be turned into a plus—comes across in a text that seems to move radically against it, but instead expresses an element that was from the beginning latent in the concept. I am referring to Gina Ferrero Lombroso's *I vantaggi della degenerazione* [The advantages of degeneration]. After stating the premise that "no clear line separates progressive characteristics from regressive characteristics in animals, that is, degeneration from evolution," she asks herself "if many of the phenomena held to be degenerate are not instead evolutionary, useful rather than damaging manifestations of the adaptation the human body makes to the conditions in which it lives."[34] Not only, but Lombroso takes another step forward that places it in a particular arrangement that lies within the immunitary paradigm. As was the case for Nietzsche in his more radical stage, this doesn't actually have an exclusionary or neutralizing character, but rather assumes and valorizes the different, the dissimilar, and the abnormal inasmuch as they are innovative and transformative powers of reality. Therefore, when Lombroso refers explicitly to the "immunity produced by the diseases suffered," she can conclude that

> the degenerates are those who fuel the sacred torch of progress; to them is given the function of evolution, of civilization. Like bacteria of fermentation, they assume the office of decomposing and reconstructing institutions; the uses that they make of their time activate the material exchange of this highly complex organism that is human society.[35]

This citation restores to degeneration all of the category's breadth as well as its paradoxical characteristics. It implies both the biological inalterability of being and its continual modification. Fixedness and movement, identity and transformation, concentration and dissemination: all are extended along a line that superimposes nature and society, conservation and innovation, immunization and communication, and they seem to rebound against themselves and to turn into their opposite, after which they once again return *[riasssestarsi]* to their initial coordinates. They oscillate from the part to the whole and back again. The idea of degeneration, which is broad enough that it includes the entire civilized world, at a certain point closes around its own sacrificial object, drastically separating it from the healthy type, pushing it toward a destiny of expulsion and annihilation. More than theories, however, artistic practices register this singular rotation of sense.[36] Already the Zolian cycle of Rougon-Macquart and the dramas of Ibsen, or in Italy De Roberto's *I viceré* or Mastriani's *I vermin*, constitute a figurative laboratory of considerable expressive depth.[37] But the works that, perhaps more than any others, account for such a semantic circuit are three texts that follow one another in the short arc of a decade, namely, Robert Louis Stevenson's *The Strange Case of Doctor Jekyll and Mr. Hyde, The Picture of Dorian Gray* by Oscar Wilde, and Bram Stoker's *Dracula*. The trajectory they seem to follow moves away from superimposition to the progressive splitting between light and shadow, health and sickness, and the norm and abnormality (all placed in a narrative framework that calls forth in detail the degenerative syndrome that was moving across the society of the time): from the scenario of a degraded and tentacle-like metropolis to the paroxysmal centrality of blood, to the battle to the death between doctor and monster.

What characterizes the three stories, however, is the growing lag between the intention of the protagonist and that of reality, which the texts both hide and allow to emerge. The more the protagonist wants to free himself from the degeneration that he carries within, projecting it outside himself, the more the result is an excess of death that bursts on the scene, swallowing him up. Thus, in Stevenson's text, Jekyll, a doctor in legal medicine, attempts to immunize himself from his own worst features through the biochemical construction of another self. "And thus fortified, as I supposed on every side, I began to profit by the strange immunities of my position."[38] But the alien creature quickly escapes from the control of its creator and takes possession of his body. It is another, but generated by the ego and so destined to reenter there. A "he," an "animal," a "brute," which, however, is impossible to isolate because he is one with himself, with his body, his blood, and his flesh:[39]

> This was the shocking thing . . . that that insurgent horror was knit to him closer than a wife, closer than any eye; lay caged in his flesh, where he heard it mutter and felt it struggle to be born; and at every hour of weakness, and in the confidence of slumber, prevailed against him and deposed him out of life.[40]

Controlled, kept, domesticated by ever larger doses of the antidote, the monstrous double (which is the same subject seen in back light) finally gains the upper hand over him who has tried to dominate him and carries him into the vortex. The degenerate is none other than the doctor himself, both his shadow and his ultimate truth. The only way to stop him is to put him to death *[dargli la morte]*, killing in the same act that self with whom he always coincides.

In the second story, that of Wilde, the divergence between self and other is accentuated. The double is no longer within the body of the subject, as was the case in Jekyll-Hyde, but is objectified in a portrait that both mirrors and betrays the original. It is what degenerates in his place—every time that he behaves in a debased way. The detachment from the real, which is to say from the constitutive alteration of the subject, is represented by the pall wrapped around the painting in order to hide it from everyone. Thus, the decay of the painted image—the projection of evil *[male]* outside itself—keeps death at a distance, ensuring the immortality of the subject. But, as in the previous case, the doubling cannot last for long. The mechanism breaks down and the image again assumes the face. The painted degeneration is in reality his own: "Upon the walls of the lonely locked room where he had spent so much of his boyhood, he had hung with his own hands the terrible portrait whose changing features showed him the real degradation of his life, and in front of it had draped the purple-and-gold pall as a curtain."[41] The final blow that Dorian delivers to the "monstrous soul-life" inevitably returns to hit him, who has already been transformed into the image of the monster.[42] It is he who lies on the ground, dead "with a knife in his heart."[43] The killing of death—the autoimmunitary dream of man—reveals itself once again to be illusory: it can't do anything except reverse itself in the death of the same killer.

With *Dracula* the relationship between reality and its mythological representation moves decisively in favor of the latter. The forces of good appear to be posed frontally against those of evil in a project of definitive immunization against disease. The demon is projected outside the mind that has created it. He encapsulates in himself all of the characteristics of the degenerate—he is no longer the other in man, but the other from man *[dall'uomo]*. Both wolf, bat, and bloodsucker, he is above all the *principle of contamination*. Not only does he live on the blood of others, but he reproduces by multiplying himself in his victims. Just as in future manuals of racial hygiene, the ultimate crime committed is the biological one of the transmission of infected blood. He carried contamination, namely, Transylvania, into London homes; he immersed the other in the same *[nello stesso]* and consigned the same to the other. The championing of contemporary degenerative theory is so absolute that the text cannot fail to cite the relevant authors: "The Count is a criminal and of criminal type. Nordau and Lombroso would so classify him."[44] Just like the degenerate, he is not a true man, but has human features. He doesn't have an image, but continually changes appearance. He is not a type but a countertype. He belongs to the world of the "non"—no longer alive, he is still and above all else "undead," repulsed by life and by death into an abyss that cannot be bridged. He is an already dead, a half dead, a living dead, just as other vampires some fifty years later will be designated with the yellow star on their arms. His killing, with a stake through the heart and the head cut off, has the characteristics of salvific death that will be shortly enlarged liberally to include millions of "degenerates." To put an end to the "man that was," to that "carnal and unspiritual appearance," to the "foul Thing," means freeing not only those whom he threatens, but also himself, giving him finally back to that death to which he belongs and which he carries within him without being able to taste it:[45]

> But of the most blessed of all, when this now Un-Dead be made to rest as true dead, then the soul of the poor lady whom we love shall again be free . . . So that, my friend, it will be a blessed hand for her that shall strike the blow that sets her free.[46]

Eugenics

The eugenics movement will take up the task of translating these kinds of literary hallucinations into reality; the movement will flare up in the opening years of the 1900s as a purifying fire across the entire Western world (countered only by the Catholic church and the Soviet Lissenkim).[47] With respect to the theory of degeneration and its folds and internal antinomies, eugenics marks both a positive result and a sharp reduction in complexity. We need only draw the necessary conclusions: if civilized peoples are exposed to progressive degeneration, the only way to save them is by reversing the direction of the process that is under way, to remove what produces the disease that corrupts it so as to reinstate it in the horizon of goodness, health, and perfection. The substitution of the positive prefix "eu" with that of the negative "de" directly expresses this reconstructive intention. But the simplicity of the move doesn't explain a dual dislocation, above all from the descriptive level (where we find degenerative semantics) to that of the prescriptive. What was understood as a given or a process becomes with eugenics a project and a program of intervention; consequently, it moves from nature to artifice. While degeneration remains a natural phenomenon, completely within the sphere of *bíos*, the eugenic procedure is characterized by the technical *[tecnica]*, which is certainly applied to life, but in a form that intends precisely to modify spontaneous development. In truth, the discourse of eugenics (more than that of nature as such) declares that it wants to correct procedures that have negatively influenced the course of nature. It begins with those social institutions and with those protective practices with regard to individuals who are biologically speaking inadequate with respect to natural selection (and which, if left to its own devices, natural selection would eliminate). The thesis variously repeated in all the texts in question is that artificial selection has no other purpose than that of restoring a natural selection that has been weakened or nullified by compensatory mechanisms of the humanitarian sort. But is it really the idea of an artificial reconstruction of the natural order that constitutes the problem—how to rehabilitate nature through artifice or how to apply artifice to nature without denaturalizing it? The only way to do so successfully is to adjust preventively the idea of nature to the artificial model with which nature wants to restore itself, rejecting as unnatural all that doesn't conform to the model. However, the negative that was to be neutralized now reappears: to affirm a good *genos* means negating what negates it from within. This is the reason that a positive eugenics (from the work of Francis Galton on), directed to improving the race, is always accompanied by a negative eugenics, one designed to impede the diffusion of dysgenic exemplars. And yet, where would the space for increasing the best exemplars be found if not in the space produced by the elimination of the worst?

The concept of "racial hygiene" constitutes the median point of this categorical passage. It represents not only the German translation of the eugenic orientation, but something that discloses its essential nervation. We can trace a significant confirmation of the change in course in Wilhelm Schallmayer's essay, *Vererbung und Auslese im Lebenslauf der Völker: Eine staatswissenschaftlich Studie auf Grund der neueren Biologie* [Heredity and selection in the vital development of nations A social and scientific study based on recent biology].[48] If we keep in mind that the same author had written a book some years earlier, dedicated to treating the degeneration of civilized nations, we can clearly see the move that German political science makes vis-à-vis biology.[49] It is true that Schallmayer doesn't adopt Aryan racism, as was the case with Ludwig Woltmann in a contemporary piece titled *Politische Anthropologie.*[50] But this makes the biopolitical approach that it inaugurates even more important. Contrary to every hypothesis put forward by the democratic left for social reform, the power of the state is tied directly to the biological health of its members. By this it is understood that the vital interest of the nation resides in increasing the strongest and checking, in parallel fashion, the weak of body and of mind. The defense of the national body requires the removal of its sick parts. In his influential manual *Rassenhygiene,* Alfred Ploetz had furnished the most pertinent key for understanding the meaning of the transformation under way: race and life are synonymous to the degree in which the first immunizes the second with regard to the poisons that threaten it.[51] Born from the struggle of cells against infectious bacteria, life is now defended by the state against every possible contamination. Racial hygiene is the immunitary therapy that aims at preventing or extirpating the pathological agents that jeopardize the biological quality of future generations.

What is sketched here is a radical transformation of the notion of politics itself, at least in the modern sense of the expression. As was the case with Francis Galton, but still more in Karl Pearson's biometrics, politics appears to be pressed among the fields of mathematics, economics, and biology. The political choices of national organisms are to be derived rigidly from a calculation of the productivity of human life with regard to its costs. If it is possible to quantify the biological capital of a nation on the basis of the vital qualities of its members, the division into zones of different value will be inferred. Nevertheless, it would be a mistake to take such a value in an exclusively economic sense. If this seems to prevail in the Anglo-Saxon and Scandinavian matrices of eugenics, it doesn't in the German case. Certainly, the reference to a differential calculus between costs and revenues isn't lacking there either, but it is always subordinated to a more profound and underlying difference relative to the typology of human life as such. It isn't man that is valued on the basis of his economic productivity, but economic productivity that is measured in proportion to the human type to which it pertains. This helps to account for the extraordinary development of anthropology in Germany in the closing decades of the nineteenth into the first half of the following century, culminating in the 1930s and 1940s, which saw 80 percent of all anthropologists in Germany join the National Socialist party. It wasn't by chance that Vacher de Lapouge wrote in his *Essais d'Anthroposociologie* on *Race et milieu social* that "the revolution that bacteriology has produced in medicine, anthropology is about to produce in the political sciences.[52] What is at stake, even before its socioeconomic implications, is the definition of the human generally and its internal thresholds.

The distinction between races, both superior and inferior, more and less pure, already constitutes the first intraspecies *clivage,* apparently confirmed by Ludwik Hirszfeld and Karl Landsteiner's contemporary discovery of different blood groups: rather than being the representative of one genus, the *anthropos* is the container of radically diverse biotypologies that move from the superman (Aryan) to the anti-man (Jew), passing through the average man (Mediterranean) and the subhuman (Slavic).[53] But what matters more is the relation between such a *clivage* within the human race and what is situated outside with regard to others. In this sense, German anthropology worked closely with zoology on the one hand and botany on the other: man is situated in a line with diverse qualitative levels that include both plants and animals. Up to this point, nevertheless, we still remain within the confines of a classic evolutionist model. The new element that brings matters to a head lies, however, in the superimposition that progressively occurs when distinguishing among the various species—in the sense that one appears contemporaneously outside and inside the other. From here a double and crisscrossed effect: on the one side, the projection of established human types in the botanical and zoological "catalog"; on the other, the incorporation of particular animal and vegetable species within the human race. In particular this second step explains not only the growing fortune of anthropology, but also the otherwise incomprehensible circumstance that Nazism itself never renounced the category of *humanitas,* on which it awarded the maximum normative importance. More than "bestializing" man, as is commonly thought, it "anthropologized" the animal, enlarging the definition of *anthropos* to the point where it also comprised animals of inferior species.[54] He who was the object of persecution and extreme violence wasn't simply an animal (which indeed was respected and protected as such by one of the most advanced pieces of legislation of the entire world), but was an *animal-man:* man in the animal and the animal in man. This explains the tragically paradoxical circumstance that in November 1933—which is to say some years before Doctor Roscher conducted experiments on the compatibility of human life with the pressure at twelve thousand meters high or with immersion in freezing water—the regime promulgated a circular that prohibited any kind of cruelty to animals, in particular with reference to cold, to heat, and to the inoculation of pathogenic germs. Considering the zeal with which the Nazis respected their own laws, this means that if those interned in the extermination camps had been considered to be *only* animals, they would have been saved. After all, in January 1937, Himmler expressed himself in similar terms when addressed the officers of the Wehrmacht: "I recently saw a seventy-two-year-old man who had just committed his seventy-third crime. To give the name animal to such a man would be offensive to the animal. Animals don't behave in such a fashion."[55] It isn't surprising that in August 1933, when Göring announced an end to "the unbearable torture and suffering in animal experiments," he went so far as to threaten to send to concentration camps "those who still think they can treat animals as inanimate property.[56]

Garland E. Allen notes how American eugenics, which was the most advanced at the beginning of the twentieth century, had its start in agriculture.[57] Its first organization was born of the collaboration between the American Breeders Association, the Minnesota Agricultural Station, and the School of Agriculture at Cornell University. Charles B. Davenport, the same Davenport who is considered to be the father of the discipline, had earlier attempted to form an agricultural company under the direction of the department of zoology at the University of Chicago in which Mendelian theories were to be experimented on domestic animals.[58] Subsequently, he turned to the Carnegie Foundation in Washington to finance a series of experiments on the hybridization and selection of plants. Finally, in 1910, with funds awarded him by the Harriman and Rockefeller families, he created a new center of genetic experimentation, the Eugenics Records Office at Cold Spring Harbor, which was committed to the study of heredity in humans. The huge success of these initiatives is largely indicative of the relation that eugenics instituted between human beings, animals, and plants. Moreover, the periodicals born in that context, in particular *The American Breeders' Magazine*, *The Journal of Heredity*, and *Eugenical News*, ordinarily published works in which one moved from the selection of chickens and pigs to the selection of humans without posing the question of continuity between them. If a farmer or a breeder wants to encourage a better reproduction of vegetables and rabbits, or conversely, wants to block a defective stock, why, the exponents of the new science asked, should it be any different with man? In 1892, Charles Richet, vice president of the French Eugenics Society and future Nobel Prize winner (in 1913), prophesized that quite soon "one will no longer simply be content to perfect rabbits and pigeons but will try to perfect humans."[59] When, some decades later, Walther Darré, Reich Minister for Nutrition will advise Himmler to "transfer his attention from the breeding of herbs and the raising of chickens to human beings,"[60] Richet's prophecy will be realized. Even in their titles, two books published a year apart, Maurice Boigey's *L'élevage humain* and Charles Binet-Sanglé's *Le haras humain*, give the sense of the general inclination of anthropological discourse toward zoology, or better, toward their complete overlapping.[61] "Let us consider coldly the fact that we constitute a species of animal," exhorts Doctor Valentino, "and from the moment that our race is accused of degenerating, let's attempt to apply some principles of breeding to its improvement: let's regulate fecundation."[62] Vacher de Lapouge had already included in his project of *Sélections sociales* the services of a "rather restricted group of absolutely perfect males."[63] But the most faithful actualization of what Just Sicard de Plauzoles called "human zootechnics" was certainly the organization Lebensborn, or "font of life," which was founded by Himmler in 1935.[64] In order to augment the production of perfect Aryan exemplars, several thousand babies of German blood were kidnapped from their respective families in the occupied territories and entrusted to the care of the regime.

If "positive" eugenics was directed to the sources of life, negative eugenics (which accompanies the positive as its necessary condition) rests on the same terrain. Certainly, it was vigilant when it came to all the possible channels for degenerative contagion: from the area of immigration to that of matrimony, which were regulated by ever more drastic norms of racial homogeneity. But "the most significant point . . . in its bio-sociological weight," as one Italian eugenicist expressed it, remained that of sterilization.[65] In addition, segregation was understood less as the restriction of personal freedom and more as the elimination of the possibility of procreation, as a sort of form of sterilization at a distance. It was no coincidence that several "feeble-minded" were given the choice between being segregated and being sterilized. The latter is the most radical modality of immunization because it intervenes at the root, at the originary point in which life is spread *[si comunica]*. It blocks life not in any moment of its development as its killer but in its own rising up—impeding its genesis, prohibiting life from giving life, devitalizing life in advance. It might seem paradoxical wanting to stop degeneration (whose final result was sterility) through sterilization, if such an antinomy, the negative doubling of the negative, wasn't an essential part, indeed the very basis of the immunitary logic itself. Therefore, on the question of sterilization the eugenicists never gave in and the Nazis made a flagship out of their own bio-thanatology. Certainly, criminals were already being castrated in 1865, but what was then considered above all else to be a punishment becomes something quite different with the development of the eugenics obsession. It concerned the principle according to which the political body had to be vaccinated beforehand from every disease that could alter the self-preserving function. Carrie Buck, a girl from Virginia who was sentenced to be sterilized after having been judged (like her mother) "weak in the mind" *[debole di mente]*, appealed her case to the County Court, the Court of Appeals, and finally to the Supreme Court. She charged that her rights had been violated under the Fourteenth Amendment (according to which no state shall deprive any person of life, liberty, or property without due process of law). Justice Oliver Wendell Holmes, a eugenicist, rejected her appeal, however, for the following reasons:

> It is better for all the world, if instead of waiting to execute degenerate offspring for crime, or to let them starve for their imbecility, society can prevent those who are manifestly unfit from continuing their kind. The principle that sustains compulsory vaccination is broad enough to cover cutting the Fallopian tubes . . . Three generations of imbeciles are enough.[66]

Defined as "poor white trash," the girl was sterilized together with another 8,300 citizens of Virginia.

If the first immunitary procedure of eugenics is sterilization, euthanasia constitutes the last (in the ultimate meaning of the expression). In a biopolitical lexicon turned into its opposite, a "good" birth or nonbirth cannot but correspond to a "good" death. Attention among scholars has recently been directed to the book, published in 1920 by the jurist Karl Binding and by the psychiatrist Alfred Hoche, with the title *Die Freigabe der Vernichtung lebensunwerten Leben* [The authorization of the destruction of life unworthy of life].[67] But such a text, which seems to inaugurate a new genre, is already the result of an itinerary that ends (at least in Germany) in another work that is no less significant. I am speaking about Adolph Jost's essay *Das Recht auf den Tod* [The right to die], which twenty-five years earlier first introduced the concept of *negativen Lebenswert,* which is to say "life without value" (which was replaced with the right to end life in the case of an incurable disease).[68] Yet the difference (also with respect to Anglo-Saxon eugenics) is the progressive shift of such a right from the sphere of the individual to that of the state. While the first preserves the right/obligation to receive death, only the second possesses the right to give it. Where the health of the political body as a whole is at stake, a life that doesn't conform to those interests must be available for termination. Furthermore, as Jost asks, doesn't this already happen in the case of war, when the state exercises its right to sacrifice the lives of its soldiers for the common good? The new element here with respect to an argument that at bottom is traditional lies in the fact that it isn't so much that medical killing falls under the category of war as that war comes to be inscribed in a biomedical vision in which euthanasia emerges as an integral part.

In relation to this framework, Binding and Hoche's essay nevertheless signals a categorical opening that is anything but irrelevant, not only on the level of quantity (from the moment that the incurably ill, as well as the mentally retarded and deformed babies are added as potential objects of euthanasia), but also on the level of argumentation. From this point of view one might say that the juridical and biological competencies that the two authors represent achieve an even greater integration, which makes the one not only the formal justification but also the content of the other. It is as if the right/obligation to die, rather than falling from on high in a sovereign decision on the body of citizens, springs from their own vital makeup. In order to be accepted, death must not appear as the negation but rather as the natural outcome of certain conditions of life. In this way, if Binding is concerned about guaranteeing the legal position of doctors engaged in euthanasia through a complex procedure of asking for the consent of those who have been judged incapable of giving it, Hoche avoids the thorny juridical question thanks to purely biological criteria: that death is juridically irreproachable not so much because it is justified by more pressing collective demands, but because the persons whom it strikes are *already* dead. The meticulous lexical research of those expressions that correspond to their diminished situation—"half-men," "damaged beings," "mentally dead," "empty human husks" *(Leere-Menschenhülsen)*, "human ballast" *(Ballast-existenzen)*—has precisely the objective of demonstrating that in their case death does not come from outside, because from the beginning it is part of those lives—or, more precisely, of these *existences* because that is the term that follows from the subtraction of life from itself. A life inhabited by death is simply flesh, an existence without life. This is the exact title of film that will later be made in order to instruct personnel working on T4, the Nazi euthanasia program: *Dasein ohne Leben* (Existence without life). Moreover, Hitler himself had juxtaposed existence and life according to an explicit hierarchy of values: "From a dead mechanism which only lays claim to existence for its own sake, there must be formed a living organism with the exclusive aim of serving a higher ideal."[69] Existence for the sake of existence, simple existence is dead life or death that lives, a flesh without body. In order to unravel the apparently semantic tension that is present in the title of Binding and Hoche's book, that of a "life unworthy of life," one need only substitute "existence" for the first term. The books are immediately balanced: the life unworthy of life is existence deprived of life—a life reduced to bare *[nuda]* existence.

The interval of value between existence and life is verified most clearly in a correlated doubling of the idea of humanity. We know the different qualitative thresholds introduced in the notion of humanity by the German anthropology of the period: *humanitas* is extended to the point of containing within it something that doesn't belong to it and indeed essentially negates it. Now, such a variety of anthropic typologies demands an analogous differentiation in the behavior of those to whom it might refer from a normative point of view. It isn't ethically human to refer to diverse types of people *[uomo]* in the same manner. Binding and Hoche had previously cautioned against "a swollen conception of humanity" and "an overevaluation of the value of life as such."[70] But against such a concept others offered a different and loftier notion of humanity, not only in relation to the collective body weakened by the unproductive weight of those of lesser worth *(Minderwertigen)*, but also to these latter ones. It was with this in mind, with the T4 Program in full operation, that Professor Lenz declared that "detailed discussion of so-called euthanasia . . . can easily lead to confusion about whether or not we are dealing with a matter which affects the safeguarding of our hereditary endowment. I should like to prevent any such discussion. For, in fact, this matter is a purely humanitarian problem."[71] Furthermore, Lenz did nothing other than fully express a reasoning that had been made long before. That euthanasia was defined as *Gnadentod,* "mercy killing," "a death with pity," or "misericordious"—which, according to Italian eugenicist Enrico Morselli, comes from "misericord," the short-bladed knife used at one time to put an end to the suffering of the dying—is the result of the conceptual inversion that makes the victim himself the beneficiary of his own elimination.[72] With birth constituting his illness, that is to say the fact of being born against the will of nature, the only way to save the defective person from such a subhuman condition is that of handing him over to death and thereby liberating him from an inadequate and oppressive life. For this reason, the book that immediately follows Binding and Hoche's text has as its title *Die Erlösung der Menschheit vom Elend* [The liberation of humanity from suffering].[73] "Free those who cannot be cured" was also the invocation on which the film *Existence without Life* concluded. In France, where state-sponsored euthanasia was never effectively practiced, Binet-Sanglé, in his *L'art de mourir* suggests carrying out the final delivery from pain through gas by injecting morphine that will transport the beneficiary to the first level of "beatitude," while Nobel Prize winner Richet holds that those killed mercifully do not suffer and that, if they were to consider it only briefly, they would be grateful to those who saved them from the embarrassment of living a defective life.[74] Even before then, Doctor Antoine Wylm had warned:

> [F]or such beings that are incapable of a conscious and truly human life, death has less suffering than life. I realize there isn't a good probability that I will be heard. As for euthanasia, which I consider to be moral, many will object with a thousand arguments in which reason will not play any role whatsoever, but in which the most infantile sentimentalism will be freely bandied about. Let us wait for the opportune moment.[75]

Genocide

That moment arrived in the opening months of 1939, when Karl Brandt, Hitler's trusted personal physician, was given the responsibility together with Philipp Bouhler, the head of the Reich Chancellery, for beginning the process of euthanasia on children younger than three years of age who were suspected of having "serious hereditary illnesses," such as idiocy, mongolism, microcephalia, idrocephalia, malformations, and spastic conditions. The ground had been meticulously prepared by the diffusion of films on the condition of the subhuman lives of the disabled, such as *Das Erbe* (Heredity), *Opfer der Vegangenheit* (Victim of the past), and *Ich klage an* (I accuse). The occasion for such steps was the request made to Hitler to authorize the killing of a baby by the name of Knauer, who was blind and was missing a leg and an arm. Just as soon as "mercy" was benevolently accorded him, a Reich's Committee was founded for assessing hereditary and serious congenital diseases, headed by Hans Hefelmann (who in fact had a degree not in medicine but in agricultural economics). Together with the committee a series of centers were set up, which were identified as "Institutions of Special Pediatrics" or even "Therapeutic Institutions of Convalescence," where thousands of children were killed by vernal injection or with lethal doses of morphine and scopolamine.

In October of the same year the decree was extended to adults as well and given the name T4 Program (from the address Tiergarten 4 in Berlin). The fact that the decree was backdated to the outbreak of the Second World War is the most obvious sign of the thanatopolitical character of Nazi biopolitics as well as the biopolitical character of modern war. Only in war can one kill with a therapeutic aim in mind, namely, the vital salvation of one's own people. Moreover, the program of euthanasia extended also geographically with the Eastern advance of German troops. Between 1940 and 1941, the Polish camps of Chelmno, Belzec, Sobibór, and Treblinka joined the six principal centers of elimination in Germany: Hartheim, Sonnensteim, Grafeneck, Bernburg, Brandenburg, and Hadamar. In the meantime, with the expansion of "special treatment" to include prisoners of war, the T4 project (which was still being implemented by doctors) was taken over by Operation I4fI3 (from the reference number in the documents of the Camp Inspectorate). This too maintained its medical outlook, but now answered directly to the SS. It was also the point of passage to outright extermination: on January 20, in the so-called Wannsee conference that had been called by Reinhard Heydrich, the final solution was decided for all Jews.

That is what is defined as "genocide." From the moment when Raphael Lemkin, a professor of international law at Yale University, coined it in 1944, the term has continued to elicit discussion (and doubt).[76] Formed from a hybrid between the Greek root *genos* and the Latin suffix *cida* (from *caedere*), the word quickly found itself linked to similar, though not identical, concepts, primarily that of "ethnocide" and "crimes against humanity." The result was a knot that was difficult to untie. What distinguishes the collective killing of the *genos* from that of *ethnos*? Is it the same thing when oppressors speak of "people" or of "race"? And what is the relation between the crime of genocide and that conceived in relation to the entire human species? Another difficulty of the historical variety was added to this first terminological problem. From the moment the subject of genocide is always a state and that every state is the creator of its own laws, it is difficult for the state that commits genocide to furnish a legal definition of the crime that it itself has committed. That said, scholars do concur that in order to be able to speak about genocide, the following minimum conditions must be met: (1) that there exists a declared intention of the part of the sovereign state to kill a homogeneous group of persons; (2) that such killing is potentially complete, that is, involves all its members; and (3) that such a group is killed insofar as it is a group, not for economic or political motives, but rather because of its biological constitution. It is clear that the genocide of the Jews on the part of the Nazis meets all these criteria. Still, to define the specificity of it is another matter, one that concerns the symbolic and material role of medicine to which we have so often drawn attention here: it involves the therapeutic purpose that is assigned to extermination from the beginning. Its implementers were convinced that only extermination could lead to the renewal of the German people. As emerges from the pervasive use of the term *Genesung* (healing) with regard to the massacre in progress, a singular logical and semantic chain links degeneration, regeneration, and genocide: regeneration overcomes degeneration through genocide.

All those authors who have implicitly or explicitly insisted on the biopolitical characterization of Nazism converge around this thesis: it is the growing implication between politics and life that introduces into the latter the normative caesura between those who need to live and those who need to die. What the immunitary paradigm adds is the recognition of the homeopathic tonality that Nazi therapy assumes. The disease against which the Nazis fight to the death is none other than death itself. What they want to kill in the Jew and in all human types like them isn't life, but the presence in life of death: a life that is already dead because it is marked hereditarily by an original and irremediable deformation; the contagion of the German people by a part of life inhabited and oppressed by death. The only way to do so seemed to be to accelerate the "work of the negative," namely, to take upon oneself the natural or divine task of leading to death the life of those who had already been promised to it. In this case, death became both the object and the instrument of the cure, the sickness and its remedy. This explains the cult of the dead that marked the entire brief life of the Reich: the force to resist the mortal infection that threatened the chosen race could only come from dead ancestors. Only they could transmit to their descendants the courage to give or to receive a purifying death in relation to that other death that grew like a poisonous fungus in the soil of Germany and the West. It was this that the SS swore in a solemn pledge that seemed to correspond to the nature and the destiny of the German people. A response was needed to the presence of death in life (this was degeneration) by tempering life on the sacred fire of death: giving death to a death that had assumed the form of life and in this way had invaded life's every space. It was this insidious and creeping death that needed to be blocked with the aid of the saving Great Death bequeathed by the German heroes. Thus, the dead become both the infectious germs and the immunitary agents, the enemies to be extinguished and the protection to be activated. Confined to this double death and its infinite doubling, Nazism's immunitary machine wound up smashed *[ingranaggi]*. It strengthened its own immunitary apparatus to the point of remaining victim to it. The only way for an individual or collective organism to save itself definitively from the risk of death is to die. It was what Hitler asked the German people to do before he committed suicide.

If this was in general terms the deadly logic of the Nazi event *[vicenda]*, what were its decisive articulations and its principal immunitary *dispositifs*? I would indicate essentially three. *Absolute normativization of life* constitutes the first. In it we can say that the two semantic vectors of immunity, the biological and the juridical, for the first time are completely superimposed according to the double register of the biologization of the *nomos* and simultaneously that of the juridicalization of *bíos*. We have already seen the growth of the influence of biology, and in particular of medicine, which took place in all of the ganglions of individual and collective experience during those years. The doctors who had enjoyed great authority and prestige in Wilhelminian and Weimar Germany became more powerful in areas that had to that point been reserved for other expertises. In particular, their presence was made felt in courtrooms, where they accompanied (and in some cases surpassed) the magistrates in the application of restrictive and repressive norms. For example, when selecting individuals to undergo sterilization, the legal commission, as well as the court of appeals, were composed of a judge and two doctors. The more the categories subjected to review were widened to include the practically unlimited field of racial deformities and social deviance, the more the power of medicine grew together with that of psychiatrists and anthropologists. The Nuremberg laws on citizenship and on the "protection of blood and the honor of the German people" further strengthened the position and power of medical doctors. When the programs of euthanasia finally began and the concentration camps came into operation, doctors became those priests of life and of death I spoke of earlier.[77]

This first side of the immunitary logic, which is attributable to the biologization of law *[diritto]*, need not, however, obscure the other side of the coin, which is to say the ever more extensive juridical (and therefore political) control of medicine. The more, in fact, the doctor was transformed into a public functionary, the more he lost autonomy with respect to the state administration on which, in the final analysis, he wound up depending. What was under way, in short, was a clear-cut transformation of the relation between patient, doctor, and state. While the relation between the first two terms was loosened, that of the second two was tightened. In the moment in which the cure (and before that still the diagnosis) was no longer a private but a public function, the doctor's responsibility was no longer exercised in relation to those who were sick, but rather to the state, the sole (and also secret) depository for archiving the conditions of the patient that before had been reserved for medicine. It is as if the role of the subject passed from the sick (who by now had become the simple object of biological definition and not of healing) to doctors, and from them in time to the state institution.[78] On the one hand, and as proof of this progressive consignment, the 1935 racial laws were not prepared by a committee of experts, as they had been the preceding year, but rather, directly by political personnel. On the other hand, if the regulations on hereditary disease still required a semblance of scientific judgment on the part of doctors, those concerning racial discrimination were assigned by pure chance. More than reflecting different biological caesuras within the population, they created them out of nothing. Doctors did nothing else except legitimate decisions with their signatures that had been made in the political sphere and translated into laws by the new legal codes of the Reich. Thus, a political juridicalization of the biological sphere corresponded to a biologization of the space that before had been reserved for juridical science.[79] To capture the essence of Nazi biopolitics, one must never lose sight of the interweaving of the two phenomena. It is as if medical power and political-juridical power are mutually superimposed over each other through alternating points that are ultimately destined to completely overlap: this is precisely the claim that life is supreme, which provokes its absolute subordination to politics.

The concentration and later the extermination camps constitute the most symptomatic figure of such a chiasmus. The term "extermination" (from *exterminare*) already refers to a terminological leak, just as the word *elimination* alludes to a moving beyond the threshold that the Romans referred to as *limes*. Naturally, the structurally aporetic character of the camp resided in the fact that the "outside" or "beyond" were constituted in the form of an "inside" so "concentrated" as to make impossible any hope of escape. It is precisely insofar as it was "open" with respect to the closed model of the prison that the camp was proven to be forever sealed off. Closed, one would say, from its own opening, just as it is destined to be interned from its own exteriority. Now, such an obviously self-contradicting condition is nothing other than the expression of the indistinction that emerges between the horizon of life and that of law that has been completely politicized. Grabbing hold directly of life (or better, its formal dimension), law cannot be exercised but in the name of something that simultaneously makes it absolute and suspends it. Against the common conviction that the Nazis limited themselves to the destruction of the law, it is to be said instead that they extended it to the point of including within what also obviously exceeded it. Maintaining that they were removing life from the biological sphere, they placed all aspects of life under the command of the norm. If the concentration camp was certainly not the place of law, neither was it that of mere arbitrary acts. Rather, it was the antinomical space in which what is arbitrary becomes legal and the law arbitrary. In its material constitution, the camp reinstates the most extreme form of the immunitary negation, not only because it definitively superimposes the procedures of segregation, sterilization, and euthanasia, but also because it anticipates all that could exceed the deadly outcome. Ordered to lock up the perpetrators of crimes that hadn't yet been committed (and therefore were not prosecutable on the basis of laws in force), the camp is configured as a form of *Schutzhaftlager* ("preventive detention"), as was written above the entrance to Dachau. What was detained in advance, which is to say what is completely lacking *[destituire]*, was life as such, subjected to a normative presupposition that left no way out.

Nazism's second immunitary *dispositif* is the *double enclosure of the body,* that is, the enclosing of its own closure. It is what Emmanuel Levinas defined as the absolute identity between our body and ourselves. With respect to the Christian conception (but also differently from Cartesian tradition), all dualism between the ego *[io]* and body collapses. They coincide in a form that doesn't allow for any distinction: the body is no longer only the place but the essence of the ego. In this sense, one can well say that "the biological, with the notion of inevitability it entails, becomes more than an object of spiritual life. It becomes its heart."[80] We know the role that the theory of the transmission of germinative plasma played in this conception and, incidental to that, of psychosomatic heredity: man is completely defined by the past that he carries and that is reproduced in the continuity between generations. The terms used by Levinas of "enchantment" *(enchaînement)* and of a "nailing" *(être rivé)* with reference to one's biological being give the material sense of a grip from which one cannot escape.[81] When faced with it, it behooves us to accept it as both destiny and responsibility rather than trying vainly to break free. And that is true both for the one whose destiny is to be condemned unremittingly (which is to say the inferior man) and for the other who recognizes in it the mark of a proclaimed superiority. In any case, it's a matter of adhering to that natural layer from which one cannot escape. This is what is meant by double enclosure: Nazism assumes the biological given as the ultimate truth because it is the basis on the strength of which everyone's life is exposed to the ultimate alternative between continuation and interruption.

This doesn't mean that it resolves itself in an absolute materialism to be identified entirely in a radicalized version of Darwinian evolution. Although the propensity of such a sort did in point of fact exist, it was accompanied and complicated by another tendency in which some have wanted to see a sort of spiritual racism, represented, for example, in Rosenberg's position. In reality, these two lines are anything but in contradiction because from the very start they share a tangential point. In none of the writings of its theoreticians does Nazism deny what is commonly defined as "soul" or spirit—only it made out of these the means not to open the body toward transcendence, but rather to a further and more definitive enclosing. In this sense, the soul is the body of the body, the enclosing of its closing, what from a subjective point of view binds us to our objective imprisonment. It is the point of absolute coincidence of the body with itself, the consummation of every interval of difference within, the impossibility of any transcendence.[82] In this sense, more than a reduction of *bíos* to *zōē* or to "bare life" (which the Nazis always opposed to the fullness of "life" understood in a spiritual sense as well), we need to speak of the spiritualization of *zōē* and the biologization of the spirit.[83] The name assumed by such a superimposition is that of race, which constitutes both the spiritual character of the body and the biological character of the soul. It is what confers meaning on the identity of the body with itself, a meaning that exceeds the individual borders from birth to death. When Vacher de Lapouge wrote that "what is immortal isn't the soul, a dubious and probably imaginary character: it is the body, or rather, the germinative plasma," he did nothing other than anticipate what Nazism will decisively elaborate.[84] The text in which this bio-theogony finds its most complete definition is Verschuer's manual of eugenics and racial heredity. Unlike in the old German state and in contemporary democracies in which one takes people to mean the sum of all citizens, which is to say, those individuals who inhabit state territory:

> [I]n the ethnic, National-Socialist state, we understand "people" or "ethnic" to be a spiritual and biological unity . . . ; the greatest part of the German people constitutes a great community of ancestors, which is to say a solidarity of blood relations. This biological unity of people is the foundation of an ethnic body, an organic structure of totalitarian character whose various parts are nothing less than the components of the same unity.[85]

This represents a further doubling or extension of that enclosure of the body on itself that Nazism placed at the center of its immunitary apparatus. Following the first operation, which remains at the level of the individual and the incorporation of the self within his own body, a second occurs by means of which every corporeal member finds himself in turn incorporated into a larger body that constitutes the organic totality of the German people. It is only this second incorporation that confers on the first its spiritual value, not in contrast to, but rather on the basis of, its biological configuration. But that is not all: connecting horizontally all the single bodies with the one body of the German community is the vertical line of hereditary patrimony "that, as a river, runs from a generation to the next."[86] It is only at this point in the biopolitical composite of this triple incorporation that the body of every German will completely adhere to itself, not as simple flesh, an existence without life, but as the incarnation of the racial substance from which life itself receives its essential form—provided, naturally, that it has the force to expel from itself all of that which doesn't belong to it (and for which reason hampers its expansive power). It is the lethal outcome that inevitably derives from the first part of the discourse. "If one begins from this notion of 'people,'" Verschuer concludes, "demographic politics is that of the *protection of the ethnic body* by maintaining and improving the healthy patrimony, the elimination of its sick elements, and the conservation of the racial character of the people."[87] In this conceptual frame, it wasn't wrong to define genocide as the spiritual demand of the German people: it is only through the removal of the infected part that that body would have experienced profoundly its enclosing on itself and through it the belonging to what is shared with every other member: "Dein Körper gehört dem Führer" (Your body belongs to the Führer) was written on posters in Berlin. When the Nazi doctor Fritz Klein was asked how he could reconcile what he had done with the Hippocratic oath, he responded: "Of course, I am a doctor and I want to preserve life. And out of respect for human life, I would remove a gangrenous appendix from a diseased body. The Jew is the gangrenous appendix in the body of mankind."[88] The German *Völkerkörper* [people's body], which was filled to the brim, couldn't live without evacuating its purulent flesh. Perhaps for this reason, another of the German doctors defined Auschwitz as *anus mundi*, anus of the world.[89]

The third Nazi immunitary *dispositif* is represented by the *anticipatory suppression of birth*, which is to say not only of life but of its genesis. It is in this extreme sense that one ought to understand the declaration according to which "sterilization was the medical fulcrum of the Nazi biocracy."[90] It isn't a simple question of quantity. Certainly, between June 1933 and the beginning of the war, more than three hundred thousand people were, for various reasons, sterilized, not to mention that in the following five years the figure would grow exorbitantly. But it isn't only a question of increased sterilization. When speaking about sterilization, Nazism had something else in mind, a kind of excess whose full sense we have yet to understand. The Nazis assumed that those numbers, which were already enormous, represented a temporary limitation with regard to what they would want to do later; for his part, Lenz declared that up to a third of the German people would have to be sterilized. Waiting for that moment to arrive, the Nazis didn't waste any time. In September of 1934, the decree on obligatory abortion was approved for degenerate parents; in June 1935, castration of homosexuals; in February 1936, it was decided that women above the age of thirty-six were to be sterilized using X-rays. We could say that deciding which method to employ keenly interested Nazi medicine. When the practice of sterilization was extended to prisoners, a real political-medical battle broke out (which is to say a thanatopolitical one) that centered on the most rapid and economical mode of operation. On the one side, there was the famous gynecologist Clauberg, the inventor of the test on the action of progesterone, who fervently supported the obstruction of the Fallopian tube. On the other side, there were Viktor Brack and Horst Schumann, who favored Roentgen rays. The result of both procedures was the atrocious suffering and death of a large number of women.

Despite the fact that both men and women were operated on without distinction, we know that it was the latter who were the principal victims of Nazi sterilization both in number (circa 60 percent) and, above all, in the frequency of death (90 percent). They were mutilated with all the pretexts in place, ones that even contradicted each other: because their husbands were psychopathic or, on the contrary, because they were unwed mothers. For those judged to be mentally deficient, the entire uterus was ablated rather than following the normal ligation of the ovarian tubes. When a number of women who had been threatened with sterilization responded with a sort of "pregnancy protest," obligatory abortion up to the seventh month of pregnancy was ordered. Moreover, in the concentration camps, maternity was punished by immediate death. To argue that all of this is the work of chance—or to obscure it in the general mechanism of extermination—would mean losing sight of the profound meaning of such an event. If we remember that the law on sterilization was in fact the first legislative measure adopted by the Nazis when in power (just as children were the first victims of euthanasia), it becomes clear that they wanted to strike at the beginning of life, life at the moment of birth. But we still haven't hit on the crux of the question. The complexity of the question will be found in the fact that these lethal measures were adopted in the midst of a pro-natalist campaign intent on strengthening the German population quantitatively as well.[91] It wasn't by chance that voluntary abortion was prohibited as a biological crime against the race, while funds were set aside for helping numerous families. How do we want to interpret such an obvious contradiction? What meaning is to be attributed to such a mingling of the production and prevention of life? How did the Nazis understand birth, and what tied birth to death?

A first response to the question lies in the distinction the Nazis wanted to make on more than one occasion between "regeneration" and "procreation." While the former, which was activated on the basis of official eugenic protocols, had to be supported at all costs, the latter (which occurred spontaneously and unexpectedly) was to be governed strictly by the state. This means that the Nazis were anything but indifferent to the biological phenomenon of birth. In fact, they gave it their utmost attention, but in a form that subordinated it directly to political command. This is the biopolitical exchange that we know so well. On the basis of the racial heredity that birth carries with it, birth appears to determine the level of citizenship in the Reich according to the principle (and also the etymology) that links birth to the nation. In nowhere more than the Nazi regime, however, did the nation seem to take root in the natural birth of citizens of German blood. In reality, here as well, what was presented as the source of power was rather derived from power, which is to say it wasn't birth that determined the political role of the living being *[vivente]*, but its position in the political-racial calculation that predetermined the value of its birth. If this living being reentered the biopolitical enclosure dedicated to breeding, it was accepted or even encouraged; if it fell outside, it had to be suppressed even before it was announced.[92] Later, when indiscriminate extermination was at hand, not even this was sufficient. Neither was it enough to prevent birth, nor simply to prompt death. It was believed necessary to superimpose the two operations, thereby subjecting birth to death. Suspending *[interrompere]* life was too little—one needed to annul the genesis of life, eliminating all posthumous traces of life. In this sense, Hannah Arendt could write: "for the status of the inmates in the world of the living, where nobody is supposed to know if they are alive or dead, is such that it is as though they had never been born."[93] They simply did not exist. This is the logical reason for which, on the one hand, they could be killed an infinite number of times in the same day and, on the other, that they were prohibited from committing suicide. Their body without a soul belonged to the sovereign. Yet, in the biopolitical regime, sovereign law isn't so much the capacity to put to death as it is to nullify life in advance.

CHAPTER FIVE
The Philosophy of *Bíos*

Philosophy after Nazism

That biopolitics experienced with Nazism its most terrifying form of historical realization doesn't mean, however, that it also shared its destiny of self-destruction. Despite what one might think, the end of Nazism in no way signaled the end of biopolitics. To hypothesize in such a way not only ignores the long genesis of biopolitics (which is rooted in modernity), but also underestimates the magnitude of the horizon they share. Nazism didn't produce biopolitics. If anything, Nazism was the extreme and perverse outcome of a particular version of biopolitics, which the years separating us from the end of the regime have proven time and again. Not only hasn't the direct relationship between life and death been moderated, but, on the contrary, the relation appears to be in continual expansion. None of the most important questions of interest to the general public (which is fast becoming ever more difficult to distinguish from the private) is interpretable outside of a profound and often immediate connection with the sphere of *bíos*.[1] From the growing prominence of ethnicity in relations between peoples and state, to the centrality of the question of health care as a privileged index of the functioning of the economic system, to the priority that all political parties give in their platforms to public order—what we find in every area is a tendency to flatten the political into the purely biological (if not to the body itself) of those who are at the same time subjects and objects. The introduction of work in the somatic, cognitive, and affective sphere of individuals; the incipient translation of political action into domestic and international police operations; the enormous growth in migratory flows of men and women who have been deprived of every juridical identity, reduced to the state of bare sustenance—these are nothing other than the clearest traces of the new scenario.[2] If we look then at the continuing indistinction between norm and exception that is tied to the stabilization of emergency legislation, we will find yet another sign of contemporary society's increasingly evident biopolitical characterization. That the obsessive search for security in relation to the threat of terrorism has become the pivot around which all the current governmental strategies turn gives an idea of the transformation currently taking place. From the politicization of the biological, which began in late modernity, we now have a similarly intense biologization of the political that makes the preservation of life through reproduction the only project that enjoys universal legitimacy.

From this perspective, however, it's opportune to recall that not only has the politics of life that Nazism tried in vain to export outside Germany—certainly in unrepeatable forms—been generalized to the entire world, but its specific immunitary (or, more precisely, its autoimmunitary) tonality has been as well. That the protection of biological life has become the largely dominant question of what now has for some time been called domestic and foreign affairs, both now superimposed on the unified body of a world without exterior (and hence without an interior), is an extraordinary acknowledgment of the absolute coincidence that has taken place between biopolitics and immunization. Fifty years after the fall of Nazism, the implosion of Soviet communism was the final step in this direction. It is as if at the end of what still saw itself as the last and most complete of the philosophies of history, life, which is to say the struggle for its protection/negation, had become global politics' only horizon of sense.[3] If during the cold war the immunitary machine still functioned through the production of reciprocal fear and therefore had the effect of deterring catastrophes that always threatened (and exactly for this reason never occurred), today, or at least beginning with September 11, 2001, the immunitary machine demands an outbreak of effective violence on the part of all contenders. The idea—and the practice—of preventive war constitutes the most acute point of this autoimmunitary turn of contemporary biopolitics, in the sense that here, in the self-confuting figure of a war fought precisely to avoid war, the negative of the immunitary procedure doubles back on itself until it covers the entire frame. War is no longer the always possible inverse of global coexistence, but the only effective reality, where what matters isn't only the specular quality that is determined between adversaries (who are to be differentiated in their responsibility and original motivations), but the counterfactual outcome that their conduct necessarily triggers—in other words, the exponential multiplication of the same risks that would like to be avoided, or at least reduced, through instruments that are instead destined to reproduce them more intensely. Just as in the most serious autoimmune illnesses, so too in the planetary conflict presently under way: it is excessive defense that ruinously turns on the same body that continues to activate and strengthen it. The result is an absolute identification of opposites: between peace and war, defense and attack, and life and death, they consume themselves without any kind of differential remainder. That the greatest threat (or at least what is viewed as such) is today constituted by a biological attack has an obvious meaning: it is no longer only death that lies in wait for life, but life itself that constitutes the most lethal instrument of death. And what else besides a fragment of life is a kamikaze, except a fragment that discharges itself on the life of others with the intent of killing them *[portarvi la morte]*?

How does contemporary philosophy position itself when confronted with such a situation? What kind of response has it furnished to the questions literally of life and death that biopolitics opened in the heart of the twentieth century and that continue to be posed differently (though no less intensely) today? Certainly, the most pervasive attitude has been to repress or even ignore the problem. The truth is that many simply believed that the collapse of Nazism would also drag the categories that had characterized it into the inferno from which it had emerged. The common expectation was that those institutional and conceptual mediations that had permitted the construction and the resistance of the modern order would be reconstituted between life and death, which had been fatally joined together in the 1930s and 1940s. One could discuss—just as one continues wearily to do so today—whether a return to state sovereignty should be applauded, a sovereignty threatened by the intrusiveness of new supranational actors, or rather whether a hoped-for extension of the logic of law to the entire arena of international relations is possible. But they are always part of the old analytic framework derived from the Hobbesian matrix, perhaps with a sprinkling of Kantian cosmopolitanism thrown in for good measure, only to discover that such a model no longer works. In other words, the model reflects almost nothing of current reality, let alone is it able to provide effective tools that might prefigure its transformation. This isn't only because of the incongruence of continuing to contrast possible options (such as those related to individual rights and sovereign power) that have from the start been reciprocally functional in the development of each from the instant that rights are not given without a sovereign power (be it national or imperial) that demands they be respected. Similarly, there doesn't exist a sovereignty that lacks some kind of juridical foundation. It's not by accident that the stunning deployment of sovereign power *[potenza]* on the part of the American imperial state is justified precisely in the name of human rights. More generally, however, the simple fact is that we can't run history backwards, which is to say Nazism (more so than communism) represents the threshold with respect to the past that makes every updating of its lexical apparatus impractical. Beginning with that threshold (which is both historical and epistemological), the biopolitical question can no longer be put off. It can, indeed needs, to be reversed with respect to the thanatological configuration that it assumed in Hitler's Germany, but not directed toward modernity, if for no other reason than because biopolitics contradictorily originates in it in both modality and intensity. This is different from the form it subsequently took in Nazi Germany.

Hannah Arendt was the person who understood early the modern roots of biopolitics, using an interpretive key that recasts its reason and even its semantic legitimacy. Contrary to the pervasive thesis that ties modernity to the deployment of politics, she not only refers it back to depoliticization, but ascribes the process to a crisis in the category of life in place of the Greek conception of the world held in common. Christianity constitutes the decisive step within such an interpretive scheme, representing in fact the original horizon in which the concept of the sacredness of individual life is affirmed for the first time (albeit inflected in an otherworldly sense). It will be sufficient that modernity secularizes it, moving the center of gravity from the celestial realm to that of the earth, to prompt that reversal in perspective that makes biological survival the highest good. From there "the only thing that could now be potentially immortal, as immortal as the body politic in antiquity and as individual life during the Middle Ages, was life itself, that is the possibility, the possibly everlasting life process of the species mankind."[4] But it is precisely the affirmation of a modern *conservatio vitae* with respect to the Greek interest for a common world that, according to Arendt, sets in motion that process of depoliticization that culminates when work that satisfies material necessities became the prevalent form of human action. Beginning from that moment,

> none of the higher capacities of man was any longer necessary to connect individual life with the life of the species; individual life became a part of the life process, and to labor, to assure the continuity of one's own life and the life of his family, was all that was needed. What was not needed, not necessitated by life's metabolism with nature, was either superfluous or could be justified only in terms of a peculiarity of human as distinguished from other animal life.[5]

It is exactly the process that Foucault will define shortly thereafter in biopolitical terms: individual life integrated in the life of the species and made distinct through a series of internal breaks in zones of different worth. But it is also the point at which Arendt's discourse tacks in a different direction, diverging from the one initiated by Foucault.[6] From the moment that the entrance of the question of life onto the scene of the modern world coincides with the withdrawal of politics under the double pressure of work and production, the term "biopolitics" (just as for the Marxian term "political economy") emerges devoid of any sense. If political activity is considered in theory to be heterogeneous to the sphere of biological life, then there can never be an experience (precisely biopolitical) that is situated exactly at their point of intersection. That such a conclusion rests on the unverified premise according to which the only valid form of political activity is what is attributable to the experience of the Greek *polis*—from which a paradigmatic separation is assumed irreflexively between the private sphere of the *idion* and the public sphere of the *koinon*—determines the blind spot that Arendt reaches concerning the problem of biopolitics: where there is an authentic politics, a space of meaning for the production of life cannot be opened; and where the materiality of life unfolds, something like political action can no longer emerge.

The truth is that Arendt didn't think the category of life thoroughly enough and therefore was unable to interpret life's relationship with politics philosophically. This is particularly surprising for the author who more than anyone else elaborated the concept of totalitarianism (unless it was precisely the specificity of what Levinas has defined as the "philosophy" of Hitlerism that eludes her or is at least hidden from her). It would have been easy to grasp its nature, to penetrate into the machine of Nazi biopolitics, beginning with a reflection on politics that is strongly marked by a reference to the Greek *polis*. The problem (relative not only to Arendt) is that such a reflection doesn't provide direct access from political philosophy, be it modern or premodern, to biopolitics. In its biocratic essence, Nazism remains mute for classical political thought. It is no coincidence that a radically impolitical thinker such as Heidegger conducted a real philosophical comparison with it (although in an implicit and reticent form). Yet he was able to attempt it, that is, to think the reverse of the question Nazism raised for world history, because his starting point, in a certain sense, was the same presupposition, which is to say the "end of philosophy," or better, its extroversion in something that can be called existence, world, or life, but which, however, cannot be comprehended in modern categories of subject and object, individual and universal, and empirical and transcendental. When in 1946 he wrote *Letter on Humanism* in the darkest moment of defeat (a defeat that was also personal), he wrestled precisely with this question. What he seeks, in the abyss that Nazi thanatopolitics had excavated, is a response capable of meeting it on its own terms, without, that is, having recourse to that humanistic lexicon that did not know how to avoid it (or even had contributed to laying the groundwork for it). Not only does his entire reflection on technology *[tecnica]* move in this direction, but also the ontological transposition of what tradition had defined each time as "subject," "consciousness," or "man" responds to the necessity of sustaining the comparison with the powers of nihilism *[potenze del niente]* on their same level. In this sense, the invitation to think against humanism is to be interpreted "because it does not set the *humanitas* of man high enough," as well as that in line with "the world historical moment," to a meditation "not only about man but also about the 'nature' of man, not only about his nature but even more primordially about the dimension in which the essence of man, determined by Being itself, is at home."[7]

Furthermore, Heidegger didn't wait for the end of the war and the fall of Nazism to undertake his reflection on the nature of man removed from that language (however humanistic) of liberal, Marxist, or existentialist ascendancy that was left undefended with regard to Nazism and the question of *bíos*. Indeed, the entire thematic of the "factical life" *(faktisches Leben)* that he took up from the beginning of the 1920s in his Freiburg courses, first in dialogue with Paul and Augustine and then with Aristotle, implied the refusal to subject the primary or concrete experience of life to the scrutiny of theoretical or objectivizing categories that were still rooted in the transcendence of the subject of knowledge—where the disruptive element with respect to the classic framework goes well beyond the results of the "philosophy of life" that authors such Dilthey, Rickert, and Bergson had elaborated in those years, to take form instead in an unsettling of both the terms and even more of the relation that binds them.[8] Not only is factical life, the facticity *[fatticità]* of life, not to be derived through a traditional philosophical investigation, but it is situated precisely in its reversal. That doesn't mean that the horizons do not intersect, namely, that the vital experience is closed to philosophical interrogation (or worse abandoned to the flux of irrationality). What it does mean is that philosophy is not the site in which life is defined, but rather that life is the primogenital root of the same philosophy:

> The categories are not inventions or a group of logical schemata as such, "lattices"; on the contrary, they are *alive in life itself* in an original way: alive in order to "form" life on themselves. They have their own modes of access, which are not foreign to life itself, as if they pounced down upon life from the outside, but instead are precisely the preeminent way in which *life comes to itself.*[9]

Already here, in this withdrawal of life from any categorical presupposition, we cannot miss seeing a connection, one that is certainly indirect, partial, and differential, with that much more immediate primacy of *bíos* that a decade later will constitute with Nazism the vitalistic battering ram against every form of philosophy. Still, this doesn't exhaust the area of the possible comparison between the thought of Heidegger and the open problem of Nazi biopolitics, not only because *bíos* echoes in the factical life that is one with its effective dimension and coincides immediately with its modes of being, but also because of the possibility or the temptation to interpret life politically (or at least negatively). If the facticity of life, which in *Being and Time* is assumed under the name of *Dasein*, doesn't respond to any external instance, from the moment that it isn't attributable to any preconstituted philosophical design, then only life is vested with its own decision of existence. But how is a life or being there *[esserci]* configured so that it can decide for itself *[su se stessa]*, or even that it *is* such a decision, if not in an intrinsically political modality? What opens the possibility of thinking *bíos* and politics within the same conceptual piece is that [first] at no point does authentic being *[poter-essere]* exceed the effective possibility of being there *[dell'esserci]*, and second that the self-decision of this being is absolutely immanent to itself. It is from this side, precisely because it is entirely impolitical, which is to say irreducible to any form of political philosophy, that Heidegger's thought emerges in the first half of the twentieth century as the only one able to support the philosophical confrontation with biopolitics.

That Heidegger faced the question of biopolitics doesn't mean that he took on its language or shared its premise, namely, the preeminence of life in relation to being in the world. Indeed, we might say that he expressed a point of view diametrically opposed to it: the biological category of life isn't the site from which the thinkability of the world opens, but is exactly the contrary. If the phenomenon of living always emerges as a living "in" or "for" or "with" something that we can indicate with the term "world," we need to conclude that "world is the basic category of the content-sense in the phenomenon, life."[10] The world isn't the container or the environment, but the content of the sense of life. It is the ontological horizon out of which only life becomes accessible to us. Thus, Heidegger distances himself both from those who, like Arendt, radically set the sphere of life against that of the world (understood as the public sphere of acting in common), and from those who reduced the world to a place for the biological deployment of life. Without being able to follow in detail the internal passages or the diachronic moments of Heidegger's discourse, one could generally trace them back to an underlying tendency to keep "factical life" apart from biology.

> Biological concepts of life are to be set aside from the very outset: unnecessary burdens, even if certain motives might spring from these concepts, which is possible, however, only if the intended grasp of human existence as life remains open, preconceptually, to an understanding of life which is essentially older than that of modern biology.[11]

Even later, when Heidegger will dedicate an entire section of his 1929–30 course to *The Fundamental Concepts of Metaphysics,* this diffidence or categorical deformity with respect to biology will not collapse. It isn't that he doesn't come into contact with some of the principal biologists of the time, as is demonstrated not by the frequent references to Driesch, Ungerer, Roux, and above all Uexküll, and by the protocols from the seminars of Zollikon, which were held specifically with a group of doctors and psychiatrists. It is precisely these protocols that allow us to see (despite the declarations of reciprocal interest) a marked communicative difficulty, if not indeed a true and precise categorical misunderstanding between conceptual lexicons that are profoundly heterogeneous. "Quite often," admits Dr. Medard Boss, who was also tenaciously involved in a complex operation of semantic loops, "the situations in the seminars grew reminiscent of some imaginary scene: It was as if a man from Mars were visiting a group of earth-dwellers in an attempt to communicate with them."[12]

Why? What are we to make of this substantial untranslatability between Heidegger's language and that of the doctors and biologists whose intention was still to be receptive? Above all, what does it suggest in relation to our inquiry? If we recall that Nazi biopolitics was characterized by the domination of the category of life as opposed to the category of existence—"existence without life" was what was given over *[destinata]* to death both in principle and in point of fact—it wouldn't be arbitrary to see in Heidegger's polemic concerning biologism a form of advance counterposition. Without wanting to homologize profoundly different terminologies (as can only be the case between the most significant philosopher of the twentieth century and the merchants selling death at a discount), we could say that Heidegger reverses the prevalent relation instituted by the latter: it isn't existence that emerges as deficient or lacking in relation to a life that has been exalted in its biological fullness, but life that appears defective with respect to an existence understood as the only modality of being in the openness of the world. Furthermore, life defined biologically doesn't have the attributes of *Dasein,* but is situated in a different and incomparable dimension with respect to the horizon of the latter. It can only be deduced negatively from *Dasein* as that which isn't it, precisely because it is "only life" *(Nur Lebenden);* as "something that only lives" *(etwas wie Nur-noch-leben):*

> Life has its own kind of being, but it is essentially accessible only in Da-sein. The ontology of life takes place by way of a privative interpretation. It determines what must be the case if there can be anything like just-being-alive. Life is neither pure objective presence, nor is it Da-sein. On the other hand, Da-sein should never be defined ontologically by regarding it as life—(ontologically undetermined) and then as something else on top of that.[13]

But the contrastive symmetry between Heidegger and Nazi biopolitics doesn't end there, not only because both for the former and the latter life and existence emerge as linked by a relation of excluding implication—in the sense that one is defined by its not being equal to the other—but in both cases the differential comparison is constituted by the experience of death. It is precisely here, nevertheless, that the two perspectives definitively diverge. While in Nazi thanatopolitics death represents the presupposition of life even before its destiny, a life emptied of its biological potentiality *[potenza]* (and therefore reduced to bare existence), for Heidegger death is the authentic *[proprio]* mode of being of an existence distinct from bare life. Certainly, the latter life dies too, but in a form lacking in meaning that, rather than a true dying *(sterben),* refers to a simple perishing, to a ceasing to live *(verenden)*. In this manner, what simply lives *[vivente]* cannot be defined in a fully mortal sense of the word, as can he who experiences his own death, but rather as the end of life, as that which from the beginning confers meaning on life. At this point, the relation between Nazi biopolitics and Heidegger's thought is delineated in all its antinomy. While in the first the sovereign structure of biopolitics resides in the possibility of submitting every life to the scrutiny of death, for the second the intentionality of death constitutes the original political form in which existence is "decided" in something that always resides beyond simple life.

Yet we can single out the point of Heidegger's greatest divergence from Nazi biopolitics in his treatment of that living specificity that is the animal. In this case as well, the point of departure is in a certain sense the same: not only what is the animal, but also how it is situated in relation to the world of man. We know how Nazism responds to such a question, in what was the culmination of a tradition born at the crossroads between Darwinian evolutionism and degenerative theory: the animal, more than a separate species from the human, is the nonhuman part of man, the unexplored zone or the archaic phase of life in which *humanitas* folds in on itself, separating itself through an internal distinction between that which can live and that which has to die. Previously in *Being and Time* (and then in a more articulated fashion in *The Fundamental Concepts of Metaphysics*) and then in the later *Contributions to Philosophy,* Heidegger travels in a different direction.[14] The question of *animalitas* is nothing but a particularly relevant specification of the relation that was already instituted between the sphere of *Dasein* and that of simple living beings. When this latter assumes the features of the animal species, the separation with respect to the one who exists in the mode of being there *[esserci]*, that is, man, becomes clearer. That the animal is defined, according to the famous tripartition, as "poor of the world" *(weltarm)*, unlike that of the stone, which is "without the world" *(weltlos)*, and then precisely of man, who is "the creator of the world" *(weltbildend)*, is in fact a way of marking an insurmountable distance in relation to human experience. It is opposed to the animalization of man, not only the one theorized but also the one the Nazis put into practice; here Heidegger situates man well on the outside of the horizon of the animal. Man is so incomparable to the animal that he is not even able to conceptualize the condition if not by inferring it as the negative of his own proper condition. The expression "poor of the world" doesn't indicate a lesser level of participation in a common nature with all living beings, including man, but an insurmountable barrier that excludes any conjugated form. Contrary to a long-standing tradition that thought man as the *rational animal*—an animal to which is added the charisma of *logos* to make him noble (according to the classic formulation of the *zōon logon echon*), man is precisely the *non*animal, just as the animal is the *non*human living being *[vivente]*. Despite all the attempts directed at tracing the affinity, symmetry, and copresence (perhaps in the existential dimension of boredom), the two universes remain reciprocally incommunicable.[15] As Heidegger writes in *Letter on Humanism:*

> It might seem as though the essence of divinity is closer to us than what is foreign in other living creatures, closer, namely, in an essential distance which however distant is nonetheless more familiar to our ek-sistant essence than is our appealing and scarcely conceivable bodily kinship with the beast.[16]

Exactly these kinds of passages, however, if they work in completely sheltering Heidegger from the thanatopolitical drift of Nazism, risk drawing him 360 degrees in the opposite direction, close to that humanism from which he had carefully distanced himself. Naturally, the entire movement of this thought (which is oriented in an ontological direction) makes impossible not only the reproposition of an anthropocentric model, but also any concept of human nature as such—autonomous from the being to whose custody man seems called. Precisely this decentering of man (or recentering of being) is connected, however, in the course of Heidegger's work, to a progressive loss of contact with the theme of "factical life" in which the semantics of *bíos* seemed inevitably implicated. It is as if the originary impulse to think life in the "end of philosophy" (or the end of philosophy in the facticity of life) slowly flows back with the effect of dissolving its same object. Wishing to trace the terms of an extremely complicated question back to an abbreviated formulation, we could say that the absolute distance that Heidegger places between man and animal is the same as that which comes to separate always in ever more obvious fashion his philosophy from the horizon of *bíos.*[17] And that is precisely because it risks entrusting *bíos* to nonphilosophy, or better, to that antiphilosophy that was terrifyingly realized in the 1930s in its most direct politicization. That it occurred exactly in that phase of Heidegger's thought, even briefly, becoming the prey of that antiphilosophy, is to be interpreted differently and in a more complex manner than it has been to now. It probably wasn't an excess of nearness but an excess of distance from both the vital and moral questions raised by Nazism that made Heidegger lose his bearings. Precisely because he didn't enter deeply enough (and not because he entered too much) into the dimension of *bíos* that is in itself political, in the rapport between qualified existence and biological life, he wound up abandoning it to those whose intention was to politicize it until it shattered. Once again the black box of biopolitics remained closed with Heidegger.

Flesh

Apparently, if we are to open the black box of biopolitics we shouldn't limit ourselves to skirting Nazi semantics, or for that matter confronting it from the outside. Something more is required and it has to do with penetrating within it and overturning one by one its bio-thanatological principles. I am referring in particular to the three *dispositifs* that I examined at the conclusion of the preceding chapter: *the normativization of life, the double enclosure of the body,* and the *preemptive suppression of birth.* Yet what does it mean exactly to overturn them and then to turn them *inside out*? The attempt we want to make is that of assuming the same categories of "life," "body," and "birth," and then of converting their immunitary (which is to say their self-negating) declension in a direction that is open to a more originary and intense sense of *communitas.* Only in this way—at the point of intersection and tension among contemporary reflections that have moved in such a direction—will it be possible to trace the initial features of a biopolitics that is finally affirmative. No longer over life but of life, one that doesn't superimpose already constituted (and by now destitute) categories of modern politics on life, but rather inscribes the innovative power of a life rethought in all its complexity and articulation in the same politics. From this point of view, the expression "form-of-life," or precisely what Nazi biopolitics excluded through the absolute subtraction of life from every qualification, is to be understood more in the sense of a vitalization of politics, even if in the end, the two movements tend to superimpose themselves over one another in a single semantic grouping.

Our point of departure, therefore, will be the *dispositif* of enclosure, or better, the *double enclosure,* of the body, which Nazism understood both as the chaining of the subject onto his own body and as the incorporation of such a body in that extensive body of the German ethnic community. It is only this last incorporation, which is radically destructive of everything that is held not to be a part of it, that also confers on the subject's body that spiritual substance that has the value of the absolute coincidence of the body with respect to itself. Naturally, this powerful ideologemme is an integral part of a biopolitical design that is already predisposed for such a paroxysmal outcome. This, however, doesn't change the fact that in it merges, or exerts an influence on, a vector of broader meaning (but also more ancient) that is part of the already classical metaphor of "political body" and, more generally, on the relation between politics and body. What I want to say is that each time the body is thought in political terms, or politics in terms of the body, an immunitary short-circuit is always produced, one destined to close "the political body" on itself and within itself in opposition to its own outside. And that is irrespective of the political orientation—either right or left, reactionary or revolutionary, monarchical or republican—to which such an operation pertains. In each of these cases, in fact, what constitutes the features either of the absolutist-Hobbesian or the democratic-Rousseauian line (without introducing genealogies even more remote in time) is the organistic model that joins every member of the body to its assumed unification. Even in contractual theories in which the political body is presented as the result of an agreement between multiple individual wills, or as the outcome of a single general will, the political body in reality is precedent to and propaedeutic to their definitions of it. It is because the political body is already inscribed in a single body that its parts can or must be consolidated in an identical figure whose object precisely is the self-preservation of the political organism as a whole. Despite all of the autonomistic, individualistic, and fragmenting impetuses that have periodically ensnared (or contradicted) this general process of incorporation, its logic has largely prevailed in the constitution and the development of nation-states, at least until modern political categories will be able to elaborate productively their own immunitary function of the negative protection of life.[18]

Then, when such a mechanism breaks down, or when the immunitary demands grow until it overflows the banks of modern mediation, totalitarianism, and in particular Nazism, produced an additional enclosure of the body on itself through a double movement. On the one side, it made absolutely coterminus political identity with the racial-biological; on the other, it incorporated into the same national body the line of distinction between inside and outside, which is to say between the portion of life that is to be preserved and what is to be destroyed. The individual and collective body—the one in the other and the one for the other—was immunized in this way, before and beyond the outside and its own surplus or lines of flight. These emerged as interrupted by a refolding of the body on itself that had the function of providing a spiritual nucleus or a surplus of meaning, to what was also considered to be absolutely biological. The concept of the political body was made functional to this direct tradition of life in politics as its antithesis, more so than to what is outside it, namely, to that part of itself judged to be not up to *[inidonea]* a similar bio-spiritual conversion. We previously saw how the first name that the Nazis gave to such an abject material was that of "existence" (because it was resistant to the double corporeal subsumption); "existence without life" is considered to be all that does not have the racial qualifications necessary to integrate ethnically the individual body with that of the collective.[19] But perhaps a more meaningful term is that of *flesh,* because it is intrinsic to the same body from which it seems to escape (and which therefore expels it). Existence without life is flesh that does not coincide with the body; it is that part or zone of the body, the body's membrane, that isn't one with the body, that exceeds its boundaries or is subtracted from the body's enclosing.

Merleau-Ponty is the twentieth-century philosopher who more than any other elaborated the notion of flesh. To recognize in his work a specific feature of the biopolitical reflection or even only an enervation of *bíos* would certainly be misleading, given the substantially phenomenological scope in which his philosophical considerations are situated.[20] This doesn't mean, however, that the theme of flesh tends precisely to exceed it in a direction not so far removed from what we brought together under the Heideggerian thematic of the "factical life." As in that case, so too the horizon of flesh *[chair]* is disclosed in the point of rupture with the traditional modality of philosophy that poses the latter in a tense and problematic relation with its own "non." When in a text titled *Philosophy and Non-Philosophy Beginning with Hegel,* Merleau-Ponty refers to the necessity that "philosophy also becomes worldly," philosophy has already shifted in a conceptual orbit in which the entire philosophical lexicon is subjected to a complete rotation on its own axis.[21] It is in this radical sense that the proposition according to which "what we are calling flesh, this interiorly worked-over mass, has no name in any philosophy" is to be understood.[22] It has no name because no philosophy has known how to reach that undifferentiated layer (and thus for this reason exposed to difference), in which the same notion of body, anything but enclosed, is now turned outside *[estroflessa]* in an irreducible heterogeneity. What this means is that the question of flesh is inscribed in a threshold in which thought is freed from every self-referential modality in favor of directly gazing on contemporaneity, understood as the sole subject and object of philosophical interrogation. From this point of view, the theme of flesh lends itself to a symptomatic reading that can also push beyond the intentions expressed by Merleau-Ponty because it is rooted therefore within the series of questions that his philosophy opened with a lexical originality at times unequaled by Heidegger himself. Without wanting in any way to propose an inadmissible comparison between the two, one could say instead that the blind point of Heidegger's analysis of *bíos* is born precisely from a missing or inadequate encounter with the concept of flesh.

Didier Franck's thesis is that Heidegger's wasn't able to think fully the notion of flesh because it is a category that is constituted spatially, and that therefore appears to be irreducible to the temporal modality that Heidegger traced in being.[23] Now, it is precisely at this point that Merleau-Ponty introduces a different perspective, beginning with an approach (but also a semantics) that is more traceable to Husserl than to Heidegger. It is from Husserl in fact that Merleau-Ponty infers not only the theme of the reversibility between sentient and felt *[senziente e sentito]*, but also that of a relation of otherness that is destined to force open the identity presupposed by the body proper. When, in a fragment from *The Visible and the Invisible*, he writes that "my body is made of the same flesh as the world (it is perceived), and moreover that this flesh of my body is shared by the world," he takes another step that brings him into a semantic range that is situated beyond both phenomenology and an existential analytic.[24] That the world is the horizon of meaning in which the body recognizes itself and which is traversed by the diversity that keeps it from being coterminous with itself, means that it has surpassed not only a Husserlian transcendentalism but also the Heideggerian dichotomy between existence and life.[25] If, for Heidegger, *bíos* does not recognize any of the modes of being that distinguish a fundamental ontology, in Merleau-Ponty it is precisely living flesh that constitutes the tissue of relations between existence and the world. Here, then, not only does the spatiality of flesh allow us to recuperate a temporal dimension, but it constitutes precisely their tangential point.

> Oppose to a philosophy of history such as that of Sartre . . . not doubtlessly a philosophy of geography . . . but a philosophy of structure which, as a matter of fact, will take form better on contact with geography than on contact with history . . . In fact it is a question of grasping the *nexus*—neither "historical" nor "geographic" of history and transcendental geology, this very time that is space, this very space that is time, which I will have

> rediscovered by analysis of the visible and the flesh, the simultaneous *Urstiftung* of time and space which makes there be a historical landscape and a quasi-geographical inscription of history.[26]

Can we read such a composite of flesh, world, and history in terms of "mondialization"? It would be imprudent to respond absolutely yes (at least considering Merleau-Ponty's personal journey). But it would be equally reductive to deny that he is the author who pushed further than others the theoretical declination of the relation between body and world. Not only, but he, before any one else, also understood that the enlargement of the body to the dimension of the world (or the configuration of the world as a singular body) would fragment the same idea of "political body," in its modern as well as in its totalitarian declensions. This is for no other reason than because, not having anything outside itself (and for that reason making it one with its own outside), such a body wouldn't be able to be represented as such—doubling upon itself in that self-identical figure, which, as we saw, constitutes one of the most terrible immunitary *dispositifs* of Nazi biocracy. For us as well as for Merleau-Ponty, the flesh of the world represents the end and the reversal of that doubling. It is the doubling up *[sdoppiamento]* of the body of all and of each one according to leaves that are irreducible to the identity of a unitary figure: "It is because there are these 2 doublings-up that are possible: the insertion of the world between the two leaves of my body [and] the insertion of my body between the two leaves of each thing and of the world."[27] That the fragment—already marked by the reference to the "thing" as the possible bridge between body and world—continues with reference to a perspective that "isn't anthropologism," further attests to the lateral move that Merleau-Ponty makes with regard to Heidegger. In the same moment in which Merleau-Ponty distances himself from anthropology (in a direction that, even if indirectly, refers to a Heideggerian ontology), he frees himself from Heidegger's ontology by assuming in the place of an object/subject not only every form of life from the human to the animal, but especially (or even) what was that "poor of the world" situated in unsurpassable remoteness from the universe of *Dasein*.[28] Again, by alluding to a "participation of the animal in our perceptive life and to the participation of our perceptive life in animality," Merleau-Ponty penetrates more deeply than Heidegger does into the most devastating imaginary of our epoch, expressing himself more forcefully against it.[29] Inscribing the threshold that unites the human species with that of the animal in the flesh of the world, but also the margin that joins the living and the nonliving,

Merleau-Ponty contributes to the deconstruction of that biopolitics that had made man an animal and driven life into the arms of nonlife.

We might be surprised that the theme of flesh, which Merleau-Ponty took up in the 1950s, remained on the margins of contemporary philosophical debates, and even more that it was treated coolly and with a certain diffidence on the part of many from whom more attention and interest might have been expected.[30] If for Lyotard the evocation of the chiasmus that flesh operates between body and world runs the risk of slipping into a "philosophy of erudite flesh," closed to the onset of the event, Deleuze sees in the "curious Fleshism" of more recent phenomenology not only a feature that deviates from what he himself defines as the "logic of sensation," but both "a pious and a sensual notion, a mixture of sensuality and religion."[31] As for Derrida, aside from the philological perplexities that he advances on the translation of the French *chair* [flesh] into the German *Leib,* he doesn't hide his fear that an immoderate use of the term can give rise to a sort of generic "globalization *[mondalisation]* of flesh": "By making flesh ubiquitous, one runs the risk of vitalizing, psychologizing, spiritualizing, interiorizing, or even reappropriating everything, in the very places where one might still speak of the nonproperness or alterity of flesh."[32] But it is perhaps Jean-Luc Nancy, to whom Derrida's texts were, however, dedicated, who expresses the most important reservation in relation to the discourse that I've traced here. This is because in the same moment in which Nancy clearly distances himself from the philosophy of flesh, he juxtaposes the urgency of a new thought of the body to it: "In this sense, the 'passion' of the 'flesh,' in the flesh, is finished—and this is why the word *body* ought to succeed on the word *flesh,* which was always overabundant, nourished by sense, and egological [égologique]."[33]

Why such a broad rejection? And to what do we owe an opposition so marked as to assume the features of a true incomprehension of what flesh signifies in the theoretical scheme I sketched above? Agitating in it certainly is an irritability on the part of contemporary French philosophy with regard to the phenomenological tradition.[34] But this particular element is not to be separated from a more general demand of differentiation in relation to the Christian conception of flesh. Indeed, one could say that it is precisely the Christian origin *[ascendenza]* (which is in no way secondary to phenomenology) that constitutes the true objective of the antiflesh polemic. If Michel Henry's most recent essay on incarnation is taken as a site of possible comparison, the terms of the question can be identified with sufficient clarity.[35] What is seen as problematic in the phenomenological (but also, eventually, in the ontological) concept of flesh is its spiritualistic connotation, which becomes evident in Henry's interpretation itself: without entering too much into the details of the question, what differentiates the flesh of the opaque and inert material of the body is its self-affectivity, which the divine Word directly transmits *[trasmessale]*. When Derrida polemicizes about an excessive fleshiness *[carnista]* that risks canceling the concreteness of the body, or when Nancy sees in incarnation a process of disembodiment and interiorization that subjects the corporeal sign to the transcendence of meaning, they do nothing other than reaffirm this spiritualistic characterization of flesh. So doing, they end up offering the same reading that Henry does, even if with the opposite intention, which is not more positive but now negative. Rather than deconstructing and overturning it in its hermeneutic effects (as one might have expected them to do), they assume the conclusions and for that reason only spurn the object. If flesh refers to the body translated into spirit, or to spirit that is introjected into the body, the path for an effective rethinking of bodies (of each body and of all bodies) moves through the definitive abandonment of the philosophy of the flesh.

Such a reasoning has its power, which rests, however, on a premise that is anything but certain—certainly, with reference to Merleau-Ponty, for whom, as we saw, *flesh* doesn't refer at all to an interiorization of the body, but if anything to its exteriorization in another body (or even in that which is not a body), but also with reference to the same Christianity, which only in exceptional circumstances links the term flesh (*sarx* or *caro)* to a spiritual dimension, which usually relates instead to the idea of body *(sōma, corpus)*. Even if the two words at a certain moment come to be partially superimposed, certainly what refers most precisely and intensely to the soul as its privileged content is the body and not flesh.[36] Flesh, for its part, finds its own specificity in the material substrate of which man is initially "made" (even before his body is filled with spirit). It is no coincidence that in Judaism (and not so differently in Greece), it is precisely the *flesh* (*basar)* that tangibly represents earthly elements and therefore suffers and is perishable. Early Christianity takes up and develops this terminology.[37] In Paul (2 Corinthians 4:11), *thnētē sarx* is the mortal existence that is exposed to pain and to sin, just as the expression "in the flesh" *(en sarki)* alludes precisely to earthly life as such, to the point where sometimes (Romans 3:20 and Galatians 2:16, in a citation from Psalms 143:2), Paul adopts the formulation *pasa sarx,* which means "every living thing *[vivente].*" It is true that the word *sōma* and then *corpus* can have analogous meanings, but more often than not it refers to the general unity of the single organism or of the collective (the church, Christianity) in which the first is positioned. As for Tertullian, the author of *De carne Christi,* he wages a difficult apologetical battle against those (Valentino, Marcione, Apelle) who argued for the spiritual or pneumatic character of Christ's flesh. His thesis instead was that while the *corpus* can be immaterial, celestial, and angelic, *caro* instead is clearly distinguished from the soul or the psyche. There does not exist a *caro animalis* [soul-flesh] or an *anima carnalis* [flesh-soul] *(nusquam animam carnem ut carnem animam)* [never soul-flesh or flesh-soul] (*De carne Christi,* XIII, 5), but only the unity, *in the body,* of two unmistakable substances that are different in and of themselves.

This notion of a material-like, inorganic, and "savage" flesh, as Merleau-Ponty would have called it, has never had a political configuration. It indicates a vital reality that is extraneous to any kind of unitary organization because it is naturally plural.[38] Thus, in Greek the term *sarx* is usually declined with the plural *sarkes,* and the expression *pasa sarx* that I noted earlier preserves a connotation of irreducible multiplicity that can be rendered with "all men" *[uomini].* So that this might set in motion the general process of constituting the Christian church, it was necessary that the diffused and dispersed flesh be reunited in a single body.[39] It so happened that we previously find in Paulian Christianity, and later in the Patristic, that the words *sōma* and *corpus* begin to displace those of *sarx* and *caro* with ever greater frequency (without ever completely replacing them). More than an expulsion of the flesh, this concerns its incorporation into an organism that is capable of domesticating flesh's centrifugal and anarchic impulses. Only the spiritualization of the body (or better, the incorporation of a spirit that is capable of redeeming man from the misery of his corruptible flesh) will allow him entrance into the mystical body of the church: "What? Know ye not that your body is the temple of the Holy Ghost *which* is in you, which ye have of God, and ye are not your own? For ye are bought with a price; therefore glorify God in your body, and in your spirit, which are God's" (1 Corinthians 6:19–20).[40] The role that the sacrament of the Eucharist had in this salvific passage from flesh to body has been noted as the double extravasation *[travaso]* of the body in Christ in that of the believer and of that of the believer in the ecclesial body. With all the variants as well as the conflicts that are derived from an initial competition, we can say that first the empire and then the nascent nation-states activated and secularized the same theological-political mechanism; but also here they did so in order to save *[riscattare]* themselves from the risk associated with "bare life," which is implicit in that extralegal condition defined as the "state of nature"—namely, the "flesh" of a plural and potentially rebellious multitude that needed to be integrated in a unified body at the command of the sovereign.[41]

The biopolitical transition that characterizes modernity advanced by this perspective didn't modify such a "corporative," as is also demonstrated on the lexical level by the long duration of the metaphor of "body politic." That the strategies of sovereign power are addressed directly to the life of subjects *[sudditi]* in all their biological requirements for protection, reproduction, and development not only doesn't weaken, but indeed further strengthens, the semantics of a body inherited by medieval political theology. There is nothing more than that body (in the individual and collective sense) that restitutes and favors the dynamic of reciprocal implication between politics and life, and this for a number of reasons. First, because of the somatic representation of legitimate citizenship prior to the growing role that demographic, hygienic, and sanitary questions began to assume for public administration. And second, because it is precisely the idea of an organic body that implicates, as necessary complement, the presence of a transcendent principle that is capable of unifying the members according to a determined functional design: a body always has a soul, or at least a head, without which it would be reduced to a simple agglomerate of flesh. Far from rejecting en masse this figural apparatus, totalitarian biopolitics (but above all Nazi biopolitics) leads it to its extreme outcome, translating what had always been considered nothing more than an influential metaphor into an absolutely real reality: if people have the form and the substance of a body, then they must be looked after *[curato]*, defended, and reinforced with instruments and a finality that are purely biological. They didn't exclude what was traditionally referred to as soul, but they understood it biologically as the carrier of a racial heredity that was destined to distinguish the healthy part from the sick part within the body—the "true" body from a flesh that lacked vital resonance and which therefore was to be driven back to death *[respingere alla morte]*. As we saw previously, this double, bio-spiritual incorporation was the final result of an immunitary syndrome so out of control that it not only destroys everything that it comes into contact with, but turns disastrously on its own body.

We noted already that such an outcome doesn't in fact mark the exhaustion or only the retreat of the biopolitical paradigm. With the end of both twentieth-century totalitarianisms, the question of life remains solidly at the center of all politically significant trajectories of our time. What recedes, however (either because of explosion or implosion), is instead the body as the *dispositif* of political identification. This process of disembodiment is paradoxically the result of an excess. It is as if the extension of the somatic surface to the entire globe makes the world the place (by way of antinomical excellence) in which inside coincides with outside, the convex with the concave, and everything with nothing. If everything is the body, nothing will rigidly define it, which is to say no precise immunitary borders will mark and circumscribe it. The seemingly uncontainable proliferation of self-identical agglomerations that are ever more circumscribed by the function of immunitary rejection of the dynamics of globalization signals in reality the eclipse of the political body in its classical and twentieth-century sense in favor of something else that appears to be its shell and proliferating substance. It is in such a substance that, perhaps for the first time with some political pregnancy, it is possible to discern something like a "flesh" that precedes the body and all its successive incorporations. Precisely for this reason it appears again when the body is in decline. That the Spinozian name of "multitude" or that of Benjamin's "bare life" can be attributed to it is also secondary with respect to the fact that in it *bíos* is reintroduced not on the margins or the thresholds, but at the center of the global *polis.*[42] What the meaning, as well as the epochal outcome, of a relation between politics and life might be (given the same material formation that escapes from the logic of immunitary) is difficult to say, also because such a biopolitical dynamic is inserted in a framework that is still weighed down by the persistence (if not by the militarization) of sovereign power. Certainly, the fact that for the first time the politicization of life doesn't pass necessarily through a semantics of the body (because it refers to a world material that is antecedent to or that follows the constitution of the subject of law *[diritto]*) opens up a series of possibilities unknown till now. What political form can flesh take on, the same flesh that has always belonged to the modality of the impolitical? And what can be assigned to something that is born out of the remains of anomie? Is it possible to extract from the cracks of *immunitas* the outlines of a different *communitas*? Perhaps the moment has arrived to rethink in nontheological terms the event that is always evoked (but never defined in better fashion) that two thousand years ago appeared under the enigmatic title "the resurrection of the flesh." To "rise again," today, cannot be the body inhabited by the spirit, but the flesh as such: a being that is both singular and communal, generic and specific, and undifferentiated and different, not only devoid of spirit, but a flesh that doesn't even have a body.

Before moving on, a final point relative to the modality of incarnation. We know that some have wanted to see in the term "incarnation" the theological bond that keeps phenomenology within a Christianity-derived semantics, and which is therefore fatally oriented toward the spiritual: penetrated by the Holy Spirit, the body of man ends up being disembodied in a dialectic that subordinates the materiality of the corporeal sign to the transcendence of meaning. The body, reduced not to signifying anything other than its own incarnated being, loses that exteriority, that multiplicity, and that opening that situate it in the real world, what in turn will refer to its anthropological, technological, and political dimension.

But is this how things really stand? Or does a similar reconstruction risk making it junior to that post-Christian or meta-Christian nucleus that it would like to deconstruct (without being able to free itself from that post-Christian or meta-Christian nucleus, which has shown through more than once in the present work)? My impression is that such a nucleus coincides in large part with the idea and the practice of incarnation. With regard to the distinction (and also the opposition) vis-à-vis the logic of incorporation: while incorporation tends to unify a plurality, or at least a duality, incarnation, on the contrary, separates and multiples in two what was originally one. In the first case, we are dealing with a doubling that doesn't keep aggregated elements distinct; in the second, a splitting that modifies and subdivides an initial identity. As the great apologetics of the first centuries after Christ argued, the Word that becomes flesh establishes the copresence of two diverse and even opposite natures in the person of Christ: the perfect and complete nature of God and the suffering and mortal nature of man. How can a God alter, disfigure, and expropriate himself to the point of really taking on the flesh of a mortal? The accent here ought to be placed on the adverb *really* because it is precisely there, on the material substantiality of a flesh that is identical to ours in all and for all, that for five centuries the Christian fathers, from Ireneus to Tertullian to Augustine, fought a difficult battle against a series of heresies (Docetism, Aranism, Monophysism,

Nestorianism), each aimed at negating the insurmountable contradiction implicit in the idea of Incarnation: to cancel either the nature of God or that of man and therefore the line they share. What appears logically unthinkable for classical culture is the two-in-one or the one-that-is-made-two through a slippage of the body out of itself, which coincides with the insertion of something within that doesn't naturally belong to it.

Given this transition, this contagion, and this denaturation, the notion of flesh needs to be rethought outside of Christian language, namely, as the biopolitical possibility of the ontological and technological transmutation of the human body. One could say that biotechnology is a non-Christian form of incarnation. What in the experience of prosthesis (of the transplant or the implant) penetrates into the human organism is no longer the divine, but the organ of another person *[uomo]*; or something that doesn't live, that "divinely" allows the person to live and improve the quality of his or her life. But that this new biopolitical feature (which inevitably is technopolitical) doesn't lose every point of contact with its own Christian archetype is witnessed in the artist who, perhaps more than any other, has placed the theme of flesh outside of the body (or of the nonorganic body) at the center of his own work. We know that classical images of the Incarnation, above all at the moment of the Crucifixion, mark a break or a rupture in the figural regime of the *mimesis* in which Christian art is framed—as if not only the Christ (for example, Dürer's), but rather also the entire order of figuration must slip into the open folds of its martyred body, damaged and disfigured, without any possibility of restoration.[43] But the flight of flesh from the body, both barely sustained and strained to the point of spasms by the structure of the bones, constitutes the center itself of the paintings of Francis Bacon, to whom I alluded above. In Bacon too this journey to the limits of the body, this slippage of flesh through its foramen explicitly refers to the ultimate experience of the Christian incarnation: "The images of the slaughterhouse and butchered meat have always struck me," Bacon remembers. "They seemed directly linked to the Crucifixion."[44] I don't know if flesh is to be related to the Nazi violence, as Deleuze would have it in his admirable comment (though the horror of that violence always remained with Bacon).[45] The fact is that in no one more than Bacon is the biopolitical practice of the animalization of man carried out to its lethal conclusion, finding a reversed correspondence perfectly in the disfigured figure of butchered flesh:

> In place of formal correspondences, what Bacon's painting constitutes is a *zone of indiscernibility or undecidability* between man and animal . . . It is never a combination of forms, but rather the common fact: the common fact of man and animal.[46]

According to all the evidence, that "common fact," that butchered, deformed, and chapped flesh, is the flesh of the world. That the painter always saw in animal carcasses hanging in butcher shops the shape of man (but also of himself) signifies that that bloody mound is the condition today of a large section of humanity. But that this recognition didn't ever lead to despair means that in it he glimpsed another possibility, tied to a different mode of understanding the relation between the phantasms of death and the power of life:

> When the visual sensation confronts the invisible force that conditions it, it releases a force that is capable of vanquishing the invisible force, or even befriending it. Life screams *at* death, but death is no longer this all-too-visible thing that makes us faint; it is this invisible force that life detects, flushes out, and makes visible through the scream. Death is judged from the point of view of life, and not the reverse, as we like to believe.[47]

Birth

The second Nazi immunitary *dispositif* to deconstruct with respect to its deadly results is that of suppressing birth. We saw how it presents itself as split in its actualization and how it is dissociated in two vectors of sense that are seemingly contradictory: on the one side, the exhibition and the strengthening of the generative capacity of the German people; on the other, the homicidal fury that is destined inevitably to inhibit it. Scholars have always seemed to have difficulty deciphering the contradiction between a politics of increasing the birthrate and the antinatalism produced first by a negative eugenics and then by the elimination en masse of pregnant mothers. Why did the Nazis commit themselves so eagerly to draining that vital fount of life that they also wanted to stimulate? The biopolitical paradigm furnishes a first response to such a question, identifying precisely the root of the genocidal discrimination in the excess of political investment on life. But perhaps a more essential motivation is to be traced in the nexus (one that isn't only etymological) linking the concepts of "birth" *[nascita]* and of "nation" in an ideological short-circuit that finds its most exasperated expression in Nazism. What kind of relationship did the Nazis institute between birth and nation? How were these superimposed in the name of Nazism—indeed, how were they constituted precisely at their point of intersection?

We know how the term "nation" is almost identical in almost all of the principal modern languages and how it derives from the Latin *natio*, which in turn is the substantive form of the verb *nascor*. Naturally, in order for the modern meaning of nation to become stable, a long process is required that doesn't leave untouched the originary relation with the concept of birth. Without entering into the details, we can say that while for the entire ancient and medieval periods the biological referent in nativity prevails over the political one that is diffused in the concept of nation, in the modern phase the equilibrium between the two terms shifts until it is reversed in favor of the latter. Therefore, if it were possible for a long period to designate as *nationes* groups of people that were joined by a common ethnic provenance (or only by some kind of social, religious, or professional contiguity), afterwards an institutional connotation prevails.[48] It is the genesis and the development of territorial states that mark this passage: in order to take on a political signification, the biological phenomenon of birth (which is impolitical in itself) needs to be inscribed in an orbit of the state that is unified by sovereign power. It was precisely in this way that a notion, which was used generically prior to that moment and often in contrasting ways—it referred to others rather to themselves, as the Roman dichotomy between uncivil and barbarian *nationes* and the *populus* or the *civitas* of Rome attests—came increasingly to assume that powerful charge of self-identification that still today connotes the national ideology. The same Declaration of Human Rights and of the Citizen (as before it habeas corpus) is to be understood in this way: as the unbreakable bond that links the bodies of subjects *[sudditi]* to that of the sovereign. In this perspective we find again the decisive reference to the category of "body." Leaving aside its monarchical, popular, as well as voluntaristic and naturalistic declinations, the nation is that territorial, ethnic, linguistic complex whose spiritual identity resides in the relation of every part to the whole, which is included in it. A common birth constitutes the thread that maintains this body's identity with itself over the course of generations. It is what joins fathers to sons and the living to the dead in an unbreakable chain. It constitutes in its continuity both the biological content and the spiritual form of self-belonging to the nation in its indivisible whole. We are dealing with a relation that isn't unlike what we saw pass between the semantics of flesh and that of the body. Just as the body constitutes the site of the presupposed unification of the anomalous multiplicity of flesh, so the nation defines the domain in which all births are connected to each other in a sort of parental identity that extends to the boundaries of the state.

With respect to this biopolitical dialectic, Nazism marks both a development and a variation; a development because it assigns a value to birth even more important in the formation of the German nation. It isn't only the unbroken line that assures the biological continuity of the people across generations, but also the material form or the spiritual material that destines the German people to dominate all other peoples (given its absolute purity of blood). But here the difference is fixed with respect to national as well as other nationalistic models that precede it. In this case, we can no longer speak of the politicization of a notion (birth, precisely) that was originally impolitical, but rather of a copresence between the biological sphere and the political horizon. If the state is *really* the body of its inhabitants, who are in turn reunified in that of the head, politics is nothing other than the modality through which birth is affirmed as the only living force of history. Nevertheless, precisely because it is invested with this immediate political valence, it also becomes the fold along which life is separated from itself, breaking into two orders that are not only hierarchically subordinated, but also rigidly juxtaposed (as are those of master and slave, of men and animals, of the living and the dead). It is from this perspective that birth itself becomes the object of a sovereign decision that, precisely because it appears to originate directly from it, transcends it, traversing it along excluding lines. This is how the ambivalence of the Nazis with regard to what was born is to be interpreted. On one side, the preventive exaltation of a life that is racially perfect; on the other side, removing the one who is assigned to death by the same statute of what is considered to be living. They could die and needed to die because they had never truly been born. Once identified with the nation, birth undergoes the same fate, as what is also held in a biopolitical clench that cannot be loosened except by collective death.

The same antinomy that characterizes the biopolitical relation between nation and birth is found at the center of the category of fraternity. For at least two centuries now (that is, from when the republican motto of the French Revolution was coined), we know that the notion of fraternity, which is originally biological or naturalistic, acquired an inevitable political resonance. Nevertheless, it is precisely the comparison with the other two truly significant words with which it is associated that reveals a *deficit* of theoretical elaboration. If liberty and equality have been analyzed, discussed, and defined at length, fraternity emerges as one of the terms least thought about by the political-philosophical tradition. Why? Why is the one that would appear to be the most comprehensible of the three concepts still unanalyzed? A first response to this question is to be sought in its originally impolitical characteristic (when not explicitly theological) that has blocked any kind of historical translation. Leaving aside their ancient roots, liberty and equality are constituted in the modern period and originate with the two great political traditions that are liberalism and socialism. This isn't the case with fraternity, whose fortune seems limited and completely consumed in the brief arc between 1789 and 1848. Indeed, with respect to the other two principles of the Revolution, fraternity is what is established later. Although previously enunciated in 1789, it only begins to appear in official documents between 1792 and 1793 when France, attacked on every side and threatened internally, needed to find words and symbols capable of calling all to the indivisible unity of the nation against its enemies. It is then that the term becomes the fundamental and founding principle with respect to the other two, which now emerge as subordinated to it both historically and logically. Only if all Frenchmen will force themselves into a single will can the nation obtain liberty and equality for itself and for those who will follow its example.[49]

Here is sketched a second and more essential motivation for the political-philosophical unthinkability of the category of fraternity.[50] Political philosophy doesn't fully grasp it not only because it is impolitical, but also because it is intensely biopolitical. This means that fraternity isn't subtracted from the concept because it is too universal, abstract, and millenarist as one might think, but, on the contrary, because it is too concrete, rooted directly in the natural *bíos.* The fact that it takes on strong national connotations in the same moment of its emergence on the political scene (as well as a nationalistic one as it appeals to the sacrality of the French nation) contrasts in some way with its supposed universalism. Unless one wants to argue (as not only Robespierre and Saint-Just did, but also Hugo and Michelet) that France represents the universal because it is the country around which the entire history of the world turns—only to discover quickly that all the people that were to be buggered with such a conviction wound up inevitably assuming the same for themselves. At stake (much more than universal abstractions of common justice) was, in reality, the reference to a self-identification founded on the consanguinity of belonging to the same nation. More than "phratry," fraternity essentially refers to the fatherland *[patria]*; it confirms the biological bond that joins in a direct and masculine lineage the brother to the father (the "motherland" *[madrepatria]* has always had symbolic connotations of virility). Now, if it is true that democracy is often referred to the idea of brotherhood, that is because democracy, like all modern political concepts, rests on a naturalistic, ethnocentric, and androcentric framework that has never been fully interrogated. What precisely is a "fraternal democracy"? Certainly, sublime accents can be heard in similar expressions: a reference to substantial values that move beyond the formalism of equal rights. Yet something different also resonates here and with a more troubling timbre. It isn't the same thing to hold that all men ought to be equal because they are brothers or that they are brothers because they are equal. Despite appearances, the category of brotherhood is more restricted and more particularistic; it is more exclusive than that of equality in the specific sense that it excludes all those who do not belong to the same blood as that of the common father.[51]

This perspective makes visible another decisive feature of the idea of brotherhood. The fact that at the moment of its maximum diffusion it was invoked *against* someone, or even all of the non-French, reveals a conflictual, when not bellicose, attitude that has been always hidden by its usual pacifist coloration. Moreover, the figure of the brother (which a long tradition from Plato to Hegel and beyond associated with that of the friend) had and has to do with the enemy, as both Nietzsche and Schmitt argued.[52] They explained that the true brother (and for that reason the true friend) is precisely the enemy because only the enemy truly puts someone to the test. The enemy confers identity through opposition; he reveals the borders of the other and therefore also one's own borders. From Cain and Abel to Eteocle and Polinice to Romulus and Remus, absolute enmity, which is to say fratricide, has always been figured through the couple of the brother, or even of twins, as René Girard demonstrates when he sees the bloodiest conflict always erupting between close relatives and neighbors.[53] One could say that blood calls forth blood. And whether metaphorically or literally, blood becomes the principle of politics, politics always risks slipping into blood.

This was Freud's conclusion, the author who perhaps more than any other decrypted the paradox of fraternity. As he tells it in *Totem and Taboo*, one day the brothers unite, oppressed by a tyrannical father.[54] They kill him and they devour his flesh, taking his place. This signifies, in the first instance (and according to a more "enlightened" interpretation), that the process of civilization is connected to the substitution of a despotic authority, indeed, to the same principle of authority, with a democratic universe in which the power that is shared by the many replaces the power of the One. In this sense, democracy emerges as both the cause and the effect of the passage from vertical domination to a horizontal one, precisely from Father to sons. But in a closer and less ingenuous analysis, Freud's allegory exhibits a more troubling truth, namely, the perpetuation of the paternal domination even inside the democratic horizon of the brothers. What else would brothers literally incorporating the dead father into themselves mean, if not that they are inexorably destined to reproduce the distinctive features (even if in a plural and domesticated form)? The fact that from such an act morality *[l'attegiamento morale]*, which is to say the sense of guilt for the homicide they have committed and the respect with regard to the Law, signifies that the act remains marked by that traumatic event, by the killing of someone who doesn't actually disappear from the scene, but is perpetually regenerated in the line of descent from brothers to sons. Once again the difference is prisoner of the repetition and the dead once again reach out and grab hold of the living.

Yet *Moses and Monotheism* is the Freudian text that most forcefully invests the biopolitical superimposition of birth and nation.[55] That it refers on several occasions to *Totem and Taboo* (following to some degree the structural schema) need not hide the *political* as well as the philosophical novelty of an essay written in three phases between 1934 and 1938: these dates are enough to indicate the adversary to whom it is addressed. It concerns Nazi anti-Semitism as it is constituted precisely along a genealogical line that joins national identity to the founding moment of its origin. In different fashion from those who refused to confront the Nazi *dispositif,* who limited themselves simply to invalidating the naturalistic presupposition, Freud met the challenge on the same terrain. In other words, he doesn't contest the connection made by Nazism between the form assumed by a people and the origin of its founder. It is true that the national community finds its own identifying foundation in the act of its birth and therefore in the birth of its most ancient Father—yet precisely because to call into question its purity and property also means to fundamentally undercut the self-identifying mechanism of the people of which it is a part. This is exactly the strategy Freud uses in *Moses and Monotheism.* He understands perfectly well the risk that he is running as is evinced by the substantial series of warnings, precautions, and reservations disseminated throughout the text as if to defend it from something close by that threatens it. When he warns in the preamble that "to deny a people the man whom it praises as the greatest of its sons is not a deed to be undertaken lightheartedly," he intends to warn the reader that he is pushing up against the adversary's position to such a degree that he risks entering in a zone of indistinction with it.[56] Why? Why precisely was it that Nazism expropriated their identity from the Jewish people, denying that they might have a form, a type, or be a race? How can one carry out this kind of expropriation, denying them even a founder by attributing a different nationality to the founder, without converging on the same anti-Semitic thesis? Why not just categorically oppose it? The opening that Freud has created is in effect rather narrow. It doesn't concern lessening the relation of the origin with regard to the Jewish people (and by extension to every people), which would mean adhering to the historicist thesis against which Nazism will have no difficulty in establishing its radical position. Rather, it concerns placing the same notion of origin under an operation of deconstruction that decenters and overturns it into its opposite: in an originary in/origin that, far from belonging solely to itself, splits from itself, divides into its own other, and thus in the other from its own *[nell'altro da ogni proprio]*.

This is the political significance of the Egyptian Moses. Freud doesn't contest that Moses founded his people; indeed, Freud supports this view with greater force than is traditionally held. But he argues that Moses was able to do it—that is, *create* a people—precisely because he did *not* belong to them, because he was impressed with the mark of the foreigner and even of the Enemy, of whom he is the natural son. It is exactly for this reason—that he was the son of the Jewish people—that he can be their Father and that he can form them according to law proper, which is to say the law of another *[di un altro]*, when not also the law of the other *[dell'altro]*.[57] However, with the relation between ethnic identity of the nation and the birth of its fathers secured (which Nazism insisted upon *in primis*), this means that that people (and therefore every people) can no longer claim the purity of their own race, which is already contaminated by a spurious origin. Not only, but no people can define themselves as the elect, as the Jewish people had first done, and then later the German people (albeit certainly in very different fashion). No people will be able to name themselves as such, that is, furnished with a national identity that is transmitted from father to son, from the moment in the archetype of Moses, in which the father is not the true father, which is to say the natural father, and whose sons are not his true sons—arriving at a point in which these Jewish sons with tremendous effort tried to free themselves from their unnatural father, killing him exactly as the brothers of the primitive hordes in *Totem and Taboo* did. Afterward, inevitably, they bowed to another law, or the law of the other, brought to them by what will be subsequently altered by Christianity. What remains in this uninterrupted sequence of metamorphosis and betrayal is the originary doubling of the Origin, or its definitive splitting in a binary chain that simultaneously unites and juxtaposes two founders, two peoples, and two religions, beginning with a birth that is itself double (just as is biologically, after all, every birth). Anything but ordered toward unifying the two (or the many in the one), birth is destined to subdivide the one (the body of the mother) into two, before the subsequent births in turn multiply those in the plurality of infinite numbers. Rather than enclosing the extraneousness within the same biological or political body (and so canceling it), birth now puts *[rovescia]* what is within the maternal womb outside. It doesn't incorporate, but excorporates, exteriorizes, and bends outside *[estroflette]*. It doesn't assume or impose but exposes someone (male or female) to the event of existence. Therefore, it cannot be used, in either a real sense or a metaphoric sense, as protective apparatus for the self-protection of life. At the moment in which the umbilical cord is cut and the newborn cleaned of amniotic fluid, he or she is situated in an irreducible difference with respect to all those who have come before.[58] With regard to them, he or she emerges as necessarily extraneous and also foreign *[straniero]*, similar to one who comes for the first time and always in different form to walk the earth. This is precisely the reason why the Nazis wanted to suppress birth, because they felt and feared that, rather than ensuring the continuity of the ethnic filiation, birth dispersed and weakened it. Birth reveals the vacuum and the fracture from which the identity of every individual or collective subject originates. Birth is the first *munus* that opens it to that in which it does *not* recognize itself. Annihilating birth, the Nazis believed that they were filling up the originary void, that they were destroying the *munus* and so definitively immunizing themselves from their traumas. It is the same reason (albeit with perfectly reversed intensity) that pushes Freud to place it at the center of his essay: not to force the multiplicity of birth into the unitary calculation of the nation, but rather to place the alleged identity of the nation under the plural law of birth.

Hannah Arendt takes the same route at war's end. We already know that her work cannot be situated within a proper biopolitical horizon (if such an expression were to connote a direct implication between political action and biological determination). The body, insofar as it is body—which is to say, like an organism it is subjected to the natural demands of protection and development of life—is radically extraneous to a politics that assumes meaning precisely by freeing itself from the order of necessity. Yet it is precisely on the basis of such an extraneousness with respect to the biopolitical paradigm that the *political* relevance that Arendt attributes to the phenomenon of birth gains more prominence. If there is a theme that recurs with equal intensity in all her texts, it is really this political characterization of birth or the "natal" features of politics. Writing against a long tradition that situates politics under the sign of death, Arendt refers precisely to the immunitary line inaugurated by Hobbes (not without an oblique glance at Heidegger's being-for-death). What she insists upon is the originary politicity of birth: "Since action is the political activity par excellence, natality, and not mortality, may be the central category of the political, as distinguished from metaphysical thought."[59] If the fear of death cannot produce anything but a conservative politics, and therefore be the negation itself of politics, it is in the event of birth that politics finds the originary impulse of its own innovative power. Inasmuch as man had a beginning (and therefore is himself a beginning), he is the condition of beginning something new, of giving life to a common world.[60]

Here Arendt seems to open a perspective in political ontology that does not coincide either with Greek political philosophy or with modern biopolitics, but refers rather to Roman usage along a line that joins the creationism of Saint Augustine to the Virgilian tradition. Birth, in a way that is different from the creation of the world (which occurred one time on the part of a single creator), is a beginning that repeats itself an infinite number of times, unraveling lines of life that are always different. It is this differential plurality that is the point in which the Arendtian political ontology is separated (or at least is placed on a different plane with respect to biopolitics). In both cases, politics assumes meaning from a strong relationship with life; but while biopolitics refers to the life of the human species in its totality or to that of a particular species of man, the object of political ontology is the individual life as such, which is to say that politics is constituted in the doubled point of divergence or noncoincidence of the individual life with respect to that of the species, as well as the single action vis-à-vis the repeated course of daily life (which is marked by natural needs).

> Yet just as, from the standpoint of nature, the rectilinear movement of man's life-span between birth and death looks like a peculiar deviation from the common natural rule of cyclical movement, thus action, seen from the viewpoint of the automatic processes which seem to determine the course of world, looks like a miracle . . . The miracle that saves the world, the realm of human affairs, from its normal, "natural" ruin is ultimately the fact of natality, in which the faculty of action is ontologically rooted. It is, in other words, the birth of new men and the new beginning, the action they are capable of by virtue of being born.[61]

At this point we cannot help but see the antinomy on which rests the entire discourse in relation to the question of *bíos*. It is clear that Arendt endeavors to keep politics sheltered from the serial repetition that tends to subject politics to natural processes and then to historical processes as well, which are ever more assimilated to the former. What is surprising, therefore, is the choice, which she often stresses, of assuming a differential element with respect to the homogeneous circularity of biological cycle, precisely a biological phenomenon that is in the final, and indeed in the first, instance, birth. It is as if, notwithstanding her refusal of the biopolitical paradigm, Arendt was then brought to use against biopolitics a conceptual instrument that was extracted from the same material—almost confirming the fact that today biopolitics can be confronted only from within, across a threshold that separates it from itself and which pushes it beyond itself. Birth is precisely this threshold. It is the unlocalizable place in space or the unassimilable moment in the linear flowing of time in which *bíos* is placed the maximum distance from *zōē* or in which life is given form in a modality that is drastically distant from its own biological bareness *[nuditità]*. That the reflection on the relationship between life and birth emerged in a monumental book on totalitarianism, which is to say in a direct confrontation with Nazism, is perhaps not unrelated to this paradox. Wanting to institute a political thought that is radically counterposed to Nazi biopolitics, Arendt, like Freud before her (but in more explicit fashion), attacks precisely the point at which Nazism had concentrated its own deadly power. As Nazism employed the production and with it the suppression of birth so as to dry up the source of political action, so does Arendt recall it in order to reactivate it. But there's more. Just as Nazism made birth the biopolitical mechanism for leading every form of life back to bare life, in the same way Arendt sought in it the ontopolitical key for giving life a form that coincides with the same condition of existence.

It has been said the perspective opened by Arendt rests on a profound antinomy relative to the theme of *bíos politikos*. It appears cut by a caesura that links the two terms in the form of a reciprocal diversity. It is true that politics, just like every human activity, is rooted in the naturalness of life, but according to a modality that assigns meaning to it precisely because of the distance from it. Birth constitutes the point at which one sees more powerfully the tension between terms united by their separation: it is the glimmering moment in which *bíos* takes up distance from itself in a way that frontally opposes it to *zōē*, that is, to simple biological life. Although birth is innervated in a process—that of conception, gestation, and parturition—that has to do directly with the animality of man, Arendt thinks birth is what distinguishes man most clearly from the animal, what exists from what lives, politics from nature. Despite all the distance she takes up from her former teacher, one can't help but sense in this political ontology a Heideggerian tonality that ends up keeping her on this side of the biopolitical paradigm. The same reference to birth doesn't appear able (except in metaphoric and literary terms) to penetrate into the somatic network between politics and life. Out of what vital layer of life is the politicity of action generated? How are the individual and genus linked in the public sphere? Is it enough in this regard to evoke the dimension of plurality without making clear beforehand its genesis and direction?

A diagonal response to this series of questions is contained in the work of an author who is less prone to directly interrogating the meaning of politics, and so precisely for this reason more likely to root it in its ontogenetic terrain. I am speaking of Gilbert Simondon, whose thematic assonance with Bergson and Whitehead (without returning to Schelling's philosophy of nature) shouldn't hide a more essential relation with Merleau-Ponty, who dedicates his essay *L'individu et sa genèse physico-biologique* to Simondon, or with George Canguilhem along a vector of sense that we will analyze shortly.[62] Without wanting to give an account of Simondon's entire system of thought, the points that have to do directly with our analysis (precisely the interrogatives that Arendt left open) are essentially two and are tightly connected between them. The first is a dynamic conception of being that identifies it with becoming and the second an interpretation of this becoming as a process of successive individuations in diverse and concatenated domains. Writing against monist and dualist philosophies that presuppose an individual that is already fully defined, Simondon turns his attention to the always incomplete movement of the individual's ontogenesis. In every sphere, be it physical, biological, psychic, or social, individuals emerge from a preindividualistic foundation that actualizes the potentialities without ever arriving at a definitive form that isn't in turn the occasion and the material for further individuation. Every individual structure, at the moment of its greatest expansion, always preserves a remainder that cannot be integrated within its own dimension without reaching a successive phase of development. And so, as the biological individuation of the living organism constitutes the continuation on another level of incomplete physical individuation, in turn psychic individuation is inscribed in a different position, which is to say in the point of indeterminacy of the biological individuation that precedes it.

What can we conclude from this with regard to our problem? First of all, we can say that the subject, be it a subject of knowledge, will, or action as modern philosophy commonly understands it, is never separated from the living roots from which it originates in the form of a splitting between the somatic and psychic levels in which the first is never decided *[risolve]* in favor of the second. Contrary to the Arendtian caesura between life and condition of existence (which is already Heideggerian), Simondon argues that man never loses his relation with his living being. He is not other from living (or more than living), but a *living human [vivente umano]*. Between the psychic and biological, just as between the biological and the physical, a difference passes through not of substance or nature but of level and function. This means that between man and animal—but also, in a sense, between the animal and the vegetal and between the vegetal and the natural object—the transition is rather more fluid than was imagined, not only by all the anthropologisms, but also by the ontological philosophies that presumed to contest them, by reproducing instead, at a different level, all their humanistic presuppositions. According to Simondon, with respect to the animal, man "possessing extensive psychic possibilities, particularly thanks to the resources of symbolism, appeals more frequently to psychism . . . but there is no nature or essence that permits the foundation of anthropology; simply a threshold is crossed."[63] Simondon defines crossing this threshold—which shouldn't be interpreted either as a continuous passage or as a sudden transition of nature—in terms of "birth." And so when he writes that "precisely speaking there is no psychic individuation, but an individuation of the living that gives *birth* to the somatic and the psychic," we need to take the meaning of that expression rather literally.[64] Every step in each phase, and therefore every individuation, is a birth on a different level, from the moment that a new "form of life" is disclosed, so that one could say that birth isn't a phenomenon of life, but life is a phenomenon of birth; or also that life and birth are superimposed in an inextricable knot that makes one the margin of opening of the other:

> The individual concentrates in himself the dynamic that gives birth to him and which perpetuates the first operation in a continuous individuation; *to live is to perpetuate a birth that is permanent and relative.* It isn't sufficient to define the living as an organism. The living is an organism on the basis of the first individuation; but it can live only if it is an organism that organizes and is organized through and across time. The organization of the organism is the result of a first individuation that can be called absolute. But the latter more than life is the condition of life; it is the condition of that perpetual birth that is life.[65]

Here Simondon completely reverses the suppression of birth that the Nazis employed as the *dispositif* for biopolitically reconverting life into death—not only by guiding all of life back to the innovative potential of birth, but by making out of it the point of absolute distinction with regard to death. If one thinks about it, life and birth are both the contrary of death: the first synchronically and the second diachronically. The only way for life to defer death isn't to preserve it as such (perhaps in the immunitary form of negative protection), but rather to be reborn continually in different guises. But the intensity of the relation that Simondon fixes between politics and *bíos,* which is to say between biological life and form of life, doesn't end here. The selfsame fact that birth is reproduced every time the subject moves beyond a new threshold, experiencing a different form of individuation, means that birth deconstructs the individual into something that was prior to, but also contemporaneously after, him. Psychic life cannot actualize the potential preindividual except by pushing him to the level of the transindividual, which is to say by translating him and multiplying him in the sociality of the collective life. The transindividual—what for Simondon constitutes the specific terrain of ethics and politics—maintains a dynamic relation with that of the preindividual, who, unable to be individualized, is precisely "placed in common" in a form of life that is richer and more complex. This means that the individual (or better, the subject) that is produced by individuating itself is not definable outside of the political relationship with those that share the same vital experience, but also with that collective, which far from being its simple contrary or the neutralization of individuality, is itself a form of more elaborate individuation. Nowhere more than here do plurality and singularity intersect in the same biopolitical node that grabs hold of politics and life. If the subject is always thought through the form of *bíos,* this in turn is inscribed in the horizon of a *cum* that makes it one with the being of man.

The Norm of Life

Nazism's third immunitary *dispositif,* in whose overturning are to be found the features of an affirmative biopolitics, is constituted by the *absolute normativization of life.* That the Nazis completely normativized life is not something current interpretation allows for. Yet couldn't one object that the uninterrupted violation of the normative order characterized Hitlerian totalitarianism and that such a distortion of natural right *[diritto]* was effectuated precisely in the name of the primacy of life over every abstract juridical principle? Actually, although both these objections contain a kernel of truth, they do not contradict (except apparently) the proposition with which I began these reflections. As to the first question—the constitutively illegal character of Nazism—and without wanting to give minimum credit to the self-interested opinions of Reich jurists, things are nevertheless more complex than they might seem at first. Certainly, from a strictly formal perspective, the never-revoked decree of February 1933 with which Hitler suspended the articles of the Weimar Constitution on personal liberty situates the twelve years that follow clearly in an extralegal context. And yet—as also emerges from the double-edged statute of the concept of the "state of exception" (which one can technically use to refer to that particular condition), a situation of extralegality isn't necessarily extrajuridical. The suspension of the effective *[vigente]* law is a juridical act, even if of the negative sort. As others have argued, the state of exception is more than a simple normative lacuna; it is the opening of a void in law intended to safeguard the operation of the norm by temporally deactivating it.[66] Moreover, not only did the Nazis formally let the complexities of the Weimar Constitution remain in force—albeit exceeding it in every possible way—but they even demanded that the Constitution be "normalized" by reducing the use of the emergency decree that had been abused by the preceding regime. This explains the cold welcome that Schmittian decisionism received on the part of the regime once it was in power. What Nazism wanted was not an order subtracted from the norm on the basis of a continuous series of subjective decisions, but, on the contrary, to ascribe them to a normative framework that was objective precisely because it originated from the vital necessities of the German people.

This last formulation takes us back to the more general question of the relationship between norm and life in the Nazi regime. Which of the two prevailed over the other to such an extent as to make it function on the basis of its own demands? Was it life that was rigidly normativized, or rather, does the norm emerge as biologized? Actually, as we saw in the preceding chapter, the two perspectives are not juxtaposed but rather integrated in a gaze that includes them both. In the moment in which one appeals to the concept of concrete, substantial, and material law against what is subjective and what belongs to the liberal matrix (but also against every kind of juridical formalism), the reference to the life of the nation appears largely to dominate. No law can be superior (or simply comparable) to that of the German community to preserve and augment its own *bíos*. From this point of view, Nazi "jurisprudence" is not attributable to a subjective or decisionistic radicalization of positive law, but, if anything, to a perverse form of natural right. Obviously, by this we understand that for "nature" is not to be understood either law expressed by the divine will or what originated with human reason, but just that biological layer in which the national order *[ordinamento nazionale]* is rooted. After all, isn't it a biological given, blood precisely, that constitutes the ultimate criterion for defining the juridical *status* of a person? In this sense, the norm is nothing but the a posteriori application of a present determination in nature: it is the racial connotation that attributes or removes the right to exist to or from individuals and peoples.

However, this biologization of law in turn is the result of a preceding juridicalization of life. If it were otherwise, where would the subdivision of human *bíos* into zones of different value be derived from, if not from such a juridical decision? It is precisely in this continual exchange between cause and effect, intention and outcome, that the biopolitical machine of Nazism is at its most lethal. In order that life can constitute the objective, concrete, and factitious reference of law, it must have already been previously normativized according to precise juridical-political caesuras. What results is a system that is doubly determined. Something else also emerges from the combined competition between the power of doctors and that of judges in the application of the biopolitical (and therefore thanatopolitical) laws of Nazism. Biology and law, and life and norm, hold each other in a doubly linked presupposition. If the norm presupposes the facticity of life as its privileged content, life for its part presupposes the caesura of the norm as its preventive definition. Only a life that is already "decided" according to a determinate juridical order can constitute the natural criteria in the application of the law. From this perspective, we can say that Nazism, in its own way, created a "norm of life": certainly not in the sense that adapted its own norms to the demands of life, but in what closed the entire extension of life within the borders of a norm that was destined to reverse it into its opposite. Directly applying itself to life, Nazi law subjected it to a norm of death, which at the same time made it absolute while displacing it.

How can this terrible thanatopolitical *dispositif* be finally broken? Or, better perhaps, how can we overturn its logic into a politics of life? If its lethal result appeared to originate from a forced superimposition between norm and nature, one could imagine that the way out might pass through a more precise separation between the two domains. Normativism and juris-naturalism—both introduced again with the fall of the Nazi regime as protective barriers against its recurring threat—followed the same path from opposite directions: in the first case, autonomizing and almost purifying the norm in an obligation always more separate from the facticity of life; in the other, deriving the norm from the eternal principles of a nature that coincides with divine will or, otherwise, with human reason. Yet the impression remains that neither of these two responses has stood the test of time, and not only because it is difficult to hypothesize the restoration of conceptual apparatuses antecedent to totalitarianism.[67] The principal reason is that neither the absoluteness of the norm nor the primacy of nature is to be considered external to a phenomenon like Nazism, which seems to be situated exactly at the point of intersection and tension of their opposing radicalizations. What else is the Nazi bio-law if not an explosive mixture between an excess of normativism and an excess of naturalism, if not a norm superimposed on nature and a nature that is presupposed to the norm? We can say that in these circumstances the "norm of life" was the tragically paradoxical formula in which life and norm are held together in a knot that can be cut only by annihilating both.

Yet this knot cannot simply be undone either, or worse still, ignored. It is here, beginning with that "norm of life," that we need today to start, not only to restore to the two terms the richness of their originary meaning, but also to invert the reciprocally destructive relation that Nazism instituted between them. We need to oppose the Nazi normativization of life with an attempt to vitalize that of the norm. But how? How should we move here and with what assumptions should we begin? I believe that the theoretical key of this passage cannot be traced to any of the grand modern juridical philosophies; nor will it be found in positivism, in juris-naturalism, in normativism, or decisionism (or at least in none of those philosophies that modernity together brought to completion and then did away with). From this point of view, not only Kelsen and Schmitt, but also Hobbes and Kant, emerge as unhelpful for thinking biopolitics affirmatively. Either they are constitutively outside its lexicon, as Kant and Kelsen are, or they are within its negative fold, as Hobbes and Schmitt are. A possible (and necessary) thread that we ought to weave is found instead in the philosophy of Spinoza—to the extent that he remains external to or lateral with respect to the dominant lines of modern juridical tradition. There is much to say (and much has been said) about the stunning force with which Spinozian philosophy destabilizes the conceptual apparatus of contemporary thought. But if we had to condense in one expression the most significant categorical step that it produces with regard to the relationship between norm and nature, between life and law, I would speak of the substitution of a logic of presupposition with one of reciprocal immanence. Spinoza doesn't negate (nor does he repress, as other philosophers do) the connection between the two domains, but deploys them in a form that situates them worlds apart from what it will assume in Nazi semantics: norm and life cannot mutually presuppose one another because they are part of a single dimension in continuous becoming.[68]

Thanks to the path he takes, Spinoza can remove himself from the formalism of the modern contract *[obbligazione]*, in particular to that of the Hobbesian variant, without, however, falling into what will be the Nazi biological substantialism. What keeps him apart from both is his refusal of that sovereign paradigm that, notwithstanding all the differences, is joined to substantialism by their same coercive tendency. When he writes in one of the most famous propositions in *Political Treatise* that "every natural thing has as much right from Nature as it has power to exist and to act," he too is thinking a "norm of life," but in a sense that rather then presupposing one to the other, joins them together in the same movement that understands life as always already normalized and the norm as naturally furnished with vital content.[69] The norm is no longer what assigns rights and obligations from the outside to the subject, as in modern transcendentalism—permitting it to do that which is allowed and prohibiting that which is not—but rather the intrinsic modality that life assumes in the expression of its own unrestrainable power to exist. Spinoza's thought differs from all the other immunitary philosophies that deduce the transcendence of the norm from the demand for protecting life and conditioning the preservation of life to the subjection to the norm. He makes the latter the immanent rule that life gives itself in order to reach the maximum point of its expansion. It is true that "each [particular thing] is determined by another particular thing to exist in a certain way, yet the force by which each one perseveres in existing follows from the eternal necessity of the nature of God," but such an individual force doesn't acquire meaning as well as possibility of success except within the internal extension of nature.[70] It is for this reason that, when seen in a general perspective, every form of existence, be it deviant or defective from a more limited point of view, has equal legitimacy for living according to its own possibilities as a whole in the relations in which it is inserted. Having neither a transcendent role of command nor a prescriptive function with respect to which conformity and deformity are stabilized, the norm is constituted as the singular and plural mode that nature every so often assumes in all the range of its expressions:

> So if something in Nature appears to us as ridiculous, absurd, or evil, this is due to the fact that our knowledge is only partial, that we are for the most part ignorant of the order and coherence of Nature as a whole, and that we want all things to be directed as our reason prescribes. Yet that which our reason declares to be evil is not evil in respect of the order and laws of universal Nature, but only in respect of our own particular nature.[71]

In nowhere more than this passage do we find the anticipated overturning that Spinoza undertakes with respect to Nazi normalization. While the latter measures the right to life or the obligation to die in relation to the position occupied with respect to the biological caesura constituted by the norm, Spinoza makes the norm the principle of unlimited equivalence for every single form of life.

It cannot be said that Spinoza's intuitions found expression and development in later juridical philosophy. The reasons for such a theoretical block are multiple. But in relation to our problem, it's worth paying attention to the resistance of the philosophy of natural right *[diritto]* as a whole to think the norm together with life: not *over* life nor *beginning from life,* but *in life,* which is to say in the biological constitution of the living organism. This is why the few heirs of the Spinozian juridical naturalism (consciously or unconsciously) are to be found less among philosophers of natural right than among those who have made the object of their research the development of individual and collective life. Or better: the moving line that runs from the first to the second, constantly translating the one into the other. As we know, it's what Simondon defines with the term and the concept of "transindividual." It is no coincidence that, beginning with Simondon, Spinoza has been interrogated, but not (as Étienne Balibar believes) because Spinoza negates individuality as such.[72] Rather, we can say that for Spinoza nothing other than individuals exists. These individuals are infinite modes of a substance that does not subtend or transcend them, but is that expressed precisely in their irreducible multiplicity; only that such individuals for Spinoza are not stable and homogenous entities, but elements that originate from and continually reproduce a process of successive individuations. This occurs not only because, as Nietzsche will later theorize, every individual body is a composite of parts belonging to other individuals and in transit toward them, but because its expansive power is proportional to the intensity and the frequency of such an exchange. Thus, at the apex of its development it finds itself part of a relation that is always more vast and complex with the environment that lets it continue to the extent that its own originary identity has been enormously reduced.

All of this is reflected in the Spinozian concept of natural right. I said earlier that the norm doesn't invest the subject from the outside because it emerges from the same capacity of existence. Not only every subject is *sui juris,* but every behavior carries with it the norm that places it in existence within a more general natural order. Considering that there are as many multiple individuals as there are infinite modes of the substance means that the norms will be multiplied by a corresponding number. The juridical order as a whole is the product of this plurality of norms and the provisional result of their mutable equilibrium. It is for this reason that neither a fundamental norm from which all the other norms would derive as consequence can exist nor a normative criterion upon which exclusionary measures vis-à-vis those deemed abnormal be stabilized. In short, the process of normativization is the never-defined result of the comparison and conflict between individual norms that are measured according to the different power that keeps them alive, without ever losing the measure of their reciprocal relation. To this dynamic, determined by the relation between individuals, is connected that relative to their internal transformation. If the individual is nothing but the momentary derivation of a process of individuation, which at the same time produces it and is its product, this indicates as well that the norms that the individual expresses vary according to his or her different composition. As the human body lives in an infinite series of relations with the bodies of others, so the internal regulation will be subject to continuous variations. More than an immunitary apparatus of self-preservation, Spinoza configures the juridical order as a meta-stable system of reciprocal contaminations in which the juridical norm, rooted in the biological norm, reproduces the latter's mutations.

It is this type of argumentation that can be ascribed to Simondon's analysis along the thread of transindividual semantics. When in *L'individu et sa genèse physico-biologique* he writes that "the values are the preindividual of the norm; they express the connection between orders of different size; born from the preindividual, they tend toward the postindividual," Simondon is negating all attempts to make absolute the normative system.[73] That such a system is likened to an individual in perpetual motion from the preindividual to the postindividual indicates that there is never a moment in which the individual can be enclosed in himself or be blocked in a closed system, and so removed from the movement that binds him to his own biological matrix. From this point of view, the only value that remains stable in the transition from the norm of one system to another is the awareness of their translatability in always more diverse and necessarily perishing forms. The most complete normative model is indeed what already prefigures the movement of its own deconstruction in favor of another that follows it: "In order for the normativity of a system of norms to be complete, it's important that there be within it both its prefigured destruction as system and its possible translation in another system according to transductive order."[74] It is true that there exists a natural tendency to imagine absolute and unchangeable norms, but that too is part of an ontogenetic process that is structurally open to the necessity of its own becoming: "The tendency to eternity becomes therefore the consciousness of the relative: this latter is no longer the will to stop becoming or to render absolute an origin and to privilege normatively a structure, but the knowledge of the meta-stability of the norms."[75] Just like Spinoza before him, Simondon also places the constitution of norms within the movement of life and makes life the primary source for the institution of norms.

If Simondon tightens norm and life in an affirmative nexus that strengthens both, the most explicit philosophical attempt to vitalize the norm is owed, however, to his teacher, Georges Canguilhem. It's certainly not the case here to consider the important passages that make up Canguilhem's resolute opposition to Nazism, many of which are biographical. Canguilhem was called in 1940 to Strasbourg to take up the chair left free by the mathematician Jean Cavaillès, a partisan who died fighting Nazism. Canguilhem also actively participated in the resistance under the pseudonym Lafont. I would say that nothing about his philosophy is comprehensible outside of this military commitment.[76] The entire conception of *bíos* to which he dedicated his work is deeply marked by it, beginning with the idea of "philosophy of biology," which in itself is counterposed to the Nazi's programmatic antiphilosophical biology. To think life philosophically, to make life the pertinent horizon of philosophy, signifies for him distancing it from an objectivist paradigm that, thanks to its alleged scientificity, ends up canceling its dramatically subjective character. But even before doing so, it's worth challenging that reduction of life to a simple material, to brute life, that Nazism precisely had pushed toward its most ruinous consequences. When he writes that "health is in no way a demand of the economic order that is to be weighed when legislating, but rather is the spontaneous unity of the conditions for the exercise of life," he can't help but refer critically and above all to Nazi state medicine, which had made that bio-economic procedure the hinge of its own politics of life and death.[77] Against it he offers the apparently tautological thesis that "the thought of what lives needs to assume its idea from the living"; here he doesn't only want to replace subjectivity at the center of the biological dimension, but also to institute a dynamic interval between life and its concept: the living is the one who always exceeds the objective parameters of life, which in a certain sense always lie beyond itself, in the median statistic on the basis of which its suitability to live and die is measured.[78] If Nazism stripped away every form of life, nailing it to its nude material existence, Canguilhem reconsigns every life to its form, making of it something unique and unrepeatable.

The conceptual instrument he adopts for such an end is precisely the category of the norm, which is assumed by juridical, as well as sociological, anthropological, and pedagogical traditions as a descriptive and prescriptive measure for valuing human behavior.[79] Canguilhem ascribes to the norm the meaning of the pure mode or state of being. In such a case, not only health but also disease constitutes a norm that is not superimposed on life, but expresses a specific situation of life. Before him, Émile Durkheim, in "Rules for the Distinction of the Normal from the Pathological," had recognized "that a fact can be termed pathological only in relation to a given species," but also that "a social fact can only be termed normal in a given species in relation to a particular phase, likewise determinate, of its development."[80] Canguilhem pushes further this "dialectical logic": what is defined as abnormal not only is included (albeit with its own fixed characterization) within the norm, but becomes the condition of recognizability and before that of existence. It is for this reason "that it is not paradoxical to say that abnormal, while logically second, is existentially first."[81] What would such a rule be that is outside the possibility of its infraction and how would it be defined? In the biological field, in fact, the normal state (as it were, of full health) is not even perceptible as such. To affirm, as the doctor Leriche does, that "health is life in the silence of organs," means that it is precisely illness that reveals to us negatively all of the physiological potentiality of the organism.[82] In order to be raised to a level of consciousness, health needs first to be lost. It is because of this second arrangement with respect to what negates it that the norm cannot be prefixed or imposed on life, but only inferred from it. Here the deconstruction is already evident that, beginning from the biological paradigm (liberated in turn from every presupposed objectivization), Canguilhem undertakes with regard to the juridical norm.[83] While this norm, which establishes a code of behavior that is anterior to its actuation, necessarily needs to foresee the possibility of the deviation of life (and therefore of sanctions with respect to it), the biological norm coincides with the vital condition in which it is manifested: "[A]n organism's norm of life is furnished by the organism itself, contained in its existence . . . a human organism's norm is its coincidence with the organism itself."[84] Once again it is the "norm of life" that is in play, but according to an order that, rather than circumscribing life within the limits of the norm, opens the norm to the infinite unpredictability of life. To the necessary negativity of the juridical norm—as Kelsen reminds us, every command can be expressed in the form of a prohibition—responds the constitutively affirmative nature of the biological.[85] Contrary to the Nazi idea that there exists a type of life which from its inception belongs to death, Canguilhem reminds us that death itself is a phenomenon of life.

Of course, it is also a negative phenomenon, like a disease that precedes and in turn determines it. But the negativity of disease (and more so death) doesn't lie in the modification of a properly original norm, as theories of degeneration would have it. It lies, on the contrary, in the organism's incapacity to modify the norm in a hold that crushes the norm on itself, forcing it into an infinite repetition. Here Canguilhem grafts the most innovative part of his proposal, situating it precisely in the point of connection and difference between normality and normativity. Derived from the Latin *norma,* both terms tend to come together in a definition that at once superimposes them while stretching them apart. Completely normal isn't the person who corresponds to a prefixed prototype, but the individual who preserves intact his or her own normative power, which is to say the capacity to create continually new norms: "Normal man is normative man, the being capable of establishing new, even organic forms."[86] It is the point of maximum deconstruction of the immunitary paradigm and the opening to a different biopolitical lexicon: the medico-biological model, employed in an intensely self-preserving key by all of modernity *[tradizione moderna]* (not to mention that of totalitarianism), is here oriented to a radically innovative meaning. As only Nietzsche of the "great health" had glimpsed, biological normality doesn't reside in the capacity to impede variations, or even diseases of the organism, but will be found rather in integrating them within a different normative material. If one interprets life according to a perspective that isn't dominated by the instinct of preservation; if, as Kurt Goldstein had argued (in a direction, by the way, that Canguilhem himself takes up and elaborates), this instinct isn't to be considered "the general law of life but the law of a withdrawn life," then disease will no longer be configured as extreme risk, but rather as the risk of not being able to face new risks, such as the atrophying of what is naturally imperiling about human nature: "The healthy organism tries less to maintain itself in its present state and environment than to realize its nature. This requires that the organism, in facing risks, accepts the eventuality of catastrophic reactions."[87] The logic of the living is capable of introducing a powerful semantic in the juridical norm against the immunitary normalization of life that is able to push beyond its usual definition.

The last work Gilles Deleuze left us is titled *Pure Immanence: Essays on a Life.*[88] A short text, in some ways elliptical and incomplete, it does, however, contain all the threads that we have woven to this point under the sign of an affirmative biopolitics. Deleuze commences with the definition of a "transcendental camp," understood as something that does not refer to an object or a subject, but rather a potentializing or depotentializing flow that moves between one sensation and another. Such a characterization is also to be contrasted with the notion of consciousness to the degree that, always focused on the constitution of a subject separated from the object proper, it ends up inevitably establishing a relationship of reciprocal transcendence. Against the latter, the transcendental field is presented as a plane of absolute immanence: it doesn't refer to anything else but itself. It is here that the category of *bíos* comes into play: "We will say of pure immanence that it is A LIFE and nothing else . . . A life is the immanence of immanence, absolute immanence: it is complete power, complete bliss."[89] Deleuze traces the conceptual genealogy in the later works of Fichte, for whom intuition of pure activity is nothing fixed.[90] It isn't a being, but precisely a life, for Maine de Biran as well (without speaking of Spinoza, Nietzsche, and Bergson, who remain the leading lights for Deleuze). Surprisingly, though, Deleuze's text introduces another unusual reference, to Dickens, and in particular to the novel titled *Our Mutual Friend* (in French *L'ami commun*), which seems to inscribe the question of *bíos* in that of *communitas* and vice versa. I would say that his "theoretical" nucleus (though we could say biophilosophical) resides in the connecting and diverging point between *the* life and precisely *a* life.[91] Here the move from the determinate article to that of the indeterminate has the function of marking the break with the metaphysical feature that connects the dimension of life to that of individual consciousness. There is a modality of *bíos* that cannot be inscribed within the borders of the conscious subject, and therefore is not attributable to the form of the individual or of the person. Deleuze seeks it out in the extreme line in which life *[la vita]* encounters *[s'incontra]* or clashes with *[si scontra]* death. It is that which happens in Dickens's text, when Riderhood, still in a coma, is in a suspended state between life and death. In those moments, in which time seems to be interrupted and opened to the absolute force of the event, the flicker of life that remains to him separates Riderhood from his individual subjectivity so as to present itself in all its simple biological texture, that is, in its vital, bare facticity:

> No one has the least regard for the man: with them all, he has been an object of avoidance, suspicion, and aversion; but the spark of life within him is curiously separable from himself now, and they have a deep interest in it, probably because it *is* life, and they are living and must die.[92]

The interest on the part of those present for this uncertain spark of life that "may smolder and go out, or it may glow and expand" is born, therefore, from the fact that in its absolute singularity, it moves beyond the sphere of the individual to be rooted in an impersonal datum—in the circumstance that, sooner or later, one dies *[si muore]*:[93]

> Between his life and his death, there is a moment that is only that of *a* life playing with death. The life of the individual gives way to an impersonal and yet singular life that releases a pure event freed from the accidents of internal and external life, that is from the subjectivity and objectivity of what happens: a "Homo tantum" with whom everyone empathizes and who attains a sort of beatitude. It is a haecceity no longer of individuation but of singularization: a life of pure immanence, neutral, beyond good and evil, for it was only the subject that incarnated it in the midst of things that made it good or bad. The life of such individuality fades away in favor of the singular life immanent to a man who no longer has a name, though he can be mistaken for no other. A singular essence, a life.[94]

A singular *[così]* life, the singularity of *a* life, Deleuze continues, is not distinguishable *[individuabile]*, that is, is not ascribable to an individual, because it is in itself generic, relating to a genre, but also unmistakable because it is unique in its genre—as that of a newborn, who is similar to all the others, but different from each of them for the tonality of the voice, the intensity of a smile, the sparkle of a tear.[95] It is constitutively improper, and for that reason common, as pure difference can be, the difference that isn't defined from anything other than from its own same differing *[differire]*. This is how the warning that appears in the section on singularity in *The Logic of Sense* ought to be understood, according to which "we can not accept the alternative . . . ; either singularities already comprised in individuals and persons, or the undifferentiated abyss."[96] The difference, which is to say the singularity, doesn't reside on the side of the individual, but rather of the impersonal—or a person that doesn't coincide with any of those [forms] in which we are accustomed to decline the subject (I, you, he), but, if anywhere, in that of the "fourth person," as Lawrence Ferlinghetti paradoxically expresses it.[97] Which is to say, in the grammar of knowledge and of power that has always excluded it:

> Far from being individual or personal, singularities preside over the genesis of individuals and persons; they are distributed in a "potential" which admits neither Self nor I, but which produces them by actualizing or realizing them, although the figures of this actualization do not at all resemble the realized potential.[98]

It is the classic and controversial Deleuzian theme of the "virtual," but at the same time of the preindividual and of the transindividual that Simondon posits.[99] Deleuze himself refers to it, citing Simondon's assertion that "the living lives at the limit of itself, on its limit," which is to say a crease in which subject and object, internal and external, and organic and inorganic are folded.[100] An impersonal singularity (or a singular impersonality), which, rather than being imprisoned in the confines of the individual, opens those confines to an eccentric movement that "traverses men as well as plants and animals independently of the matter of their individuation and the forms of their personality."[101]

In such a move we can glimpse something that, while still not tracing the figure of an affirmative biopolitics, anticipates more than one feature. If we superimpose the pages of Dickens to which reference was already made, we perceive that these kinds of features emerge once again from the reversal of Nazi thanatopolitics: the life that qualifies the experience of Riderhood, depersonalizing it, is, as in the Nazi laboratory, in direct contact with death. What Dickens calls "outer husk" or a "flabby lump of mortality" has not a little to do with the "empty shells" and "life unworthy of life "of Binding and Hoche—with Treblinka's flesh of the ovens—yet with a fundamental difference that has to do with a change in orientation; no longer from life seemingly to death, but from death seemingly to a life in which Riderhood awakes.[102] When Deleuze speaks of a "sort of beatitude" as a condition that lies beyond the distinction between good and evil (because it precedes, or perhaps because it follows, the normative subject that places it in being), he is also alluding to "a norm of life" that doesn't subject life to the transcendence of a norm, but makes the norm the immanent impulse of life. The appeal to the impersonal as the only vital and singular mode isn't unrelated to the going beyond a semantics of the person that has been represented from the origin of our culture in its juridical status (at least insofar as the law was and continues to function in relation to the intangible individuality of the person). It is this biojuridical node between life and norm that Deleuze invites us to untie in a form that, rather than separating them, recognizes the one in the other, and discovers in life its immanent norm, giving to the norm the potentiality *[potenza]* of life's becoming. That such a unique process crosses the entire extension of life without providing a continuous solution—that any thing that lives needs to be thought in the unity of life—means that no part of it can be destroyed in favor of another: every life is a form of life and every form refers to life. This is neither the content nor the final sense of biopolitics, but is a minimum its presupposition. Whether its meaning will again be disowned in a politics of death or affirmed in a politics of life will depend on the mode in which contemporary thought will follow its traces.

Notes

Translator's Introduction

1. Roberto Esposito, *Communitas: Origine e destino della comunità* (Turin: Einaudi, 1998).

2. Or, when not "exposing" presumed terrorists, force-feeding them so as to protect their lives. See Luke Mitchell, "God Mode," *Harper's* 313 (August 2006): 9–11. The American business in question (though by no means the only one) is Wal-Mart, where, in order to discourage "unhealthy job applicants, it was suggested that Wal-Mart arrange for 'all jobs to include some physical activity (e.g., all cashiers do some cart-gathering)'" ("Wal-Mart Memo Suggests Ways to Cut Employee Benefit Costs," *New York Times*, October 26, 2005).

3. Niklas Luhmann, *Social Systems*, trans. John Bednarz Jr. with Dirk Baecker (Stanford, Calif.: Stanford University Press, 1995); Donna Haraway, "Biopolitics and Postmodern Bodies," in *Simians, Cyborgs and Women: The Reinvention of Nature* (New York: Routledge, 1991), 203–30; and Jean Baudrillard, *The Transparency of Evil: Essays on Extreme Phenomena* (New York: Verso, 1991), 85. Compare as well Robert Unger's discussion and problematization of "immunity rights" and radical democracy in *False Necessity: Anti-Necessitarian Social Theory in the Service of Radical Democracy* (Cambridge: Cambridge University Press, 1987), 513–17, 530. My thanks to Adam Sitze for drawing my attention to Unger's important contribution to immunity theory.

4. Agnes Heller and Ferenc Fehér, *Biopolitics* (Brookfield: Aldershot, 1994); Agnes Heller and Sonja Puntscher Riekmann, eds., *The Politics of the Body, Race, and Nature* (Averbury: Aldershot, 1996); and idem, *Theory of Modernity* (Malden, Mass.: Blackwell, 1999). For Mark C. Taylor, see *Nots* (Chicago: University of Chicago Press, 1993), as well as *Hiding* (Chicago: University of Chicago Press, 1998).

5. Jacques Derrida, "Faith and Knowledge: The Two Sources of Religion," in *On Religion*, ed. Jacques Derrida and Gianni Vattimo (Stanford, Calif.: Stanford University Press, 1998); *The Politics of Friendship*, trans. George Collins (New York: Verso, 1997); "Autoimmunity: Real and Symbolic Suicides," in *Philosophy in a Time of Terror: Dialogues with Jürgen Habermas and Jacques Derrida*, ed. Giovanna Borradori (Chicago:

University of Chicago Press, 2003); and *Rogues: Two Essays on Reason* (Stanford, Calif.: Stanford University Press, 2005).

6. Michel Foucault, *"Society Must Be Defended": Lectures at the Collège de France, 1975–1976,* ed. Mauro Bertani and Alessandro Fontana, trans. David Macey (New York: Picador, 2003). See also his lectures from 1978 to 1979, collected in *Naissance de la biopolitique: Cours au Collège de France (1978–1979),* under the guidance of Alessandro Fontana (Paris: Seuil, 2004).

7. Giorgio Agamben, *Homo Sacer: Sovereign Power and Bare Life,* trans. Daniel Heller-Roazen (Stanford, Calif.: Stanford University Press, 1998); *Remnants of Auschwitz: The Witness and the Archive,* trans. Daniel Heller-Roazen (New York: Zone Books, 1999); and *The Open: Man and Animal,* trans. Kevin Attell (Stanford, Calif.: Stanford University Press, 2004).

8. Michael Hardt and Antonio Negri, *Empire* (Cambridge: Harvard University Press, 2000) and *Multitude: War and Democracy in the Age of Empire* (New York: Penguin Press, 2004).

9. See in this regard Esposito's earlier works on political philosophy: *Vico e Rousseau e il moderno Stato borghese* (Bari: De Donato, 1976); *La politica e la storia: Machiavelli e Vico* (Naples: Liguori, 1980); *Ordine e conflitto: Machiavelli e la letteratura politica del Rinascimento italiano* (Naples: Liguori, 1984); *Categorie dell'impolitico* (Bologna: Il Mulino, 1988); *Nove pensieri sulla politica* (Bologna: Il Mulino, 1993); and *L'origine della politica: Hannah Arendt o Simone Weil?* (Rome: Donzelli, 1996).

10. Judith Butler, *Precarious Life: The Powers of Mourning and Violence* (London: Verso, 2004) and *Giving an Account of Oneself: A Critique of Ethical Violence* (New York: Fordham University Press, 2005); Keith Ansell-Pearson, *The Viroid Life: Perspectives on Nietzsche and the Transhuman Condition* (New York: Routledge, 1997) and *Germinal Life: The Difference and Repetition of Deleuze* (New York: Routledge, 2000); Jürgen Habermas, *The Future of Human Nature* (London: Polity Press, 2004); and Ronald Dworkin, *Life's Dominion: An Argument about Abortion, Euthanasia, and Individual Freedom* (London: Vintage Books, 1994) as well as *Sovereign Virtue: The Theory and Practice of Equality* (Cambridge: Harvard University Press, 2000).

11. Karl Binding and Alfred Hoche, *Die Freigabe der Vernichtung lebensunwerten Leben: Ihr Mass und ihre Form* (Leipzig, 1920). Selections from the work were translated into English in 1992. See "Permitting the Destruction of Unworthy Life," in *Law and Medicine* 8 (1994): 231–65.

12. Roberto Esposito, *Immunitas: Protezione e negazione della vita* (Turin: Einaudi, 2002).

13. Esposito, *Communitas,* xii.

14. Émile Benveniste, *Indo-European Language and Society,* trans. Elizabeth Palmer (Coral Gables, Fla.: University of Miami Press, 1973), and Marcel Mauss, *The Gift: The Form and Reason for Exchange in Archaic Societies,* trans. W. D. Halls (London: Routledge, 2002).

15. Esposito, *Communitas,* xiii.

16. Ibid., xiv.

17. Cf. the chapter in *Communitas* dedicated to guilt: "Community is definable only on the basis of the lack from which it derives and that inevitably connotes it precisely as an absence or defect of community" (33).

18. See "Immunity" in chapter 2.

19. Ibid.

20. Ibid.

21. Ibid.

22. What Esposito has done, it seems to me, is to have drawn on Nancy's arguments in *The Inoperative Community* regarding precisely the excessive nature of community vis-à-vis the metaphysical subject. Nancy writes that "community does not weave a superior, immortal, or transmortal life between subjects . . . but it is constitutively, to the extent that it is a matter of 'constitution' here, calibrated on the death of those whom we call, perhaps, wrongly, its 'members' (inasmuch as it is not a question of organism)." Esposito demonstrates instead that the calibration of which Nancy speaks doesn't just involve the future deaths of the community's "members," but also revolves around the mortal threat that the other members represent for each other. It is precisely this threat and the calls for immunization from it that explain why so many have in fact made the question of community "a question of organism." Or better, it is precisely the unreflected nature of community as organism that requires deconstruction. Only in this way will the biopolitical origins of community be made clear via community's aporia in immunity (Jean-Luc Nancy, *The Inoperative Community*, ed. Peter Connor, trans. Peter Connor, Lisa Garbus, Michael Holland, and Simone Sawhney [Minneapolis: University of Minnesota Press, 1991], 14).

23. See "Immunity" in chapter 2.

24. Rossella Bonito Oliva's analysis of the immunization paradigm is apropos: "The route of a mature modernity unbinds the originarity of the relation [between *zoon* and the political] and makes immanent the reasons of living with *[cum-vivere]*, which is always assumed as a subsequent and therapeutic step for the condition of solitude and the insecurity of the individual" ("From the Immune Community to the Communitarian Immunity: On the Recent Reflections of Roberto Esposito," *Diacritics* 36:2 (summer 2006).

25. Michel Foucault, "Governmentality," in *The Foucault Effect: Studies in Governmentality*, ed. Graham Burcell, Colin Gordon, and Peter Miller (Chicago: University of Chicago Press, 1991), 103.

26. Foucault, *"Society Must Be Defended,"* 253.

27. Butler, *Precarious Life*, 24. See as well Butler's discussion of the opacity of the subject: "The opacity of the subject may be a consequence of its being conceived as a relational being, one whose early and primary relations are not always available to conscious knowledge. Moments of unknowingness about oneself tend to emerge in the context of relations to others, suggesting that these relations call upon primary forms of relationality that are not always available to explicit and reflective thematization" (Butler, *Giving an Account of Oneself*, 20).

28. Butler, *Precarious Life*, 20.

29. Butler does come close to Esposito's position when describing the violent, self-centered subject: "Its actions constitute the building of a subject that seeks to restore and maintain its mastery through the systematic destruction of its multilateral relations . . . It shores itself up, seeks to reconstitute its imagined wholeness, but only at the price of denying its own vulnerability, its dependency, its exposure, where it exploits those very features in others, thereby making those features 'other to' itself" (ibid., 41).

30. Roberto Esposito, "Introduzione: Termini della politica," in *Oltra la politica: Aantologia dell'impolitico* (Milan: Bruno Mondadori, 1996), 1. Lest I appear to reduce their respective positions to a Hobbesian declension of biopolitics in Esposito and a Hegelian search for recognition in subject positions in Butler, each does recognize the need to

muster some sort of new understanding of the changing conditions of what qualifies as life. For Butler, that search is premised on the need to enlarge "the differential allocation of grievability that decides what kind of subject is and must be grieved"; hence the importance she places on narratives of multilateralism and changing the normative schemes of what is or isn't human proffered by the media (Butler, *Precarious Life,* xiv). For his part, Esposito chooses to focus on the process of individualization that occurs at both the individual and collective level, arguing that "if the subject is always thought within the form of *bíos,* this in turn is inscribed in the horizon of a *cum* [with] that makes it one with the being of man" (See "Philosophy after Nazism" in chapter 5.). The title *Bíos* comes into its own here as a term that marks the vital experiences that the individualized subject shares and has "in common" politically with others. Esposito's excursus on life as a form of birth that he elaborates in chapter 5 may in fact be read as a necessary preface for the kind of changed recognition protocols related to grieving that Butler herself is seeking.

31. Derrida, "Faith and Knowledge," 44.

32. Ibid. Cf. in this regard the pages Foucault devotes to the theme in *The Hermeneutics of the Subject: Lectures at the Collège de France 1981–82,* trans. Graham Burchell (New York: Palgrave, 2005), 120–21, 182–85. My thanks to Adam Sitze for pointing out the important connections between biopolitics and these later seminars.

33. Derrida, "Faith and Knowledge," 51.

34. Ibid.; emphasis in original.

35. Ibid. In this regard, see A. J. P. Thomson's "What's to Become of 'Democracy to Come?'" *Postmodern Culture* 15:3 (May 2005).

36. Derrida, *The Politics of Friendship,* 35; emphasis in original.

37. "Thus Deleuze's ultimate response to Hegel's argument against the 'richness' of immediacy is that the significance of the singular—'this,' 'here,' 'now'—is only grasped within the context of a problem, a 'drama' of thought that gives it sense, in the absence of which it is effectively impoverished" (*Gilles Deleuze: Key Concepts,* ed. Charles J. Stivale [Montreal: McGill-Queen's University Press, 2005], 47).

38. Derrida, *Rogues,* 33.

39. Ibid., 36.

40. Ibid., 36–37.

41. Derrida, "Autoimmunity," 95; emphasis in original.

42. With that said, it is also true that with a different set of texts in hand a different reading of Derrida emerges, namely, *Specters of Marx: The State of the Debt, the Work of Mourning, and the New International,* trans. Peggy Kamuf (New York: Routledge, 1994), as well as Derrida's later texts on hospitality, in particular *On Hospitality,* trans. Rachel Bowlby (Stanford, Calif.: Stanford University Press, 2000). Hent de Vries analyzes Derridean thought and hospitality as well in the last chapter of his *Religion and Violence: Philosophical Perspectives from Kant to Derrida* (Baltimore: Johns Hopkins University Press, 2002). My thanks to Miguel Vatter for pointing out these other more "communist" texts.

43. Esposito, *Immunitas,* 170.

44. See "The Norm of Life" in chapter 5.

45. See Andrea Cavalletti's recent *La città biopolitica,* where he implicitly invokes the life of the city as one requiring protection (Andrea Cavalletti, *La città biopolitica: Mitologie della sicurezza* [Milan: Bruno Mondadori, 2005], esp. 20–27). See as well my interview with Esposito in *Diacritics* 36:2 (summer 2006).

46. See too the recent, brilliant contributions of Simona Forti to discussions of biopolitics originating in Italy. In addition to her groundbreaking work from 2001 titled *Totalitarianismo* (Rome: Laterza, 2001), her stunning "The Biopolitics of Souls: Racism, Nazism, and Plato" recently appeared in English (*Political Theory* 34:1 [February 2006]: 9–32). There she examines "the ambivalences that connect some of the assumptions of our philosophical tradition to Nazi totalitarianism" (10).

47. Foucault, *"Society Must Be Defended,"* 246–47.

48. Ibid., 246.

49. Ibid., 247.

50. Ibid., 246.

51. Ibid., 259.

52. See especially Paolo Virno, *The Grammar of the Multitude,* trans. Isabella Bertoletti (New York: Semiotext[e], 2004); *Governing China's Population: From Leninist to Neoliberal Biopolitics,* ed. Susan Greenhalgh and Edwin A. Winckler (Stanford, Calif.: Stanford University Press, 2005); *Lessico di biopolitica,* ed. Renata Brandimarte, Patricia Chiantera-Stutte, et al. (Rome: Manifestolibri, 2006); and Antonella Cutro, *Biopolitica: storia e attualità di un concetto* (Verona: Ombre Corte 2005).

53. Agamben, *Homo Sacer,* 1.

54. On this note see Laurent Dubreuil's "Leaving Politics: Bíos, Zoé, Life," *Diacritics* 36:2 (summer 2006).

55. Carl Schmitt, *Political Theology: Four Chapters on the Concept of Sovereignty,* trans. George Schwab (Cambridge: MIT Press, 1985). Agamben discusses at length the relation among Schmitt, Benjamin, and the state of exception in *State of Exception,* trans. Kevin Attell (Chicago: University of Chicago Press, 2005).

56. Agamben, *Homo Sacer,* 7.

57. Ibid., 8.

58. Ibid., 174. In this sense I agree with Eric Vogt's view that Agamben "corrects" Foucault's analysis. See his "S/Citing the Camp," in *Politics, Metaphysics and Death: Essays on Giorgio Agamben's Homo Sacer* (Durham, N.C.: Duke University Press, 2005), 74–101.

59. Agamben does take up his analysis of modern biopolitics in *The Open,* where what he calls the anthropological machine begins producing "the state of exception" so as to determine the threshold between the human and the inhuman. Yet to the degree the optic moves along the horizon of the state of exception, modernity and with it a nineteenth-century anthropological discourse remain wedded to a political (and metaphysical) aporia. "Indeed, precisely because the human is already presupposed every time, the machine actually produces a kind of state of exception, a zone of indeterminacy in which the outside is nothing but the exclusion of an inside and the inside is in turn only the inclusion of an outside" (Agamben, *The Open,* 37).

60. Marco Revelli, *La politica perduta* (Turin: Einaudi, 2003).

61. Hardt and Negri, *Empire,* 421.

62. See Paolo Virno's previously cited *Grammar of the Multitude* as well as Michael Hardt and Antonio Negri's edited collection of essays on Italian radical thought *Labor of Dionysus: A Critique of the State-Form* (Minneapolis: University of Minnesota Press, 1994).

63. Hardt and Negri, *Multitude,* 348.

64. Certainly, the Deleuzian optic is crucial in accounting for Hardt and Negri's positive vision of biopolitics, as they themselves readily admit. A new sense of the communal

based on the multitude and cooperation makes clear the illusory nature of modern sovereignty. See in this regard Negri's *Kairòs, alma venus, multitudo: Nove lezioni impartite a me stesso* (Rome: Manifestolibri, 2000): "The teleology of the common, inasmuch as it is the motor of the ontological transformation of the world, cannot be subjected to the theory of sovereign mediation. Sovereign mediation is always in fact the foundation of a unit of measure, while ontological transformation has no measure" (127).

65. Hardt and Negri, *Multitude,* 206.

66. In a recent essay, Esposito pushes his reading of Foucault to a global reevaluation of the term "totalitarianism": "Recognizing the attempt in Nazism, the only kind of its genre, to liberate the natural features of existence from their *historical* peculiarity, means reversing the Arendtian thesis of the totalitarian superimposition between philosophy of nature and philosophy of history. Indeed, it means distinguishing the blind spot in their unassimilability and therefore in the philosophical impracticability of the notion of totalitarianism" (Roberto Esposito, "Totalitarismo o biopolitica: Per un'interpretazione filosofica del Novecento," *Micromega* 5 [2006]: 62–63).

67. See "Regeneration" in chapter 4.

68. We ought to note that much of Esposito's critique of Foucault also holds true for Agamben. But where Foucault links socialism to Nazism via racism, Agamben joins a Nazi biopolitics to modern democracies through the state of exception. The result is, however, the same: to highlight Nazism's shared biopolitical features with contemporary democracies and so to lessen its singularity.

69. In this regard, see the entry for sovereignty in Esposito's *Nove pensieri sulla politica* (Bologna: Il Mulino, 1993), 87–111.

70. "One can speak of the Nazi state as a 'biocracy.' The model here is a theocracy, a system of rule by priests of a sacred order under the claim of divine prerogative. In the case of the Nazi biocracy, the divine prerogative was that of cure through purification and revitalization of the Aryan race." Lifton goes on to speak of biological activism in the murderous ecology of Auschwitz, which leads him to the conclusion that the "Nazi vision of therapy" cannot be understood apart from mass murder (Robert Jay Lifton, *The Nazi Doctors: Medical Killing and the Psychology of Genocide* [New York: Basic Books, 1986], 17, 18).

71. See "Regeneration" in chapter 4. In *Immunitas* Esposito makes explicit his attempt to fold the notion of exception into that of immunization. Alluding to Agamben, Esposito notes that "the irreducibly antinomical structure of the *nomos basileús*—founded on the interiorization or better the 'internment' of an exteriority—is especially evident in the case of exception that Carl Schmitt situates in the 'most external sphere' of law" (Esposito, *Immunitas,* 37). Here Esposito attempts to think immunity through a Benjaminian reading of law and violence, but elsewhere he notes that such a method is in fact Bataillian. See his *Categorie dell'impolitico* for the debt such a methodology owes Georges Bataille and the term *partage,* or the liminal copresence of separation and concatenation (Esposito, *Categorie dell'impolitico,* xxii).

72. See "Poliitics over Life" in chapter 1.

73. See in particular the 2001 roundtable discussion among Esposito, Negri, and Veca ("Dialogo sull'impero e democrazia," *Micromega* 5 [2001]: 115–34), as well as Esposito's recent elaboration of biopolitical democracy ("Totalitarismo o biopolitica: Per un'interpretazione filosofica del Novecento," *Micromega* 5 [2006]: 57–66).

74. Hardt and Negri, *Multitude,* 206.

75. "Interview with Roberto Esposito," *Diacritics* 36:2 (summer 2006). With more time, it would be of great interest to trace how Esposito's early work on the Italian avant-garde informs his later reflections on immunity and biopolitics. See in this regard his analysis of the poetry of Nanni Balestrini in *Ideologia della neo avanguardia* (Naples: Liguori, 1976) and the resemblance between *communitas* as a vital sphere with that of Balestrini's poetics.

76. Hardt and Negri, *Multitude,* 356. But we shouldn't assume that the contact implicit in a network doesn't risk precisely the kind of autoimmunitary deficiencies that Baudrillard, for instance, sees as the principal feature of current politics. He writes: "All integrated and hyperintegrated systems—the technological system, the social system, even thought itself in artificial intelligence, and its derivatives—tend towards the extreme constituted by immunodeficiency. Seeking to eliminate all external aggression, they secrete their own internal virulence, their own malignant reversibility" (Jean Baudrillard, "Prophylaxis and Virulence," in *The Transparency of Evil: Essays on Extreme Phenomena,* trans. James Benedict [London: Verso, 1993], 62).

77. See "Property" in chapter 2.

78. See "Flesh" in chapter 5; emphasis in original.

79. What he will later say about Deleuze's final text, "Pure Immanence: A Life . . . ," is a shorthand for his own analysis: *bíos* is inscribed in the question of *communitas* and vice versa (See "The Norm of Life" in chapter 5).

80. See "Philosophy of Nazism" in chapter 5.

81. In this sense, Esposito's conception of biopolitics differs from Donna Haraway's. Haraway, we recall, leans directly on the immunitary paradigm as a model for interaction. If she doesn't sing its praises, she does recognize in it the postmodern mode by which "the semi-permeable self [is] able to engage with others (human and non-human, inner and outer), but always with finite consequences" (Haraway, "Biopolitics and Postmodern Bodies," 225). Significantly, these include "situated possibilities and impossibilities of individuation and identification; and of partial fusions and dangers." In short, only when immunized is every member capable of interacting with every other. *Bíos* moves the accent off of the individual and the body, the individual body, to a notion of life, one that cannot be traced back to a specific individual, but rather to the dynamic motor of the virtual and the singularities that precede the genesis of individual selves. In other words, to *communitas* as the preindividualizing mode of having and being in common.

82. More similarities between Butler and Esposito's reading of the subject emerge here. "Do we want to say that it is our status as 'subjects' that binds us all together even though, for many of us, the 'subject' is multiple or fractured? And does the insistence on the subject as a precondition of political agency not erase the more fundamental modes of dependency that do bind us and out of which emerge our thinking and affiliation, the basis of our vulnerability, affiliation, and collective resistance?" (Butler, *Precarious Life,* 49).

83. Of particular importance for Esposito is the category of flesh appropriated from Merleau-Ponty, and its usefulness for scrambling and eliding previously inscribed immunitary borders. Flesh, for Esposito, offers the possibility of thinking a politicization of life that doesn't move through a semantics of the body, as flesh refers to a "worldly material that is antecedent to or that follows the constitution of the subject of law" (See "Flesh" in chapter 5.). The distinctively anti-immunitary features of flesh make

it possible to countenance the "eclipse of the political body," and with it the emergence of a different form of community in which contagious exposure to others gives way to constituitive openness. Flesh will then name what is common to all, a being that is "singular and common" (ibid.).

84. Giorgio Agamben, *The Coming Community,* trans. Michael Hardt (Minneapolis: University of Minnesota Press, 1993), 64.

85. Esposito, *Communitas,* 139. In this regard, see Adriana Cavarero's compelling reading of speech and politics in the thought of Hannah Arendt, to which Esposito's understanding of the relation between community and communication is indebted: "According to her [Arendt], speech—even it is understood as *phone semantike*—does not become political by way of the things of the community that speech is able to designate. Rather, speech becomes political on account of the self-revelation of speakers who express and communicate their uniqueness through speaking—no matter the specific content of what is said. The political valence of signifying is thus shifted from speech—and from language as a system of signification—to the speaker" (Adriana Cavarero, *For More Than One Voice: Toward a Philosophy of Vocal Expression,* trans. Paul A. Kottman [Stanford, Calif.: Stanford University Press, 2005], 190). For the relation Bataille draws between the individual and communication, see his *On Nietzsche,* trans. Bruce Boone (New York: Paragon, 1992), esp. 18–19.

86. Cf. Judith Butler's gloss of Laplance's "Responsibility and Response" in *Giving an Account of Oneself:* "The other, we might say, comes first, and this means that there is no reference to one's own death that is not at once a reference to the death of the other" (75).

87. Georges Bataille, "The College of Sociology," in *Visions of Excess: Selected Writings, 1927–1939,* trans. Alan Stoekl (Minneapolis: University of Minnesota Press, 1985), 251.

88. Esposito, *Communitas,* 142.

89. See "The Norm of Life" in chapter 5.

90. Ansell-Pearson, *Viroid Life,* 182, 189.

91. Esposito, *Immunitas,* 205.

92. See "Birth" in chapter 5.

93. Ibid.

94. See Andrew Fischer's helpful summary of the debate, "Flirting with Fascism: "The Sloterdijk Debate," *Radical Philosophy* 9 (January/February 2000): 20–33.

95. Habermas, 14.

96. Ibid., 10.

97. Esposito, "Totalitarismo o biopolitica," 63–64; emphasis in original.

98. Dworkin, *Life's Dominion,* 76–77.

99. Dworkin, "Playing God," in *Sovereign Virtue,* 452.

100. Ibid., 449.

101. Dworkin, "Liberal Community," in *Sovereign Virtue,* 227.

102. Dworkin, "Playing God," 452.

103. See "The Norm of Life" in chapter 5.

104. Ibid.

105. Cf. Esposito's reading of Gehlen in *Immunitas:* "For Gehlen, the other, more than an alter ego or a different subject is essentially and above all else a non-ego; the 'non' that allows the ego to identify with the one who is precisely other from its own other" (123).

106. Gilles Deleuze, *Pure Immanence: Essays on A Life,* trans. Anne Boyman (New York: Zone Books, 2001), 28–29.

107. See "The Norm of Life" in chapter 5.

108. Ibid.

109. I wish to thank Miguel Vatter for the terminology. For a discussion of the difference between biopower and biopolitics, which seems to me implicit in this distinction, see Maurizio Lazzarato, "From Biopower to Biopolitics," available at http://www.generation-online.org/c/fcbiopolitics.htm (accessed October 10, 2007): "Foucault's work ought to be continued upon this fractured line between resistance and creation. Foucault's itinerary allows us to conceive the reversal of biopower into biopolitics, the 'art of governance' into the production and government of new forms of life. To establish a conceptual and political distinction between biopower and biopolitics is to move in step with Foucault's thinking."

110. "Transcript, "President Bush Discusses the War on Terror," National Endowment for Democracy, October 5, 2005 (available at www.whitehouse.gov).

111. "As Americans, we believe that people everywhere—everywhere—prefer freedom to slavery, and that liberty once chosen, improves the lives of all" (ibid.).

112. Cf. Achille Mbembe's discussion of the individual as opposed to the person in discussions of African societies: "Finally, in these societies the 'person' is seen as predominant over the 'individual,' considered (it is added) 'a strictly Western creation.' Instead of the individual, there are entities, captives of magical signs, amid an enchanted and mysterious universe in which the power of invocation and evocation replaces the power of production, and in which fantasy and caprice coexist not only with the possibility of disaster but with its reality" (Achille Mbembe, *On the Postcolony* [Berkeley: University of California Press, 2001], 4). My thanks to Adam Sitze for pointing out the deep connections between Esposito and Mbembe.

113. See "Flesh" in chapter 5.

Introduction

1. Roberto Esposito, *Communitas: Origine e destino della comunità* (Turin: Einaudi, 1998) and *Immunitas: Protezione e negazione della vita* (Turin: Einaudi, 2002).

1. The Enigma of Biopolitics

1. See in this regard the collection *Biopolitik,* ed. Christian Geyer (Frankfurt, Suhrkamp, 2001).

2. Karl Binding, *Zum Werden und Leben der Staaten: Zehn Staatsrechtliche Abhandlungen* (Munich and Leipzig: Duncker & Humblot, 1920); Eberhard Dennert, *Der Staat als lebendiger Organismus: Biologische Betrachtungen zum Aufbau der neuen Zeit* (Halle [Saale]: C. E. Müller, 1920); and Edward Hahn, *Der Staat, ein Lebenwesen* (Munich: Dt. Volksverlag, 1926).

3. Rudolph Kjellén, *Stormakterna: Konturer kring samtidens storpolitik* (Stockholm: Gebers, 1905).

4. Rudolph Kjellén, *Staten som Lifsform* (Stockholm: Hugo Geber, 1916).

5. Rudolph Kjellén, *Grundriss zu einem System der Politik* (Leipzig: Rudolf Leipzig Hirzel, 1920), 3–4.

6. Jakob von Uexküll, *Staatsbiologie: Anatomie, Phisiologie, Pathologie des Staates* (Berlin: Verlag von Gebrüder Paetel, 1920).

7. Ibid., 46.

8. Ibid., 55.

9. Morley Roberts, *Bio-politics: An Essay in the Physiology, Pathology and Politics of the Social and Somatic Organism* (London: Dent, 1938).

10. Ibid., 153.

11. Ibid., 160.

12. Aroon Starobinski, *La biopolitique: Essai d'interprétation de l'histoire de l'humanité et des civilisations* (Geneva: Imprimerie des Arts, 1960).

13. Ibid., 7.

14. Ibid., 9.

15. Edgar Morin, *Introduction à une politique de l'homme* (Paris: Éditions du Seuil, 1969).

16. Ibid., 11.

17. Ibid., 12.

18. Edgar Morin, *Le paradigme perdu: La nature humaine* (Paris: Éditions du Seuil, 1973).

19. André Birré, "Introduction: Si l'Occident s'est trompé de conte?" *Cahiers de la biopolitique* 1:1 (1968): 3.

20. Antonella Cutro also discusses this first French production in biopolitics in her *Michel Foucault. Tecnica e vita. Biopolitica e filosofia del "Bíos"* (Naples: Bibliopolis, 2004), which constitutes the first, useful attempt to systematize Foucauldian biopolitics. More generally on biopolitics, see *Politica della vita,* ed. Laura Bazzicalupo and Roberto Esposito (Milan: Laterza, 2003), as well as *Biopolitica minore,* ed. Paolo Petricari (Rome: Manifestolibri, 2003).

21. *Research in Biopolitics,* ed. Stephen A. Peterson and Albert Somit (Greenwich, Conn.: JAI Press). The volumes, in order, are *Sexual Politics and Political Feminism* (1991); *Biopolitics in the Mainstream* (1994); *Human Nature and Politics* (1995); *Research in Biopolitics* (1996); *Recent Explorations in Biology and Politics* (1997); *Sociology and Politics* (1998); *Ethnic Conflicts Explained by Ethnic Nepotism* (1999); and *Evolutionary Approaches in the Behavioral Sciences: Toward a Better Understanding of Human Nature* (2001).

22. Lynton K. Caldwell, "Biopolitics: Science, Ethics, and Public Policy," *Yale Review,* no. 54 (1964): 1–16; and James C. Davies, *Human Nature in Politics: The Dynamics of Political Behavior* (New York: Wiley, 1963).

23. Roger D. Masters, *The Nature of Politics* (New Haven and London: Yale University Press, 1989).

24. Walter Bagehot, *Physics and Politics, or, Thoughts on the Application of the Principles of "Natural Selection" and "Inheritance" to Political Society* (Kitchener, Ont.: Batoche, 2001).

25. Thomas Thorson, *Biopolitics* (Washington, D.C.: University Press of America, 1970).

26. See, on this point, D. Easton, "The Relevance of Biopolitics to Political Theory," in *Biology and Politics,* ed. Albert Somit (The Hague: Mouton, 1976), 237–47, as well as before that William James Miller Mackenzie, *Politics and Social Science* (Baltimore:

Johns Hopkins University Press, 1967), and H. Lasswell, *The Future of the Comparative Method,* in *Comparative Politics* 1 (1968): 3–18.

27. Warder C. Allee's volumes on the animal are classic: *Animal Life and Social Growth* (Baltimore: Williams & Wilkins Company and Associates in Cooperation with the Century of Progress Exposition, 1932) and *The Social Life of Animals* (Boston: Beacon Press, 1958). Also of interest are Lionel Tiger, *Men in Groups* (New York: Vintage Books, 1970) and Desmond Morris, *The Human Zoo* (New York: Dell, 1969). For this "natural" conception of war, see especially Quincy Wright, *A Study of War* (Chicago: University of Chicago Press, 1942), and Hans J. Morgenthau, *Politics among Nations: The Struggle for Power and Peace* (New York: Alfred A. Knopf, 1948). More recently there is V. S. E. Falger, *Biopolitics and the Study of International Relations: Implication, Results, and Perspectives,*in *Research in Biopolitics* 2: 115–34.

28. Albert Somit and Stephen A. Peterson, *Biopolitics in the Year 2000, Research in Biopolitics* 8: 181.

29. In this direction, compare Carlo Galli, "Sul valore politico del concetto di 'natura,'" in *Autorità e natura: Per una storia dei concetti filosofico-politici* (Bologna: Centro stampa Baiesi, 1988), 57–94, and Michela Cammelli, "Il darwinismo e la teoria politica," *Filosofia politica,* no. 3 (2000): 489–518.

30. An acute historical-conceptual analysis of sovereignty, if from another perspective, is that proposed by Biagio De Giovanni, "Discutere la sovranità," in Bazzicalupo and Esposito, *Politica della vita,* 5–15. See as well Luigi Alfieri's "Sovranità, morte, e politica," in the same volume (16–28).

31. For an analytic reconstruction of the problem, see Alessandro Pandolfi, "Foucault pensatore politico postmoderno," in *Tre studi su Foucault* (Naples: Terzo Millennio Edizioni, 2000), 131–246. On the relation between power and law, I refer the reader to Lucio D'Alessandro, "Potere e pena nella problematica di Michel Foucault," in *La verità e le forme giuridiche* (Naples: La città del sole, 1994), 141–60.

32. Michel Foucault, *"Society Must Be Defended": Lectures at the Collège de France, 1975–1976,* ed. Mauro Bertani and Alessandro Fontana, trans. David Macey (New York: Picador, 2003), 239–40.

33. Michel Foucault, "Crisis de un modelo en la medicina?" in *Dits et Écrits,* vol. 3 (Paris: Gallimard, 2001), 222.

34. Michel Foucault, *Discipline and Punish: The Birth of the Prison,* trans. Alan Sheridan (New York: Vintage Books, 1977).

35. Michel Foucault, *Abnormal: Lectures at the Collège de France 1974–1975,* trans. Graham Burchell (New York: Picador, 2003).

36. Michel Foucault, *The History of Sexuality,* vol. 1: *An Introduction,* trans. Robert Hurley (New York: Vintage Books, 1978), 89.

37. Ibid., 145.

38. Michel Foucault, "Return to History," in *Aesthetics, Method, and Epistemology,* ed. J. Faubion (New York: New Press, 1998), 430–31.

39. Michel Foucault, "The Crises of Medicine or the Crises of Anti-Medicine," *Foucault Studies,* no. 1 (December 2004): 11.

40. Michel Foucault, "Human Nature: Justice versus Power" (Noam Chomsky and Michel Foucault), in *Michel Foucault and His Interlocutors,* ed. A. I. Davidson (Chicago: University of Chicago Press, 1997), 110. Cf. Stefano Catucci's *La 'natura' della natura umana: Note su Michel Foucault,* in Noam Chomsky and Michel Foucault, *Della natura umana: Invariante biologico e potere politico* (Rome: Derive Approdi, 2004), 75–85.

41. Foucault, *History of Sexuality,* 143.

42. Michel Foucault, "Bio-histoire et bio-politique," in *Dits et Écrits, 1954–1988,* vol. 3 (Paris: Gallimard, 1994), 97.

43. Foucault, *History of Sexuality,* 143.

44. Ibid.

45. Ibid.

46. Foucault, *"Society Must Be Defended,"* 35; my emphasis.

47. Ibid., 36.

48. Ibid.; my emphasis.

49. Foucault, *History of Sexuality,* 138.

50. On the processes of subjectivization, cf. Mariapaola Fimiani, "Le véritable amour et le souci commun du monde," in *Foucault: Le courage de la vérité,* ed. Frédéric Gros (Paris: Presses universitaires de France, 2002), 87–127, and Yves Michaud, "Des modes de subjectivationaux techniques de soi: Foucault et les identités de notre temps," *Cités,* no. 2 (2000): 11–39. Fundamental for the theme remains Gilles Deleuze, *Foucault,* trans. Seán Hand (Minneapolis: University of Minnesota Press, 1988).

51. Michel Foucault, "The Subject and Power," *Critical Inquiry* 8:4 (summer 1982): 781.

52. Michel Foucault, "'Omnes et Singulatim': Towards a Critique of Political Reason," in *Power,* ed. James Faubion (New York: New Press, 1997), 321.

53. Ibid., 322.

54. Ibid.

55. Foucault, *History of Sexuality,* 95.

56. Ibid., 144–45.

57. I am alluding to Michael Hardt and Antonio Negri's *Empire* (Cambridge: Harvard University Press, 2000), esp. 22–41, but also to the group headed by the French journal *Multitudes.* See in particular the first issue of 2000, dedicated precisely to *Biopolitique et biopouvoir,* with contributions by Maurizio Lazzarato, Éric Alliez, Bruno Karsenti, Paolo Napoli, and others. It should be said that the theoretical-political perspective is in itself interesting, but only weakly linked to that of Foucault, who inspires it.

58. See, on this point, Valerio Marchetti, "La naissance de la biopolitique," in *Au risque de Foucault* (Paris: Éditions du Centre Georges Pompidou: Centre Michel Foucault, 1997), 237–47.

59. Michel Foucault, "The Political Technology of Individuals," in Faubion, *Power,* 405.

60. Marco Revelli has recently discussed the relation between politics and death in a vigorously ethical and theoretical essay, *La politica perduta* (Turin: Einaudi, 2003). See as well his earlier *Oltre il Novecento* (Turin: Einaudi, 2001).

61. Foucault, *"Society Must Be Defended,"* 241; my emphasis.

62. Ibid., 36.

63. Ibid., 253–54.

64. Ibid., 254.

65. Cf. Michael Donnelly, "On Foucault's Uses of the Notion 'Biopower,'" in *Michel Foucault Philosopher,* ed. Timothy Armstrong (New York: Routledge, 1992), 199–203, as well as Jacques Rancière, "Biopolitique ou politique?" *Multitudes* 1 (March 2000): 88–93.

66. Foucault, "The Subject and Power," 779.

67. This is the outcome that Giorgio Agamben coherently arrives at in *Homo Sacer: Sovereign Power and Bare Life,* trans. Daniel Heller-Roazen (Stanford, Calif.: Stanford University Press, 1998).

2. The Paradigm of Immunization

1. On the communitarian motif in Hegel, see in particular Rossella Bonito-Oliva's *L'individuo moderno e la nuova comunità* (Naples: Guida, 1990), esp. 63–64.

2. Émile Durkheim, *The Rules of Sociological Method,* trans. W. D. Halls (New York: Free Press, 1982), 73.

3. Max Scheler, *Problems of a Sociology of Knowledge,* trans. Manfred Frings (London: Routledge, 1980) and *Person and Self-Value: Three Essays,* trans. M. S. Frings (Boston: Kluwer Academic Publishers, 1987); Helmuth Plessner, *Conditio humana* (Frankfurt am Main: Suhrkamp, 1983) and *Limits of Community: A Critique of Social Radicalism,* trans. Andrew Wallace (New York: Humanity Books, 1999); and Arnold Gehlen, *Urmensch und Spätkultur: Philosophische Ergebnisse und Aussagen* (Bonn: Athenäum-Verlag, 1956) and *Man, His Nature and Place in the World* (New York: Columbia University Press, 1988).

4. Plessner, *Conditio humana,* 72.

5. Gehlen, *Urmensch und Spätkultur,* 44–45.

6. Norbert Elias, *The Civilizing Process,* trans. Edmund Jephcott (Oxford: Blackwell, 1994), 453.

7. For this reading of Parsons, see as well Stefano Bartolini, "I limiti della pluralità: Categorie della politica in Talcott Parsons," *Quaderni di teoria sociale* 2 (2002): 33–60.

8. Niklas Luhmann, *Social Systems,* trans. John Bednarz Jr. with Dirk Baecker (Stanford, Calif.: Stanford University Press, 1995), 371–72.

9. [Esposito deals more at length with Luhmann and immunity, particularly in the juridical sense, in *Immunitas: Protezione e negazione della vita* (Turin: Einaudi, 2002), 52–61.—*Trans.*]

10. Luhmann, *Social Systems,* 374.

11. See in this regard A. D. Napier's *The Age of Immunology* (Chicago: University of Chicago Press, 2003).

12. Dan Sperber, *Explaining Culture: A Naturalistic Approach* (Oxford: Blackwell, 1996), and Donna Haraway, "The Biopolitics of Postmodern Bodies: Determinations of Self in Immune System Discourse," in *Simians, Cyborgs, and Women: The Reinvention of Nature* (London: Routledge, 1991), 204.

13. See Odo Marquard, *Aesthetica und Anaesthetica: Philosophische Überlegungen* (Paderborn: F. Schöning, 1989).

14. On this last point, see Alain Brossat, *La démocratie immunitaire* (Paris: Dispute, 2003), and Romano Gasparotti, *I miti della globalizzazione: "Guerra preventiva" e logica delle immunità* (Bari: Dedalo, 2003). On globalization more generally, see the works of Giacomo Marramao, which have been collected in *Passaggio a Occidente: Filosofia e globalizzazione* (Turin: Bollati Bolinghieri, 2003).

15. In this regard, see my *Immunitas,* as well as *Communitas: Origine e destino della comunità* (Turin: Einaudi, 1998). Giuseppe Cantarano has recently written as well on some of these same themes. See his *La comunità impolitica* (Troina: Città Aperta, 2003).

16. Bruno Accarino has drawn attention to the opposing bipolarity of *Belastung/Entlastung* (debt/exoneration) in *La ragione insufficiente: Al confine tra autorità e razionalità* (Rome: Manifestolibri, 1995), 17–48.

17. With regard to the aporia and the potentialities of this dialectic (or nondialectic) between immunity and community, see the intelligent essay that Massimo Donà

has dedicated to the category of immunization using a key that productively pushes it toward a different logic of negation: Massimo Donà, "Immunity and Negation: On Possible Developments of the Theses Outlined in Roberto Esposito's *Immunitas,*" *Diacritics* 36:3 (2006).

18. See the section "Politics of Life" in chapter 1.

19. [Esposito is clearly referring to Giorgio Agamben's discussion of paterfamilias. See Agamben's *Homo Sacer: Sovereign Power and Life,* trans. Danile Heller-Roazen (Stanford, Calif.: Stanford University Press, 1998), esp. 81–90.—*Trans.*].

20. Plato, *Republic,* trans. Robin Waterfield (Oxford: Oxford University Press, 1993), 173.

21. See in this regard the invaluable essay by Simona Forti, "The Biopolitics of Souls," *Political Theory* 34:1 (2006): 9–32.

22. Joachim Günther, *Hitler und Platon* (Berlin and Leipzig: W. de Gruyter, 1933) and *Hitlers Kampf und Platons Staat: Eine Studie über den ideologischen Aufbau der nationalsozialistischen Freiheitsbewegung* (Berlin: W. de Gruyter, 1933); A. Gabler, (Berlin and Leipzig: W. de Gruyter, 1934); and Hans F. K. Günther, *Platon als Hüter des Lebens: Platons Zucht- und Erziehungsgedanken und deren Bedeutung für die Gegenwart* (Munich: J. F. Lehmann, 1928). In the same direction as Günther see *Humanitas* (Munich: J. F. Lehmann, 1937). For Wilhelm Windelband, see his *Platon* (Stuttgart: F. Frommann, 1928). The following are the texts that Günther cites in the third edition of his book on Plato (1966, 9–10): Alfred E. Taylor, *Plato: The Man and His Work* (New York: Dial Press, 1927); Julius Stenzel, *Platon der Erzieher* (Leipzig: F. Meiner, 1928); Paul Friedländer, *Platon* (Berlin and Leipzig: W. de Gruyter, 1928–30); Constantine Ritter, *Die Kerngedanken der platonischen Philosophie* (Munich: E. Reinhardt, 1931); Werner Wilhelm Jaeger, *Paideia: Die Formung des Griechischen Menschen* (Berlin and Leipzig: W. de Gruyter, 1936); Léon Robin, *Platon* (Paris: F. Alcan, 1935); Gerhardt Krüger, *Einsicht und Leidenschaft: Das Wesen des platonischen Denkens* (Frankfurt am Main: V. Klostermann, 1939); and Ernst Hoffmann, *Platon* (Zurich: Artemis-Verlag, 1950).

23. Plato, *Republic,* 174.

24. Aristotle, *The Politics,* trans. Trevor J. Saunders (New York: Penguin Classics, 1981), 88.

25. In addition to Mario Vegetti's recent *Quindici lezioni su Platone* (Turin: Einaudi, 2003), see in particular "Medicina e potere nel mondo antico" in the forthcoming *Biopolitiche.* With regard to these problems and with an implicit attention to the immunitary paradigm, there is the recent publication of the important essay by Gennaro Carillo, *Katechein: Uno studio sulla democrazia antica* (Naples: Editoriale Scientifica, 2003).

26. With regard to Peter Sloterdijk, one ought to keep in mind the three important volumes that appeared under the title *Sphären* (Frankfurt: Suhrkamp, 2004) in which the author traces the lineaments of a true and actual "social immunology."

27. This reading of modernity has for some time been the object of discussion for Paolo Flores d'Arcais. See his important essay *Il sovrano e il dissidente: La democrazia presa sul serio* (Milan: Garzanti, 2004) and the debate that ensued in *Micromega* 2–3 (2004).

28. Martin Heidegger, "The Age of the World Picture," in *The Question concerning Technology and Other Essays,* trans. William Lovitt (New York: Harper and Row, 1977), 149–50.

29. Thomas Hobbes, *Leviathan,* ed. Francis B. Randall (New York: Washington Square Press, 1976), 87.

30. Ibid.

31. Ibid., 87–88.

32. Thomas Hobbes, *De Cive* (London: R. Royston, 1651; 1843), 158; Thomas Hobbes, *The Elements of Law* (London: Tönnies, 1889), 178; Hobbes, *Leviathan,* 240.

33. See in this regard Carlo Galli's "Ordine e contingenza: Linee di lettura del *Leviatano,*" in *Percorsi della libertà: Scritti in onore di Nicola Matteucci* (Bologna: Il Mulino, 1996), 81–106; Alessandro Biral, *Hobbes: La società senza governo,* in *Il contratto sociale nella filosofia politica moderna,* ed. Giuseppe Duso (Milan: FrancoAngeli, 1993), 51–108; and Giuseppe Duso, *La logica del potere* (Rome-Bari: Laterza, 1999), 55–85.

34. I am referring in particular to Roman Schnur, *Individualismus und Absolutismus: Zur politischen Theorie vor Thomas Hobbes, 1600–1640* (Berlin: Duncker & Humblot, 1963).

35. Michel Foucault, *"Society Must Be Defended": Lectures at the Collège de France, 1975–1976,* ed. Maruo Bertani and Alessandro Fontana, trans. David Macey (New York: Picador, 2003), 90.

36. Hobbes, *Leviathan,* 149.

37. Ibid., 150.

38. John Locke, *Two Treatises of Government* (Cambridge: Cambridge University Press, 1967), 224.

39. Ibid., 223.

40. John Locke, *Epistola de Tolerantia: A Letter on Toleration,* trans. J. W. Gough (Oxford: Clarendon Press, 1968), 67. Cf. the following: "And 'tis not without reason, that he seeks out, and is willing to joyn in Society with others who are already united, or have a mind to unite for the mutual *Preservation* of their Lives, Liberties, and Estates, which I call by their general name, *Property*" (Locke, *Two Treatises,* 368).

41. Locke, *Two Treatises,* 324.

42. With regard to the dialectic of property in modern political philosophy, I have drawn important insights from Pietro Costa, *Il progetto giuridico: Ricerche sulla giurisprudenza del liberalismo classico* (Milan: Giuffrè, 1974), and Francesco De Sanctis, *Problemi e figure della filosofia giuridica e politica* (Rome: Bulzoni, 1996). Paolo Grossi's *Il dominio e le cose: Percezioni medievali e moderne dei diritti reali* (Milan: Giuffrè, 1992) remains crucial for understanding the premodern tradition.

43. Locke, *Two Treatises,* 305–6.

44. Ibid., 316–17.

45. Karl Marx, *Economic and Philosophic Manuscripts of 1844,* ed. Dirk J. Struik, trans. Martin Milligan (New York: International Publishers, 1964), 128–29.

46. See, on this point, Pietro Barcellona, *L'individualismo proprietario* (Turin: Bollati Boringhieri, 1987).

47. On this transformation, see Adriana Cavarero's "La teoria contrattualistica nei *Trattati sul governo* di Locke," in *Il contratto sociale nella filosofia politica moderna,* ed. Giuseppe Duso (Bologna: Il Mulino, 1987), 149–90.

48. Immanuel Kant, *The Philosophy of Law: An Exposition of the Fundamental Principles of Jurisprudence as the Science of Right,* trans. W. Hastie (Edinburgh: T. & T. Clark, 1887), 64–65.

49. [I have chosen to translate the Italian *libertà* with "liberty" (and not "freedom"), not only because the passages Esposito cites from Locke include the term, but also to mark the assonances that Esposito will hear between liberty, deliberation, *libertates,* and, of course, liberalism.—*Trans.*].

50. Cf. Dieter Nestle, *Eleutheria: Studien zum Wesen der Freiheit bei den Griechen und im Neuen Testament* (Tübingen: Mohr, 1967); Émile Benveniste, *Indo-European Language and Society*, trans. Elizabeth Palmer (London: Faber, 1973); and Richard. B. Onians, *The Origins of European Thought about the Body, the Mind, the Soul, the World, Time, and Fate: New Interpretations of Greek, Roman and Kindred Evidence Also of Some Basic Jewish and Christian Beliefs* (Cambridge: Cambridge University Press, 1988).

51. In this regard, see Pier Paolo Portinaro's dense postface to the translation of Benjamin Constant's *La libertà degli antichi, paragonata a quella dei moderni* (Turin: Einaudi, 2001).

52. Isaiah Berlin, "Two Concepts of Liberty," in *Four Concepts of Liberty* (New York: Oxford University Press, 1970), 130; my emphasis.

53. Martin Heidegger, *The Essence of Human Freedom: An Introduction to Philosophy*, trans. Ted Sadler (New York: Continuum, 2002), 13.

54. [Esposito is punning here on the assonance between *alterità* (otherhood) and *alterazione* (alteration).—*Trans.*]

55. Niccolò Machiavelli, *Discourses on Livy*, Oxford World's Classics, trans. Julia Conaway and Peter E. Bondanella (Oxford: Oxford University Press, 1997), 64.

56. Thomas Hobbes, "Of Liberty and Necessity," in *The English Works of Thomas Hobbes*, vol. 4 (London: John Bohn, 1890), 273.

57. Hobbes, *Leviathan*, 37.

58. Locke, *Two Treatises*, 302.

59. Ibid., 289.

60. Charles de Scondat, Baron de Montesquieu, *Spirit of Laws*, trans. Thomas Nugent (Kitchener, Ont.: Batoche Books, 2001), 206; Jeremy Bentham, *Rationale of Judicial Evidence*, in *The Works of Jeremy Bentham*, vol. 7 (Edinburgh: John Bowring, 1843), 522.

61. Jeremy Bentham, *Manuscripts* (University College of London), lxix, 56. See the doctoral thesis of Marco Stangherlin, "Jeremy Bentham e il governo degli interessi" (University of Pisa, 2001–2).

62. Michel Foucault, "La questione del liberalismo," in *Biopolitica e liberalismo: Detti e scritti su potere ed etica 1975–1984*, trans. Ottavio Marzocca (Milan: Medusa, 2001), 160.

63. Hannah Arendt, "What Is Freedom?" in *Between Past and Future: Eight Exercises in Political Thought* (New York: Viking Press, 1961), 155.

64. Ibid., 150.

65. Michel Foucault, *Technologies of the Self: A Seminar with Michel Foucault*, ed. Luther H. Martin (Amherst: University of Massachusetts Press, 1988), 152.

66. Luis Dumont, *Essays on Individualism: Modern Ideology in an Anthropological Perspective* (Chicago: University of Chicago Press, 1986).

67. For the figure of the *homo democraticus* I refer to the reader to Massimo Cacciari's important observations in *L'arcipelago* (Milan: Adelphi, 1997), 117–18. See too Elena Pulcini, *L'individuo senza passioni: Individualismo moderno e perdita del legame sociale* (Turin: Bollati Boringhieri, 2001), 127–28. On Tocqueville more generally, cf. Francesca Maria De Sanctis, *Tempo di democrazia: Sulla condizione moderna* (Naples: Editoriale Scientifica, 1986).

68. Alexis de Tocqueville, *Democracy in America*, ed. Francis Bowen, trans. Henry Reeve, vol. 2 (Cambridge: Sever and Francis, 1862), 121, 124.

69. Ibid., 169.

70. Friedrich Nietzsche, *Twilight of the Idols, or, How to Philosophize with a Hammer*, trans. Duncan Large (Oxford: Oxford University Press, 1998), 68; 64.

3. Biopower and Biopotentiality

1. [The term Esposito uses in the chapter title is *biopotenza*, which connotes both power and a potentiality for producing and undergoing change. Since Esposito intends it as a necessary step on the way to thinking an affirmative biopolitics, I have translated it as potentiality unless otherwise indicated.—*Trans.*]

2. [See the introduction to Esposito's 1998 preface to *Categorie dell'impolitico* (Bologna: Il Mulino, 1988) for further thoughts on the "impolitical."—*Trans.*]

3. Karl Löwith, "European Nihilism: Reflections on the European War," in *Martin Heidegger and European Nihilism*, trans. Gary Steiner (New York: Columbia University Press, 1995), 206; Georges Bataille, "Nietzsche and the Fascists," in *Visions of Excess*, trans. Allan Stoekl (Minneapolis: University of Minnesota Press, 1985), 24.

4. Michel Foucault, "Nietzsche, Genealogy, History," in *Aesthetics, Method and Epistemology*, ed. J. Faubion (New York: New Press, 1998), 369–91.

5. Friedrich Nietzsche, *On the Genealogy of Morals: A Polemic: By Way of Clarification and Supplement to My Last Book, Beyond Good and Evil*, trans. Douglas Smith (Oxford: Oxford University Press, 1997), 66.

6. Friedrich Nietzsche, *Frammenti postumi (1885–1887)*, in *Opere complete di Friedrich Nietzsche*, vol. 8 (Milan: Adelphi, 1992), 139. [As no complete edition of Nietzsche's posthumous works exists in English, I have cited the Italian and where possible the German.—*Trans.*].

7. Friedrich Nietzsche, *Ecce Homo*, trans. Walter Kaufmann (New York: Vintage Books, 1967), 311.

8. Friedrich Nietzsche, *Twilight of the Idols, or, How to Philosophize with a Hammer*, trans. Duncan Large (Oxford: Oxford University Press, 1998), 65.

9. Nietzsche, *Frammenti postumi (1888–1889)*, 408.

10. On the complex relationship between Nietzsche and Darwinism and more generally with the biological sciences, see especially Éric Blondel, *Nietzsche, le corps et la culture: La philosophie comme généalogie philologique* (Paris: Presses universitaires de France, 1986); H. Brobjer, *Darwinismus*, in *Nietzsche-Handbuch* (Stuttgart-Weimar: Metzler, 2000); Barbara Stiegler, *Nietzsche et la biologie* (Paris: Presses universitaires de France, 2001); Gregory Moore, *Nietzsche, Biology and Metaphor* (Cambridge: Cambridge University Press, 2002), as well as Andrea Orsucci, *Dalla biologia cellulare alle scienza dello spirito* (Bologna: Il Mulino, 1992).

11. Nietzsche, *Frammenti Postumi (1881–1882)*, 432–33.

12. I am referring, of course, to Martin Heidegger, *Nietzsche*, trans. David Farrell Krell (San Francisco: HarperSanFrancisco, 1991).

13. Friedrich Nietzsche, *The Gay Science*, trans. Walter Kaufmann (New York: Vintage Books, 1974), 34–35.

14. For this relation see especially Remo Bodei's chapter dedicated to Nietzsche in his important work *Destini personali: L'età della colonizzazione delle coscienze* (Milan: Feltrinelli, 2002), 83–116, as well as Ignace Haaz, *Les conceptions du corps chez Ribot et Nietzsche* (Paris: L'Harmattan, 2002).

15. In this sense the work contemporary with Nietzsche of the greatest importance is Wilhelm Roux's *Der Kampf der Theile im Organismus* (Leipzig, 1881). For more on Roux, see Wolfgang Müller-Lauter, "Der Organismus als innere Kampf: Der Einfluss von Wilhelm Roux auf Friedrich Nietzsche," *Nietzschen Studien* 7 (1978): 89–223.

16. Nietzsche, *Frammenti postumi (1884–1885)*, 238.

17. Nietzsche, *Ecco Homo*, 231–32.

18. Nietzsche, *Frammenti postumi (1885–1887)*, 77–78.

19. "It should be considered symptomatic when some philosophers—for example Spinoza who was consumptive—considered the instinct of self-preservation decisive and *had* to see it that way; for they were individuals in conditions of distress" (Nietzsche, *The Gay Science*, 292).

20. Ibid., 291–92.

21. Nietzsche, *Thus Spake Zarathustra: A Book for All and None*, trans. Thomas Wayne (New York: Algora Publishing, 2003), 87.

22. Friedrich Nietzsche, "History in the Service and Disservice of Life," in *Unmodern Observations*, trans. Gary Brown (New Haven: Yale University Press, 1990), 89.

23. Friedrich Nietzsche, *Beyond Good and Evil: Prelude to a Philosophy of the Future*, trans. Judith Norman (Cambridge: Cambridge University Press, 2001), 153, 154.

24. The reference here is to W. H. Rolph's *Biologische Probleme zugleich als Versuch zur Entwicklung einer nationalen Ethik* (Leipzig: Wilhelm Engelmann, 1882).

25. "Uncommon is the highest virtue and useless, luminous it is and gentle in its brilliance: a bestowing virtue is the highest virtue" (Nietzsche, *Thus Spake Zarathustra*, 57).

26. Friedrich Nietzsche, *Human, All Too Human: A Book for Free Spirits*, trans. R. J. Hollingdale (Cambridge: Cambridge University Press, 1986), 376.

27. See in this regard Umberto Galimberti's *Gli equivoci dell'anima* (Milan: Feltrinelli, 1987).

28. Friedrich Nietzsche, "On Truth and Lies in a Nonmoral Sense," in *The Nietzsche Reader*, ed. Keith Ansell-Pearson and Duncan Large (Oxford: Blackwell Publishing, 2006), 122.

29. Nietzsche, *Human, All Too Human*, 89.

30. Friedrich Nietzsche, *Twilight of the Idols*, trans. Duncan Large (Oxford: Oxford University Press, 1998), 78.

31. Nietzsche, *Human, All Too Human*, 113.

32. Friedrich Nietzsche, *Daybreak: Thoughts on the Prejudices of Morality*, trans. R. J. Hollingdale (Cambridge: Cambridge University Press, 1982), 52.

33. Nietzsche, *Frammenti postumi (1888–1889)*, 214.

34. Nietzsche, *On the Genealogy of Morals*, 105.

35. I am referring to Gilles Deleuze, *Nietzsche and Philosophy*, trans. Hugh Tomlinson (New York: Columbia University Press, 2006).

36. Nietzsche, *Frammenti postumi (1885–1887)*, 283, 289.

37. Ibid., 93.

38. For the theme of decadence, see Giuliano Campioni, "Nietzsche, Taine et la décadence," in *Nietzsche: Cent ans de réception française*, ed. Jacques Le Rider (San-Denis: Éditions Suge, 1999), 31–61.

39. Nietzsche, *Ecce Homo*, 233–34. [The Italian translation of the German differs widely from the English. For "unclean" (*Lauterkeit* in German), one reads "contaminated" *(contaminate)* and for "cleanliness" (*Reinheit* in German), purity *(purezza)*. Given Esposito's emphasis on the themes of integrity and purity, I have chosen to add the German in brackets.—*Trans.*]

40. Nietzsche, *Frammenti postumi (1888–1889)*, 217.

41. Ibid., 377.

42. [See in particular Michel Foucault's *"Society Must Be Defended": Lectures at the*

Collège de France, 1975–1976, ed. Mauro Bertani and Alessandro Fontana, trans. David Macey (New York: Picador, 2003), esp. the seminars of February 18 and 25, 1976.—*Trans.*]

43. Nietzsche, *Beyond Good and Evil*, 152.

44. I am referring to Domenico Losurdo's important and debatable book *Nietzsche, il rebello aristocratico, biografia intellettuale e bilancio critico* (Turin: Bollati Boringhieri, 2002).

45. Nietzsche, *The Gay Science*, 100.

46. Rather important in this direction is Alexander Tille, *Vom Darwin bis Nietzsche: Ein Buch Entwicklungsethik* (Leipzig: C. G. Naumann, 1895).

47. Cf. Alfred Espinas, *Des sociétés animales: Étude de psychologie comparée* (Paris: G. Baillière, 1877), and two texts from Georg Heinrich Schneider: *Der Tierische Wille* (Leipzig: Abel, [188?]) and *Der menschliche Wille vom Standpunkte der neueren Entwicklungstheorien (des "Darwinismus")* (Berlin: F. Dummlers, 1882). The texts of Espinas and Schneider were part of Nietzsche's library.

48. Nietzsche, *Twilight of the Idols*, 61.

49. Ibid., 68.

50. Nietzsche, *The Gay Science*, 177.

51. Ibid.

52. In this direction, see Marco Vozza, *Esistenza e interpretazione: Nietzsche oltre Heidegger* (Rome: Donzelli, 2001). On the metaphor of illness, see Patrick Wotling, *Nietzsche et le problème de la civilisation* (Paris: Presses universitaires de France, 1995), 111ff.

53. Nietzsche, *Human, All Too Human*, 99.

54. Nietzsche, *The Gay Science*, 346.

55. Nietzsche, *Human, All Too Human*, 107.

56. Ibid., 108.

57. Nietzsche, *Beyond Good and Evil*, 96.

58. Georges Bataille, *On Nietzsche*, trans. Bruce Boone (New York: Paragon, 1992), 8, 25.

59. Cf. Furio Semerari, *Il predone, il barbaro, il giardiniere* (Bari: Dedalo, 2000), 145ff.

60. Massimo Cacciari dedicates intense pages to this theme in *L'arcipelago* (Milan: Adelphi, 1997), 135–54.

61. Nietzsche, *Frammenti postumi (1884–1885)*, 317.

62. Nietzsche, *Daybreak*, 149.

63. Nietzsche, *Frammenti postumi (1881–1882)*, 348.

4. Thanatopolitics

1. Michel Foucault, *"Society Must Be Defended": Lectures at the Collège de France, 1975–1976*, ed. Mauro Bertani and Alessandro Fontana, trans. David Macey (New York: Picador, 2003), 258–63.

2. See the section titled "Politics over Life" in chapter 1.

3. Foucault, *"Society Must Be Defended,"* 258.

4. Ibid., 260.

5. Alain Brossat, *L'épreuve du désastre: Le XXe siècle et les camps* (Paris: Albin Michel, 1996), 141ff.

6. Simona Forti offers an exemplary profile of the relation between totalitarianism and philosophy in her *Il totalitarianismo* (Rome-Bari: Laterza, 2001).

7. Robert Lifton, *The Nazi Doctors: Medical Killing and the Psychology of Genocide* (New York: Basic Books, 1986), 31.

8. Erwin Baur, Eugen Fischer, and Fritz Lenz, *Grundriss der menschlichen Erblichkeitslehre und Rassenhygiene* (Munich: J. F. Lehmann, 1923), 417–18.

9. Rudolf Ramm, *Ärztliche Rechts und Standeskunde: Der Arzt als Gesundheitserzieher* (Berlin: W. de Gruyter, 1943), 156.

10. Hans Reiter, "La biologie dans la gestion de l'État," in *État et santé* (Paris: F. Sorlot, 1942). Other contributions include L. Conti, "L'organisation de la santé publique du Reich pendant la guerre"; F. von Verschuer, "L'image héréditaire de l'homme"; E. Fischer, "Le problème de la race e la législation raciale allemande"; A. Scheunert, "La recherche et l'étude des vitamines au service de l'alimentation nationale.

11. Hans Weinert, *Biologische Grundlagen für Rassenkunde und Rassen Hygiene* (Stuttgart: Ferdinand Enke Verlag, 1934).

12. Cf. Benno Müller-Hill, *Murderous Science: Elimination by Scientific Selection of Jews, Gypsies, and Others, Germany 1933–1945*, trans. George R. Fraser (Oxford: Oxford University Press, 1988), 94.

13. Joachim Mrugowsky, "Einleitung," in *Das ärztliche Ethos*, ed. Christoph Wilhelm Hufeland (Munich and Berlin: J. F. Lehmann, 1939), 14–15. See in this regard Lifton, *The Nazi Doctors*, 32.

14. Robert N. Proctor, *The Nazi War on Cancer* (Princeton, N.J.: Princeton University Press, 1999), 55.

15. In addition to the work of Lifton cited above, see too in this connection the relevant work of Rafaella de Franco, *In nome di Ippocrate: Dall'olocausto medico nazista all'etica della sperimentazione contemporanea* (Milan: F. Angeli, 2001).

16. K. Blome, *Arzt im Kampf: Erlebnisse und Gedanken* (Leipzig: Johann Ambrosius Barth Verlag, 1942).

17. Andrzej Kaminski, *Konzentrationslager 1896 bis heute: Eine Analyse* (Stuttgart: W. Kohlhammer, 1982), 145.

18. Primo Levi, *Survival in Auschwitz: The Nazi Assault on Humanity*, trans. Stuart Woolf (New York: Touchstone Books, 1996), 40.

19. Kaminski, *Konzentrationslager 1896 bis heute*, 200.

20. Adolf Hitler, *Libres propos sur la guerre et la paix recueillis sur l'ordre de Martin Bormann*, vol. 1 (Paris: Flammarion, 1952), 321.

21. Cf. Christopher R. Browning, *The Path to Genocide: Essays on Launching the Final Solution* (Cambridge: Cambridge University Press, 1992), 153–54.

22. Otto Helmut, *Volk in Gefahr: Der Geburtenrückgang und seine Folgen für Deutschlands Zukunft* (Munich: J. F. Lehmann, 1933), and Friedrich Burgdörfer, *Völker am Abgrund* (Munich: J. F. Lehmann, 1936).

23. On transformations in the concept of "degeneration," compare Georges Paul Genil-Perrin, *Histoire des origines et de l'évolution de l'idée de dégénérescence en médecine mentale* (Paris, 1913), as well as R. D. Walter, "What Became a Degenerate? A Brief History of a Concept," *Journal of the History of Medicine and the Allied Sciences* 11 (1956): 422–29.

24. Benedict-Augustin Morel, *Traité des dégénérescences physiques, intellectuelles et morales de l'espèce humaine et des causes qui produisent ces variétés maladives* (Paris: J. B. Baillière; New York: H. Baillière, 1857).

25. Valentin Magnan and Paul Maurice Legrain, *Les dégénérés, état mental et syndromes épisodiques* (Paris: Rueff, 1895), 79.

26. Morel, *Traité des dégénérescences physiques*, 5; Giuseppe Sergi, *Le degenerazioni umane* (Milan: Fratello Dumolard, 1889), 42.

27. Edwin Ray Lankester, *Degeneration: A Chapter in Darwinism* (London: Macmillan, 1880), 58.

28. On Italian degenerative theory, see A. Berlini, "L'ossessione della degenerazione: Ideologie e pratiche dell'eugenetica," diss., Ist. Orientale di Napoli, 2000, and more generally Maria Donzelli, ed., *La biologia: Parametro epistemologico del XIX secolo* (Naples: Liguori Editore, 2003).

29. Prosper Lucas, *Traité philosophique et physiologique de l'hérédité naturelle* (Paris: J. B. Baillière, 1847–50), and Théodule Ribot, *L'hérédité: Étude psychologique sur ses phénomènes, ses lois, ses causes, ses conséquences* (Paris: Ladrange, 1876). On Ribot, see Remo Bodei, *Destini personali: L'età della colonizzazione delle coscienze* (Milan: Feltrinelli), 65ff.

30. Eugène Apert, *L'hérédité morbide* (Paris: E. Flammarion, 1919), 1.

31. Lucas, *Traité philosophique et physiologuqe de l'hérédité naturelle*, 5.

32. André Pichot, *La société pure, de Darwin à Hitler* (Paris: Flammarion, 2000), 80–85.

33. Max Nordau, *Degeneration*, introduction by George L. Mosse (Lincoln: University of Nebraska Press, 1993), 22.

34. Gina Ferrero Lombroso, *I vantaggi della degenerazione* (Turin: Bocca, 1904), 56, 114.

35. Ibid., 185.

36. For the literary references that I take up and elaborate in the following pages I am indebted to the directions that Daniel Pick provides in *Faces of Degeneration: A European Disorder, 1848–1918* (Cambridge: Cambridge University Press, 1989), 155–75. On the concept of degeneration, see as well J. Edward Chamberlin and Sander L. Gilman, eds., *Degeneration: The Dark Side of Progress* (New York: Columbia University Press, 1985).

37. Émile Zola, *His Excellency* (London: Elek Books, 1958); Federico De Roberto, *I viceré* (Milan: Garzanti, 1970); Francesco Mastriani, *I vermi* (Naples: M. Milano, 1972).

38. Robert Louis Stevenson, *The Strange Case of Doctor Jekyll and Mr. Hyde* (New York: Viking Penguin, 2002), 60–61.

39. Ibid., 67, 68, 66.

40. Ibid., 69.

41. Oscar Wilde, *The Picture of Dorian Gray* (Oxford: Oxford University Press, 1998), 115.

42. Ibid., 183.

43. Ibid., 184.

44. Bram Stoker, *Dracula* (Toronto, Ont.: Broadview Press, 1998), 383.

45. Ibid., 279, 252, 252.

46. Ibid., 253.

47. For a detailed (and positive) review of eugenic institutions and practices in the first decades of the last century, see Marie-Thérèse Nisot, *La question eugénique dans les divers pays* (Brussels: G. Van Campenhout, 1927–29).

48. Wilhelm Schallmayer, *Vererbung und Auslese im Lebenslauf der Völker: Eine staatswissenschaftlich Studie auf Grund der neueren Biologie* (Jena: G. Fischer, 1903).

49. Wilhelm Schallmayer, *Über die drohende körperliche Entartung der Kulturmenschheit und die Verstaatlichung des ärztlichen Standes* (Berlin: L. Heuser, 1891).

50. Ludwig Woltmann, *Politische Anthropologie* (Eisenach and Leipzig: Thüringische Verlags-Anstalt, 1903).

51. Alfred Ploetz, *Die Tüchtigkeit unserer Rasse und der Schutz der Schwachen: Ein Versuch über Rassenhygiene und ihr Verhältnis zu den humanen Idealen, besonders zum Socialismus* (Berlin: Fischer Verlag, 1895).

52. Georges Vacher de Lapouge, *Race et milieu social: Essais d'anthroposociologie* (Paris: M. Rivière, 1909).

53. See in this regard the essays collected in M. B. Adams, *The Wellborn Science: Eugenics in Germany, France, Brazil and Russia* (Oxford: Oxford University Press, 1990).

54. Reference has already been made to the success of Alfred Espinas's *Des sociétés animales: Étude de psychologie comparée* (Paris: G. Baillière, 1877), 13–60. The most relevant sections for our discussion are perhaps the initial ones on parasites (distinguished in "parasites, commensals, and mutualists").

55. Joël Kotek et Pierre Rigoulot, *Le siècle des camps: Détention, concentration, extermination, cent ans de mal radical* (Paris: Lattès, 2000).

56. Proctor, *The Nazi War on Cancer*, 129.

57. Garland E. Allen, "Chevaux de course et chevaux de trait: Métaphores et analogies agricoles dans l'eugénisme américain 1910–1940," in *Histoire de la génétique: Pratiques, techniques et théories*, ed. Jean-Louis Fischer and William Howard Schneider (Paris: Créteil, 1990), 83–98.

58. On the figure of Davenport, see in particular his *Heredity in Relation to Eugenics* (New York: Henry Holt and Company, 1911).

59. Charles Richet, "Dans cent ans," *La Revue scientifique* (March 12, 1892): 329.

60. Lifton, *The Nazi Doctors*, 279.

61. Maurice Boigey, *L'élevage humain* (Paris: Payot, 1917), and Charles Binet-Sanglé, *Le haras humain* (Paris: Albin Michel, 1918).

62. Charles Valentino, *Le secret professional en médecine, sa valeur sociale* (Paris: C. Naud, 1903).

63. Vacher de Lapouge, *Sélections sociales* (Paris: A. Fontemoing, 1896), 472–73.

64. Just Sicard de Plauzoles, *Principes d'hygiène* (Paris: Éditions Médicales, 1927).

65. A. Zuccarelli, "Il problema capitale dell'Eugenica," *Nocera Inferiore* (1924): 2.

66. In *Buck v. Bell*, 274 U.S. 200 (1927). Cf. Amedio Santuosuosso, *Corpo e libertà: Una storia tra diritto e scienza* (Milan: R. Cortina, 2001). On American biopolitics and its close relations with Nazi Germany, see Stefan Kühl, *The Nazi Connection: Eugenics, American Racism and German National-Socialism* (Oxford: Oxford University Press, 1994).

67. Karl Binding and Alfred Hoche, *Die Freigabe der Vernichtung lebensunwerten Leben: Ihr Mass und ihre Form* (Leipzig: Meiner, 1920).

68. Adolph Jost, *Das Recht auf den Tod* (Göttingen: Grunow & Co., 1895).

69. Lifton, *The Nazi Doctors*, 17.

70. Binding and Hoche, "Ärztliche Bemerkungen," in *Die Freigabe der Vernichtung lebensunwerten Leben, 61–62.*

71. Müller-Hill, *Murderous Science*, 40.

72. Enrico Morselli, *L'uccisione pietosa* (Turin: Bocca, 1928), 17.

73. Ernst Mann (pseudonym of Gerhard Hoffmann), *Die Erlösung der Menschheit vom Elend* (Weimar: F. Fink, 1922).

74. Charles Binet-Sanglé, *L'art de mourir: Défense et technique du suicide secondé* (Paris: Albin Michel, 1919); Richet, "Dans cent ans," 168.

75. Antoine Wylm, *La morale sexuelle* (Paris: Alcan, 1907), 280.

76. Raphael Lemkin, *Axis Rule in Occupied Europe: Laws of Occupation, Analysis of Government, Proposals for Redress* (Washington, D.C.: Carnegie Endowment for International Peace, Division of International Law, 1944). On the vast literature related to genocide, I direct the reader only to *Genocide: A Critical Bibliographic Review* (New York: Facts on File Publications, 1988), as well as to Y. Ternon, *L'état criminel* (Paris: Seuil, 1995).

77. See the section titled "Regeneration" in this chapter.

78. See, on this point, Anne Carol, *Histoire de l'eugénisme en France* (Paris: Seuil, 1995).

79. In addition to Paul Weindling's *Health, Race and German Politics between National Unification and Nazism 1870–1945* (Cambridge: Cambridge University Press, 1989), which is a rich source on the relation between medicine and politics from Wilhelminian to Nazi Germany, see too Michel Pollak, "Une politique scientifique: Le concours de l'anthropologie, de la biologie et du droit," in *La politique nazie d'extermination*, ed. François Bédarda (Paris: Albin Michel, 1989), 75–99.

80. Emmanuel Levinas, "Reflections on the Philosophy of Hitlerism," *Critical Inquiry* 17:1 (fall 1990): 69.

81. The impossibility of escape *[evasione]* is at the center of Levinas's *On Escape*, trans. Bettina Bergo (Stanford, Calif.: Stanford University Press, 2003). It seems to me that no one has noted that Brieux, in his play titled precisely *L'évasion*, takes up the identical theme, at first affirming and then contesting the idea that a hereditary disease cannot be cured (Eugène Brieux, *L'évasion, comédie en 3 actes* [Paris: Stock, 1914]).

82. On the dialectic of incorporation, cf. Claude Lefort, "L'image du corps et le totalitarisme," in *L'invention démocratique* (Paris: Fayard, 1981).

83. This dual procedure of the biologization of the spirit and the spiritualization of the body constitutes the nucleus of Nazi biopolitics. See, in this regard, the chapter titled "Politique biologique" of the *Anthologie de la nouvelle Europe*, which was published in occupied France by Alfred Fabre-Luce (Paris, 1942). It includes contributions from Gobineau, Chamberlain, Barrès, Rostand, Renan, and Maurras, alongside those of Hitler.

84. Vacher de Lapouge, *Sélections sociales*, 306. Cf. Pichot's *La société pure*, 124.

85. Otmar von Verschuer, *Manuel d'eugénique et hérédité humaine* (Paris: Masson, 1943), 114. I am citing the French version and not the original, *Leitfaden der Rassenhygiene*, in the following paragraphs.

86. Ibid.

87. Ibid., 115.

88. Lifton, *The Nazi Doctors*, 16.

89. Ibid., 147.

90. Ibid., 27.

91. Cf. Gisela Bock, "Il nazionalsocialismo: Politiche di genere e vita delle donne," in *Storie delle donne in Occidente: Il Novecento* (Rome-Bari: Laterza, 1992), 176–212. See as well her *Zwangssterilisation im Nazionalsozialismus: Studien zur Rassenpolitik und Frauenpolitik* (Opladen: Westdeutscher Verlag, 1986); more generally, on women under Nazism, see Claudia Koonz, *Mothers in the Fatherland: Women, the Family and Nazi Politics* (New York: St. Martin's Press, 1987).

92. In his text on female fertility, *Fruchtbarkeit und Gesundheit der Frau*, which opens with the Nazi slogan that "the genus and the race are above the individual," Dr. Hermann Stieve holds that the value of women is measured by the state of their ovaries.

To prove such a thesis, he himself conducted experiments on the degree to which ovaries could suffer lesions under bouts of terror until they atrophied. On this, compare the third chapter of Ernst Klee's *Auschwitz, die NS-Medizin und ihre Opfer* (Frankfurt am Main: S. Fischer, 1997).

93. Hannah Arendt, *The Origins of Totalitarianism* (New York: Harvest Book, 1968), 444.

5. The Philosophy of *Bíos*

1. On new biopolitical emergencies, compare the exhaustive survey by Laura Bazzicalupo, "Ambivalenze della biopolitica," in *Politica della vita: Sovranità, biopotere, diritti*, ed. Laura Bazzicalupo and Roberto Esposito (Rome-Bari: Laterza, 2003), 134–44. See as well Bazzicalupo's *Governo della vita: Biopolitica ed economia* (Rome-Bari: Laterza, 2006).

2. For further discussion of these aspects, see Alessandro Dal Lago, *Non-persone: L'esclusione dei migranti in una società globale* (Milan: Feltrinelli, 2002); Salvatore Palidda, *Polizia postmoderna: Etnografia del nuovo controllo sociale* (Milan: Feltrinelli, 2000); and, more generally, Sandro Mezzadra and Petrillo Agostino, *I confini della globalizzazione: Lavoro, cultura, cittadinanza* (Rome: Manifestolibri, 2000).

3. In this sense, see Agnes Heller, "Has Biopolitics Changed the Concept of the Political? Some Further Thoughts about Biopolitics," in *Biopolitics: The Politics of the Body, Race, and Nature*, ed. Ferenc Fehér and Agnes Heller (Aldershot: Avebury, 1996), as well as Heller and Fehér's *Biopolitics* (Aldershot and Brookfield, Vt.: Avebury, 1994).

4. Hannah Arendt, *The Human Condition* (Chicago: University of Chicago Press, 1958), 55.

5. Ibid., 22.

6. In this direction, cf. Leonardo Daddabbo, *Inizi: Foucault e Arendt* (Milan: B. A. Graphis, 2003), esp. 43–46.

7. Martin Heidegger, "Letter on Humanism," in *Basic Writings from Being and Time (1927) to The Task of Thinking (1964)* (New York: HarperSanFrancisco, 1977), 210, 225. Interesting elaborations of these reflections are contained in the reading of "Letter on Humanism" as well as Heidegger's entire thought by Peter Sloterdijk in *La domestication de l'être: Pour un éclaircissement de la clairière*, a paper given at the Centre Pompidou in March 2000 (Paris: Mille et une Nuits, 2000).

8. For such a tonality of Heidegger's thought, and more generally on the early Heidegger, see Eugenio Mazzarella, *Ermeneutica dell'effettività: Prospettive ontiche dell'ontologia heideggeriana* (Naples: Guida, 2002).

9. Martin Heidegger, *Phenomenological Interpretations of Aristotle: Initiation into Phenomenological Research*, trans. Richard Rojcewicz (Bloomington: Indiana University Press, 2001), 66.

10. Ibid., 65.

11. Ibid., 62.

12. Medard Boss, "Preface to the First German Edition of Martin Heidegger's *Zollikon Seminars*," in *Zollikon Seminars: Protocols-Conversations-Letters*, ed. Medard Boss, trans. Franza Mayr and Richard Askay (Evanston, Ill.: Northwestern University Press, 2001), xviii.

13. Martin Heidegger, *Being and Time,* trans. Joan Stambaugh (Albany: State University of New York Press, 1996), 46.

14. Martin Heidegger, *The Fundamental Concepts of Metaphysics: World, Finitude, Solitude,* trans. William McNeill and Nicholas Walker (Bloomington: Indiana University Press, 1995); *Contributions to Philosophy: From Enowning,* trans. Parvis Emad and Kenneth Maly (Bloomington: Indiana University Press, 1999). Luca Illetterati accurately analyzes this itinerary in *Tra tecnica e natura: Problemi di ontologia del vivente in Heidegger* (Padova: Paligrafo, 2002).

15. [Esposito's obvious target is Giorgio Agamben's discussion of boredom and the animal in *The Open.*—*Trans.*]

16. Heidegger, "Letter on Humanism," 206.

17. See, on this point, the persuasive essay by Marco Russo, "Animalitas: Heidegger e l'antropologia filosofica," *Discipline filosofiche* 12:1 (2002): 167–95.

18. Cf. Jacob Rogozinski, "Comme les paroles d'un homme ivre . . . : chair de l'histoire et corps politique," *Les Cahiers de Philosophie,* no. 18 (1994–95): 72–102.

19. See the section titled "Degeneration" in chapter 4.

20. Nonetheless, see Antonio Martone, "La rivolta contro Caligola: Corpo e Natura in Camus e Merleau-Ponty," in Bazzicalupo and Esposito, *Politica della vita,* 234–43.

21. Maurice Merleau-Ponty, "Philosophy and Non-Philosophy since Hegel," in *Philosophy and Non-Philosophy since Merleau-Ponty* (New York: Routledge, 1988; reprinted, Evanston, Ill.: Northwestern University Press, 1997), 63.

22. Maurice Merleau-Ponty, *The Visible and the Invisible,* trans. Alphonso Lingis (Evanston, Ill.: Northwestern University Press, 1968), 147.

23. See Didier Franck, *Heidegger et le problème de l'espace* (Paris: Éditions de Minuit, 1986).

24. Merleau-Ponty, *The Visible and the Invisible,* 248.

25. The work that has excavated this terrain the most deeply and with innovative results is Lisciani Petrini's *La passione del mondo: Saggio su Merleau-Ponty* (Naples: Edizioni Scientifiche Italiane, 2002).

26. Merleau-Ponty, *The Visible and the Invisible,* 258–59.

27. Ibid., 264.

28. See again Petrini, *La passione del mondo,* 119.

29. Maurice Merleau-Ponty, *The Nature: Course Notes from the Collège de France,* ed. D. Seglard, trans. Robert Vallier (Evanston, Ill.: Northwestern University Press, 2003), 103. In this regard, see too the chapter that Élisabeth de Fontenay dedicates to Merleau-Ponty in *Le silence des bêtes: La philosophie à l'épreuve de l'animalité* (Paris: Fayard, 1998), 649–60.

30. Maurizio Carbone has reconstructed the reasons, tracing in turn a twentieth-century genealogy of the theme of flesh in "Carne: Per la storia di un fraintendimento," in *La carne e la voce: In dialogo tra estetica ed etica,* ed. Maurizio Carbone and David M. Levin (Milan: Mimesis, 2003).

31. François Lyotard, *Discours, figure* (Paris: Klincksieck, 1971), 22; and Gilles Deleuze and Félix Guattari, *What Is Philosophy?,* trans. Hugh Tomlinson and Graham Burchell (New York: Columbia University Press, 1994), 178.

32. Jacques Derrida, *On Touching—Jean-Luc Nancy,* trans. Christine Irizarry (Stanford, Calif.: Stanford University Press, 2005), 236, 238.

33. Jean-Luc Nancy, *The Sense of the World,* trans. Jeffrey S. Librett (Minneapolis: University of Minnesota Press, 1997), 149. I have previously anticipated these critical

reflections in "Chair et corps dan la déconstruction du christianisme," in *Sens en tous sens: Autour des travaux de Jean-Luc Nancy,* ed. Francis Guibal and Jean-Clet Martin (Paris: Galilée, 2004), 153–64.

34. Davide Tarizzo provides a descriptive map of contemporary French philosophy in *Il pensiero libero: La filosofia francese dopo lo strutturalismo* (Milan: Cortina Raffaello, 2003).

35. Michel Henry, *Incarnation: Une philosophie de la chair* (Paris: Éditions du Seuil, 2000).

36. Jérôme Alexandre, *Une chair pour la gloire: L'anthropologie réaliste et mystique de Tertullien* (Paris: Beauchesne, 2001), 199ff.

37. Cf. E. Schweizer, F. Baumgärtel, and R. Meyer, "Flesh," in *Theological Dictionary of the New Testament,* ed. Gerhard Kittel, trans. Geoffrey W. Bromiley (Grand Rapids, Mich.: W. B. Eerdmans, 1985).

38. Cf. the neophenomenological perspective of Marc Rihir in *Du Sublime en politique* (Paris: Payot, 1991).

39. Cf. Xavier Lacroix, *Le corps de chair, les dimensions éthique, esthétique et spirituelle de l'amour* (Paris: Éditions du Cerf, 1992). On the theme of flesh in Saint Paul, see as well J. A. T. Robinson, *Le corps, étude sur la théologie de Saint-Paul* (Lyon: Éditions du Chalet, 1966).

40. *The Bible: The Authorized King James Version* (Oxford: Oxford University Press, 1997).

41. I previously introduced these themes in *Immunitas: Protezione e negazione della vita* (Turin: Einaudi, 2002), 78–88 and 142–44. A seemingly different reading of the body is present in the ample frame that Umberto Galimberti offers in *Il corpo* (Milan: Feltrinelli, 1987).

42. See in particular Aldo Bonomi, *Il trionfo della moltitudine* (Turin: Bollati Boringhieri, 1996); Paolo Virno, *A Grammar of the Multitude: For an Analysis of Contemporary Forms of Life* (Cambridge, Mass.: Semiotext[e], 2003); Antonio Negri, "Approximations: Towards an Ontological Definition of the Multitude," *Multitudes,* no. 9 (2002) (available at http://www.nadir.org/nadir/initiativ/agp/space/multitude.htm); and Augusto Illuminati, *Del Comune: Cronache del general intellect* (Rome: Manifestolibri, 2003). As interesting and diverse as these perspectives are, the risk ultimately is that the reading of biopolitics that results may be, if not economistic, then minimally productivistic or workerist, and therefore impolitical. Compare, on this point, the observations of Carlo Formenti, *Mercanti di futuro: Utopia e crisi del Net Economy* (Turin: Einaudi, 2002), 237ff.

43. Cf. Georges Didi-Huberman, *Confronting Images: Questioning the Ends of a Certain History of Art,* trans. John Goodman (University Park: Pennsylvania State University Press, 2005).

44. David Sylvester, *Entretiens avec Francis Bacon* (Geneva: A. Skira, 1996), 29.

45. Gilles Deleuze, *Francis Bacon: The Logic of Sensation,* trans. Daniel W. Smith (New York: Continuum, 2003), 67. On the relation between Deleuze and Bacon, see Ubaldo Fadini, *Figure nel tempo: A partire da Deleuze/Bacon* (Verona: Ombre Corte, 2003).

46. Deleuze, *Francis Bacon,* 21.

47. Ibid., 62.

48. For a lucid genealogy of the concept of "nation," see Francesco Tuccari, *La nazione* (Rome-Bari: Laterza, 2000), not to mention Étienne Balibar's "History and Ideology: The Nation Form," in *Race, Nation, Class: Ambiguous Identities,* ed. Étienne Balibar and Immanuel Wallerstein (London: Verso, 1991), 88–106.

49. On the notion of "fraternity," with particular reference to France, see Marcel David, *Fraternité et Révolution française: 1789–1799* (Paris: Aubier, 1987), as well as his *Le Printemps de la Fraternité, Genèse et Vicissitudes, 1830–1851* (Paris: Aubier, 1992).

50. Eligio Resta critically interrogates the possibility of a fraternal right in *Il diritto fraterno* (Rome-Bari: Laterza, 2002).

51. On the relation among friend-enemy-brother, see also Jacques Derrida, *The Politics of Friendship,* trans. George Collins (London: Verso, 1997).

52. For the need of a fraternal brother in Nietzsche, see especially *Thus Spoke Zarathustra,* trans. Thomas Wayne (New York: Algora Publishing, 2003), 42–43, 46–47, 161. For Carl Schmitt, see *Ex captivitate salus: Erfahrungen der Zeit 1945/47* (Cologne: Greven Verlag, 1950).

53. René Girard, *Violence and the Sacred,* trans. Patrick Gregory (Baltimore: Johns Hopkins University Press, 1977).

54. Sigmund Freud, *Totem and Taboo: Resemblances between the Psychic Lives of Savages and Neurotics,* trans. A. A. Brill (Amherst, N.Y.: Prometheus Books, 2000).

55. Sigmund Freud, *Moses and Monotheism,* trans. Katherine Jones (New York: Vintage Books, 1967). On this theme, see as well my *Nove pensieri sulla politica* (Bologna: Il Mulino, 1993), 92–93, as well as *Communitas: origine e destino della comunità* (Turin: Einaudi, 1998), 22–28.

56. Freud, *Moses and Monotheism,* 3.

57. Cf. Phillipe Lacoue-Labarthe and Jean-Luc Nancy, "Il popolo ebraico non sogna," in *L'altra scena della psicoanalisi: Tensioni ebraiche nell'opera di Sigmund Freud,* ed. David Meghnagi (Rome: Carucci, 1987).

58. Compare this reading of the mother–son relation with Angela Putino, *Amiche mie isteriche* (Naples: Cronopio, 1998).

59. Arendt, *The Human Condition,* 9.

60. Cf. Eugenia Parise, ed., *La politica tra natalità e mortalità: Hannah Arendt* (Naples: Edizioni scientifiche italiane, 1993).

61. Arendt, *The Human Condition,* 246–47.

62. Cf. A. Fagot-Largeault, "L'individuation en biologie," in *Gilbert Simondon: Une pensée de l'individuation et de la technique* (Paris: Albin Michel, 1994). See as well the other anthology of essays titled *Simondon,* ed. Pascal Chabot (Paris: J. Vrin, 2002).

63. Gilbert Simondon, *L'individu et sa genèse physico-biologique* (Paris: J. Millon, 1995), 77.

64. Gilbert Simondon, *L'individuazione psichica e collettiva* (Rome: DeriveApprodi, 2001), 84; my emphasis.

65. Ibid., 138.

66. Giorgio Agamben, *State of Exception,* trans. Kevin Attell (Chicago: University of Chicago Press, 2005).

67. The insurmountable aporia in which the polemics between normativism and natural right take place are in plain view in the joint publication of two essays, the first by Ernst Cassirer, "Vom Wesen und Werden des Naturrechts," *Zeitschrift für Rechtsphilosophie in Lehre und Praxis* 6 (1932–34): 1–27, and Hans Kelsen, "Die Grundlage der Naturrechtslehre," *Österreichische Zeitschrift für öffentliches Recht* 13 (1963): 1–37. In 2002, the Italian journal *Micromega,* in its second issue, published a number of essays by Angelo Bolaffi, Stefano Rodotà, Sergio Givone, Carlo Galli, and myself precisely on this theme.

68. For this juridical philosophical interpretation of Spinoza, see above all the relevant essay by Roberto Ciccarelli, *Potenza e beatitudine: Il diritto nel pensiero di Baruch Spinoza* (Rome: Carocci, 2003).

69. Baruch Spinoza, *Political Treatise,* in *Complete Works,* ed. Michael L. Morgan, trans. Samuel Shirley (Indianapolis: Hackett Publishing Company, 2002), 683.

70. Baruch Spinoza, *Ethics,* trans. G. H. R. Parkinson (Oxford: Oxford University Press, 2000), 154.

71. Spinoza, *Political Treatise,* 685.

72. Étienne Balibar, *Spinoza and Politics,* trans. Peter Snowdon (London: Verso, 1998), 64–68.

73. Simondon, *L'individu et sa genèse physico-biologique,* 295.

74. Simondon, *L'individuazione psichica e collettiva,* 188.

75. Ibid.

76. Canguilhem's metapolitical reflections were already expressed in his *Traité de Logique et de Morale,* published in Marseille in 1939. See in particular the last two chapters, "Morale et Politique" and "La Nation et les Relations internationales" (259–99).

77. Georges Canguilhem, "Une pédagogie de la guérison est-elle possible?" in *Écrits sur la médicine* (Paris: Éditions du Seuil, 2002), 89.

78. Georges Canguilhem, *La connaissance de la vie* (Paris: Librairie Hachette, 1952), 12.

79. Cf. Guillaume Le Blanc, *Canguilhem et les normes* (Paris: Presses universitaires de France, 1998).

80. Émile Durkheim, "Rules for the Distinction of the Normal from the Pathological," in *The Rules of Sociological Method,* trans. W. D. Halls (New York: Free Press, 1982), 92.

81. Georges Canguilhem, "New Reflections on the Normal and the Pathological," in *The Normal and the Pathological,* trans. Carolyn R. Fawcett (New York: Zone Books, 1991), 243.

82. René Leriche, "Introduction générale. De la santé à la maladie. La douleur dans les maladies. Où va la médicine?" in *Encyclopédie Française,* vol. 6, 16-I; quoted in Canguilhem, *The Normal and the Pathological,* 91.

83. Cf. Pierre Macherey, "Pour une histoire naturelle des normes," in *Michel Foucault philosophe* (Paris: Éditions du Seuil, 1989), 203–21.

84. Canguilhem, "New Reflections," 258–59.

85. Hans Kelsen, *General Theory of Norms,* trans. Michael Hartney (Oxford: Oxford University Press, 1991), 158–62. On the complex theme of the norm, I will limit my references to Alfonso Catania, *Decisione e norma* (Naples: Jovene, 1979), as well as *Il problema del diritto e dell'obbligatoreità: Studio sulla norma fondamentale* (Naples: E.S.I., 1983). More recently, see also Fabio Ciaramelli, *Creazione e interpretazione della norma* (Troina: Città Aperta, 2003).

86. Canguilhem, *The Normal and the Pathological,* 139.

87. Ibid., 199. For the reference to Goldstein, see Kurt Goldstein, *The Organism: A Holistic Approach to Biology Derived from the Pathological Data in Man* (New York: Zone Books, 1995).

88. Gilles Deleuze, *Pure Immanence: Essays on a Life,* trans. Anne Boyman (New York: Zone Books, 2001).

89. Ibid., 27. See as well René Schérer's "Homo tantum, L'impersonnel: Une politique," in *Gilles Deleuze: Une vie philosophique,* ed. Éric Alliez (Le Plessis-Robinson: Institut Synthélabo pour le progrès de la connaissance, 1998), 25–42, and Giorgio Agamben,

"Absolute Immanence," in *Potentialities: Collected Essays in Philosophy,* trans. Daniel Heller-Roazen (Stanford, Calif.: Stanford University Press, 1999), 220–39.

90. Deleuze, *Pure Immanence,* 27.

91. [English usage doesn't require the determinate article, "the," with any regularity, thus "life" and not "the life," but I have retained the article as Esposito's analysis makes little sense without it. The interested reader is also directed to the closing pages of *Immunitas* in which Esposito discusses at length in a different setting the use of the determinate article preceding self as in "the self."—*Trans.*]

92. Charles Dickens, *Our Mutual Friend* (New York: Alfred A. Knopf, 1994), 443.

93. Ibid., 444.

94. Deleuze, *Pure Immanence,* 29.

95. [I have translated *cosí* as "singular" following the English translation of Deleuze. Thus Deleuze writes: "The singularities and the events that constitute *a* life coexist with the accidents of *the* life that corresponds to it" (29).—*Trans.*]

96. Gilles Deleuze, *The Logic of Sense,* trans. Mark Lester with Charles Stivale (New York: Columbia University Press, 1990), 103.

97. Lawrence Ferlinghetti, "Il," in *Un regard sur le monde* (Paris: C. Bourgeois, 1969), 111.

98. Deleuze, *The Logic of Sense,* 103.

99. For the problematicity of the virtual in Deleuze, in relation to the logic of immanence, see the intense and acute monograph that Alain Badiou dedicates to it in *Deleuze: The Clamor of Being,* trans. Louise Burchill (Minneapolis: University of Minnesota Press, 1999).

100. Simondon, *L'individu et sa genèse physico-biologique,* 260; quoted in Deleuze, *The Logic of Sense,* 104.

101. Deleuze, *The Logic of Sense,* 107.

102. Dickens, *Our Mutual Friend,* 443, 444.

Index

Roberto Esposito teaches contemporary philosophy at the Italian Institute for the Human Sciences in Naples. His books include *Categorie dell impolitico, Nove pensieri sulla politica, Communitas: Orgine e destino della comunità*, and *Immunitas: Protezione e negazione della vita.*

Timothy Campbell is associate professor of Italian studies in the Department of Romance Studies at Cornell University and the author of *Wireless Writing in the Age of Marconi* (Minnesota, 2006).

THE SACRED COMPLEX

THE SACRED COMPLEX

On the Psychogenesis of *Paradise Lost*

William Kerrigan

Harvard University Press
Cambridge, Massachusetts
and London, England
1983

Printed in the United States of America
10 9 8 7 6 5 4 3 2 1

This book is printed on acid-free paper, and its binding materials have been chosen for strength and durability.

Publication of this book has been aided by a grant from the Andrew W. Mellon Foundation

LIBRARY OF CONGRESS CATALOGING IN PUBLICATION DATA

Kerrigan, William, 1943-
The sacred complex.

Includes bibliographical references and index.
1. Milton, John, 1608-1674. Paradise lost.
2. Psychoanalysis and literature.
I. Milton, John, 1608-1674. Paradise lost.
II. Title.
PR3562.K4 1983 821'.4 83-4367
ISBN 0-674-78500-2

For my parents,
Barbara and Wally

PREFACE

WORK ON THIS BOOK started during a leave of absence from the University of Virginia in 1974-75, when I began studying Renaissance medicine, hoping to discover Milton's conception of his blind body, and simultaneously began studying Freud at the Forum on Psychiatry and the Humanities of the Washington Institute of Psychiatry, hoping to be able to interpret Milton's conception. The second venture proved more fateful than the first. Gradually I was given to know that the benign demon of this book wanted me to read and meditate rather than pursue the sort of literary "research" conventional in Renaissance studies. I am grateful to have been supported by three grants from the Research Committee of the University of Virginia. Much of the actual writing was completed during another leave of absence in 1979-80, partially financed by my appointment as a Sesquicentennial Associate of the Center for Advanced Studies.

It will be evident to readers of my earlier book, *The Prophetic Milton* (1974), that the concerns I now bring to Milton, if unchanged in some respects, have undergone considerable revision. I associate the difference with James W. Earl of Fordham University. For a long time—though not long enough—we were able to share our minds about everything we read and experienced, or wanted to read and experience. No one has better exemplified for me a spirit of free inquiry unencumbered by dogma or egotism.

My largest intellectual debt is to the work of Paul Ricoeur. This patient philosopher has guided me through psychoanalysis, phenomenology, metaphor, symbolism, and to some extent through religious myth and theology. I do not possess his faith in Christian myth, and in those few places where his work makes me uneasy I feel that he is assuming a new version of the oldest argument for the existence of God—namely, that God exists because he is symbolized. But, like Ricoeur, I am convinced that there remains an indissoluble core of meaning unique to religion after every negative critique has had its say. My broadest question arises from this conviction: what could be said about human nature were Freudian psychoanalysis and Miltonic Christianity to be allowed their truth? *Freud and Philosophy*, with its sharp

distinction between a "space of fantasy," where symbols repeat and reanimate, and a "space of culture," where symbols initiate and transcend, provides the conceptual background for this interpretation of Milton. My book searches for the border between fantasy and culture—the points at which psychoanalysis is all at once confirmed and led beyond itself.

The organization is less eccentric than it may appear to be in the first two chapters, and I trust that by the end of the second its rationale will be clear. But the reader may appreciate some preliminary guidance. I approach my theory of the "sacred complex" gradually and within the context of Milton's life. After a brief chapter anchoring questions posed generally in the Introduction, I consider the problematic temptation of *Comus* (Chapter 2) and the paradigmatic temptation of *Paradise Regained* (Chapter 3) as indications, respectively, of the false settlement overcome in the second phase of Milton's artistic career and the abstract formula of this reversal. The remainder of the book explores the religious symbolism of *Paradise Lost*. What do these symbols tell us about the creation of the poem? How can the meanings of these symbols be thought to escape their genesis in Milton's psychic history? Criticism must somehow abbreviate when faced with the bulk and almost unremitting sublimity of the epic. Hoping to combine the up-close of local interpretation with the back-far of theoretical reflection, I have focused on passages dense with its major themes: the invocation to light (Chapter 4): the dialogue on nourishment, matter, and accommodated speech (Chapter 5); the *felix culpa* of the final books and the six great lines at our exit from the epic (Chapter 6). For an aesthetic justification of this procedure the reader may turn to the discussion of the enfolded sublime in Chapter 5.

Dr. Joseph Smith, the teacher of my first course in Freud, has been indispensable during my labor on this project. I feel honored to serve as his colleague and collaborator in the interdisciplinary activities of the Forum on Psychiatry and the Humanities. He will recognize the seminal importance for me of his articles on the absent object, primitive ideation, the Freudian and Heideggerian ideas of affect, and the translation through deferred mourning of separation anxiety into castration anxiety. My father-in-law, Dr. Donald Burnham, listened with customary generosity to rough formulations of many of the ideas in this book. They have benefited from his native shrewdness as well as his professional expertise. It almost goes without saying that these two psychoanalysts share responsibility for my felicities, not my lapses. I would be pleased if my own pages were to betray something

of their ability, born of a wisdom beyond system, to release unsuspected resources in familiar psychoanalytic concepts.

My notes fail to indicate the degree to which I have been instructed by Milton scholarship. Among the early Miltonists, I have learned much from the "complicated Richardson," as the eighteenth-century authors, father and son, of an enduring achievement in Milton studies called themselves. The general atmosphere of the Romantic Miltonists, the sense that here was a poetic Titan struggling with the forces sustaining the universe, remains an inspiration; in the age of Freud one can learn from the banal chatter of a television talk show that life is the constant management of conflict, but one must go back at least as far as Romanticism to appreciate the size of the conflicts Milton set in motion (Satan and God, death and justice, authority and liberty, temptations of all kinds) and the power he needed in order to compose them. Like all Miltonists, I am indebted to the three figures that towered over this field in the first half of our century: Denis Saurat, who addressed the thought of Milton, not as an artifact to be nailed to sources and traditions, but as the living speculation of a man profoundly in touch with the logic of Christian myth at just the moment when its salvation history was about to mutate into transcendental idealism; James Holly Hanford, who illuminated everything he saw in Milton and saw an aspect of nearly everything; and E. M. W. Tillyard, whose rich and unpredictable wit was often more insightful than his measured judgments. Today their legacy is being handsomely overseen by William Hunter, Edward Le Comte, Barbara Lewalski, Michael Lieb, and others.

But with respect to key issues of interpretation that cannot be decided on the basis of local passages alone, depending instead on a sense of the entirety of Milton, on what Saurat termed the "man and the thinker," my relationship to contemporary Milton studies is to some degree an adversary one. While I find little to applaud in the schoolboy debunking of Waldock and Empson, I also find little farcical in the portrayal of Satan. The treatment of this creation by many modern critics strikes me as an instance of a distinct literary syndrome, The Denial of the Most Interesting Crux, on a par with claiming that Hamlet does not delay or that Cleopatra dies to reassure us that *cupiditas* truly is a folly. (What sort of discipline is it, to paraphrase Durkheim's reaction to a similar situation, whose principal discovery is that the main problem it was created to solve does not exist?) I am unable to read *Paradise Lost* as a comfortably orthodox poem, however intricate the rhetorical means devised for eliciting our stock responses

or reconverting us to commonplace positions, without a feeling of impoverishment. Nor am I able to understand how an artistic career ending with the "rousing motions" of *Samson Agonistes,* which command from within the will of an otherwise impervious hero, and the "inward Oracle" of *Paradise Regained,* which renders books and other doctrines superfluous, can be characterized as an undeviating crusade to the Holy Land of Reason. In these areas of interpretation my major benefactors include W. B. C. Watkins, whose *An Anatomy of Milton's Verse* remains the best introduction to the poet, and Harold Bloom, whose extraordinary disclosure of the Milton encountered by the finest poetic minds in our language should prompt us to reassess the Milton encountered by our scholarship.

Edward Tayler read drafts of two of these chapters and strengthened my grasp on things I had already not quite said. I am grateful for the timely encouragement of Philip Gallagher and Robert Fallon. Among my colleagues, James Nohrnberg often shared with me, literally at a moment's notice, the current state of that vast coherence he seems always to be elaborating. Ralph Cohen prevented me from concluding prematurely that I understood the invocation to light. Ernest Gilman put me in the way of an idea about riddling in Milton. Leo Damrosch's excellent seminar in Puritanism reacquainted me with the commanding importance of theodicy in *Paradise Lost.* Gordon Braden was incalculably helpful when it came to revising and rearranging. When it came to indexing, I served as assistant to the meticulous Patricia Welsch. Leaving unnamed many students who have broadened and clarified my thinking, I must at least mention, with more warmth than a list can suggest, the contributions of Jon Harned, Reginald Bell, John Peter Rumrich, Emily Miller, William Fuller, Regina Schwartz, Nahla Hashweh, and Beth Ash.

Amelia Burnham Kerrigan is everything a man could want in a therapist. To begin with the letter, her ministrations permitted this book to be written: Brian and Ann were not always napping, nor Christopher always occupied, when the scholar plied his elsewhere slighted trade. To close with the spirit, the long education and sometimes punishing practice of a psychiatrist have not blunted her regard for the mystery of Being. It was the fullness of our love, more than anything else, that permitted this book to be thought.

I have dedicated it to the couple who will always live within me.

W. K.

CONTENTS

THE SACRED COMPLEX

Hermeneutics seems to me to be animated by this double motivation: willingness to suspect, willingness to listen; vow of rigor, vow of obedience. In our time we have not finished doing away with *idols* and we have barely begun to listen to *symbols.* It may be that this situation, in its apparent distress, is instructive: it may be that extreme iconoclasm belongs to the restoration of meaning.

Paul Ricoeur, *Freud and Philosophy*

INTRODUCTION

Milton gave heartfelt answers to some astonishing questions as he designed a universe in which he could understand as well as dramatize the fatal human decision of Christian myth. How might the Son be related to the Father? Where does matter come from? Do angels eat? Is there time in heaven? Why does death, the sentence of fallen nature, exemplify justice? After three centuries of progressively astringent epistemology, including the cumulative religious disillusionment wrought by Feuerbach, Marx, Durkheim, Nietzsche, and Freud, living on a world that has been photographed from outer space, it is impossible for me to read Milton wholly inside the worldview that made these questions urgent and their answers plausible.

If never uniformly and instantaneously, fundamental change happens. One of its effects is to render unmistakable the symbolic, projected, even illusory character of what was, for the age now known as previous, the merely real. This effect is initially devastating in the case of religion, producing a reclassification of religious thought that no amount of historical empathy can undo entirely. When Milton, feeling licensed by inspiration to reject traditional solutions, ranged through the theological issues embedded in the images of Christian myth, he moved in a realm where the coherence exacted by reason "from above" had few safeguards to protect itself from secret accommodation with the imperatives of the unconscious arising "from below." This truth, our truth, seems foreshadowed in the very procedures of his mythmaking, waiting there for our recognition. For Milton did not seek an independent and purified reason—just the opposite. As a theologian he attributed his positions to the agency of the Spirit, aware that extrarational forces inclined his conviction to this harbor or that, and as a poet he knew himself to be writing unpremeditated verse, given sometimes in dreams and likened to a bird's song.

Some critics regard the assumption of fundamental change as either exaggerated or, if true, then only for the worse, seeking to persuade their readers and students that the heroism proper to a modern intellectual consists in mocking with great shows of learning the hubris implicit in the idea of modernity itself. They tend to call any approach

to literature that does not begin and end in terms accessible to the author or his first readers "decadent," but if decadence is the refusal to encounter the pressing issues of a current situation, it is not difficult to reverse this indictment. Others engage the problem abstractly. An immense and growing body of literary theory exists to advise readers about whether they should or should not reconceive the meaning of past texts on the grounds of what is credible today. Were it not for a widespread sense of fundamental change, this advice would have no point—hence the folly of trying to adjudicate such a matter in theoretical terms: someone for whom the change has occurred *must* reconceive the achievement of past authors on this basis or else he cannot, reverting to the author at issue here, think Milton and at the same time think consequentially. The effort to do both, one would suppose, declares the value of Milton's accomplishment.

Or perhaps creates it. As the givens of a past ontology become manifestly self-given, cultural rather than natural, we undertake the labor of historical explanation, addressing the "how" and "why" of this old order. Almost everyone agrees that a literary work arising from and referring back to a discarded image can be studied as such without becoming obsolete. One may argue in several ways that the literary work possesses an autonomy, escaping the fate of propositional language through an inherent acknowledgment of its own createdness or symbolicity; whether it meant to or not, literature displays the true character of the discarded image, its fiction, and is therefore exempt from the reclassification that discards. One may argue in more complicated and satisfying ways that the literary work enjoys a special mode of existence, not timeless, monadic, and "in itself," but temporal, diverse, and "for us," subject by its essence to endless reinterpretation and free in this respect from the fate of its initial horizons. I prefer the second kind of argument because the first would secure the unique invulnerability of the work by sheer prescription, as if art's immortality were given in its nature rather than won, time after time, by its power to occasion the enlightenment of disparate audiences. Whatever our theoretical strategy for immunizing texts, literature will not escape from the iconoclasm of change by virtue of how it is defined. The definition is prelude to a task, and this task will be pretty much the same for readers of every theoretical persuasion. The survival of literature as anything more than artifact depends on our ability to extend its original reference into a genuinely revelatory description—not, let me emphasize, a retracing of one or another orthodox description—of the world we inhabit now.

Adopting this task, we face the threat of an irreparable cleveage in

the ongoing activities of literary study. Does not the explanatory function of historical scholarship, which appears to concede the pastness of the work in its search for intentional meaning, oppose the effort to extend or renew its reference? The problem is somewhat ameliorated by the recognition that some of what passes for historical scholarship committed to restoring or recollecting intentional meaning in fact levels off this meaning by giving undue weight to artificially constructed "traditions" and "conventions," forcing the sense of great texts into a simplistic redundance of past commonplaces—a result comparable to the meanings achieved in clumsy efforts at modernization. Still, by endeavoring to set a psychoanalysis of Milton's symbolism alongside a reflection on his symbolism, this book must of necessity wrestle with its own version of the threatened cleavage between historical placement and contemporary renewal.

Psychoanalysis is often classed uncritically, by opponents and practitioners alike, with so-called "ahistorical" approaches to literature. The grouping is false, first of all, to the procedures of psychoanalytic interpretation. Although he broadens the usual notion of intentional meaning to include the unconscious, the psychoanalytic critic cannot find this extended intentionality without first having found the customary one: the intended meaning, the minimal possession of all literary study, is the meaning he interprets. The locale of this "second-order" interpretation, this meaning behind intentional meaning, continues to be historical. The psychoanalytic critic is a literary historian. More strictly than any other mode of literary history, psychoanalysis appears to limit meaning to its origins. One cannot overstate the genetic, anti-teleological bias of a theory that discovers the very energy of psychic life coiled into the structure of a wish (the optative, the fundamental mood of childhood), then defines wish as the desire to return to an earlier state of affairs. As a type of literary history, psychoanalysis comes up against the problem of the pastness or fixity of meaning in an especially radical form. Banishing incredible gods from the imagination, it contributes answers to our interrogation of the past. How did you come to be? Why did you disguise your creation as another's? As you served your order, how did your order serve you? But psychoanalysis purchases its explanations at the price of apparent damage to the semantic potency of the work.

Psychoanalysis is not a "method" in the Renaissance sense. It will not place us beyond ourselves, like the compass that permits a fool to outgo Leonardo working freehand. If brought to literature with respect for the cultural life of symbols, its principles can disclose new configurations in the structure of a work or a career. Its presence as a

possible interpretive stance has repercussions for all the other stances: psychoanalysis questions, as we will see, the concept of aesthetic unity prevalent in Milton studies. More important, since these explanatory principles assume a developmental history, psychoanalysis can (unlike structuralism, for example) *animate* its interpretation, allowing one to generate a narrative of the creative act oriented toward its "second life" as the source of our pleasure and our wisdom. In the terms of Paul Ricoeur, one can try to capture meaning in its flight from the "space of fantasy" to the "space of culture."[1]

But when the Freudian narrative comes upon the finished composition from which, as the evidence for inferences, it began, the barrier appears. The reference of psychoanalytic discourse is pinned to the fantasy side of the text, unable to break loose from archaic materials in the mind of the author. There are many confused or bogus reasons for disappointment over psychoanalytic interpretations, but this one is pertinent: how can psychoanalysis contribute to the ongoing vitality of the work? It does not help matters to shift the locus of discourse from the author and the text to the text and the audience, purporting to show, for instance, that the work activates universally conflictual ideas, then pacifies us through the defensive transformations of aesthetic form. For here, too, albeit on the other side of the text, meaning is funneled back into childhood. Is literature but a series of ornate scenarios for our happy adjustment to the unbearable?

So the problem of confining meaning to the past, which may beset any historical approach to literature, belongs to psychoanalysis intrinsically. This psychology shares with linguistics, sociology, and anthropology—disciplines whose advancement in recent years has been achieved by adopting models of synchronous order, trivializing change—a difficulty in accounting for, even in admitting as data to which theorizing must be answerable, the fact of significant innovation. Analysts have often noted that Freud's idea of sublimation, which is one of the few notions in classical psychoanalysis to acknowledge the futural thrust of psychic life, was never made consistent with his metapsychology. One can point to the sublimation of the artist or scientist. One can describe it: whereas a symptom replaces the object of an instinct, a sublimation would imply, beyond this change in objects, a redirection of the *aim* of the instinct. But it is difficult to comprehend how, given the rest of psychoanalytic theory, such a process could come about.[2]

The ego psychology of Anna Freud, Hartmann, Kris, Rapaport, and Erikson is sometimes said to have overcome the predominant "ar-

chaeology" of Freud. This supposed victory has proven useful for psychoanalytic students of the humanities in defending their work and its assumptions against the charge of reductionism. Looking carefully, however, one notices that the independent strength of the ego is more often asserted than investigated in detail. I suspect that the reasons for this vagueness lie in the theories being drawn from. By stressing the adaptational role of fantasy, defense, and regression in general, the ego psychologists have offered a partial relief at best from Freud's powerful and knowingly reductive emphasis on the conservatism of human desire, our many ruses for holding fast to our earliest objects and satisfactions: they have simply deflected the problem of meaningful innovation into a realm of function, and although meaning has a function, meaning itself is not a function. One intuits that the functionalism of ego psychology applied to literature, like the functionalism of social anthropology applied to myth and ritual, falls short of addressing the sort of meaning that, in the end, gives these creations value.

To avoid the entrapment of meaning in either repetition or function I have tried to let my psychoanalysis of the life and character of Milton interact with the parabolic openness of his art. This art is the master expression of a religious vision coherent, sizable, and for all its antecedents, original enough to be termed "Miltonic Christianity." There is nothing to be gained by hoodwinking his vision in the name of some presupposed clarity. I have no interest in yet another demonstration, annexing this poet for the colonial empire of "applied" psychoanalysis, wherein the principles of explanation borrowed from psychoanalysts are assumed to be sturdy enough to eliminate the uncertainties of other fields. Often they are, and I have profited from many such endeavors. But how can one be confident of having solved a problem psychoanalytically when the psychoanalytic principles of explanation, on which all results depend, have not been thematized? I could not refrain altogether from this imperfect interdisciplinary mode without producing an unreadably restless and self-referential book, but I have tried here and there to move my discourse toward psychoanalysis itself, which is, after all, part of the world Milton may illuminate for us. "A reduction from one system to another," Nelson Goodman has argued, clearing up some familiar confusions about this matter, "can make a genuine contribution to understanding the interrelationships among world-versions."[3] If, however, we are not to undertake this reduction for its own sake as a formal exercise, we must proceed in a fully interdisciplinary attitude, putting the antagonistic

assumptions of these world-versions—in this instance, Freudian psychoanalysis and Miltonic Christianity—at risk. My aim is to allow the current of explanation to flow from psychoanalysis to Milton *and* from Milton to psychoanalysis, thus enabling the poet to lend his significance to the theory that elucidates him.

The area of psychoanalysis specifically at risk here is the theory of the superego. One cannot begin to think psychoanalytically about Milton without realizing that the oedipus complex is the generative center of his character and his art. From the viewpoint of psychoanalysis, of course, this could be predicated of all authors who are not psychotic; but among them only Milton created a religious poem so great as *Paradise Lost* and so intensely concentrated on the issues of the complex. Concerns prominent in his epic—the justification of the ways of the Father, the freedom found in serving the Father, the failed heroism of Satan's rebellion, the tragedy of human disobedience, the reconciliation to death, the reworking of the Trinity into an uncreated Father and an indebted Son—impress the situation of an oedipal child on a religion that already derived our salvation from an identification with a divine and ideally obedient Son whose love was manifested to us through a sacrificial death demanded by paternal justice. In this sense Milton all but delivers up his religion to the secular verdict of oedipal interpretation. Yet in another and no less significant way, psychoanalytic readers engaging his epic have the opportunity to submit their theory of the superego to the intelligence and sublimity of Milton's faith.

The importance of Freud in the history of religious thought does not rest on the coincidence of his intellectual authority and his atheism, but on the fact that he proposed one of the few compelling explanations for the universal appeal and tenacity of religious belief. Comments on religion are frequent in his works, and if we group *Totem and Taboo* and *Civilization and Its Discontents* with *The Future of an Illusion* and *Moses and Monotheism*, he devoted four books to the subject. Unlike some inheritors of his science, Freud recognized how foolish it would be first to construct a psychology and then to ask what this psychology can make of religion; a psychology completed without reference to religion and other cultural phenomena has called itself into question by this very decision. Freud began rather than concluded his investigations with the understanding that religious experience is illusory, which is not to say merely false, for his new discipline was in large measure a hermeneutics of the manifold agencies of illusion in our experience. Certain it is that the first psychoanalyst

did not have the last word on religion when, like a final antitype of the Moses with whom he wrestled all his life, he brought us the laws of illusion. Nevertheless, his negative revelation is nobler, even in its provocative way more inspirational than one would gather from apologists who pass him by with a few quick gibes.

Atheism made Freud uneasy. Religions had always helped mankind to renounce instinct and cooperate with inhibiting rules. To know them as illusory was to add to warfare and revolution yet another sign of how precarious an undertaking civilization was. At the bottom of this uneasiness stood suppositions about the amorality of the ego. Having exposed rewarding and punishing gods as regal creatures of the superego, Freud could entrust the future of morality only to some lessened superego: there was no impetus in the ego proper to attempt ethical self-determination. When Freud spoke of the "categorical imperative" of the primitive superego, he was ironically taunting the autonomous rationality of Kant's ethics while at the same time laying the groundwork for an eventual solidarity with Kant. Through education the maker of gods might descend into this world, lending its authority to the ego as maker of sciences. Heir to an irrational drama of fantasy, symbol, and affect, the bare imperative of the early superego—"Thou shalt" and "Thou shalt not"—might one day serve no higher power than "impersonal" respect for mankind.[4]

In the wake of Freud, psychoanalysts have tended to think of ethical maturity as the gradual assimilation of the superego into the ego. There is nothing esoteric about the sources of this attitude. The clinical literature is full of sorrow and violence produced by the easy translation between primitive conscience and deity. History itself in this regard is a clinical literature. The oedipal child yields to a superior will; the institution of law in psychic development occurs as far as possible from rational assent. Our happiest strategy for domesticating the archaic voluntarism of the superego, in commonsense as in the mainstream of psychoanalytic thought, is to forge an explicit ethics, absorbing the regulatory function of the superior will by making our duties both concrete and internally consistent: a mature ego can command itself, manage its own economy of value.[5] Religion rarely enters into the discussion of morality in contemporary psychoanalysis with the problematic force it possessed for Freud.

I would like to recover something of the original urgency of the encounter between religion and psychoanalysis. If psychoanalysis would guard us against the primitive illusions of religion, perhaps religious affirmations of the superego would guard us against the civilized illu-

sions of a self-authenticating ego. For the superego is not, like the unconscious, accessible only through an interpretive process. It is immanent: a complement to desire, an other who prods and forbids, requites and accuses, inhabits my consciousness. Putting aside the question of a God beyond the mind that senses, signifies, and attests to God, my study of Milton has led me to posit the existence of a "sacred complex" refashioning the superego and reworking the layers of mental structure implicit in the superego. The acquisition of this first and secular conscience, from the beginning a scandal in the eye of the ego, is the condition in psychic history for the religious experience of Milton, or of Paul, Augustine, Luther, and Kierkegaard. Without concealing from our view the anxieties generated by a religious expansion of conscience, these figures speak to us of what may become precious and creative in the intensification of our singular internal relatedness. The superego may make us ill, may make us ordinary, or may fade into an ego ready for solitary self-regulation: but there is another and uniquely religious vicissitude of the superego. Traced to the turmoil of the oedipus complex, God appears to be illusion. Such an illusion, due in part to the process of excavation by which it has come into our ken, seems at first overwhelmingly conservative—a strategy for preservation. Yet it may have a future. The dynamics of the sacred complex turn against the oedipal settlement they also preserve, liberating the ego to some degree from its secular submission. Through religious symbols Milton was able to empower himself in a manner directly opposed to the dissolution of superego into ego. The way beyond the complex began inside the complex. The destination was a poem freed from the past, not only as an authentic gift to our culture, but as an act of creation transcending the psychological conditions of its possibility.

• 1 •

BEGINNING WITH A TRUE POEM

But such as are good men can give good things.

—The Lady of *Comus*

FAMILIES, too, have their myths. Elizabeth Milton told John Aubrey, an interviewer who delighted in these summary anecdotes, a story she had from her late husband, author of *Paradise Lost* and defender of regicide. It was probably an oft-told tale in the family mythology—a story of origins, a way for the sometimes recklessly opinionated man at the head of the family to explain why Miltons should be as he was. A generation back there had been a break in the history of his lineage reflecting as in a microcosm the conflict throughout all Europe between old law and new spirit. One day the grandfather of the poet, Richard Milton, had discovered his son John reading an English Bible in their Catholic home. An argument ensued and John was banished from the family.[1] We have in one brief domestic tragedy the drama of the Reformation: a Protestant son claims his right to search the divine book, guided by an internal Spirit whose prompting exempts him from paternal authority. For this new line of Miltons, the myth implies, religious conscience should take precedence over every other form of obligation.

When in time the son of this ousted Protestant, bearing his father's name, became himself a family man, he acquired a family Bible. He chose the King James version of 1612, a copy that had passed through the homes of previous owners. Life would be nearer to psychic life if he had inherited the same book his father had dared to examine in the house of his grandfather. For Milton justified the spirit of his father, devoting his prose to an uncompromising defense of the right to interpret this book and his poetry to a representation of the book he rightfully interpreted. The absolute claim of religious conscience made him blind, endangered, impoverished—and made him what he is to our culture, the major proprietor in English poetry of the Christian Bible.

He made the 1612 Bible his own. On the blank page before the opening of Genesis he recorded his beginning. "John Milton was born the 9th of December 1608 die Veneris half an howr after 6 in the morning." The hour mattered. We possess the horoscope he had cast for himself.[2] A son of dawn, the young poet began his Nativity Ode, a "birthday present" for Christ, on the "first light" of Christmas morning 1629, and crowned the consolation of "Lycidas" with the finest rendition in our language of the everyday metaphor of sunrise as rebirth; the inspiration for *Paradise Lost* would strike the mature visionary either in dreams or "when Morn / Purples the East."[3] Beneath his own birth he inscribed the birth of his younger brother Christopher (but not the hour), preserving in this sacred setting the origins of the two male offspring of his father. Anne, their older sister, is represented by her two male children, Milton's nephews Edward and John Phillips. There follow the births of his children by Mary Powell—Anne in 1646, Mary in 1648, John in 1650, Deborah in 1652. Another hand records the aftermath of Deborah's birth for the blind owner of the book: "my wife hir mother dyed about 3. days after. And my son about—6. weeks after his mother." The final entry announces the birth in 1657 of "Katherin my daughter," followed by her early death, and in a subordinate clause, the related death of "Katherin my second wife." The entries suggest a bias in the book, the culture of the book, and the owner of the book. His daughters are born, but his wives merely die. His brother is born, but his sister joins the family record only through her male children. This is the bookkeeping of a patriarch.

The practice of inscribing births and deaths in a family Bible was conventional in Protestant homes and remains common to this day. But it will repay an admirer of *Paradise Lost* to open this old Bible in his mind and reflect at some length on what he sees there. On one page the birth of John Milton and the history, brutal as a tombstone, of his wives and children and nephews; on the next page the birth of the world and the history of the family of man. The last date on the first page is 1657. *Paradise Lost*, begun about this time, hovers ten years in the future, when Milton will publish for fit audience his own version of the births and deaths registered on the next page under the name of a creating God. It might be said that the title page of this new Genesis is a third creation combining, in the provisional eternity of art, the two contiguous leaves of the family Bible.

In the family Bible we behold *Paradise Lost* unmade, disassembled. The disparities implicit in the great harmony of the poem could not be

clearer. Long before the birth of Milton the Bible had become an unrepeatable book of books or, put another way, a book that, as the closure of prophecy, could only be repeated—our access to the sacred, God-given. It was not repeated uniformly. As Milton argued in *Areopagitica*, the full meaning of the book lay dispersed in innumerable commentaries, theologies, works of art, heteronomous acts of interpretation. From the birth of Milton, who will one day gather and recorder the discordant pieces of this significance, depends a bare and finite record. We note the premature deaths, imagine his sorrow; we infer his blindness from the presence of an alien script. The family Bible poses the mystery of composition as a wonder and a challenge to psychology. How was this man able to extend his paternity to the recollected coherence of the Word of God? How was he able to wrest from the finite resources of his existence the three dimensions of the sacred mastered in *Paradise Lost*: the demonic, the divine, and the human? We may wonder how the son of a country glovemaker could locate in himself the evil of Macbeth or the tenderness of Cordelia. But the narrative voice of *Paradise Lost* enacts its victory over finitude, emerging from the rooted circumstances of the invocations to manifest sacred spaces lying long flights away from these roots. There is no poem in our language that so openly represents the complete dilation of a human being to the uttermost reach of his desire and his capacity.

The patriarch is not, as his Satan will try to be, self-fathered. A pattern derived from the previous generation has been repeated here: like his own father, Milton named his first daughter Anne and his first son John. No less than any man's, Milton's is a contingent beginning, and his creations proceed down the page under the influence of a prior history—an unmentioned mother and a father who need not be mentioned, since to name the son is to name him. "John Milton was born . . ." Human genesis demands a passive verb. When we turn the page we move to the absolute beginning of God—but not, suggestively, as Milton read this book. In recreating the divine genesis, he will present the absolute beginning of traditional biblical exegesis as a contingent and therefore heretical beginning—not the creation of God untroubled by priority, but the work of a Son who is himself, precisely like the patriarch on the previous page, the created image of a prior and invisible Father. "In the beginning God created." For Milton, the "God" of this primal sentence signified Christ, who like every son could receive in full propriety the name of his father. As the juxtaposed pages begin to seem, from the perspective of *Paradise Lost*, more similar than contrary, the emblem of the family Bible supplies a

first context for our inquiry into the psychogenesis of Milton's epic.

Particulars represent universals, instances become exemplary. Endemic to language, structures of typification also inhere in familiar acts of mind such as memory and perception, where any one view of a tree may serve to represent the whole tree or the whole class of trees. There is, like the tie between signifier and signified in language, an arbitrary relationship between the partial and the entire in these cognitive acts. The class may be indicated by one or another of its members, the maxim "do not underestimate your enemy" illustrated by one or another cautionary tale drawn from history. But neither religious nor (usually) poetic meaning is given in this fashion. Here specific symbols engender specific meanings, rather than being arbitrarily attached to meanings discoverable without them; even the apparent exceptions, allegory and fable, tend as literary forms to become nonillustrative, their symbolic sense inextricable from the thickened texture of their "naive," "literal," or lexical sense. Assuming as an axiom that "father" and "son" will never be purely logical ideas, cleansed of psychic history, for any man, the necessary bond between symbol and symbolized directs our attention to the psychological genesis of a poem such as *Paradise Lost*: the nature of *this* God depends on the status of *these* symbols. In the sphere of logic the representative and the represented are found together in the same signifying intention. Socrates is, being Socrates, a man; political indolence, recognized as such, is already exemplary. Yet it is by no means clear that the father Milton repeated in naming his children belongs to the category of divinity. The logical necessity of this connection cannot be guaranteed, nor its status assessed, through a linguistic analysis of the symbolism of the poem. We must descend beneath language to the birth of symbols in life. The semantic event in which "father" and "son" generate sacred meaning rests on a prior emergence at the level of psychic history. If transcendent or existential truths dwell in Milton's religious symbolism of fatherhood and sonship, rather than a mystified repetition of his psychological past, we must search for evidence of an epigenetic process, dramas of transformation and structural realignment, in the character of the poet—a lived meaning that authenticates the signifying intentions of *Paradise Lost*.

"But such as are good men can give good things." Despite the central position of psychoanalysis, this project answers to a familiar concern. The said, it has often been asserted, lacks weight or authenticity unless consonant with the being of the speaker. "*Language* most shewes a man," Jonson wrote: "speak that I may see thee."[4] In the

concluding stanza of "The Windows," George Herbert voiced the ideal sought and the failure feared throughout his poetry:

> Doctrine and life, colours and light, in one
> When they combine and mingle, bring
> A strong regard and aw: but speech alone
> Doth vanish like a flaring thing,
> And in the eare, not conscience ring.[5]

Speech should be the confession of existence. Augustine, who wrote the chronicles of time and eternity, narrated his life as well—and did so partly to educate his audience in the difference between humility and humiliation, teaching them to know the self hidden by the authoritative voice of the theologian as the ground of this authority, not its undermining; for such reasons, I think, he closed his autobiography with a commentary on Genesis. Certainly Milton never doubted that the idea of a worthy poem was nonsense apart from the worth of the poet.

He declared this principle in his most admired pronouncement about the art of poetry:

> And long it was not after, when I was confirm'd in this opinion, that he who would not be frustrate of his hope to write well hereafter in laudable things, ought him selfe to bee a true Poem, that is, a composition, and patterne of the best and honourablest things; not presuming to sing high praises of heroick men, or famous Cities, unless he have in himselfe the experience and the practice of all that which is praiseworthy.[6]

The correspondence between poet and poem is not primarily synchronic. It is an achievement of personal history. Heroic poetry assumes the prior existence of the praiseworthy in the poet as well as his subject. What the poet says he must first be: a poem must somehow inhere in the "composition, and patterne" of "things." When interest shifts, as it does here, from the interpretation of what is written to the act of the writer, clear distinctions between moral and literary value, deed and composition, character and statement, soon evaporate. Milton acted on this principle, as if expecting posterity to be curious about him. He welcomed scrutiny almost in the spirit of a dare; there is an element of aggressive invitation about this and similar passages: "search me, I am pure." Partly as a result of his public intimacies, we

know more about Milton than about any other English poet who preceded him. The materials, if not as full as one could wish, are plentiful nonetheless. Occupied with "things" as well as texts, a criticism answerable to this aesthetic should be able to interpret the "true Poem" fulfilled in the meaning of *Paradise Lost*, while also interpreting the meaning of *Paradise Lost* as brought forth in the "composition, and patterne" of the poet's life. Wary of this confluence, we normally separate the true poem from its verbal successor, assigning the flyleaf of the family Bible to the biographer of Milton and the opening of Genesis to the critic or scholar of *Paradise Lost*.

Our inquiry into the psychic work of the poet, the genesis of meaning "in himself," begins with temptation. Among the early poems *Comus* contains the fullest exposition of virtue as both a moral condition and a power; in the next chapter I will try to discern, as other Miltonists have, the private declaration implied in the victorious resistance of the Lady. When Milton returned to poetry in the last fifteen years of his life, he organized the "composition, and patterne" of the sacred history represented in his religious epics around temptation, and the trials of garden and desert, though subdued, give shape to the final day of Samson. Milton emphasized "the best and honourablest things" in the moral aesthetic of *An Apology*, but he understood from the beginning that virtue arose in combat with its adversary:

> And perhaps this is that doom which *Adam* fell into of knowing good and evill, that is to say of knowing good by evill. As therefore the state of man now is; what wisdome can there be to choose, what continence to forbeare without the knowledge of evill? He that can apprehend and consider vice with all her baits and seeming pleasures, and yet abstain, and yet distinguish, and yet prefer that which is truly better, he is the true warfaring Christian. I cannot praise a fugitive and cloister'd vertue, unexercis'd & unbreath'd, that never sallies out and sees her adversary, but slinks out of the race, where that immortal garland is to be run for, not without dust and heat. Assuredly we bring not innocence into the world, we bring impurity much rather: that which purifies us is triall, and triall by what is contrary.
>
> (*CP* II, 514–515)

This famous passage bears signs of the conflict it urgently defends. A few sentences away Milton committed a well-known parapraxis, telling us that Spenser taught these truths by having the palmer

accompany Guyon into the Cave of Mammon, when in fact Guyon journeyed alone.[7] It is, at least for an instant, comparably odd to be told that we bring impurity with us into the world, but gain virtue in being tried by what is contrary—by purity? The rebound of contraries that momentarily overturns the intended sense, making virtue into our adversary and tempter, can stand as a vague prefiguration of the difficulties to be found in *Comus.* Here I wish to emphasize that in the moral epistemology of this passage the Miltonic knower is a tempted man.

Like all religions, Christianity designates certain moments of our experience as sacred. These moments, posited for all time in etiological myths, are renewed historically through layer after layer of replicas in image, legend, exegesis, and the lives of holy men. The formal temptation of a solitary individual is probably among the few such moments that serve to distinguish the religious visions derived from the Bible, for little of what we know about the religions of Mesopotamia, Egypt, Greece, or Rome anticipates the steady emphasis accorded to the solitude of choice, the moral agent outside of social boundaries, in Judaism and Christianity. This is the moment that Milton, among Christian poets, claimed as his own. Man knows good and evil, and knows himself, in the heat of forbearing, abstaining, preferring: temptation opens to him, in a single experience, each of the dimensions of sacred space explored by the poet of *Paradise Lost*. But where has Milton centered us? What is the psychological or existential core out of which this symbolism evolves?

In *Comus* and the two epics temptation occurs between the representatives of opposing communities. Someone who knows the law of God is now alone, away from the counsel and support of others who know this law. His isolation is crucial. Many of the ideals Milton expounded in his political works can be understood as attempts to conceive of a nation whose coherence would still allow this essentially asocial encounter with evil and alien ways to transpire; the quarrel over solitary labor in Book 9 of *Paradise Lost* contains the whole lesson of liberty in seminal form.[8] Since there could be for Milton no meaningful virtue in pure knowledge of the law, in forced or untried obedience, the secret ordeal of temptation is our sole initiation into the moral life. So the tempter appears. He seeks to replace the community as a source of counsel and support. He offers enticing rewards, but to join the community of transgressors who enjoy them, one must break the law. God watches unseen while the tempted chooses between homes.

The community that tells this myth wants to be preserved even in its absence. First allegiance should be the only allegiance. What the tempter offers is a negative of the moment of conversion: there would have been no dramatic climax to the *Confessions* had Augustine remained in the bosom of Monica. But for Augustine as for Milton, a fire of existential revelation burns through the superficial didacticism of the moment. The liminal state of the tempted, suspended during his trial between clear social determinations, is the state of the soul—forever choosing in this life, forever between God and Satan, forever in the restless present between memory and anticipation. Every man is solitary. In his solitude every man finds conflicting allegiances. A myth urging the immortality of the community became, in the exegetical tradition, one of the best explanations for its inevitable discontents.

The interior communities disclosed in the moment of temptation are those of law and desire. Through the enticements of the tempter, with his alien possibilities, the tempted comes to know desire as the yearning for what he lacks, must not do, cannot be; and knowing desire, he comes to know the law *as law*. Like the Creation that mastered Chaos, laws govern what would be lawless without them. This conflict betrays the differences between man and God in voluntarist theologies. Law makes sense in another world: here one obeys because of the prestige of divine authority. In the moderately rationalist theology Milton professed, conflict results from the divided nature of the law-bearer as well as from the incomprehensibility or transcendent source of the law. Given to reason in order to govern passions and appetites, law discovers a lesion in the soul; God and nature bid the same in man only when his nature overcomes itself. In *Paradise Lost* Milton endeavored to minimize the trauma of this conflict. God provides in abundance, enormous bliss, and there is seemingly no opening for the appearance of an unsanctified desire. When disobedience occurs nonetheless, the poet dwells on the fraudulence of the tempter and the deceived reason of Eve. But Milton knew that the hunger for righteousness could not be all of hunger. No one reads of the evil hour in *Paradise Lost* without realizing, by the time the apple goes down, that Milton has exposed a will to deceit in Eve, an interior fraud that opens out to welcome her tempter, clothing itself in images of power. Breaking the law means being a man, a queen, an empress, an angel—"nor was Godhead from her thought." What do we desire? To be without the law or, what is the same thing, to be its source. Marking our otherness from God, law unveils a thwarted divinity in ourselves.

"All men are made in the image of God," Arnold Stein has re-

marked, "but"—correcting piety with the sting of wit—"only Milton was made in the image of Milton's God."[9] Unlike angels and essences, however, or an idiosyncratic God, the knowledge given in the moment of trial cannot be thought away. We *are* between these communities. At the end of his *Two Treatises,* William Perkins devised a Ramist diagram of the possible relations between flesh and spirit in human choice.[10]

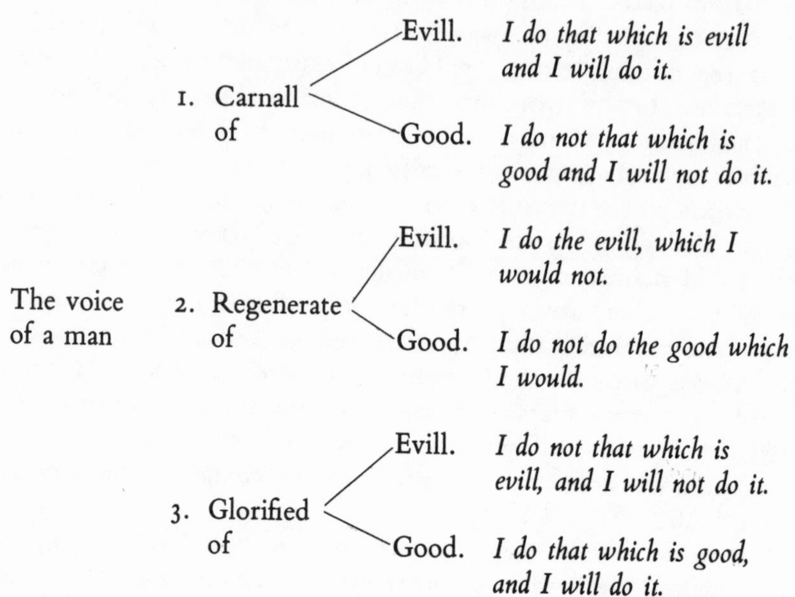

The voice of a man
1. Carnall of
Evill.
I do that which is evill and I will do it.
Good.
I do not that which is good and I will not do it.
2. Regenerate of
Evill.
I do the evill, which I would not.
Good.
I do not do the good which I would.
3. Glorified of
Evill.
I do not that which is evill, and I will not do it.
Good.
I do that which is good, and I will do it.

This voice speaks every language. Insofar as Milton reimagined his religion about the events of temptation, he pitched his art on the ground of our existence. His Christian moment reveals the life beneath particular conceptions of the moral life, for ethics begins in internal division, where the self cannot be wholly coincident with the self. The myth of the destined tempter and the single definitive trial contracts into one experience our slow and fragmentary effort to choose ourselves amidst the disruptions of psychic conflict. It tells of our abiding homelessness. The final image of Adam and Eve between Eden and the world, choosing their place of rest, brings the symbolism of temptation to its basis in our history.

The myth announces that our fundamental posture toward experience must be vigilance—an assumption shared by Freud, whose metaphors of censors and watchmen posit, beyond our awareness, a tense diplomacy that guards the ego from the unmediated countenance

of desire. "Everyone," Milton wrote, "must one day render to him [God] an account of his actions, good and bad alike" (*CP* VI, 132). Believing that the interior side of every action has two witnesses who, in the ideal case, approve the same action for the same reasons, many of his contemporaries would have subscribed to this statement and nodded, as they had been nodding all their lives, to its warning call for an explicit self-justification. But the man who owned the myth of temptation placed his assessments in the public record. God was not audience enough for his awakened conscience. Milton exhibited his vigilance, offering to the public not just his choices and self-delineations, but his own contentious estimates of the state of his soul. As the birthday sonnet, the two sonnets on blindness, and the autobiographical sections of the prose give us to know, indeed demand that we know, the scrivener's son prepared all his life for the day when he and God would match accounts, continually revising the ledger, fitting old desires to new circumstances, resisting the pressure of repentance issuing from alternative ways of reckoning his spiritual estate: the accounting itself was often enough a drama of temptation. In a classic article on the "temptation motive" in Milton, Hanford traced a line of conflicts ascending from sensuality to distrust of the deity.[11] One wonders if, given the insistent publicity of his conscience, temptation was not a motive in another sense, compelling the very act of creation.

Milton did not lack opportunities to perfect this dramaturgy of moral action. He lived in unsettled times, when a long and bloody revolution demanded that men take account of themselves in making fateful decisions of allegiance. Examining such moments in our own lives, we catch the rough sketch of the temptation scene—our existence reaching, as it were, toward a sacred articulation. We feel both profoundly alone and yet somehow observed by a recording presence as we wrestle with the large possibilities of self-determination. When the decision is made, we feel, whatever our religious beliefs, that we stand now *sub specie aeternitatis*. The philosopher Hans Jonas connects this aspect of choosing with a particular order of religious symbolism:

> In moments of decision, when our whole being is involved, we feel as if acting under the eyes of eternity . . . . We may say, for instance, that what we do now will make an indelible entry in the "Book of Life," or leave an indelible mark in a transcendent order; that it will affect that order, if not our own destiny, for good or for evil; that we shall be accountable for it before a

timeless seat of justice, or—if we are not there for the accounting, because we have flowed down the river of time—that our eternal image is determined by our present deed, and that through what we do to that image of ours here and now, we are responsible for the spiritual totality of images that evergrowingly sums up the record of being and will be different for our deed. Or, less metaphysically, we may say that we wish to act so that, whatever the outcome here in the incalculable course of mundane causality—whether success or miscarriage—we can live with the spirit of our act through an eternity to come, or die with it the instant after this. Or, that we are ready to see ourselves, in an eternal recurrence of all things, when our turn comes round, and again we stand poised as now, blind as now, unaided as now—to see ourselves making the same decision again, and ever again, always passing the same imaginary test, endlessly reaffirming what is yet each time only once. Or, failing that certainty of affirmation, that at least the agony of infinite risk may be rightfully ours. And in this, eternity and nothingness meet in one: that the "now" justifies its absolute status by exposing itself to the criterion of being the last moment granted of time. To act as if in the face of the end is to act as if in the face of eternity, if either is taken as a summons to the unhedging truth of selfhood.[12]

"Tarry awhile," Goethe's Faust would have said to the moment of pleasure, "thou art so fair!" But to the *nunc stans* of fateful decision we say, "Remain forever, thou art so difficult and so just!" Whether or not they mythologize temptation, all religions possess a symbolism responsive to our sense of the "monumental" choice that leaves an objective record or icon of itself—a system of animal doubles perhaps or, in totemic cultures, the soul, which faithfully takes on the personality and even the physical characteristics of its host.[13] Students of medieval and Renaissance Christianity know this doubling or reiteration, not only from the figure of the Book of Life but also in the familiar appeal, possibly the single most persuasive ploy in the repertoire of Christian rhetoric, to an eternal existence in the mind of God, where the exhorted soul, congregation, or nation was already suffering for its sins or hearing hosannas for its saintliness.

What the poet says, he must first be. Milton created through poetry effigies of his self-determination. In "Lycidas," the anxious contemplation of a premature death that visits poets just when they "think to

burst out into sudden blaze" can be stilled by thinking of the "*Fame*" (etymologically, "the spoken") kept before the "perfect witness of all-judging *Jove*" because, adjusting the moral aesthetics of *An Apology* to a sacred context, artistic works transcribe true poems and true poems have already been transcribed in the regard of God. Poetry is a human version of the divine record, its "composition, and patterne" shaped first of all in the chronicles of conscience. Milton made this motive, the need to provide a public revelation of will, central to his heretical godhead, wherein Christ issues forth as the visible manifestation of a paternal will otherwise secret, "doubling" or "fulfilling" that will in the creation and management of objectivity. But despite the parallel between poet and Father, Milton was not the source of the law; and despite the identification we will study, he was not the Son, truest poem, the supreme embodiment of the law. One can scarcely imagine a more arduous conception of poetry. The entire process of artistic creation, inseparable from the account to be rendered to God, takes place under the shadow of ultimate judgment.

Comus, the first formal temptation in Milton's art, shows how the mastery of desire by law results in a "hidden strength." The allegiances in conflict here wind back through the history of the soul to childhood and even to birth: this temptation occurs in the "navel" of the woods. We sense from "L'Allegro" and "Il Penseroso," stylized as they are, that Milton hoped to conclude his youth with a gesture of great finality, abandoning one sort of life in the act of embracing another. Critics speak sensibly of the necessary reconciliation of the two goddesses in any reasonable life, including Milton's, but the disjunctive foundation of the poems tells its own tale. Fidelity to one goddess requires a scornful exorcism of the other. After drawing up a long optative catalogue of expected rewards, the poet twice offers his contract: if you give me my wish, I in return will make my home with you. Although their content is rearranged somewhat, these mutally exclusive pledges shape the temptation resolved in *Comus*. The Lady of the work is a monument left by a choice, an early crystallization of Milton's identity as man and poet.

Following in the train of critics who have found strange doctrines and troubled symbolism in the masque, I can produce no new evidence for believing that Milton concluded his youth by prolonging it, attaching his poetic ambitions to a vow of chastity. I will offer a different interpretation of the evidence we have. Law and desire are not haphazardly related: they constitute each other. The law obeyed in *Comus* forbids the sexual life, and everything that supports obedience

to this law—the family of the Lady, her spiritual guardians, her internal virtue, the transcendent existence toward which she aspires—cannot be understood apart from sexual desire. A work about virginity, *Comus* is also and inevitably a work about eroticism.

Milton could not abandon the elements of psychic life composed in the flawed decision of his masque as he could a material possession. Thirty years later they will reappear, composed anew, in the resolute poet of the final masterpieces. But in studying the psychological structure of desire's negation, we can appreciate why Milton had to pass through this early delineation of himself, falling under and then breaking the spell of the Lady's virtue. The cruelest temptations have the power to wring a permutation from the voice of our volition that William Perkins, like Aquinas before him, did not consider possible: *I do the good which I would not; I do not do the evil which I would.*

• 2 •

THE ROOT-BOUND LADY

THE OCCASION for the entertainment now known as *Comus* was the presidency of John Egerton. But after the opening speech installs his "new-entrusted Scepter" within the cosmic hierarchy of government, the masque proceeds to celebrate his authority and his fitness to rule within the realm of domestic politics. Families, too, have their rituals, and *Comus* is a ritual for the unification of a family, its plot derived from the actual journey of three children to the "wish't presence" (950) of their parents.[1] When the siblings arrive at the parental seat, they are certified as Egertons in the manner of a ceremony of investiture:

> Noble Lord, and Lady bright,
> I have brought ye new delight,
> Here behold so goodly grown
> Three fair branches of your own. (966–969)

These "fair branches" belong to the family tree by merit as well as birthright. Unlike the herd of Comus, who "all their friends and native home forget" (76), the Egerton children have remembered their native home, and because they have not lost their "human count'nance" (68), they can in turn be remembered *at* their native home. The particular question decided en route to this reunion is whether the scepter of the Lord President commands obedience and legitimacy in the conscience of his daughter when she is, for a time, the captive of a tempter's magic. In line with the general movement from political to familial authority, the symbolism of virtue in *Comus* situates conscience at its origins in childhood.

Virtue bears its own "bright day," able to see in dark places, whereas vice "Benighted walks under the midday Sun" (381–384). The tempter cannot, his captive assures him, "touch the freedom of my mind" (663). "Love virtue, she alone is free" (1019). Such passages have the ring of ancient sentiment, a wisdom almost global, and we know before we check that maxims like these will have ricocheted through the diverse minds of Plato, Plutarch, Cicero, Seneca, and

Augustine. But what is this wisdom about virtue and freedom that has flowed into Ludlow Castle from so many sources? Dancing in the night, the tempter clarifies his opposition:

> Rigor now is gone to bed,
> And Advice with scrupulous head,
> Strict Age, and sour Severity,
> With their grave Saws in slumber lie. (108–110)

Rigor, Advice, Age, and Severity are personifications of parental authority, ways in which an older generation transfers its "grave Saws" to a new generation. Now the parents of the world are in bed, eyes closed—or so the tempter thinks. "Tis only daylight that makes sin," Comus continues, equating darkness with secrecy ("these dun shades will ne'er report") and light with tattletales ("the blabbing Eastern scout," "the telltale sun"). Daylight is the symbol of being observed by the older generation, exposed to the watch of their external authority. If an observed person harbors scenes of orgiastic disobedience within him, meditating all that he would do in the absence of his observer, then this man "Benighted walks under the midday Sun; / Himself is his own dungeon" (383–384). He lives invisibly, chained in the dungeon of his privacy, where there is no authority and he enjoys the indulgence of a perpetual midnight. Why does he behave, limiting himself to fantasy? The gaze of authority. But because he would not behave in its absence, the gaze of authority is his jailer.

Virtue has no jailer: it is its own. Acting without regard to external conditions, the same in public as when alone, the virtuous man has the authority within him, where it is always bright noon. "Virtue could see to do what virtue would / By her own radiant light, though Sun and Moon / Were in the flat Sea sunk" (373–375). He sustains the conditions of external observation in the well-lit privacy of his mind, keeping watch over himself: the daylight that literally constrains the evil man becomes, as a symbol, the definition of the virtuous man. He has consented to a tradition, and will in time take his place with Age, Rigor, and Advice. But the only way for someone to tell a free man from a constrained man, the virtuous from the vicious, is to let him out of his sight, which is why the midnight encounter with Comus permits the Lady to be recognized as a fair branch of the family tree. She has demonstrated that the "Noble Lord, and Lady bright" have been enthroned within her, a presence even in their absence.

The Masque of the Superego

Freud's theory of the superego is a continuation of the venerable wisdom informing the ethics of the masque. It seeks to explain how the regulation of desire, which begins for all of us in the scrutinies and restraints of external authority, metamorphoses into a self-regulating conscience, wherein the ego can accuse itself and impose the psychic punishment or self-aggression of guilt. The state of being revealed to us by the temptation myth achieves a formal organization in our lives through the turbulence of the oedipus complex. The love of the child, focused for the first time on a distinct object, encounters law; all of the earlier conflicts between felt omnipotence and experienced finitude coalesce in this illegal wish. At the resolution of the complex, the two relationships of love and rivalry, desire and law, "are dissolved, destroyed internally and something novel, psychic structure, comes into being."[2] As a monument to the sacrifice of desire, formed primarily from an identification with the law-bearer, the superego establishes at the beginning of moral existence an intrinsic connection between an ideal and a prohibition. This is precisely the moral structure celebrated by the temptation myth: one becomes ideal by prohibiting oneself, by refusing and renouncing, by saying "no." The oedipal child is the first hero of temptation.

Certain features of *Comus* acquire a new lucidity when read from this perspective. No doubt there is a world of difference between an observed child listening to the grave saws of its elders and the luxuriously furnished conscience of the Lady, replete with myth, magic, religion, song, and the charm of divine philosophy; her parents have not gone to bed this night, and the very fact of the masque exonerates them from the charge of sour severity. Yet the symbolism of virtue, representing the mature state in terms of the primitive one, keeps this world of difference before us. The symbolism, furthermore, has its counterpart in the form of the masque, which represents concretely the etiology of conscience preserved in this symbolism.

The definitive character of the genre lies in its attachment to an ephemeral occasion. Art is embedded in event. In masques the relation of actor to role, usually incidental in the playhouse, is supplied (for some roles, at least) with a motive; fiction and occasion interpenetrate through winking allusions to the stage machinery (as in ll. 221–225 and 331–335) or to the nobility present, waiting for the plot to glide into the festive dancing of a party. Always original in his ways with genre, Milton has informed the special ontology of the masque with

the life of an idea. For the parents of Alice Egerton look on from their thrones as an unacknowledged audience—as if absent, as if invisible—while the residue of their spectatorship in the mind of this child is being tried. "Eye me blest Providence," the Lady declares as she leaves with the false shepherd Comus (dark inverse of the Attendant Spirit, also disguised as a shepherd), knowing that she must act, like her author, "As ever in my great task-Master's eye" (Sonnet VII). In the darkness of the woods, seemingly, *no one would know.* But in fact everyone would know: the inward gaze of conscience would see, the unacknowledged eyes of the parents would see, the invisible watch of providence would see. All of the stages of conscience—external and internal, secular and religious—are present at the exposure of her virtue, having become analogons of each other through a symbolic use of the nature of a masque.

Milton recovers in this way the primordial experience of conscience. The gaze and the light inside the virtuous mind of the Lady have been dismantled and reexternalized in her "Noble Lord, and Lady bright." Surely the most common metaphors for the vigilance of the superego are the sun and the eye, often merged in iconographical traditions—which is one of the cheaper lessons to be learned from a dollar bill. But what do we learn from these metaphors? Reflexivity, the great theme of philosophies of consciousness from Plotinus to Husserl, is binary. Self-awareness demarcates the boundary between me and not-me, setting the knowing ego against known objects. With the institution of the superego, however, a third term occupies our mental space, transforming self-awareness into the triangular structure of moral existence. Jonas' description of the choice that engraves its image is one instance of this transformation: the figure of authority has come to rest within us, sensed in the observation of our conduct and heard in the "call" of conscience. The sacrifice instituting the superego *does* leave an indelible impression, completing the structure of mind with a unique otherness—the burden of the law. This new reflexivity is not addressed to the ego as reason, determining objects in the realm of knowledge, but to the ego as will, determining itself in the realm of value. Its expansion and exploration occur in the symbolism of the sacred.

We could not speak feelingly of this otherness without making symbols of the scene of observation. It is a tendency of all symbolism to represent the higher by the lower, what is futural and mysterious by what is previous and overcome. The biblical vocabulary of evil, Ricoeur has shown, draws on the everyday facts of the spot, the

weight, and the deviating path.[3] If we say "God is the sun" as a metaphor, this statement depends for its evocative power on our ability to imagine, and be moved by imagining, that God *is* the sun. Symbolism has a regressive component, an internal dynamic that pulls symbol and meaning back toward idol and image. It is not surprising, then, that the pattern we have discovered in the tropes of virtue and the form of the masque will be found elsewhere in *Comus*. But the representation of a complex or ideal state by a simpler or earlier one is so predominant here that questions arise concerning the freedom of these symbols. To what extent is the meaning of this work "root-bound," captive to a regressive movement from which there is no significant return? This question is accentuated for a psychoanalytic reading, but many interpreters of *Comus* have struggled in their own terms with a problem of this sort. For at the center of the entertainment chastity narrows to virgin chastity, a state of the soul evolving from, and unable to detach itself from, a precondition of the body.

Virginity's Logic

The theme of solitude leads directly to this lapse. It is already ironic that the manifestation of virtue requires solitude, yet as we learn from the brothers, this stipulation poses a more serious difficulty in the case at hand. Their exchange turns on two consequences of aloneness. The Younger Brother fears that his sister may be in the "direful grasp" of "Savage hunger" or "Savage heat" (357–358), vulnerable to rape. Somewhat obtusely, the philosophical brother replies that his sister is not "unprincipl'd in virtue's book," taking solitude to be the prerequisite for moral trial and limiting the consequence that worries his brother to a parenthesis (370). When corrected, the Elder Brother offers his peculiar doctrine of the "hidden strength" (416), the "complete steel" (421) and "arms of Chastity"(440), symbolized in ancient myth by the bow of Diana and the shield of Minerva: more than internal, chastity has the kind of strength a savage would respect. This assertion arises solely because the virtue at issue rests on virginity. A faithful, hopeful, charitable, wise, prudent, courageous, patient, or temperate Lady could not have her virtue snatched away by force: she could say to her tempter, "Thou canst not touch the freedom of my mind" (633), and that would be that. But the virtue of the Lady is not only in her mind, and if it is to be tempted (as it must) in solitude (as it must), then guardian forces not of this world must protect it from unchosen ruin (as they do).

The moral logic of the masque's religious universe identifies virginity as a virtue of singular prestige:

> Against the threats
> Of malice or of sorcery, or that power
> Which erring men call Chance, this I hold firm;
> Virtue may be assail'd but never hurt,
> Surpris'd by unjust force but not enthrall'd,
> Yea even that which mischief meant most harm
> Shall in the happy trial prove most glory.
> But evil on itself shall back recoil,
> And mix no more with goodness, when at last
> Gather'd like scum, and settl'd to itself,
> It shall be in eternal restless change
> Self-fed and self-consum'd; if this fail
> The pillar'd firmament is rott'nness,
> And earth's base built on stubble. (586–599)

The Elder Brother places the trial of his sister in the context of a theodicy, for assuming that she is indeed virtuous, her captivity puts the moral direction of the universe on trial. His transition from the Lady's ordeal to the end of time, "when at last" the contraries will be disengaged, evil contained and virtue purified, is not at all gratuitous: the vulnerable base on which her virtue is built calls forth an intervention, a sign of the full theophany to come. Feeble to the degree that it rests on a bodily state, virginity makes heaven stoop. Its exemplars, when tried by force as well as fraud, will come into contact with transcendent beings.

As virginity is connected with heaven, so is it disconnected from nature. Like all tempters, Comus tries to appall his captive with the extremity of her virtue: how can you deny yourself this without denying yourself everything? So after his appeal to the positive benefits of his liquor has failed, Comus attacks the refusal itself. He accuses the Lady of advocating "The lean and sallow Abstinence" (709), and then, making explicit the sexual advance symbolized by the liquor, taunts her with the specific abstinence of "that same vaunted name Virginity" (738). The Lady can extricate herself from the blunt either/or of the first charge by invoking the "holy dictate of spare Temperance" (767): to partake or not partake, that is *not* the question, but when, with whom, how much, what, and in what attitude. A symmetrical answer to the taunt of virginity would be chastity, widely believed in the

Renaissance, especially among Protestants, to include marriage. As temperance to abstinence, so Protestant chastity to virginity. But the Lady refuses to voice even a hint of compromise with the second, specifically sexual charge of abstinence, and in a gesture that has provoked much critical debate, accepts the name given to her by her extempter. Her chastity is virginity, transvalued from a defective extremity to a "sublime notion and high mystery" (786).

Is the doctrine of virginity so strange that even critics who would prefer to treat literature as the storehouse of intellectual history must fall back on a biographical explanation? This is the conception of the problem that the most influential critic of the masque, A. S. P. Woodhouse, responded to. In his view, the Lady's reply, dramatically apt for the fifteen-year-old Alice, need not implicate the doctrines of the author in any way whatsoever. Woodhouse contended that Milton held chastity able to survive the marriage bed, but did not bother to make this "qualification" explicit: "Nor is it necessary, surely, for any rational mind, that he should."[4] I suppose there must be rational minds for whom the difference between living with or without sexuality amounts to a dispensable "qualification," and I suppose that these minds would not be changed by the fact that the five best poems written during the first half of Milton's life are "An Ode on the Morning of Christ's Nativity" (the Virgin Birth of a virgin God), "Il Penseroso" (a virgin goddess, whose literary tastes do not include the Hymen masques enjoyed by L'Allegro), *Comus*, and the two elegies for virgins, "Lycidas" and "Epitaphium Damonis." But to take the matter in context, Milton could have written a speech conveying both the Lady's allegiance to virginity *and* her awareness that Comus professes a concept of chastity as exaggerated and impoverished as his concept of temperance. The last words of the taunting Comus, "you are but young yet" (755), actually invite a reply of this kind, all but asking the Lady to concede a relationship between her sexual continence and her youth. Why did our notably contentious poet forbear in this instance? Was it his practice to pass up opportunities to correct his tempters? It is odd for a reader of *Comus* to hear that no rational mind could mistake the conventionality of the Lady's chastity, since she calls virginity and virginity alone a "mystery," "sage" as well as "serious," which presumably means that the virginity here professed is not, like temperance or quotidian chastity, a cardinal virtue derived from reason. The common ways of this world, we are told in the first speech of the masque, do not aspire toward "high mystery."

If we find something more esoteric in *Comus* than the trite allegory

whose appreciation Woodhouse made into a test of our rationality and, in other places, our spirituality, we need not adopt the equally odd position of those who reason that, because virtue tightens into mystical virginity during the temptation scene, *Comus* is a disunified work.[5] As may be learned from the Platonic One, the Cartesian piece of wax, or the identity of any man, unity is a complex idea, arguably the most complex idea. Language such as "tied together," "linked," and "bound up" suggests an undercurrent of tyranny in commonplace descriptions of a unified text. I do not understand why conflicts of affect or proposition in a work should be prima facie evidence of "disunity," or if so, why disunity should be considered prima facie a flaw. To require of art that its ambivalences and ambiguities be "resolved" is to ask for a dangerous variety of aesthetic illusion—one set against life. We can try to update the neoplatonic *concordia discors* by insisting that conflicts become incorporated into the coherent sense of a text when and only when they are intentionally produced. But this type of solution, whose consequence is the division of literature into mature texts able to manage their own subversive tendencies and immature texts given over to biography and psychology, seeks to contain the idea of the unconscious prematurely. Conflict is inherent in coherence. Freud proposed that the concept of "no," emerging from primitive repression, makes judgment possible by allowing the repudiated to be thought without being repressed: intentional meaning, like psychic unity, depends on the achievement of contradiction (S.E. XIX, 235–239).[6] One may of course debate with Freud; those appealing to the self-sufficiency of intentional meaning should be doing precisely that. But *Comus* is a work about temptation—the myth of the heroism of "no"—and it seems reasonable to imagine that here, if anywhere, meaning will exude its own adversary.

Discussions about the success or failure of this work have been marked for some time by the adherence of both sides to an idea of coherence more suitable to symbolic logic than to symbolic literature. Enid Welsford adduced the standards of Ben Jonson to devalue *Comus*. Subsequent critics, who seem always to be rediscovering the masqueness of the masque, have been quick to dismiss her belief that Hymen is the presiding deity of the genre, which turns Milton into a Puritan at the Maypole, but they have not questioned seriously whether Jonson's prescriptive analysis of the form exhausts its possibilities. It would be highly improbable, perhaps a unique case in the history of literature, if Jonson had laid down The Rule for a good masque when he wrote of "one entire body, or figure . . . with that general har-

mony so connexed, and disposed, as no one little part can be missing to the illustration of the whole."[7]

Working from this principle, critics have generally approached *Comus* with something close to the following model of what would constitute a perfect interpretation: find its theme, conceived of in scholastic fashion as a single substantive, and then demonstrate that the meaning of the masque is a set of mutually consistent predicates for this nominated noun. If such a model guided the composition of *Comus,* Milton deliberately chose to withhold predicates from the moment at which his Lady, all the eyes of conscience upon her, declares the magic substantive:

> Thou hast nor Ear nor Soul to apprehend
> The sublime notion and high mystery
> That must be utter'd to unfold the sage
> And serious doctrine of Virginity,
> And thou art worthy that thou shouldst not know
> More happiness than this thy present lot.
> . . . Thou art not fit to hear thyself convinc't;
> Yet should I try, the uncontrolled worth
> Of this pure cause would kindle my rapt spirits
> To such a flame of sacred vehemence,
> That dumb things would be mov'd to sympathise,
> And the brute Earth would lend her nerves, and shake,
> Till all thy magic structures rear'd so high,
> Were shatter'd into heaps o'er thy false head.
> *Comus.* She fables not, I feel that I do fear
> Her words set off by some superior power. (784–789; 792–801)

At this critical juncture the Lady's theme is *only* a substantive; the "whole," in Jonson's terms, is entirely missing to the illustration of this "part." Her strength is "hidden" in the sense that a solitary Lady does not appear to be a formidable foe, but also in the sense that her doctrine of virginity remains undivulged, itself virginal, as if speech this intimate would be equivalent to the sexuality her virtue forbids—the exhibition of the self to the other.

Although doubly hidden, the mere threat of strength's disclosure proves its mettle, vindicating the architect of this pillared firmament. Virginity is indeed a "vaunted name" in the masque: at no time in her competent exposition of temperance did the Lady seem ready to become an Orpheus ("dumb things would be mov'd to sympathise") ac-

complishing the feat of a Samson ("Till all thy magic structures rear'd so high, / Were shatter'd into heaps o'er thy false head"); the doctrine of virginity, not the doctrine of temperance, is able to demonstrate that "earth's base" is not "built on stubble" by literally calling to its aid "the brute Earth." But what is the substance that would lend this name, if uttered, its superior power? In order to couple predicates of any kind to this unforthcoming theme, we must first examine the cosmos whose laws protect the "happiness" of a virgin.

Wright and Arthos have confirmed the decisive influence of Plato.[8] Inspired by the myth of the two earths in the *Phaedo,* Milton set his trial in the cosmos of the Orphic tradition. The "daemon," as the Attendant Spirit is called in the Trinity MS., introduces the masque with a severe contrast between "My mansion" and the "dim spot" called "Earth," where frail and feverish men are "Confin'd and pester'd in this pinfold" (1–11). This imagery strongly suggests that to live at all is to have undergone the animal metamorphosis produced by Circe and her son, for the archetype of their ability to transform a man into "some brutish form" is the imprisonment of soul in body. Virtue begins, as it has always begun in the Orphic cosmos, with the gnosis of this estrangement, teaching us that we are the same as the soul and other than the body. What the potion of Comus offers is the familiar opponent of salvation in the many philosophies and religions founded on this cosmology:

> And they, so perfect is their misery,
> Not once perceive their foul disfigurement,
> But boast themselves more comely than before,
> And all their friends and native home forget,
> To roll with pleasure in a sensual sty. (73–77)

His crew have *forgotten.* They represent life in ignorance of the difference.

Within Christian thought this Greek cosmos has usually been interpreted by reference to the Pauline dualism of "flesh" and "spirit." Here, as in *Paradise Lost*, the difference to be remembered is inside the soul. Including passions and appetites, "flesh" is a diagnostic category in the pathology of the will; the body as a thing belonging to an alien world of things becomes, not the dead fact against which the soul is known, but a symbol in moral psychology.[9] The virtue of temperance expounded by the Lady answers to this interpretation of the flesh. There is nothing wicked in needing to eat or in the food consumed,

for wickedness can only arise in the volition of the soul. But the mystery of virginity she next announces can be associated with a partial reversion, felt elsewhere in the masque, from the Christian to the Greek dualism.

Descending from purity, the Attendant Spirit finds "rank vapors" and a "sin-worn mold." On this night "black usurping mists" (337) obscure the stars of his homeland. Given these allusions to the bad odors and thickened atmosphere of our pigsty earth, the "unhallow'd air" (757) of the enchanter's den, where the Lady rues the necessity of unlocking her lips to speak, seems the concentrated distillate of air itself. In Fulke Greville's haunting and truly Pauline sonnet on the terrifying fantasies we have in the dark, these hellish images "but expressions be of inward evils."[10] A "thousand fantasies" (205) beset the lost Lady of *Comus* and, as her brother assures us, a "thousand liveried Angels" (455) attend the chaste soul—an angel for every fantasy. Yet these superior powers appear to guard her, not from interior impurity, but from contamination by the tainted physicality of the world, "Driving far off each *thing* of sin and guilt" (456; my italics). These are not fantasies expressive of inward evil, but fantasies indicative of external danger. In the older conception of the difference, virtue was the soul's defense against the invasions of matter. This idea dominates the symbolism of *Comus.*

Before learning of the difference, an inhabitant of the Orphic cosmos will have drawn breath and eaten food, and unless virtue is suicide (as indeed it was for Sabrina), he will continue to do so. The one taint of the flesh subject to absolute denial is sexuality. Providing our fullest exposition of the sage and serious doctrine, the Elder Brother equates lust with the intrusion into the soul of an incarnate evil:

> So dear to Heav'n is Saintly chastity,
> That when a soul is found sincerely so,
> A thousand liveried Angels lackey her,
> Driving far off each thing of sin and guilt,
> And in clear dream and solemn vision
> Tell her of things that no gross ear can hear,
> Till oft converse with heav'nly habitants
> Begin to cast a beam on th' outward shape,
> The unpolluted temple of the mind,
> And turns it by degrees to the soul's essence,
> Till all be made immortal: but when lust
> By unchaste looks, loose gestures, and foul talk,
> But most by lewd and lavish act of sin,

Lets in defilement to the inward parts,
The soul grows clotted by contagion,
Imbodies and imbrutes, till she quite lose
The divine property of her first being.
Such are those thick and gloomy shadows damp
Oft seen in Charnel vaults and Sepulchers,
Lingering and sitting by a new-made grave,
As loath to leave the body that it lov'd,
And link't itself by carnal sensuality
To a degenerate and degraded state.
Second Brother. How charming is divine Philosophy!
(453–476)

The magic of the poetry is simply splendid: *lust, looks, loose, lewd, lavish, lets in, lose, lingering, loath, leave, lov'd,* and *link't* lead us to imagine that one consonant was given to the lolling tongue of man in order that his gross ear could be attuned to the death knell of the soul, then taken unawares by the redemptive chime of *charnel, carnal, charming.* Like a wish, this enchantment is intense enough to suspend unwelcome knowledge. For if the "unpolluted temple of the mind"—which means the body, in apposition to "th'outward shape"—becomes "the soul's essence," making "all" immortal, could there *be* any virgins in a graveyard? Milton first expressed his refusal of this fact in "On the Death of a Fair Infant":

Yet can I not persuade me thou art dead
Or that thy corse corrupts in earth's dark womb,
Or that thy beauties lie in wormy bed,
Hid from the world in a low delved tomb. (29–32)

In the invented myth that opens this poem, death comes to a "Virgin Soul" (21) from sexual assault. *Comus* is a magic fortress to preserve the wish whose earlier formula was "can I not persuade me." Spiritual guardians make sexual assault impossible. Divine philosophy proves that, if we retain the virginity we are born with, the "divine property" of our "first being," our bodies will never "lie in wormy bed."

Virginity being the supreme virtue, lust emerges as the supreme sin. In keeping with the pattern of regression we discovered first in the symbolism of conscience and now discover in the history of moral dualism, the Elder Brother represents the sin of unchastity by eating and drinking. Foreshadowing the liquor of the temptation scene, the act

"most" opposed to chastity lets in, clots, and imbrutes. One can understand why, in the Orphic cosmos, virginity might present itself as a strategy of integrity. Let us suppose that, as the Elder Brother implies, body and soul are posed against each other as the profane against the sacred. Life would involve a constant warfare of the soul against the body, its inescapable other; linked to the contagion of profane flesh, regularly absorbing impure substance, the soul would be enmeshed in what Durkheim termed the most absolute dualism "in all the history of human thought."[11] Through choosing virginity, however, the soul can transform the brute condition of the body, moving flesh inside the shelter of its sacredness. Become "soul's essence," the virgin body is no longer other to the soul, and although the dualism persists, it has been happily resituated: a unified person confronts the profanity of the world across the boundary of the body. Hence the loss of virginity would institute the more divisive form of the dualism, consigning body to the "contagion" of a material world constantly brought inward to the seat of the soul by nourishment. Bad food, therefore, symbolizes sexuality.

The new partnership of soul and virgin body, through the ministrations of "heav'nly habitants," somehow bypasses the moment of death. In the passage of the *Phaedo* from which Milton drew his vision of the graveyard, Socrates says that "the soul which is deeply attached to the body . . . hovers around it and the visible world for a long time, and it is only after much resistance and suffering that it is at last forcibly led away by its appointed guardian spirit. And when it reaches the same place as the rest, the soul which is impure *through having done some impure deed, either by setting its hand to lawless bloodshed or by committing other kindred crimes which are the work of kindred souls*, this soul is shunned and avoided by all."[12] The Miltonic version replaces murder as the crime of lingering souls with the self-murder of lust. Socrates proves immortality before going to his death, but *Comus* positions this subject before sexuality, claiming through this shift to have abolished the death sentence for body as well as soul.

In a general way the emotions that underlie this wishful transposition of Plato are not obscure, nor were they obscure during the Renaissance—witness the semantic range of the word "death." Perhaps, as Frye has noted of the threatened virgins of romance, the loss of virginity has meant for many people, male and female, a metaphorical death, a fall into the ordinary, a submission to the processes of life that hurry us toward its end.[13] Christians have traditionally represented salvation as a new virginal wholeness, a purification through rebirth; to

this extent, Woodhouse was right to argue that virginity in *Comus* is an "illustration" and a "symbol" of chastity in the order of grace.[14] But Milton's determinedly physical elaboration of the sage doctrine does not permit the symbol to be detached from the illustration. The argument of *Comus* is trying to include the body, not just what the body means, in the order of grace. Its divine philosophy asserts that only virginity can accomplish this inclusion, producing a sublimation of the flesh "by degrees." Clearly the reverse of Comus' transformation, this magical undoing of birth has a narrative echo in the "quick immortal change" (841) of Sabrina, "a Virgin pure" (826). The water god revives her corpse by pouring "Ambrosial Oils" into "the porch and inlet of each sense" (839–840).[15] Her perfect chastity, it would appear, has prevented the orifices that admit the external world, whether as substance or as sensation, from becoming "clotted." The symbolic significance of her myth is predicated, in some detail, on the fact of her virginity.

According to divine philosophy, the "quick immortal change" of the Sabrina myth will happen to the Lady "by degrees." More than critics have realized, the opposition between a bestial and a virginal physiology defines the alternatives presented during the temptation scene. As there are spirits from the transcendent world that attend aspiring mortals, so there are "spirits" in the blood that "Strive to keep up" our "frail and feverish being." Comus wants the Lady to see his "cordial Julep" as a simulacrum of lively blood able to produce this liveliness, when imbibed, by sympathetic magic:

> See, here be all the pleasures
> That fancy can beget on youthful thoughts,
> When the fresh blood grows lively, and returns
> Brisk as the *April* buds in Primrose-season.
> And first behold this cordial Julep here,
> That flames and dances in his crystal bounds
> With spirits of balm and fragrant Syrups mixt. (668–674)

The use of words such as "spirits" and "cordial" in the modern vocabulary of alcohol derives from the belief, presumed in this passage, that intoxicants quicken the three kinds of spirits distributed by the blood.[16] Like a Renaissance physician, Comus prescribes a glass of dancing spirits to refresh his tired captive. Given the known effect of this cordial, there is an implicit pun here on "animal spirits," and given the known desires of the tempter, betrayed in the word "beget,"

it might not be untoward to remember that in the Galenic tradition the pneumata of the blood distilled a potent essence known as "seminal spirits."[17] But the Lady already contains a magic potion. Were she to expound the doctrine of virginity, the internal effect would be holy inebriation: the cordial that "flames and dances" in the crystal cup of the magus has its sacred complement in her "rapt spirits" kindled "To such a flame of sacred vehemence." The last words of Comus conclude this battle of opposed physiologies with the high comedy of a projection:

> I must not suffer this, yet 'tis but the lees
> And settlings of a melancholy blood;
> But this will cure all straight, one sip of this
> Will bathe the drooping spirits in delight
> Beyond the bliss of dreams. Be wise, and taste. (809–813)

He unmistakably diagnoses his own depression. Far from being weighed down by unvivified spirits, hers is a sacred melancholy, inflamed and enraptured like the "sad virgin" of "Il Penseroso." The sage doctrine of the masque concerns a virginity ambitious enough to have generated a quasi-science.

But the highest mystery of *Comus* is a symbol that questions the efficacy and coherence of this doctrine. Because of the inexperience of the brothers, Comus retains his wand, and he makes good apparently on his initial threat to render the Lady "as Daphne was, / Root-bound." Her body is now inert; the lips that were reluctant to speak, now locked indeed. "We cannot free the Lady that sits here / In stony fetters fixt and motionless" (818–819). Wilkenfeld has analyzed brilliantly the symbolic character of movement in *Comus*, where the freedom of virtue has a spatial correlate in rising, descending, and dancing.[18] Disrupting this metaphorical structure, the temporary paralysis of the Lady invites us to reinterpret the virtue we have pieced together from philosophy and cosmology. Comus has restated his diagnosis in the very flesh of his opponent. She never spoke the mysterious doctrine. She will not speak again in the masque. Motionless before us, her virgin body has become the emblem of itself, or half an emblem, an emblem without words, as if the doctrine and privileges and expectations of virginity had been reduced by the enchanter to nothing more than a stasis in the body, bound to their root. Within the artistic frame of *Comus*, the author of this subversive new symbol is the hero of the anti-masque, and its meaning need not be fettered by the divine

philosophy that leads to Jove's court. We must not rush to the theologians: we are instead invited to "revise" the Lady, to look again at her virtue, submitting the temptation of the masque to a second and open critique. The issue of unified meaning arises at this point, but as it does, it ceases to be a question directed toward the pure formalism of meaning, dissolving into other and more pressing questions about the authenticity of the values intended in the work. Why should the Lady, having resisted her tempter and given proof of "some superior power" at her command, still be enchanted by the power of the wand? Is virtue *not* free? How strong is a virtue that would exchange the sexual life in order to gain a tinge of realism for the wish to be exempt from death? Might her paralysis at the scene of temptation symbolize her deliverance to the equal but contrary forces of law and desire? Has Comus exposed through symbolism a deeper knowledge of the high mystery than the one the Lady refused him?

On at least one occasion the foremost interpreter of *Comus* left evidence of his own uncertainty over the strength of its virtue. Milton, toward the end of his travels abroad, placed two inscriptions beneath his name in the guest book of the Cerdogni family of Geneva: *Coelum non animum muto dum trans mare curro* (Though I travel across the sea, I do not change my mind with my skies); and the last words of *Comus*, "if virtue feeble were / Heaven itself would stoop to her."[19] His mind is steadfast in alien lands, unmoved by foreign counsel, guarded by the impenetrable shield of its own "no," but if virtue feeble were... The symbol created by the enchanter hints that this uncertainty over the strength or self-sufficiency of virtue conceals and perpetuates the true problem. *Because* virginity *is* fixed and unmoved, strengthened by magical delusions, it is feeble. In the second stage of his poetic career Milton will also make a strength of weakness—of blindness. But the possibility of this stronger future lay in the affirmation of the antimasque.

Miltons and Egertons

Milton inscribed three images of himself in the Ludlow masque. The first is an encoded signature. Following precedent, he recorded his authorship in the cryptography of pastoral romance:

> Care and utmost shifts
> How to secure the Lady from surprisal,
> Brought to my mind a certain Shepherd lad

Of small regard to see to, yet well skill'd
In every virtuous plant and healing herb
That spreads her verdant leaf to th'morning ray.
He lov'd me well, and oft would beg me sing,
Which when I did, he on the tender grass
Would sit, and hearken even to ecstasy,
And in requital ope his leathern scrip,
And show me simples of a thousand names,
Telling their strange and vigorous faculties;
Amongst the rest a small unsightly root,
But of divine effect, he cull'd me out . . . .
He call'd it *Haemony*, and gave it me,
And bade me keep it as of sovran use
'Gainst all enchantments, mildew blast, or damp,
Or ghastly furies' apparition. (618–630; 638–641)

Milton as shepherd lad is the friend of Lawes, devoted to song and to acquiring the knowledge expected of poets. He is modestly subordinate to "*Meliboeus* old," normally identified as Spenser, "The soothest Shepherd that ere pip't on plains" (822–823), who has taught the Attendant Spirit the higher magic of invoking Sabrina. But he belongs in this company, and his "unsightly" yet powerful herb is the embodiment of his own unrecognized worth, "Of small regard to see to, yet well skill'd." A student of herbals, old languages, and ancient geographies, he mixes these to germinate *Haemony*, a plant as original as it is traditional, whose botanical and etymological roots continue to occupy modern scholars; our pursuit testifies to his ability to enter traditions by inventing uniquely. The shepherd lad is Milton, we might say, as he was born through his own imagination into the literary and cultural history presupposed by his masque.

His posture of studious leisure, hearkening in ecstasy, like the speakers in the companion poems, to the beauty of song (Il Penseroso also meditates on "every Herb that sips the dew"), reminds us that Milton's work on the text of *Comus* from its performance in 1634 to its published version in 1637 is almost coextensive with the five years he spent, between Cambridge and the Italian journey, at his father's country homes. I think the stasis of this period—an educated man still living at home, gathering knowledge of diverse kinds for a creative achievement in the future—has much to do with the paralyzed virgin of *Comus*. Insofar as the Lady gains heavenly consult, high doctrine,

and enraptured speech from her virtue, chastity must be associated, as Tillyard and Sirluck have maintained, with Milton's artistic ambition; pure diet and a chaste youth distinguish the divine poet in Elegy VI.[20] Like other Miltonists, Sirluck relates the hiatus of these years to a crisis in the choosing of a profession, a long struggle over whether to become a poet or a clergyman, arguing on this basis that the sacrifice of sexuality was Milton's way of atoning for his refusal of Orders (which would make better sense if Milton had been a Catholic) by elevating poetry to the dignity of the priesthood. Yet were these in the same category, equally vocations in the seventeenth century?

Donne, Herbert, Herrick, and countless others felt no disjunctive tension between the pastoral and poetic lives—nor does Milton in his pastoral eulogy for Edward King. Few men earned a living by writing poetry; poetry was written as a means to a living, the preferment of a secretaryship, a court appointment, a career in tutoring. How much contempt an aristocratic patron might harbor for his hired versifier is well illustrated by Donne's catastrophe while serving the Egerton family, the same people that Milton was flattering three decades later with the sublime notions of *Arcades* and *Comus*. Publishing the masque in 1645, Milton was proud of the work; he included a commendatory letter from Henry Wotton, who had read the anonymous text of 1637. Yet that version bore an epigram from Vergil in which a miserable shepherd, failing to attend to his business, speaks metaphorically of a wind that destroys his flowers.[21] The Egertons had butchered his text in 1634. Surely Milton must have realized that writing masques and poems of flattery, such as his "Epitaph on the Marchioness of Winchester," would buy him a precarious life at the expense of a debased art: he sought almost from the beginning a "Celestial Patroness" whose bounty was "unimplored." If he chose against the church for the reason he later named—his unwillingness to bind his conscience to the oaths of men (*CP* I, 822–823)—then he chose against "poetry" for the same reason. Virginity did not become a doctrine out of Milton's guilt over preferring a poetic to a priestly career: I think he preferred, as a career or means to a career, neither, and his inability to conceive of *any* desirable career, rather than his difficulty in choosing between two of them, probably goes further in explaining the caesura in his life after graduating from Cambridge. In the end he wandered into a congenial solution by declaring, through his prose, the very principles that made social life problematic for him, securing in this way the preferment of the Latin Secretaryship. He was concerned that no one think he had solicited this position (*CP* IV, 627; *EL*, 29), and often charged his op-

ponents with the prostitution of writing for financial reward. That Milton was wise not to become a professional poet on the model of Ben Jonson, inventor of the English court masque whose cankered muse was ever wary of her entrapment in the web of patronage, can be seen from the judgment that John Egerton, the Elder Brother of *Comus*, would one day scrawl on the flyleaf of the *Defensio: Liber igni, Author furcâ, dignissimi* (The book deserves the fire, the author the gallows).[22]

Ad Patrem and the letter containing Sonnet VII suggest that Milton did feel, throughout his period of retirement, a tension between poetry and gainful employment, whether in the church or not. It took time to become a poet of the highest sort. Because of the nature of this poetry, it would bring no worldly rewards. The intensely private ambition of divine poetry, like virginity in a woman, removed one from the system of exchange that constitutes social order. From this perspective, *Comus* was a trial step in the direction of Jonson, an attempt to write sage and serious poetry that nobility would nonetheless support. But why, if the shepherd lad was trying to concede something to the ways of the world, did he produce such an esoteric and personal statement? Part of the answer lies in the second of his self-portraits.

He is also the Elder Brother. To my knowledge, it has not been noticed that the family on the stage of *Comus* is the near image of Milton's own. Alice Egerton was fifteen, her brothers eleven and nine. Sharing a Christian name with John Egerton, Milton grew up in a household comprised of an older sister (Anne, how much older we do not know) and a younger brother (Christopher, seven years his junior). Given what we know of the artistic imagination, this is not a negligible parallel. Lawrence Stone has contended that, because of the separation of children from parents, the rough economics of marriage, and the convention of hiring outsiders for nursing, training, and tutoring children, the bonds of affection that we think of as intrinsic to the family were looser and dispersed among employees during the seventeenth century.[23] Milton seems to have been exceptional in this respect. He often expressed devotion for his father and gratitude for the education provided him; about his mother we will speak presently. No doubt relations with his brother, who followed his father into the networks of London commerce and became a Royalist Catholic, were sometimes strained, but Milton protected him during the Interregnum, entered into investments with him (unprofitably, and at the price of much litigation), and Christopher was the executor of his will. Perhaps he was just the sort of fellow who, like the Second

Brother of *Comus,* would compare his lost sister to a miser's treasure left in the open while John was applying philosophy to the situation and determining the relevant classical myths.

About Milton's feeling for Anne there can be no question. She is almost certainly the bereaved mother addressed in "On the Death of a Fair Infant." His first ambitious poem in the English language presents her dead child as the reluctant savior of the world, and promises its mother that, if patience triumphs over lamentation, her name will live forever (a promise Isaiah tendered to eunuchs). Subsequent births were more fortunate. Her two boys lived some while with Milton, receiving his meticulous tutoring, in the first home that he established for himself. Edward Phillips, the oldest nephew, wrote a loving biography of his uncle and is thought to have incorporated opinions drawn from the poet's conversation into his *Theatrum Poetarum Anglicorum;* Edward received all of Milton's papers at his death. Anne's child—"that he brought up," as Aubrey put it (*EL*, 4)—seems to have been a surrogate for the son that Milton himself was not destined to enjoy. The reiteration of Milton's first family by the masquing family goes some way toward explaining how the genre of the masque, which in its subordination of the poet to the public honor of his patron would appear to lie at the pole farthest from self-expression, became in the instance of *Comus* a vehicle for intimate revelation.

Milton limns himself in the figure of the philosophical brother with affectionate irony. As an ardent justifier of the ways of chastity, John Egerton is, if you will, the shepherd lad decoded—one who also knows and loves the "artful strains" of Thyrsis (494–496), an apologist for the truth of poetic fables, a young master of polemic and disputation whose abiding posture is that of the champion. One expression of this attitude in the character of Milton was his love for swords (see *EL*, 32); he recommended fencing in *Of Education* and wrote of his youth in the *Defensio Secunda:* "I was not ignorant of how to handle or unsheathe a sword, nor unpracticed in using it each day. Girded with my sword, as I generally was, I thought myself equal to anyone, though he was far more sturdy, and I was fearless of any injury that one man could inflict on another" (*CP* IV, 583). Is not walking through the streets in this frame of mind the social embodiment of the stance of fearless rectitude and partisanship we find in much of his prose and poetry? As Tillyard noted, the Elder Brother drops the charming strains of neoplatonic philosophy for the heroic rhetoric of Shakespeare's sword-carrying noblemen when he sees the Attendant Spirit approaching:[24]

> That hallo I should know; what are you? Speak;
> Come not too near, you fall on iron stakes else. (490–491)

One need only imagine Lawes standing there in his shepherd's garb while this fustian is squeaked out by an eleven-year-old to appreciate the fine irony with which the poet, in inventing a masque identity for the oldest Egerton, is remembering his own premature and embarrassingly solemn imitations of Renaissance ego-ideals. But Milton also liked to think of himself as continuous, keeping his life on a steady course of dedication: whatever he was doing, it was something he had always done or always prepared himself to do. The Elder Brother discloses more about the high mystery of chastity than his tempted sister, unveiling the cosmological and theological postulates that legitimize her virtue, and all of his contentions are validated by the subsequent action of the masque—with the exception of the "heretical" symbol of temporary paralysis. In the framework of intentional meaning as customarily understood, the Elder Brother is the privileged interpreter of *Comus.*

Milton is also the Lady. More than the sobriquet "The Lady of Christ's," which nice habits and delicate appearance earned the poet at Cambridge, supports this identification. I will begin with the famous digression of *An Apology*, often quoted by critics of the masque because it reveals the layers of cultural experience in Milton's idea of chastity and alludes to the Circe myth. Romance was the teacher of chastity in the register of literature:

> Next, (for heare me out now Readers), that I may tell ye whether my younger feet wander'd; I betook me among those lofty Fables and Romances, which recount in solemne canto's the deeds of Knighthood founded by our victorious Kings; & from hence had in renowne over all Christendome. There I read it in the oath of every Knight, that he should defend to the expence of his best blood, or of his life, if it so befell him, the honour and chastity of Virgin or Matron. From whence even then I learnt what a noble vertue chastity sure must be, to the defence of which so many worthies by such a deare adventure of themselves had sworne. And if I found in the story afterward any of them by word or deed, breaking that oath, I judg'd it the same fault of the Poet, as that which is attributed to *Homer*; to have written undecent things of the gods. Only this my minde gave me that every free and gentle spirit without that oath ought to be borne a Knight, nor needed to expect the guilt spurre, or the laying of a

> sword upon his shoulder to stirre him up both by his counsell, and his arme to secure and protect the weaknesse of any attempted chastity. (*CP* I, 890–891)

He identified with the knights, champions of chastity like the Elder Brother. When these "gods" were less than ideal, this anxiously virtuous reader could not separate their moral lapse from that of the author. The lesson had to be unadulterated. Since romance itself is counsel in the form of stories about arms, teaching "every free and gentle spirit" to "secure and protect" feeble chastity "both by his counsell and his arme," the author of a knight who breaks his oath has transgressed the duty sworn by this oath—and threatens to spread this crime among readers. In the most primitive terms, you are what you represent, and you urge the imitation of what you represent. One presumes that this young reader did not exercise a comparable vigilance with other genres teaching other virtues, since otherwise it would be hard to see how he could have approved of anything but hagiography.

After discussing the influence of Greek thought, Milton shows in Christian terms the ultimate logic of his engagement with this literary kind:

> This that I have hitherto related, hath been to shew, that though Christianity had bin but slightly taught me, yet a certain reserv'dnesse of naturall disposition, and morall discipline learnt out of the noblest Philosophy was anough to keep me in disdain of far less incontinences then this of the burdello. But having had the doctrine of holy Scripture unfolding those chaste and high mysteries with timeliest care infus'd, that *the body is for the Lord and the Lord for the body*, thus also I argu'd to my selfe; that if unchastity in a woman whom Saint *Paul* termes the glory of man, be such a scandall and dishonour, then certainly in a man who is both the image and glory of God, it must, though commonly not so thought, be much more deflouring and dishonourable. In that he sins both against his owne body which is the perfeter sex, and his own glory which is in the woman, and that which is worst, against the image and glory of God which is in himselfe.
> (*CP* I, 892)

When interpreted spiritually, the literary drama of deflecting unchaste eventualities from women belongs more aptly to "deflouring" lapses that threaten honorable men. Being a knight is the same thing as being

the lady he is sworn to protect, for the surest way to guard the chastity of another is to swear an oath of chastity within oneself, becoming one's own lady. The double identification with knight and lady that structured Milton's reading of romances has been projected onto the plot of *Comus* in the kinship of Elder Brother and Lady. They are vested in each other: the armed brother says that his chaste sister has the arms and armor of a knight, while the sister announces a high mystery that the brother expounds.

The self-portrait of the Lady conveys what Erikson termed, speaking of young Luther, "a secret furious inviolacy"—Milton as he was known, perhaps, only to Diodati.[25] Studying the passions bound in her virtue, we can reconstruct the etiology of his temperament.

Like many men born into the marketplace Protestantism of Renaissance cities, Milton possessed in his fashion all three of the traits linked by psychoanalytic theory in the notion of the anal character: thrift, orderliness, and stubbornness. Of the last it is unnecessary to speak at length—not one divorce pamphlet, but four; when news arrives that the author of the pamphlet under attack is Moulin rather than More, Milton persists in abusing More, since "one of them was as bad as the other" and "he having writ it, it should go into the world" (*EL*, 15). As for the others, both poet and early biographers stress the temperate habits and frugal ways (extravagant only in books) that enabled him, despite heavy financial setbacks, to survive for a lifetime primarily on his father's estate. But I am not concerned with how he apportioned his money, organized his daily round, or stubbornly bullied his daughters. What fascinates is Milton's translation of the anal character into ideas, systems of metaphor, a power of artistic design and economy, and, not least, into the psychological drama of his career. Jon Harned has argued that this sublimation, like the character itself, was the legacy of his father, who deliberately raised his oldest son and namesake to achieve a sacred restatement of his own success, at once justifying and transcending the financial affairs that had occupied his own life.[26] Christopher Milton, Edward Phillips wrote, was "principally designed for the study of the Common-Law of England," while the oldest son "was destin'd to be the Ornament and Glory of his Countrey" (*EL*, 52–53)—*he* was the one whose portrait was twice painted (at ten and twenty-one), who was given special tutors and a nurse to attend him when he studied late, who was allowed to spend five years at home after matriculating, then to travel abroad in the old style of the aristocracy. Milton represented, as Kermode remarks, "his father's chief investment."[27]

The investment paid off, although not—and this was probably inevitable, were there to be a genuine return—in the manner foreseen by the father. "You may pretend to hate the delicate Muses," Milton wrote in *Ad Patrem*, "but I do not believe in your hatred. For you would not bid me go where the broad way lies wide open, where the field of lucre is easier and the golden hope of amassing money is glittering and sure; neither do you force me into the law and the evil administration of the national statutes" (p. 84). So much for the life of the father and the destiny contrived for Christopher! Money is a "morbid preference" (p. 85).

But if Milton detested hirelings and intellectual profiteers, he brought to the financial metaphors of Christianity, particularly the gospel parables of work, an acute and moving sense of *debt*, which is one of the master tropes of his life and art. More than any Christian poet, Milton opens to us emotionally the fundamental religious assertion that life is a gift, obligating us to the veneration of thankfulness and the ambition of wholehearted endeavor, "all passion spent." God lends us the capital of time, and we repay this endowment in significant action. For Milton the action was to be an inspired song: the gift of life, represented in Genesis by the breath that animates the dust of Adam, would be repaid in kind. All his life he was vulnerable to the tense convictions that he was premature and at the same time belated.[28] Subsumed into spiritual finance, thrift had become waiting, preparing, concealing the talent. Orderliness, on the other hand, became transmuted into the belief in a divine schedule to which his significant action must be answerable. Between the thrift of unspent talent and the orderliness of being on schedule, stubborness mediated, allaying the pressures of anxious self-assessments with the "no" of his unwavering rectitude. "All is, if I have the grace to use it so, / As ever in my great task-Master's eye." The usury of the father will be made sublime. But the burden of that Egyptian eye awaiting payment gives us some intimation of why his great poem should contain our most fascinating version of the figure who takes but recognizes no debt—Satan, the thief of paradise, who after voicing his hatred for the illumination of the sun, locates the motive for his rebellion in the desire to "quit / The debt immense of endless gratitude, / So burdensome still paying, still to owe" (4.51–53).

The author of *Comus* planted the Pauline virtue of virginity in the pigsty of an excremental world alien to the moral psychology of Paul himself. Matter is rank, contagious, and fatal. Keeping this world outside, the perfection of temperance is the retention of a "divine prop-

erty." Comus opposes the unworldly scheduling of this virtue in pervasively financial terms. Nature is a lender. Having been given "dainty limb," the Lady must repay in "gentle usage" (680–681). His attack on her abstinence proclaims the wealth and industry of a capitalist nature. When he shifts to the "vaunted name," he identifies virginity with greed and miserliness:

> Beauty is nature's coin, must not be hoarded,
> But must be current, and the good thereof
> Consists in mutual and partak'n bliss,
> Unsavory in th'enjoyment of itself. (739–742)

But her "*due* steps" aspire to the "Golden Key / That opes the Palace of Eternity" (as it also does in "Lycidas," l. 111). She hoards and shuts fast in this life to gain entrance to the treasures of the next; here the metaphorical structure of the masque seems to be a solemn inversion of conventional ribaldry such as this, from *The Revenger's Tragedy*:

> Virginity is paradise lock'd up.
> You cannot come by your selves without fee,
> And 'twas decreed that man should keep the key! (II.i.153–155)

Precisely by remaining "lock'd up," the undefiled Lady moves by degrees and due steps to the key of a transcendent paradise. The one talent that is death not to hide must be lodged in her useless, a divine property gripped by the guardian passions of the anal character.

This in itself is not interesting, psychoanalytically or artistically. If the anal period results in a characteristic organization of attitudes toward pure and impure, keeping in and letting out, the genesis tells us nothing about the value of the objects of these emotions. To recognize, for example, that the familiar representation of Hell in Christian myth or the familiar conception of matter in the Orphic tradition draw upon a primal experience of disgust is hardly to refute, even to question, the moral and philosophical bearing of these systems. The three virtues of anality may in fact be virtues. But in *Comus* the perfect opposition between virtue and vice on the subject of chastity reproduces the defensive ethics epitomized by the anal child. For this is the reaction formation par excellence: feces—the first creation, the first gift—are devalued as the sign of shame and incompetence, and a new virtue is fashioned from the denial of an old one. When they remain at the level of the original conflict, reaction formations are pseudo-vir-

tues, depending for their strength on a continued attachment to the opposing wish. There are many oppositions in *Comus* that bespeak at least the structure of reaction formation; Haemony itself, a dark and unregarded plant in this neck of the woods, trod on by shepherds every day, yet bearing in another country "a bright golden flow'r" (633), seems an excellent emblem for the psychic reversal that creates a golden ideal out of something denied. As is not the case with temperance and immoderation, which exist in a relationship of degree, the virtue of unblemished chastity *could* be produced by idealizing the logical opposite of a prohibited sexual wish. Negate the tempter, dress the negation in the imagery of Christian Platonism, and one beholds the heroine. Reverse the spell of Comus, we are told, and the Lady would be free to rise. Yet for a time she sits, as if caught from behind by the childhood origin of her ethical strategy. According to Sabrina, the "venom'd seat" has been "Smear'd with gums of glutinous heat" (916–917), and "glutinous," Le Comte reminds us, derives from *gluteus*, "buttock."[29] Because of the sexual interests of her tempter, we recognize these hot gums as a symbol for sperm; but the dominant iconography of paralysis on a throne is anal. The pattern of regression found in the imagery of conscience and the exposition of cosmic dualism now extends to the phases of psychic history.

Sabrina is summoned. Although the wand is missing, the formula of release is still reversal:

> Brightest lady look on me,
> Thus I sprinkle on thy breast
> Drops that from my fountain pure
> I have kept of precious cure,
> Thrice upon thy finger's tip,
> Thrice upon thy rubied lip;
> Next this marble venom'd seat
> Smear'd with gums of glutinous heat
> I touch with chaste palms moist and cold.
> Now the spell hath lost his hold. (910–919)

Comus disparaged the "clear stream" (722) when offering his "luscious liquor." Risen from such a stream, Sabrina consecrates the act of eating. Her first gesture purifies the breast, earliest source of food. Then, moving from hand to lips, she enacts the gesture of mature consumption. Finally she turns to the site of food's exit, opposing the moist and hot of lust or excretion with the moist cold of chaste

water. Her magic may well symbolize baptism, as is often claimed. But why would the rite of baptism, the first antidote to original sin, be appropriate in this context? *Her ritual of undoing implies that the Lady, having figuratively drunk the potion of her tempter, is guilty.*

Were *Comus* Alice Egerton's fantasy, the origins of this guilt would be fairly obvious. The tempter with his wand, his throne, and his desire to make Alice "my queen" (265) represents a projection of her father with his "new-entrusted Scepter" (36), his throne, and his "Lady bright." Before the witness of her superego, she relives the oedipal temptation, invited to usurp the throne of her mother. She chooses virginity, fortifying this choice with the lock of anal passions. However, the power of the wand persists: because her virtue is bound to a repressed wish, she is fixed at the stasis of primitive superego formation, a captive of old love and old fear that have not dissolved into the mobile self-possession of intrapsychic structure. She must be forgiven by her rival. So the virgin Sabrina appears, a pre-oedipal mother with no sexual entanglements, to give her life once again through the sanctification of nourishment.

This pretense yields a plausible and coherent interpretation because Milton had a fine intuitive understanding of women *in certain situations*. Besides the fulsome rendering of virginity in *Comus*, Milton also wrote a gentle and intelligent sonnet, "Lady that in the prime," in which he plays the role of counselor and confessor to a lady similar to his half-invented heroine of 1634; considering the significance of a baby within the psychic history of many women, there is also a deftness in consoling the grieving mother of "On the Death of a Fair Infant" with a promise given originally to eunuchs. One way men acquire such intuition is from their own oedipal settlements. Milton's father was, we have seen, a willful man who designed the destinies of his sons. The conscience he instilled in his oldest boy deployed its ego with a sometimes heedless rigor: "My father destined me in early childhood for the study of literature, for which I had so keen an appetite that from my twelfth year scarcely ever did I leave my studies for my bed before the hour of midnight. This was the first cause of injury to my eyes, whose natural weakness was augmented by frequent headaches. Since none of these defects slackened my assault upon knowledge, my father took care that I should be instructed daily both in school and under other masters at home" (*CP* IV, 612). Sacrifice to the father's plan was obviously rewarded. The thought of a pubescent boy studying past midnight, suffering already from headaches and eye-strain, might bring smiles of approval from the rod-wielding schoolmasters of

the Renaissance, but they, too, were extensions of the superego.

Erikson has said that the fundamental conflict within the oedipal child is between initiative and guilt.[30] Rephrased for our poet, the urgent question stemming from this period in his life might have been this: how do I venerate my father, acknowledging his creation of me, his laws, his plans for me, while still clearing a space for my own ambitions? In the oedipus complex proper, the male child normally sacrifices love to wholeness. Fearing the talionic law of retaliation, he abandons the fantasy of supplanting his father and "identifies with the aggressor": men gain their psychological maleness, assimilating themselves to the ideal type of their own sex, only in the face of a death threat to this maleness—an irony to be observed in the ensuing period of latency, that long paralysis of the penis, when masturbation and sexual fantasy cease until puberty. Should his castration anxiety be severe, the oedipal boy may try to solve his dilemma by entering into the structure of the "negative" complex. Here he identifies with his mother (or sister), extricating himself from a hostile relationship to the father by presenting himself as an object of love.

This could not have been the primary solution adopted by Milton, for he was not a homosexual. Although his friendship with Diodati has some traces of homoerotic feeling—the letter of 1637 (*CP* I, 325–328) contains passages that might have been the work of a jealous woman, or a man imitating one while half-aware of his tone— Shawcross' belief in a real sexual adventure runs contrary to the regularities of his character manifested by the bulk of his writings.[31] This was not a timid man. When discussing issues of personal probity, which he was often compelled to do in poetry as well as prose, he hurled disdain at those time-servers who adjust their behavior to any kind of external coercion, from unreflective prejudice to reasoned creed, and rose in prideful tones to declare his own freedom from such constraints. Self-justification being almost an obsession with him, he did not lack the courage to indulge inclinations so fundamental as sexual preference. Had Milton found himself primarily and continuously attracted to men, he would have acted on this attraction at some point in his life, and if he had, sodomy would not have been routinely listed among the acts "opposed to chastity" in the *Christian Doctrine* (*CP* VI, 726, 757). Was he ever wrong in such matters? Instead of searching Christian tradition to justify the sexual love of men for men, he evolved a humane philosophy of divorce (managing to explain away the contrary opinion of Christ himself) and defended biblical polygamy. But a feminine identification in the oedipal context does illuminate his experience

with the genre of romance, his unconscious understanding that a baby is a phallus, his ability to represent the positive complex of Alice Egerton by reimagining his own negative one, and in his subsequent work, such apparent anomalies as the transposed genders in the myths of *Areopagitica* (Psyche as Adam, female Truth as Osiris, Britannia as a strong and long-haired Samson), which have an echo in several odd parallels linking the major figures of *Paradise Lost*. This feminine identification became for Milton a private sphere of power, the locus of energies and resources that his earthly father had *not* given him. We will see that transformations of the lady in Milton are in effect the history of his poetic identity from *Comus* to *Paradise Lost*.

When Milton left Cambridge, his father had retired from business and moved out of London. A beloved father having completed his work, a son with his work before him: it might have been a time of consolidation for the young man. Sons often make a lasting peace with their old rival when they come to replace him; time, in the normal course of things, awards the oedipal triumph to the young. But Milton remained at home for five years, and rather than pursue the "potent inclination" of "the desire for house & family of his owne" (*CP* I, 319), he developed a mysticism of virginity. The doctrine proclaimed in *Comus* repeats as a choice the trauma of oedipal submission, prolonging his latency well into manhood.[32]

This archaic submission can be discerned in one of its most puzzling passages:

> These thoughts may startle well, but not astound
> The virtuous mind, that ever walks attended
> By a strong siding champion Conscience.—
> O welcome pure-ey'd Faith, white-handed Hope,
> Thou hov'ring Angel girt with golden wings,
> And thou unblemish't form of Chastity,
> I see ye visibly. (210–216)

Out of the conscience of the Lady as she moves toward her trial come the three theological virtues, with the famous substitution of chastity for the third and greatest of these—the only one whose direct object is our activity in this world. Chastity is already virginity in these lines. If "unblemish't" modifies "form" alone, the Lady could be speaking platonically of the blemish of particularity. But it would be peculiar for her to call "unblemish't," in the philosophical sense, a form seen "visibly." When we add the context of the masque as a whole, the weight of the evidence is that the adjective qualifies "form

of Chastity," implying that married chastity is a blemished form.

The psychoanalytic model of symptom formation offers one solution to the crux. Chastity is repressed, and in its place a symptom is installed—the high mystery of virginity. What is the relation between the excised virtue and its successor? In Renaissance iconography, charity is commonly portrayed as a mother suckling her children. Spenser's Charissa, perpetually giving birth, keeps "Her necke and breasts . . . ever open bare," ready at any time to suckle her babies:

> A multitude of babes about her hong,
> Playing their sports, that ioyd her to behold,
> Whom still she fed, whiles they were weake and yong,
> But thrust them forth still, as they wexed old.[33]

Diekhoff's collection of Milton's autobiographical passages runs to nearly 275 pages; in all of this material, a discourse of the self more generous than that of any other poet of the period, there are but two allusions to Sarah Milton, one of them being: "my mother [was] a woman of purest reputation, celebrated throughout the neighborhood for her acts of charity" (*CP* IV, 612).[34] Reenacting the genesis of the superego, the virtue of maternal love disappears from the text of *Comus*, replaced by the form of its sterile and unblemished opposite. Virginity is the response to a mother's love that, suspending his maleness, makes a lady of a son.

Comus can create the symbol interpreting this root-bound virtue because Milton has represented, in the figure of his tempter, an overcoming of the oedipal temptation. Gradually released from the original universality of the father's law, men arrive through "potent inclination" at a mature object choice modeled, to some degree, on the lost love of childhood. In this sense, Comus feels a wholly unexceptional attraction to the Lady. She is his mother regained:

> I have often heard
> My mother *Circe* with the Sirens three
> Amidst the flow'ry-kirl'd *Naiades*,
> Culling their Potent herbs and baleful drugs,
> Who as they sung, would take the prison'd soul,
> And lap it in *Elysium*; *Scylla* wept,
> And chid her barking waves into attention,
> And fell *Charybdis* murmur'd soft applause:
> Yet they in pleasing slumber lull'd the sense,
> And in sweet madness robb'd it of itself,

But such a sacred and home-felt delight,
Such sober certainty of waking bliss,
I never heard till now. I'll speak to her
And she shall be my Queen. (252–265)

Perverse rites aside, the spokesman for sexuality in the masque (who also betrays an unblushing interest in the gems "hutch't" in the "loins" of Mother Nature) has transcended incestuous love in the only way men do, choosing his "Queen" out of a "home-felt delight" that identifies her with his mother, yet also valuing the difference and uniqueness of his choice ("never heard till now"). The first love, after all, was a dream, whereas its restoration is the "sober certainty of waking bliss." In the complement to this moment of desire, we find the shepherd lad listening to the ravishing songs of his father's friend Henry Lawes. His magic plant counteracts the "potent herbs" of sensual experience that Comus has retained from the women of his childhood.

The virginity of the masque preserves the "first being" of a person in a psychological sense as well. In the plot of psychic development, the oedipus complex stands as a rude critique of the omnipotent ego shaped in the early years of life. The child must concede the grandiosity of his desire, obey the law of a superior, postpone his nascent sexuality in the most arduous test of the reality principle, and come to understand death, not merely as an intolerable flood of anxiety in the absence of gratification, a death *in* the ego, but through castration as a death *of* the ego.[35] The high mystery of virginity seeks to compensate for the loss of sexuality by eluding the cleansing humiliation of these narcissistic injuries. "I choose chastity, mastering the trauma I once suffered passively by restaging it as my active, heroic decision. I will not be the rival of my father, nor will I envy his prerogatives. This vow will protect me from my hostility and his wrath: I forego." But as the Lady sits immobile, two spiritual guardians gather about her, a sacred prefiguration of the domestic reunion at the end of the masque. The Attendant Spirit and Sabrina, like the fantasy parents of the family romance, nourish illusions surviving from the "first being" of the preoedipal ego—a body that cannot die, an interior unstained by excremental impurity, a self-sufficiency free of earthly attendants, an innate superiority to "bestial" people. The high mystery, then, is a strategy for making a triumph of submission, a virtue of necessity. Unable to take the other as a sexual object, the virgin ego takes itself as an object; the replacement of charity with unblemished chastity that obeys the "no" of the father's law also enacts the compensatory gesture of nar-

cissism, substituting the virtue of cherishing oneself for the virtue of cherishing the other.[36]

Thus, we have found another explanation in depth psychology for Milton's reluctance to leave home. He was unable to enjoy the fruits of his obedience, passing beyond sonship to become himself a husband and father. But he had also transformed this oedipal sacrifice into a retrieval of the unlimited ego formed during the early symbiosis of infant and mother. Just as in infancy her absence was the primary threat to this ego, so in maturity her continued presence might well have nurtured its continued survival. Castration anxiety, to be technical, fused with separation anxiety, doubling his attachment to his first family. Perhaps this fusion of oedipal with preoedipal wholeness shows us why the phallic magic of Comus should reside in the navel of the woods, filling the scar that marks our original dismemberment.

The final and sweetest bounty of this "hidden strength" lies in its secret assertion of the very hostility being renounced. I have maintained that the virgin soul attempts to make of its incarnation a sacred existence—in Christian terms, to extricate itself from the lineage of Adam and Eve by refusing to participate in the act by which original sin is propagated. But the other side of this sanctification of the self is an indictment of progenitors. If one makes a high mystery of virginity, the fact that one exists points to a crime. The Lady is guided to the two thrones, set side by side, on which her "Noble Lord, and Lady bright" recline, professing by their union that they no longer know the sage and serious doctrine in its unblemished form: *she* is the "Brightest lady," as her magical mother has proclaimed. Absolving the ego from oedipal crime, virginity returns the accusation. *The superego is guilty*. Implicit in the virtue of *Comus* is a suppressed indictment of the community of parents who, guilty themselves, impose on their children obedience to the law they defile. This subtext in the doctrine of virginity stands opposed to the theodicy on the surface of the masque. Justifying the ways of its new master, the oedipal ego must confront the implacable truth that the superego always derives from those who violate its "no"—that good and evil proceed, in this sense, from the same source. Through extreme obedience, virginity exploits an irony discovered in the rudiments of theodicy.

As the intimate core of Milton's poetic identity, virginity was a sign of his separateness (the poet is found "far from the sight of profane eyes" in *Ad Patrem*), his hope for the divine favor of inspiration, his desire to surpass his predecessors, and his wish to author a deathless poem. But from the symbol left by the enchanter we can surmise that virginity also symbolized his reluctance to begin and his fear of ex-

posure before his audience. The conflict that Milton is trying to solve during his years at home is not between poetry and some other vocation: it is within poetry itself. The poet in the act of composition is solitary and, if he is inspired, engaged in communication with transcendent forces. When, however, he publishes his composition, he divulges the hidden strength of his narcissistic esteem, purchased at the cost of sexuality, to the judgment of others. Given a source of strength, but forbidden to reveal his source of strength—the dilemma of Samson could be taken as a parable about this artistic conflict. Poetry externalizes conscience, and only a fit audience, willing to concede immortality to a poetic identity shaped about a pure and immortal self, could be entrusted with the sacred songs of Milton. In the epigram that accompanied his gift of the 1645 *Poems* to the Oxford librarian John Rouse, he imagines his book "scraped by the dirty, calloused hand of an illiterate dealer" (p. 147). Insofar as a text contains "a potencie of life," "the pretious life-blood of a master spirit" (*CP* II, 492–493), its availability to soiled hands and unclean thoughts recalls the Elder Brother's charged account of the invasion of impurities into the "inward parts" of our being. A poem is the corpus of the soul. The published poet is, like the Lady, fixed in the gaze of his audience, vulnerable to outrage.

All writers know in some fashion the fear of narcissistic injury, but I think Milton knew this fear primitively. His fascination with the fate of Orpheus is perhaps the most succinct expression of his ambivalent calling to the sacred art. Did not the exhibition of his talent cause the dismemberment of Orpheus—and at the hands of women devoted to the father of Comus, virginity's opponent? In Milton's imagination, the desire to exhibit can bring down a terrible vengeance: it is just when "we think to burst out into sudden flame" that the blind Fury "slits the thin-spun life."[37] "You ask me what I am thinking of?" Milton wrote to Diodati in 1637, the year Edward King drowned and *Comus* was published anonymously, "So help me God, an immortality of fame" (*CP* I, 327). But the fulfillment of this wish aroused watchful furies with sharp instruments because it was born of energies thwarted by the formation of the superego—including sexual energies. The fusion of separation and castration anxiety had as its complement a fusion at the roots of his poetic ambition of narcissistic and oedipal desire.

The hoarded power symbolized by virginity placed Milton *before* the consummation of his ambition, owing rather than paying. The Lady does not speak her doctrine; she promises that she can. *What she says about the revelation of her mystery*—I shall not now, but could; not to

you, but only to someone worthy—*would have been the reply appropriate to her sexual temptation*, symmetrical to her exposition of temperance. How, given the implied equations of virginity with the power to speak and sexuality with its actualization in speech, can the virgin poet fulfill himself without losing his power? As paralysis precludes action, so concealment is the safeguard against exposure. Milton wrote in *Ad Patrem* that his debt to his father was such that "my greatest gifts could never repay" (p. 82). Immense pride often accompanies the piety of the unpayable or infinite debt, for to be infinitely indebted, one must be infinitely endowed. Spending his talent, Milton would risk its defilement, the realization of its finitude, and, most of all, the judgment of his debtor. Instead, he exhibited his beforeness, letting the promise of youth endure throughout most of his life. "Neither doe I think it shame to covnant with any knowing reader, that for some few yeers yet I may go on trust with him toward the payment of what I am now indebted" (*CP* I, 820). Remaining beholden kept the energies of his "first being" intact. For who covenants with us? Nor was Godhead from his thought.

In a theological reading of the masque, the Lady's paralysis might be interpreted as a Protestant acknowledgment of the inability of man to will his own merit, all virtues requiring for their completion the gift of grace. If it is permissible to offer a psychological description of this human insufficiency, and if the doctrine of grace implies for those in need of it a yearning for a new or transformed existence that, unable to be imagined precisely, cannot therefore be sought through an ordinary act of volition, then the theological reading of this symbol is not inconsistent with my own. Comus leaves behind a body whose posture is the emblem of its soul. Hers is a root-bound virtue, caught in a reaction formation to oedipal temptation. It is not free. It is in bondage to the desire denied. The virgin "parents" who come to her aid, generated from Orphic myth, simplified Platonism, and impossible physiology, can release her only in the sense that they compensate for her essential imprisonment. Their superiority to the real parents awaiting her at Ludlow manifests a strategy of idealization, endeavoring to make of her stasis a spiritual aspiration. Comus has shown us the need for grace. When articulated in the epilogue of the masque, this grace contains in some indistinct way the recovery of his rejected vice.

The evidence brought together here is subject to diverse constructions within the framework of psychoanalysis. In particular, I have emphasized the maternal direction of Milton's incest wish, which is clearly in defiance of the chaste silence of the poet himself. Given the reiteration of his family in the masquing family, would not the passion

of brother for sister seem more probable? Watkins found "a faint suggestion of latent jealousy" in the Elder Brother's insistence that his sister be untouchable, "clad in complete steel" (421), and Milton's subsequent enthusiasm as an uncle would lend support to the idea that the lady he protected when reading romances was Anne.[38] A strong current of sister incest would also allow us the gratuitous pleasure—in some ways, the best kind—of squaring literary history with psychic history, putting even the unconscious of our poet in the romantic line. But sister incest almost always derives from a primary attachment to the mother or to whomever occupies the slot "mother."[39]

I am also bolstered in my opinion by the other of Milton's two allusions to his mother—a statement collaborated by Wood and Phillips (*EL*, 36, 55):

> When I had occupied five years in this fashion, I became desirous, my mother having died, of seeing foreign parts, especially Italy, and with my father's consent I set forth.
>
> (*CP* IV, 614)

The death of Sarah Milton was the first step in lifting the spell that bound him to his home. It was followed by the father's consent, travel abroad, residence in London, the profession of tutoring, the early pamphlets, the declaration in *An Apology* that "marriage must not be called a defilement" (*CP* I, 893), and the hasty union of 1642. The epilogue to *Comus* uproots the symbolism of the masque, freeing its meaning to "set forth" into a future that could possibly hold *Paradise Lost*.

The Promise of Jove

Mortals that would follow me,
Love virtue, she alone is free,
She can teach ye how to climb
Higher than the Sphery chime;
Or if Virtue feeble were,
Heav'n itself would stoop to her. (1018–1023)

"Here at last," Robert Martin Adams has written, expressing in this instance the critical consensus, "is the Christian heaven, unmistakably."[40] Critics have tended to interpret the epilogue as an allegorical progression toward the final acknowledgment of a realm "Higher than

the Sphery chime," retrospectively translating the entire speech into Christian terms. But can we make this translation from symbol to doctrine with unqualified confidence? Unmistakably Christian, yes—but distinctly Christian? As John Arthos maintains, "it is certain that as an allegory the Epilogue is incomplete. Whatever the context and whatever the sources, the Epilogue itself does not work out in an explicitly schematic way the relationship of each of the realms it names, and the characters of the beings in them, to a defined philosophical system. The suggestions are everywhere, the definitions are lacking."[41] Emerging from a fictive philosophy that exists only fragmentarily in the masque itself, the epilogue seems a pseudo-allegory, the announcement of a mysterious knowledge into which none of the "mortals" on this stage, whether as characters or as people, have been fully initiated. I suspect that here, as maybe elsewhere in the mythological poetry of the Renaissance, we foreclose the passage by assuming from the onset that its meaning can be stated without remainder in Christian terms. Why, we might ask instead, is this passage set in the myths of a dead culture? What is being said here that could not be said in conventional delineations of Christian doctrine? At this stage in Milton's career the polytheism of imagination could address problems left unresolved, or rendered all the more acute, by the monotheism of reason.

Whatever the significance of its allegory, the rhetorical purpose of the epilogue *is* unmistakable: a distinct departure for the journey of interpretation. Dropping the disguise of his subservience to the Egerton family, the Attendent Spirit, like a good tempter, beckons chaste mortals—those with "true" ears (997) unclotted by "foul talk" (464) or "barbarous dissonance" (550)—to their reward in a great good place beyond the dim spot of this earth. Tillyard believed that the epilogue cut beneath the oppositions of the temptation scene, showing "the lady partly wrong and Comus partly wrong" by proposing marriage as the mediate state between virginity and lust.[42] Although his reading leaves out of account the otherworldly locale of this marriage, his intuition responds to a genuine lure. The return of erotic imagery concedes that sexuality is, if not in the end, in the end of this work at least, the protean lord of our desire, and worlds we yearn for will be marked as desirable by erotic symbolism, all the more so if we must forfeit sexuality in order to reach them. The sacred tempter speaks directly to desire—to the desire of the young, who must negate desire in the gaze of those who indulge it.

Assuming that Spenser's Garden of Adonis was the primary source

for this passage, Milton has produced several telling revisions. Whereas in Spenser the two sets of lovers inhabit the same paradise, Milton splits the good place into two tiers, making Cupid and Psyche reign over Venus and Adonis. He also denies to the subordinate pair the blissful intercourse they enjoy in the *Faerie Queene*[43] without dread of "that foe of his":

(List mortals, if your ears be true)
Beds of *Hyacinth* and Roses
Where young Adonis oft reposes,
Waxing well of his deep wound
In slumber soft, and on the ground
Sadly sits th' *Assyrian* Queen. (997–1002)

In the sky above this garden, Cupid and Psyche embrace:

But far above in spangled sheen
Celestial *Cupid* her fam'd son advanc'd,
Holds his dear *Psyche* sweet entranc't,
After her wand'ring labors long,
Till free consent the gods among
Make her his eternal Bride,
And from her fair unspotted side
Two blissful twins are to be born,
Youth and Joy; so *Jove* hath sworn. (1003–1011)

This vision of the transcendent world could be interpreted as the climactic expression of the psychological subordination we have discerned in the virtue of virginity. The genital love of Venus and Adonis is not being consummated; meanwhile Cupid and Psyche, Love and Mind, revel in an apotheosis of narcissism. But if we associate Psyche with the Lady, comparing her "labors" in the masque with her promised destiny, the epilogue sketches a new strategy for liberating her from immobility.

The oath of Jove does justice to the young, satisfying their implied indictment of the community of parents. Here the older generation is separated. Venus opposed the love of Cupid and Psyche, yet in the Adonis myth she herself was guilty of the crime of loving a mortal. In Milton's garden a "deep wound" still afflicts her chosen consort, and if he is "waxing well," he remains fixed in the posture of sleep while the Queen "Sadly sits"—both of them motionless, the paralysis of chaste

children having been transferred to their guilty oppressors. Far above, where the "free consent" of the gods has legitimized sexuality, the young embrace before their marriage. The issue is Youth and Joy indeed, for this is the oedipus complex tuned to the triumph of children: as a deep wound of castration divides the parental generation, their sexual privileges are "advanc't" to Cupid and Psyche. The concluding vision of the masque releases the erotic and aggressive wishes whose denial bound virtue to its root. Could the young Milton have achieved this clarification of desire within the terms of Christian Myth?

It is characteristic of Milton's temptations, both here and in the epics of his maturity, that whatever the law forbids will be salvaged for righteousness in another and transmuted form. But the just restitution of the sexual life in a transcendent experience must have posed difficulties for an imagination nurtured by Protestant thought. In the Catholic tradition one could eroticize heaven through metaphors such as the soul as bride of Christ, the "Celestial Cupid": these metaphors of recovery had been earned for everyone by the renunciations of a celibate priesthood and their monastic communities. Abolishing the ecclesiastical law against sexuality, the reformers gave the erotic heaven a prominent place in their general critique of images. Since imagination was by nature an idolator, the symbolism of a luxurious heaven surreptitiously expressed the human in the guise of signifying the divine.[44] This Protestant uneasiness can be felt in Milton's two great attempts to supply appropriate consolation for dead virgins. His scriptural warrant was Rev. 14, where the 144,000 virgins stand closest to the throne at the marriage of the lamb. Thus elevated, the subjects of "Lycidas" and "Epitaphium Damonis" participate in a celebration, feasting and making music, yet they are not married. The last line of the poem for Diodati, *Festa Sionaeo bacchantur et Orgia Thyrso,* reminds us of the explicitly sexual consolation of the masque, for Comus is the son of Bacchus, and the name taken by the Attendant Spirit, Thyrsis, puns on the Dionysian symbol of the thyrsus.[45] Awaited by the gods of desire, the Lady is the only one of Milton's gallery of virgins to be promised the full recompense of intercourse. She will be "sweet entranc't" in the arms of Love. Through classical myth, Milton was able to write the text of desire suppressed in the body of the masque. His climactic symbol for the Lady's spiritual destination brings the prohibited wish into consciousness, a precondition for dismantling the reaction formation of virginity.

The regressive movement initiated by this symbol is, within the psychic structure we have articulated, a progressive one. Read as naive

statement, before undertaking an allegorical translation, the journey of Psyche leads away from primitive configurations toward intercourse, marriage, and parenthood. The pattern is broken by reversal: an archaic desire (the wishful apotheosis of narcissism) is here represented by an "advanc't" state.

Luxuriating in the surface of the passage, we discover with the word "entranc't" an extraordinary redemption of the lewd magic of Comus. The tempter, we recall, wanted us to forget; his sorcery induced his followers to wallow in sensuality, ignorant of their souls. Now the good entrancement of Psyche provides an occasion to recognize what has been implicit throughout—that the sexual act is the epitome of this forgetfulness. Kierkegaard and Heidegger have probed the distinctive structure of anxiety among human affects. Love is love-of, fear fear-of, but anxiety has no intentional correlate; we are anxious about "nothing," which is why Heidegger places anxiety at the threshold of the knowledge of our homelessness in this world, exhorting us to achieve the state of authentic resolution he terms Being-toward-death.[46] The phenomenology embodied in the everyday language of the Renaissance understood that orgasm, too, was a "death"—the happy inverse of anxiety, we might say. Freud pointed to the experience underwriting this metaphor in a passage about hysteria that seems particularly suggestive in the context of *Comus*, with its arresting interlude of paralysis: "The loss of consciousness, the '*absence*,' in a hysterical attack is derived from the fleeting but unmistakable lapse of consciousness which is observable at the climax of every intense sexual satisfaction, including auto-erotic ones" (*S.E.* IX, 233). Whereas anxiety violates intentionality by effacing the object-pole of our awareness, sexual experience culminates in a moment of oblivion that shatters the I-pole of consciousness. The soul really is, in this sense, destroyed by sexuality. If virginity is the effort to transform the body into "soul's essence," orgasm is the longing of the soul to be wholly incarnate. In mysteriously reclaiming the power of Comus, the vision at the end of the masque, before drifting off into nebulous allegory, bears intimations of a new wisdom concerning this life.

The magic rite of intercourse engenders the child, visitor from Non-Being. Excised from the conscience of the Lady, the maternal virtue of charity is restored, iconographically at least, in the promised motherhood of Psyche. Her vagina will not be "used" for birth, a displacement that continues the *noli me tangere* of virginity, but like "entranc't," the phrase "fair unspotted side" suggests an imaginative victory over the psychic origins of this virtue.

Probably "unspotted" is meant to signal the flesh's escape from the contagious matter of "this dim spot," realizing at last the preoedipal wishes of divine philosophy. The birth of twins from the "side," however, returns us to the family now reunited on this earth. In the Echo Song the two brothers were described as "a gentle Pair / That likest thy Narcissus are." How can two be like one unless they are, like Narcissus, twinned? This gentle pair, Comus told the Lady, "left your fair side all unguarded" (283), and in her hour of trial the place of these youthful champions was taken by the sterile virtue of her "strong siding champion Conscience." Insofar as the fate of Psyche foretells the destiny of the Lady, the brothers are again at her side in the symbolism of transcendence. The child of Cupid and Psyche in Spenser is the traditional Voluptas. This last rearrangement of Milton's "pre-text" completes with perfect finality the oedipal conquest announced in the separation and subordination of Venus and Adonis. Mother of her earthly siblings, wife to love himself, the "Brightest lady" will occupy the place of her rival—a daughter-become-mother begetting brothers-become-sons. She will be married, furthermore, to a son-become-father. Within the polytheism of a discredited, therefore malleable religion that Renaissance poets inherited as a province of their imaginations, Milton was able to play out the male and female complexes (or the positive and negative of the male complex) to the ideal consummation of "Youth and Joy."

Releasing the hostility and desire frozen by the virtuous choice of the temptation scene, the epilogue gestures toward a new negotiation of the oedipal settlement. It tries to achieve what the children and their spiritual guardians could not: the capture of the magician in the navel of the woods and the appropriation of his power. Imagination finds what vigilance lost. And Jove vouchsafes this ambition: the father at his most sublime, at the pinnacle of a cosmic hierarchy including the Lord President and all other trustees of earthly scepters, decrees the wished advancement of youth. Milton was endeavoring to recreate his superego as a consenting as well as denying force. The grace of a new fit between wish and defense would answer those abiding questions that the feeble superiority of virginity was retreating from, an "excrementall" virtue in the language of *Areopagitica*, "fugitive and cloister'd" despite its willingness to be tempted (*CP* II, 515–516). How does a son shaped by the will of his father become independent? How can this son fulfill erotic and literary ambitions, yet retain his power?

For the divine poet, the answer lay in translating the vision of the

epilogue into Christianity without puncturing its assertion of triumphant filial desire. I believe that Milton found this transit from classical to Christian deeply problematic. Jove deposed, and in some Renaissance versions of the myth actually castrated, his father: could not this god endorse filial hopes more readily than God the Father? Christianity also deifies a son. Unlike the victorious Jove, however, Christ—virgin son of a virgin mother—manifests his Godhead in an act of oedipal submission, absorbing the wrath of his Father; "the stroke of death he must abide," as Milton wrote in "The Passion." This fragmentary lyric foreshadows its own failure in metaphors of paralysis: "These latter scenes confine my roving verse, / To this Horizon is my *Phoebus* bound"; "And here though grief my feeble hands up-lock." We will find abundant evidence to prove that identification with the Son, on which all forms of Christian salvation rest, was for Milton vital and shaping, not merely imputed and gratuitous. But during the first stage of his artistic life the dark mysteries of the sacrificed Son stood (in the words affixed to his arrested poem of arrested grief) "above the years he had." For someone as obedient to his superego as the youthful Milton, the crucial moment of his religion might well have seemed, at a depth of his being immune to the "must" of doctrine, unbearably unjust. He would never really warm to the image of the Crucifixion. It is represented in ten moving but unextraordinary lines in *Paradise Lost* (12.411–420), and is kept at a distance of prefiguration in *Paradise Regained* and *Samson Agonistes*, whose prophetic climaxes sweep through the Passion to the unambivalent triumph of the Last Judgment. Of course we can tally up the many allusions to the Cross in his works; he was a Christian.[47] But when, in the second burst of his poetic career, Milton completed a monumental translation of "so *Jove* hath sworn" to the ways of the Christian Godhead, myth and theology underwent heretical transformations in expressing and assimilating the hard truths of sonship deposited in this religion.

For the man, the answer lay in pulling the vision of the epilogue down to his earthly future. I have maintained that the death of his mother prompted his first step in this direction. When Milton returned from Europe, he set himself up as a tutor, and the presence of his sister's children must have intensified what was already fatherly about a profession that allowed him to impose elaborate schemes (just how elaborate we can surmise from *Of Education*, as well as from the testimony of Edward Phillips in *EL*, 60–62) on the development of children. But the last step, his actual decision to marry, coincides more

importantly with his enthusiastic participation in the early years of the prelacy controversy: licensed rage was the other key to his release. The blocked aggression inscribed in the epilogue to *Comus*, having endured "wand'ring labors long," embraced an appropriate object in the profession for which his father had destined him. His violence as a controversialist, after allowances are made for the rude rancor conventional in public disputes of the time, remains astonishing.

The passion of his first work, *Of Reformation*, despite a generous ladling of rich invective (the devotion of ministers under episcopacy "gives a Vomit to GOD himselfe"—*CP* I, 537), is in the main dignified and erudite, as if his reason had decided to enter the debate by appealing to the image of itself in his audience. The peroration, however, divulges pure force, making of language an instrument of the will that grants utter beatitude to those who agree and lays a primitive curse of damnation on the unconvinced. Those plying the strange trade of writing argumentative prose know that, brooding in solitude over the minds that will peruse the text, one now and again becomes so convinced of what one is saying that merely saying it begins to seem insufficiently assertive, and one is tempted to add some extrarational, magical proclamation of the might of these invincible thoughts. But with the conclusion of his first pamphlet Milton has outdone most of the zealots of disputation history in this topos of narcissistic speech. Commanding the will of God, he makes the eternal destinies of his readers proceed from their evaluation of his statement:

> Where they undoubtedly that by their *Labours*, *Counsels*, and *Prayers* have been earnest for the *Common good* of *Religion* and their *Countrey*, shall receive, above the inferiour *Orders* of the *Blessed*, the *Regall* addition of *Principalities*, *Legions*, and *Thrones* into their glorious Titles, and in supereminence of *beatifick Vision* progressing the *datelesse* and *irrevoluble* Circle of *Eternity* shall clasp inseparable Hands with *joy*, and *blisse* in over measure for ever.
>
> But they contrary that by the impairing and diminution of the true *Faith*, the distresses and servitude of their *Countrey* aspire to high *Dignity*, *Rule* and *Promotion* here, after a shamefull end in this *Life* (which *God* grant them) shall be thrown downe eternally into the *darkest* and *deepest Gulfe* of HELL, where under the *despightfull controule*, the trample and spurne of all the other *Damned*, that in the anguish of their *Torture* shall have no other ease then to exercise a *Raving* and *Bestiall Tyranny* over them as their *Slaves* and *Negro's*, they shall remaine in that plight for ever,

the *basest*, the *lowermost*, the *most dejected*, most *underfoot* and *down-trodden Vassels* of *Perdition*. (CP I, 616–617)

The champion is in the field. Although, as he will soon assess his strength, "I have the use, as I may account it, but of my left hand" (*CP* I, 808) – a figure drawn from wrestling or fencing? – the left is indisputably in the right. *Caveat lector*.

A few months later, after his warning had gone unheeded and Bishop Hall had dared to disagree with the Smectymnuans, Milton produced one of the most furious vituperations of the century, "a vehicle of unrestrained bitterness," the Yale editor remarks, "unequaled as yet among the serious Puritan sallies against the prelates" (*CP* I, 123). Like the enchanted Lady of *Comus*, the Remonstrant of *Animadversions* must sit helpless while his "inside nakednesse" is displayed "to public view" (I, 688). At times the stupidity of this captive is a mental body fetid with sores and filth. Metaphors of biting, gnawing, spitting, vomiting, defecating, and whoring burn their paths through the text, often joined to worms, vipers, and other creatures; when some fecal badinage prompts the Remonstrant to note that "Wanton wits must have leave to play with their owne sterne," Milton replies that his knowledge of ass-licking must have been learned "from one of the *Archiepiscopall Kittens*" (696). Now and then he turns from his imprisoned puppet to his readers, confirming their pleasure in this exposure. He suspends the abuse for a gorgeous prayer to Christ (705–707), requesting a speedy end to the world, then resumes his brilliant rant and rave. In reply to one of the Remonstrant's defenses, he passes beyond speech altogether: "Ha ha ha" (726). Although Milton claims that his task has been undertaken with "a sad and willing anger . . . without all private and personall spleene" (*CP* I, 663), and although some studies of his political rhetoric have approached this "well heated fervencie" as a cunningly crafted style imitated from noble precedents, it is absurd to imagine Milton deciding, in the calm center of his humanist intelligence, that because a language which symbolically disrobes, smears, spurns, wounds, dismembers, murders, damns, and tortures its opponents for all eternity would be efficacious in the good cause, he would therefore work up a facsimile of violent intent. Nor can we dismiss these "subliterary tactics," in the phrase of Michael Lieb, as beneath his genius and essentially unrelated to the sublimity of *Paradise Lost*, concentrating instead on a "true Milton" found on those islands of prayer, meditation, and autobiography that protrude here and there from the seethe of detestation released during his political years.[48]

In the preface to *Animadversions*, we learn that "those two most rationall faculties of humane intellect, anger and laughter, were first seated in the brest of man" (664) to aid him in combating false prophets. Milton's animation of these "seated" powers was a form of self-healing, purging inactivity from himself by projecting its image onto another, then heaping scorn on this disguised icon of himself. *Animadversions* recreates the subversive paralysis of *Comus*, but now it is the root-bound Remonstrant who will not grow up, will not extricate himself from bondage to authority and risk the adventure of reformation. Milton discovered in the pamphlet wars a realistic channel for filial hostility. His omnipotent self came forth into the world, clad in the complete steel of invincible truth: his accusers were the guilty ones. Seeking to reform church and country, Milton was reforming from within the structure of his superego.

The introspective pamphleteer of *The Reason of Church Government*, analyzing his conscience in a passage thick with financial metaphors, reasons that "those few talents" God has "lent" him must be spent in these controversies lest he be cursed forever with his "own brutish silence" (*CP* I, 804–805). He is not spending the most precious talent, which still remains locked in the stasis of the past; assuming the correctness of the traditional dating of *Samson Agonistes*, all the poetry written during his public career is short and in some fashion occasional. Yet payment on this noble if lesser debt appears to him as the precondition for any future speech. Milton has severed himself. The positive achievement of poetic ambition has been left where it was during the years of country retirement, under the ban of primitive fears—of leaving the ego sullied and depleted, exhibited before vengeful powers—while virginity has been detached from this ambition, allowing Milton to cast himself into the processes of life, where he can through revolutionary politics attack the bond of obedience between the ego and the law. The accomplishment of this negative and destructive aim will stir the talent "Lodg'd with me useless" (Sonnet XIX) from its long quiescence. (I reserve for the next chapter the position of God and religion generally in this self-transformation.)

The England exhorted in these early pamphlets reflects the image of a reawakened Milton:

> Why else was this nation chos'n before any other, that out of her as out of *Sion* should be proclaim'd and sounded forth the first tidings and trumpet of Reformation to all *Europ*. And had it not bin the obstinat perversnes of our Prelats against the divine and

> admirable spirit of *Wicklef*, to suppresse him as a schismatic and *innovator*, perhaps neither the *Bohemian Husse* and *Jerom*, no nor the name of *Luther*, or of *Calvin* had bin ever known: the glory of reforming all our neighbours had bin compleatly ours. But now, as our obdurat Clergy have with violence demean'd the matter, we are become hitherto the latest and the backwardest Schollers, of whom God offer'd to have made us the teachers.
>
> (*CP* II, 552–553).

As Kermode suggests, grouping Milton with the psycho-heroes of Erikson, "his personal crisis melted into the national."[49] Numerous sympathies must have facilitated this alchemy, including the inescapable association between father and king that Milton attacked so vehemently in *The Tenure of Kings and Magistrates* and *Eikonoclastes*, but I wish to emphasize the coincidence of the historical moment of England with the personal moment of her prophet. Milton, too, possessed a "chos'n" self subjugated by the advent of a superego, and he, too, was ready to burst forth from "obstinat perversnes" and begin to claim its promise. His zealous proddings of these "latest" and "backwardest Schollers" must have seemed incontestibly righteous to him in part because he was simultaneously undertaking the exorcism of that static obedience memorialized in the Lady of *Comus*.

In June of 1642 he departed from London for the country, a journey that brought him close to the home where his father and brother now resided. The anonymous biographer stresses the brisk efficiency of the groom: "hee in a moneths time (according to his practice of not wasting that precious Talent) courted, marryed, and brought hom from Forrest-hall near Oxford a Daughter of Mr. Powell" (*EL*, 22). The perceptive metaphor of spiritual finance had a secular reach in this instance, for as Parker has shown, the reason for the journey was almost certainly monetary. Milton's father had made several investments for him, among them a loan of £300 to Mr. Powell; badly in debt in 1642, he had probably defaulted on the payments.[50] The eager son-in-law was thirty-three years old, all the time that had been allotted to the Son of God. He came to collect a debt, and perhaps the daughter he received from arrangements made between fathers was, in due time, the payment owed him for a lifetime—*the* lifetime—of celibacy. The process begun with the death of his mother had converted "his practice of not wasting that precious Talent" from virginal hoarding to abrupt risk. One can well imagine his frustration at hav-

ing chosen, as if bound by some curious destiny to the backward timidity of his own youth, a daughter of Eve unable to leave her home.

Virtue Renewed

Something changes, as Descartes learned from the piece of wax. Fixity and innovation, the polarities of *Comus*, are synthesized in every sort of change, including the development of a man. This man displayed the poles of change radically.

He owned what Le Comte has termed an "unchanging mind" (the mind he brought to Geneva), loyal to the same attitudes, causes, and ideas, even to the same metaphors and turns of phrase.[51] He knew his ambition from the beginning, and he fulfilled himself in the end: as temptation remained his myth, obedience remained his virtue. When his opponent called him a "mushroom . . . lately sprung from the earth," Milton was cut to the quick: "You err, More, and know not me. To me it was always preferable to grow slowly, and as if by the silent lapse of time" (*CP* V, 819). Yet in his prose he tried out almost every posture of writing, acting as orator, satirist, panegyrist, historian, scholar, logician, educator, philosopher, theologian, epistolist, and editor. The poet went from genre to genre, leaving at least one innovative work in each of them. Milton suffered the abandonment of Mary Powell, the deaths of two wives and two children, blindness, and political failure, adapting to these crises without losing his sense of a destiny on schedule. Although his enemies were always branded as children unwilling to liberate themselves from tyrannical authority, his political opinions shifted dramatically, making him an Independent and then a sect of one. Against a base of constancy, we perceive his changes as he probably perceived them himself—moments in an unfolding unity whose particular shapes may be surprising, but always appear, in retrospect, to have been foretold. Miltonists from Madsen to Tayler have demonstrated the centrality of "typological structure" or "proleptic form" in his poetry, and insofar as the theory of time to which he gave aesthetic shape conceives of repetition as advent ("Yet once *more*," Tayler would pin the nuance), the movement of his career appears no less shaped than his poems.[52] *Comus* is the Old Testament of the Miltonic canon, at once annulled and enlarged by his future works.

Milton did not outgrow the need for converse with transcendent beings. When a network of spiritual guardians fails in "Lycidas," he

converts the evidence of their impotence into a sign of new presence and efficacy; the drowned poet has become, at the end of the elegy proper, "Genius of the shore," tending beneficently "To all that wander in that perilous Flood"—a saltwater, male, contemporary, Christian Sabrina. Similarly, in the *Second Defense* Milton proclaims that his blindness, which enemies have taken as evidence of desertion, is really proof of divine protection (*CP* IV, 588). But in his blindness the old preoccupation with being defended against malign powers broadens to include a new intimacy with the figure of the guide:

> What supports me, dost thou ask?
> The conscience, Friend, to have lost them overplied
> In liberty's defense, my noble task,
> Of which all Europe talks from side to side.
> This thought might lead me through the world's vain mask
> Content though blind, had I no better guide. (Sonnet XXII)

The guardians of the masque shepherd the immaculate Lady through dark woods toward a symbolic restoration, in the next life, of what her conscience has sacrificed, yet the "guide" of blind Milton will reward his conscience with prophetic vision in the "vain mask" of this life. Guarding the poet from depression and death during the composition of his epic (7.32–39; 9.41–47), the Heavenly Muse of *Paradise Lost* completes the line of supernatural protectors. She is summoned by desire, not prohibition. Her ministrations, indistinguishable from the free mobility of "thoughts, that voluntary move / Harmonious numbers" (3.37–38), no longer comprise a *feeble* compensation for the enormity of the ego's obedience. She is strength manifested, a transcendence emerging from the unconscious sources of the creative act.

There are traces of the Orphic cosmos in *Paradise Lost*. As we will see, disobedience has physiological effects in the epic, inducing a "forgetfulness" in the organism as it descends toward relatively spiritless or involuntary matter; fallen angels and fallen men are imprisoned in coarsened bodies. Moreover, the poet includes escape from the thickness of material existence among the good destinies of unfallen Adam and Eve. Miltonic man still wishes to become, as in the divine philosophy of *Comus*, "soul's essence." Yet the theme of confinement in matter is transfigured by the heretical promise that matter will be encountered, in modes continuous with the earthly ones, in the transcendent world. Again, we can see a vestige of the excremental earth of *Comus*, not only in the fate of the fallen cosmos but in Milton's replacement of the "oral" Creation *ex nihilo* with the "anal"

Creation *ex deo*, where matter preexisting in the Godhead is externalized to produce the universe. Yet as conceived in the vitalist ontology of the epic, this matter is neither dead nor, in its primal state, contagious:

> Thus God the Heav'n created, thus the Earth,
> Matter unform'd and void: Darkness profound
> Cover'd th' Abyss: but on the wat'ry calm
> His brooding wings the Spirit of God outspread,
> And vital virtue infus'd, and vital warmth
> Throughout the fluid Mass, but downward purg'd
> The black tartareous cold Infernal dregs
> Adverse to life. (7.232–239)

In the first of many separations, the creator purges away the fecal "dregs / Adverse to life." This prelude in the anal mode guarantees an essential alliance between man and the material world. The sublimation once made possible by chastity now appears as the universal instinct of the "first being" of matter, which retains in "various degrees" the spiritual animation of the creator, "from whom / All things proceed, and up to him return / If not deprav'd from good" (5.469–471). The epic celebrates fecundity. Excremental conceptions of matter intrude into the cosmos from hell and will be gathered, in an apocalyptic repetition of the first purgation, back into this "Infernal" realm, while the created universe seems the offspring of the deity "advanc't" from the deep model of defecation to that of parturition. Suffused with the divine animation, matter released from God can "up to him return," assured like the good children in *Comus* of being recognized and welcomed by its begetter.

Paradise Lost firmly repudiates the high mystery of the masque. The copulating angels of its erotic heaven are, Hanford noted many years ago, "the celestial counterpart of an experience already known on this earth."[53] When God, withholding playfully, suggests a life of virgin solitude for the first man, Adam manifests his freedom and independence by desiring against this apparent opposition the "nice and subtle happiness" of a mated man (8.399). The Father consents:

> This turn hath made amends; thou hast fulfill'd
> Thy words, Creator bounteous and benign,
> Giver of all things fair, but fairest this
> Of all thy gifts, nor enviest. (8.491–494)

Stein is correct to infer something ominous or fallen from the sentiment of "nor enviest," for Milton has contracted into a conversation the plot of the oedipus complex.[54] A superego has "made amends" when, after a time of envious denial, a son at last receives permission from his conscience to realize "Thy wish, exactly to thy heart's desire" (451). This is one of several evocations of oedipal peace in the epic, the last of which occurs inside the reader of the final lines. As we will discover in the closing pages of this study, it is far—it is in fact as far as one can go—from the filial vengeance expressed at the end of *Comus*.

More clearly, even, than the conversation that precedes her birth, the psychic history of Eve exposes the falsity of the doctrine of the masque. Her brief concern with virginity, as she recoils from her first sight of Adam, is explicitly identified with her earlier narcissistic infatuation with the beauty of her own body, "Less winning soft, less amiably mild" (4.479) than the momentarily threatening angularity of Adam. To have turned back to the form in the pool would have been paralysis: "there I had fixt / Mine eyes till now, and pin'd with vain desire" (465–466). In place of the regressive currents of the masque, we find in the epic a vector of maturation. The "marked sexual overevaluation" characteristic of the state of being in love, Freud argued, "is doubtless derived from the child's original narcissism and thus corresponds to a transference of that narcissism to the sexual object" (*S.E.* XIV, 88). The natural marriage of Adam and Eve renders this evolution with remarkable clarity. A divine voice interrupts the reverie of Eve. She must love, not the image of herself, but "hee / Whose image thou art" (4.471–472)—a directive that substitutes Adam for the image in the pool by identifying Eve as the living image of her beloved. Her entrancement with her form is transferred, as it were, to Adam, lifting the figure reflected in the mirror of narcissism into the higher dialectics of mutual love. In the end, to be sure, the self-love bound inevitably into their love will result in tragedy. The tempter reinflames Eve's adoration for her own image, and she offers the fatal meal to Adam in an inversion of the initial terms of her love, recreating the man as an image of herself, while Adam, too, remains under the spell of his own version of narcissistic idolatry, "fondly overcome with Female charm" (9.999). Indispensable to the success of *Paradise Lost*, the authority of its temptation scene depends on the entanglement, in the tempted, of narcissism and mutual love. Because Eve has left her first desire to the extent that she enjoys sexual love before her trial, the tempter of the epic can reveal, as Comus does not, the hidden scope of

this first desire. And although the fundamental human lapse stems from the same root as love, the last books of the poem deliberately confront and conquer the wish to retreat from this difficult bliss, gradually reconstituting both a marriage and the value of marriage. I will propose in my final chapter that vulnerability to the beauty of Eve has a positive, ultimately theodicial aspect. It cannot be limited to a "marked sexual overevaluation," the persistence of narcissism.

In the hymn to marital love, Milton declares the hypocrisy of a tradition in the imagining of Eden that would have the ideal marriage unconsummated or the hymen magically unbroken. Augustine supposed that the erection and orgasm of unfallen Adam must have been volitional, causing no disturbance in his reason. The unwilled behavior of the fallen penis is the wage of disobedience paid in the flesh; man can no more command his unruly member than God could prevent Adam and Eve from falling.[55] Milton's Adam, however, experiences unfallen erotic passion as a "Commotion strange" (8.531). Suggesting an anodyne for this disruption, Raphael instructs him to meditate on one of the meanings of the Circe myth in *Comus*:

> But if the sense of touch whereby mankind
> Is propagated seem such dear delight
> Beyond all other, think the same voutsaf't
> To Cattle and each Beast. (8.579–582)

But Adam knows something of the magic of Eros, an entrancement accessible decades before only in symbolism pointing to another world. As he moves by negation toward the supreme delights of spiritual companionship, he pauses in a parenthesis to dispute the angel's assumption:

> Neither her out-side form'd so fair, nor aught
> In procreation common to all kinds
> (Though higher of the genial Bed by far,
> And with mysterious reverence I deem)
> So much delights me, as those graceful acts. (8.596–600)

Virginity, the high mystery of *Comus*, has yielded to the "mysterious reverence" for a sensation "higher . . . by far" than the coupling of animals.

When Adam asks about angelic touch, Raphael's famous reply would seem to validate this reverence. In a proportional cosmos, the

superiority of angelic to human sexual pleasure argues the superiority of human to animal sexuality (as is also implied at 4.751–752). To live in the body is not to be in Circe's train, for even at the level of involuntary appetites, man is not brutish. The moral life does not require a law of absolute abstinence. "Our maker bids increase, who bids abstain / But our Destroyer, foe to God and Man?" (4.748–749) Milton's tendentious hymn to the "Rites / Mysterious"—the term Adam used to upbraid an angel—can be understood as a late purgation of his own youth, associating the convergence of psychic forces in his doctrine of virginity with the bidding of Satan. I will maintain in subsequent chapters that Satan is in truth the vehicle of urges condensed in this first, imperfect symbol of Milton's poetic power.

Aside from the fact of it, what can be known of Milton's change? We should not expect to explain his renovation in the strong sense of offering an explanation that invites no further explanation. But we can attempt to infer, from the vantage point of the mature works, how the données of this psyche might have been newly aligned, fresh resources appearing from old rigidities.

If the Lady of *Comus* is the effigy of a youthful self lived through triumphantly, the Christ of *Paradise Regained* is Milton's schematic representation of this internal metamorphosis. Here the poet creates another unmoved hero, one who has seemed to many readers as paralyzed as the heroine of the masque. Like her, he is asked to join a dark and inverted religion; like her, he meets his tempter in the wilderness, away from family and community, then returns home after successfully completing his trial. The hero is thirty years old—as Kermode has noted, the age at which Milton left home and country for foreign lands.[56] In his temptation we can begin to study the psychic structure, translated into Christianity, of the new order adumbrated at the end of the masque. I count it as one measure of the freedom of this "greater man" that, while we have been able to interpret *Comus* without straining the principles of psychoanalysis, this discipline must stretch to encompass in a minimally satisfactory way the mysteries unsealed in *Paradise Regained*.

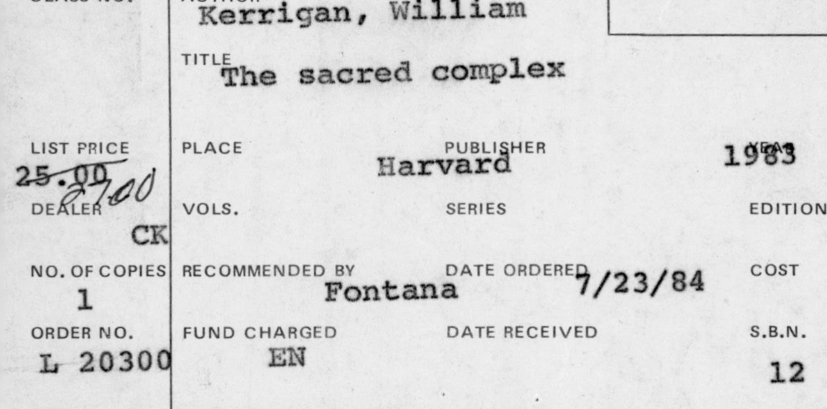

CLASS NO.

AUTHOR Kerrigan, William

L. C. CARD NUMBER

TITLE The sacred complex

LIST PRICE	PLACE	PUBLISHER	YEAR
~~25.00~~ 27.00		Harvard	1983

DEALER	VOLS.	SERIES	EDITION
CK			

NO. OF COPIES	RECOMMENDED BY	DATE ORDERED	COST
1	Fontana	7/23/84	

ORDER NO.	FUND CHARGED	DATE RECEIVED	S.B.N.
L 20300	EN		12

XAVIER UNIVERSITY LIBRARY
CINCINNATI, OHIO 45207

0-674-78500-2

• 3 •

NEW ADAM:
SACRED HISTORY AS PSYCHIC HISTORY

Most of us carry in our hearts the Jocasta, who begs Oedipus for God's sake not to enquire further; and we give way to her, and that is the reason why philosophy stands where it does.

—Schopenhauer to Goethe, November 11, 1815

THE RELIGIOUS ELEMENT in some lives—stiffening defensive habits and amplifying conflicts, encouraging cultural narcissism, arrogating divine sanction for every variety and degree of intolerance—appears to be reducible without much leftover to early psychic arrangements. My thesis, however, is that other people, not all of whom have written books, undergo through religion an authentic reshaping of the self that psychoanalysis has yet to recognize adequately.[1] The process I have in mind stems from a regeneration of the oedipus complex, and might be described as a mature fulfillment of the complex.

It is currently fashionable in psychoanalysis and literary psychoanalysis alike to interpret the cultural life in terms of primitive representations of a unified self. Whether or not we are a narcissistic generation, we are unquestionably a generation seeking to wield theories of narcissism. Some of these admit patches of sunny optimism, while others are shadowed by an undisguised scorn for the concept of the ego in psychoanalysis congruent with a similar scorn among certain contemporary philosophers for the concept of the subject in traditional metaphysics. It is as if the oedipal explanations favored by a previous generation of Freudian intellectuals, like the idealist systems constructed by the heirs of Descartes, hid from us the most withering truths about our cultural life and its history. And the most withering truths are, for many today, the most interesting truths—truths to be driven through to ultimate formulations, then held there in a posture compounded about equally of bravery and weariness. The fascination of narcissism is consequence as well as cause: great energies are devoted to contemplating it, whereas the

oedipal drama seems by comparison drab and unmysterious, played out.

Psychoanalytic theorists would agree that the distinction between narcissistic and oedipal interpretations of a person or a work of art cannot be absolute. But, given the recent emphasis on how oedipal themes are foreshadowed in the earlier period, it is well to remember that narcissistic themes also move forward as the complex works over, organizes, and transforms preoedipal configurations. Milton reclaimed an ensemble of narcissistic entitlements through his lifelong internalization of religious symbolism, among them something comparable to the "basic trust" Erikson located in the heart of Protestant faith; he would not venture into "What in me is dark" and "what is low," releasing Satan from the depths of perdition, without the accompanying wing of a Heavenly Muse.[2] In each of the three late works, however, the way back is the way forward. For Milton the recovery of a "secondary naiveté" in powers and privileges answerable to narcissistic desires—the theophany of Christ; the renewed external strength and divinely implanted inward eyes of Samson; the boldness and inspiration of the poet—results from a demonstration that one is master of desire, whether in temptation, like Christ, or like Samson and the poet, in the ability to suffer irksome fate without condemning the Father. Anyone looking at Milton psychoanalytically will discern narcissistic themes, yet the full appreciation of his poetic vitality depends on comprehending how, in every case, even that of Satan, these themes are located within and governed by an oedipal structure. Like Eve, Milton became distant enough from the mothering image in the pool to relocate its lure in a maturer framework. He and his late heroes exist in a different order of coherence.

The oedipus complex does not just "whittle down" or lessen the degree of narcissistic esteem. There is, if all goes well, a change in kind in the structure of selfhood, generating a unity peculiar to human existence. There are no perfect models for this integrity in mathematics or linguistics; we speak of it by making metaphors of other unities. The boy's desire for unrivaled possession of his mother announces his victory over the narcissistic position, in that the mother is sufficiently separate to be united with and sufficiently absent to be yearned for, but includes a regressive component also, in that this *re*union would preserve the original narcissistic dyad. With respect to the regressive current of oedipal desire, the castration threat acquires a positive aspect: the father sanctions and completes the process of separation from the preoedipal mother; like a clarification, castration cuts at the

lingering dependence joining the self to its first object. It threatens precisely that false image of integrity adored by the primitive ego: the self as a reflection of the maternal gestalt, an image in space vulnerable to division. Just here, between the threat of castration and the inflexible desire for narcissistic entitlement, we find the Lady of *Comus*, whose paralysis is an exemplary instance of what Jacques Lacan has called "the inertia characteristic of the formations of the I."[3] If the boy is able to accept the dismemberment of his narcissistic image, he gains a new "image" under the eye of the superego. Kierkegaard, the great prophet of the self, disclosed a higher and more exacting integrity, the unity of a task undertaken, a self "relating itself to its own self and . . . willing to be itself."[4]

Conscience enjoins this relation. The boy adopts the active task of creating himself in the light of the father, now become, more than law merely, the representative of his own futurity. The lost mother is refashioned and repositioned in the new and futural dimension of hope. The complex has a prophetic cast: the law that bars the child from first desire disposes him toward a restoration to come. This is the genesis in psychic life of Heidegger's Dasein, a finite self who must choose an unchosen situation in order to free its possibilities, the ultimate one, foretold by castration, being death. No longer spatial, the being of the post-oedipal ego is temporal through and through. Its essence lies in volition's grasp on possibility. Its unity is the unity dispensed to Milton by "Time . . . and the will of Heav'n" in Sonnet VII—"All," lifetime now.

Where is God, the "great task-Master's eye," in the typology of psychic history? Initially he is found in culture, which is where we are thrust by the complex.

If, as Lévi-Strauss asserts, the incest taboo, by demanding exogamy and therefore generating social structure, is the difference par excellence between nature and culture, the complex affects our entrance at the level of psychology into the cultural order. Lacan has gone so far as to identify one manifestation of the oedipal father, the so-called "Name-of-the-Father," with language, kinship, custom, the *socius* in general.[5] At first, perhaps, God to the child is no more than a name invoked when people sing, fall silent before eating, congregate and act reverently, become angry and act aggressively. One supposes that, coincident with the differentiation of the child, this Name-of-God is gradually clarified as a means by which his family aligns and distinguishes itself in the social field: to know oneself as child of God is to know oneself inside a web of mundane classifications. But for God

to be owned and chosen, events must occasion a *search*; I choose this term in memory of the first words Milton's God speaks to Adam in paradise, which also contain the first, perhaps the profoundest, of his attempts to define himself to man: "Whom thou sought'st I am" (8.316). Normally a social group will provide the occasion in a ritual, such as catechizing or prophesying, though in the absence of (or in addition to) the ritual, a trauma of some sort, a death, anxiety, or guilt, would be likely to prompt the search. During the search cultural symbols bearing the idea of God will be articulated in regions of the mind distant from abstract thought, "constructed" there from self-representations and object-representations. A relationship will be struck. It is the beginning of a process, since throughout life there will be occasions to search the heart. But these later searches will be shaped by the fixed articulation of the initial ones. The psychic incarnation of God repeatedly stressed by Freud was the oedipal father at the threshold of the superego, an omnipotent giver and taker, our master and definer of our ideal, who demands a tribute of devotion that we human beings are notoriously inconsistent in giving him. I will follow Freud, but to a different conclusion.

Attached to God, a certain arrangement of early life becomes in a sense immutable. It is remembered in that deep way Freud called "repetition"—an unknowing reproduction of the past. This concept should not be deployed in a wholly pejorative manner, for daily experience affords numerous examples of nonpathological repetition (the exclamation that greets the arrival of dinner) and I see no reason why elements of religious behavior (the urge to give thanks to one greater than oneself, the making rapturous of the transcendent) should not be classed among them. By virtue of repetition, presumably, religion "makes sense" before and after rational intelligence mounts its critique. Yet the success of the religious search may also allow for that other way of activating the past Freud termed, in the psychoanalytic treatment, "working through."[6] The internalization of the sacred—and along with it, necessarily, of the profane—bestows on the early arrangement a potential flexibility in the very act of preserving or "remembering" it in religious symbolism. I think of religion in this regard as "conscious fixation," a psychoanalytic paradox. What might otherwise have remained wholly private, severed from both the subject and his culture, has been reached. No doubt the primitive organization of the psyche affixed to religious symbols accounts for the fact that they have so often proved congenial to neurotic and psychotic people. Religion opens a pathway between culture and the repressed. A

specific language of symbols, eventually to be enriched by theology, has been set aside for the curative, repetitious, or degenerative survival of the past.

In Protestantism the liberty of religion's "once more" may be particularly adventitious in light of the nature of the superego. This latecomer to the psychological trinity, Freud maintained, is the vehicle of culture. It grows toward the normative: as teachers and other authorities come to write their lessons after the "Thou shalt" and "Thou shalt not" of primitive conscience, the superego tends to fall into the ordinary, enjoining the same relation of self to self prominently enjoined by the community at large. God may never leave the field of culture in which he was initially discovered, evolving into the God of Durkheim, society itself. But if the searching child has attached the Name-of-God, with a reverence indistinct from fear and trembling, to the father at the threshold of the superego, an unpredictable figure of pure power whose authority stands prior to any apprehension of law as such, religious devotion may permit a creative "immaturing" of the fallen superego.[7] One may regain through such a voluntarist God eccentric possibilities obscured by the evolution of the superego, and if society is to some degree tolerant of these possibilities, a dimension of the indefinite, of radical futurity, may appear in the cultural order. Milton's pamphlets, notably *Areopagitica,* attest to the excitement of this dimension—England's second chance or "reformation"—and the mood of his public career can be traced in the expansions and contractions of this aperture, as disillusionment progressively closed over the vision of hope. But Milton was in touch with a God whom, as Lewalski has argued in her superb study of Donne's *Anniversaries,* one searched for "by occasion," attuned to the flow of personal situation rather than the calendar of the community.[8] God generated lifetime in fundamentally asocial decisions, such as the postponement of Milton's vocation, which might be considered the equivalent in Protestant individualism to monastic retirement. The superego not only watched, but watched over. Milton had discovered a future through an apparently fugitive and cloistered affirmation of his inviolate selfhood, and he would discover another in the creation of a great prophetic poem: he transferred his futurity to the symbols that made it possible. The subject of the rest of this book is how, by reenacting the oedipus complex in the sphere of religion, Milton composed the terrors and ecstasies bound to his ambition while simultaneously he gave life to the symbols of his faith.

This man mastered, as no other literary figure of the period did, his

unchosen circumstances. One result of this accomplishment is that the plot of his "true poem" manifests to us with exceptional clarity the impositions peculiar to his historical moment. "Father" and "Son" comprise the abiding anthropomorphism in the Christian Godhead, whose emotional origins, if almost invisible in some theologies, have always been evident in scenarios of devotion. But in Renaissance culture the first male relationship must have been tense and pervasive, resonant to a degree we can scarcely imagine. This was, after all, a culture that with few exceptions educated men only. Men had their own languages, and these were the languages of culture at its grandest—of diplomacy and international exchange, of the thought-world inhabited by Renaissance intellectuals, where wisdom regularly took the form of classical allusion. A suggestive fable lurks in those stories that have come down to us about Milton's daughters being compelled to read to their father, stumblingly, in languages they could not fathom. Many were his idiot daughters: could anyone have situated himself in the high culture of the Renaissance, even in that part of it given over to the mother tongue, without knowing Latin? The visual half of the emblem was often said to constitute a natural "language" known to everyone, yet Latin mottoes were commonly strewn across its disjointed landscapes, and many of the jokes of Shakespeare, that middling classicist whom we like to think of as a "popular" artist, aim for an educated ear. The possession of male tongues was more than a sign of class, more than prestige: without them, in situation after situation, one could not have *gotten the point.* Fathers handed down to sons, not only the world, but the sense of the world, the complete orientation. The former Latin Secretary drew from male languages the matter for his double simile on the pinnacle of *Paradise Regained,* a dense riddle about the destiny of sons and the history of culture that will occupy us for much of this chapter.

Milton's is an exemplary career in the sense that he completed, grasped, and finally broke outside the paradigm of a humanist lifetime. During the Renaissance the association I have posited between culture and the complex was firmly demarcated, for it was normally at the age of six or seven, soon after the institution of the superego, that boys were taken from home, women, and the mother tongue to the masculine world of the schoolroom. First truly and then figuratively, ambitious speech emerged from this place—the initial locus of the Name-of-the-Father. To honor the poet, or to defend his fiction-making, Englishmen called him "teacher." The traces of classroom disputation in Renaissance literature, in the style of all Renaissance

discourse—habitually contentious, regularly intent on producing an opponent—have been well documented. Before they competed, boys followed and imitated, assimilating the model speech of immortal predecessors into what I have elsewhere termed a "linguistic ego."[9] As an initiation into the ways of men, Latin rewarded oedipal sacrifice, and no one made more of the reward than Milton. *Paradise Lost* was the creation of the finest Latinist in the country (as even the government realized), a man who hired himself out as a tutor, wrote about education, compiled a Latin thesaurus, assembled textbooks on grammar and logic, and hoped to be doctrinal to a nation; his epic is shaped about pedagogical relations between the muse and himself ("Instruct me, for thou know'st"), Raphael and Adam, Michael and Adam. When he is not being disobedient, the Miltonic Adam is above all a student, and Pope was characteristically observant about social scenery when he remarked that God presides over heaven like a "School-Divine."[10] It was in the classroom, we can be sure, that Milton conceived the cultural form of his ambition: the linguistic consummation of the linguistic ego, the insemination of the mother tongue with the monumentality of the father tongues. By the time Milton brought his studies home from Cambridge, the original association between sacrificing the feminine and access to culture had blossomed into the mystique expressed in *Comus*: where we would expect to find a woman in the oedipal dimension of hope, we find instead powerful speech, a great poem, sexual hope having been deferred beyond this life.

There was another displacement in the humanist classroom. As women were absent, so in crucial ways was God, whose native tongue received far less attention than the languages of Zeus and Jupiter. Whatever the yield in cultural vitality, it seems an odd society that would educate its men in the imaginative splendors of another, and in their regard, incredible religion. It *is* odd: how many parallels does history offer us? The formal separation of culture from religion, of profane from sacred culture, or in some few humanist classrooms, of one sacred culture from another, occasioned another search. For a boy who, like Milton, felt himself destined to the priesthood by his father, the Renaissance effort to "find" God in the male culture, making unity of apparent disparity, must have been particularly urgent. One prevalent way of connecting classical and Christian, born in medieval exegesis but given its profoundest development in poetry, was a quasi-typology. The male culture of the schoolroom foreshadows Christian culture, as when the infant Christ of the Nativity Ode assumes the

likeness of the infant Hercules even as he shows "his Godhead true" by routing the "damned crew" of his predecessors. Grown up with his type, Christ is again a hercules at the climax of *Paradise Regained.* Milton's prolonged attempt to reach Christianity through classical gods and heroes represents the visible surface of a slow reworking of his oedipal settlement. Acts of signification that made of the male culture a *figura* of Christian truth had as their counterpart a psychological movement in which the complex itself became, not a permanent structure symbolized *by* his art, but a symbol *of* its new religious form in his art. If the first complex thrusts us from nature into culture, the second transforms the profane into the sacred culture—which is not, as we will see by the end of this chapter, encultural in the usual way for Milton. The poet and his late heroes encounter the sacred in solitude, discovering through a private relationship to God an identity importantly untouched by the roles and ascriptions borne into our lives with the Name-of-the-Father.

"On the Morning of Christ's Nativity," "Upon the Circumcision," "The Passion"—these poems are unique among the youthful works of Milton in their direct engagement with the surface of Christian myth, and the two of them he was able to complete deal with an infant. Elsewhere Christianity is the signified beyond the text, promised in the text and to that degree unrealized. The persistence in Milton of a narcissistic integrity attracting him to the self-display of authorship, but also leading him to dread the desecration or dismemberment of his "true poem" through its exhibition in a text, helped to produce the hiatus in his poetic career. This first oedipal situation is enmeshed in the culture of the schoolroom—Saturn and Jove, Venus and Cupid, the Circe tale in Homer. The profuse episodes and diverse cast of the old mythology may serve to clarify desire, as in the epilogue to *Comus*; they do not, however, solve the interrelated problems of law, death, and narcissistic ambition. "Tomorrow to fresh Woods, and Pastures new." The clear difference in the three late poems is that Milton possesses his faith. Remembered in similes and allusions, the schoolboy gods have been de-centered by the immediate illumination of sacred history. The revelation is no longer "above the years he had," nor must it be approached, as it is in "Lycidas," through the partialities of the male culture. Milton passed beyond the shadowy types of himself. This was not the work of a moment, or of a year. The labor I have in mind was rather comparable to the one performed by "the sad friends of Truth" in *Areopagitica* (II, 549)—many signifying acts, all of them drawing their sense from a single form no one of them can signify entirely.

As Christ unfolds his wisdom in *Paradise Regained,* Milton represents in an abstract and ideal fashion his own entrance into a religious existence capable of presenting this wisdom. Tillyard proposed that Christ in this poem is "partly Milton himself imagined perfect."[11] Were the proposition to be proven in detail, some readers might feel the poem devalued. But all religions offer their adherents one form or another of merger with deity, and Christianity hinges on the promise of identification with Christ. In Protestantism this identification has two currents: a projective one, in which Christ "takes on" our sins, absorbing the violence of man against man and man against God; and an introjective one, in which we receive from Christ a postlapsarian innocence, ending the violence of God against man (the Atonement). There must be a psychological substrate beneath these theological channels. In his early searches for God, I believe that Milton attached Christ to the ego-ideal, his vision of an ego perfectly responsive to, and therefore fully rewarded by, the superego. The most important heresy of his mature theology—subordinationalism, making Christ truly a Son—projects this transaction onto the Godhead. Lucifer becomes Satan in *Paradise Lost* when he envies the power and position of the exalted son; the structure of oedipal resolution is already there at the moment the rebel appears. In sacred history as in psychic history, the militant ego-ideal at the conclusion of the oedipus complex retrieves and reaffirms in a tragic context a prior love between the recent antagonists. Consequently, the several triumphs over Satan in sacred history display a power preexistent in the bond between Father and Son, and these displays or theophanies recreate in sacred terms the genesis of the superego. Read as a Freudian psychomachia, the climactic defeat of Satan in *Paradise Regained* enacts the peripetea of the oedipal drama, in which the ego as rival to the father falls into repression, while the ego-ideal, obediently correlated with the ways of the father, ascends to its sovereign position in the structure of conscience. But there is advent here as well.

Christ is the best teacher because he is the best pupil. On the pinnacle, where his quotation of a law is simultaneously the mutation of a law, he fuses in a single oracle these contrary stances toward truth. In precisely this way, Milton's mature faith emerges from a doubling back on the profane complex, which is both repeated and altered, symbolized and made a symbol. Because it must have occurred in order to be repeated, the first complex is the condition of Miltonic Christianity—our purchase on his language of the sacred. Its persistence grounds the affective power of the religious symbolism within which the second complex transpires. The interlaced metaphors of *Paradise*

Regained—"falling," "standing," and "being led," for instance—can receive such an investment of emotional force only because all of us have been children, and any adult who has lent his hand to guide a toddling child will have sensed in the strength of that tiny grip the origins of Milton's definitive trope for the spiritual conscience. The same logic holds in a more comprehensive way for the relation between the sacred history Milton reimagines and the psychic history he bears within him. As we forge symbols that at once retain and pass beyond a literal or lexical sense, so we may forge ourselves. Repeated, the dynamics of the complex are also unwound and rewoven into a strange negation of their first arrangement, as if it were the genius of religion to outwit the plots of secular obedience by conceding their success. The most radical wisdom of this Christ, we know from the Athenian temptation, could never be heard in a humanist schoolroom. But without the schoolroom, no temptation, no poem, no Christ, no Milton.

I have raised questions about how lifetime situates the self and how the sacred is situated in lifetime, how cultures and individuals change, how God informs culture and how religious symbols enter into psychological processes. All of them await us at the end of *Paradise Regained.* But the intellectual pain sometimes provoked by the effort to release meaning from its original circumstance might be alleviated in this instance if it could be shown that Milton's own text empowers us to address these matters with ideas beyond its historical scope. I think it does, several times over, and I am convinced that one of the good futures for Milton studies—not in the sense that we must stop doing what we do, but in the sense that we might also do some things we are not—lies in the assimilation of this truth.

The Riddle Posed

Milton riddled the climax of his brief epic with enigmas of such power that they occasion us to reflect in unfamiliar ways about his life, his art, and his religion:

> There on the Highest Pinnacle he set
> The Son of God, and added thus in scorn.
> There stand, if thou wilt stand; to stand upright
> Will ask thee skill; I to thy Father's house
> Have brought thee, and highest plac't, highest is best,
> Now show thy Progeny; if not to stand,

Cast thyself down; safely if Son of God:
For it is written, He will give command
Concerning thee to his Angels, in thir hands
They shall up lift thee, lest at any time
Thou chance to dash thy foot against a stone.
   To whom thus Jesus. Also it is written,
Tempt not the Lord thy God; he said and stood.
But Satan smitten with amazement fell
As when Earth's Son *Antaeus* (to compare
Small things with greatest) in *Irassa* strove
With *Jove's Alcides,* and oft foil'd still rose,
Receiving from his mother Earth new strength,
Fresh from his fall, and fiercer grapple join'd,
Throttl'd at length in th'Air, expir'd and fell;
So after many a foil the Tempter proud,
Renewing fresh assaults, amidst his pride
Fell whence he stood to see his Victor fall.
And as that *Theban* Monster that propos'd
Her riddle, and him who solv'd it not, devour'd,
That once found out and solv'd, for grief and spite
Cast herself headlong from th'*Ismenian* steep,
So struck with dread and anguish fell the Fiend,
And to his crew, that sat consulting, brought
Joyless triumphals of his hop't success,
Ruin, and desperation, and dismay,
Who durst so proudly tempt the Son of God.
So Satan fell; and straight a fiery Globe
Of Angels on full sail of wing flew nigh,
Who on their plumy Vans receiv'd him soft
From his uneasy station, and upbore
As on a floating couch through the blithe Air,
Then in a flow'ry valley set him down
On a green bank, and set before him spread. (4.549–587)

The last simile of what some take to be his last work knits together the two most famous sons in Western culture at just the moment when each of them, Oedipus to Thebes and Christ to the temple, returns in triumph to his father's house. It was the fate of the Greek son to answer two riddles, the first with "man," the word for everyone, and the second with "Oedipus," the word for him alone. Freud's extraordinary unriddling of his story exposed the link between

the two: man is the animal who bears the oedipus complex, and therefore culture, which together constitute his fate. Heir to Freud as well as to Milton and Sophocles, my reflections on this passage spring from my own position in our ongoing fatality. But I hope those readers prone to feel, with Merritt Hughes, that "psychoanalysis can tell us very little that is not obvious about the forces which went into the creation of the hero of *Paradise Regained*, for they are better understood in the light of general, cultural history," can at least be made willing to concede that the Oedipean simile represents, by all the measures we bring to literary study, a moment of summary radiance in this poem and this career.[12]

It encapsulates, and reveals in retrospect, one of the structural principles that has informed the temptations of the poem. Like the shapeshifter of the Greek riddle, Satan appears three times. First he is "an aged man in rural weeds" (1.314), then "a man" (2.298) of urban refinements. Finally "Out of the wood he starts in wonted shape" (4.449), which with MacKellar I assume to mean "in his usual shape," the satanic form that he began to acquire in the infancy of his evil.[13] The end of *Paradise Regained* is unmistakably eschatological, "The dreadful Judge in middle Air shall spread his throne" from the Nativity Ode being one gloss on the action. It also recapitulates the first type of the end in Miltonic Christianity. The poet concludes his negative parallel to the fall of man (this Adam triumphs over temptation) with a positive parallel to Christ's appearance in the heavenly warfare of *Paradise Lost*; at the instant of victory in *Paradise Regained* things have wound back to the primal expulsion, when Satan for the first time threw himself "headlong" (6.864) from the Father's territory to avoid the terror of the Son. Hearing this echo, we may remember that Satan lay "Prone on the Flood" (1.195) after his first fall. Antaeus, Sphinx, and Satan create an imaginative equation between the outstretched body of the dead or defeated and the four legs of childhood. However we choose to interpret it, one of the secrets disclosed on the pinnacle is that Satan has enacted the riddle of the Sphinx *backward*.

The simile also encapsulates the theme of the poem. In fact, it is a simile *for* the theme of the poem. Above the ancient heroes of martial combat, Christ will "vanquish by wisdom hellish wiles" (1.175), as in the end he has, smiting Satan with the intellectual blow of "amazement." That Antaeus and Hercules have received more attention from critics than Oedipus and the Sphinx would appear to be, within the express values of the poem, wrongheaded. For the wrestling of the first simile speaks primarily to the smiting, the physicality of the

climax, whereas the outwitting of the second simile concerns the *meaning*, the ferocious wisdom that Christ has brought into the world, before which bedazzled Satan falls. The simile of intellectual combat identifies the climax of the poem with the defeat of question by answer, enigma by revelation.

If, in the few references to Oedipus scattered about his work, Milton shows scant fascination with this myth, the last simile of *Paradise Regained* may serve to remind us that the Christianity of his late masterpieces supplies an interesting and perhaps original role for riddling broadly considered as the unexpected solution to an apparent conundrum. In *Paradise Lost* the foreseen victory of Satan over mankind confronts God with a theodical antinomy. Because he cannot intervene without suspending freedom, man will freely fall, and the Creation will then belong to Satan, whose ways are injustice itself. Should the Father subvert this triumph by withholding his judgment from erring man, the diety would also become unjust: "Die hee or Justice must" (3.210). Put in simplest terms, the victory of Satan, in which the spoils of this world are given as a meal to Death, coincides with the just sentence of God already included in the law, since on the day they eat of the fruit, they shall die. With a hint from the Father, the Son solves the riddle by inventing Christianity (3.210–343). The answer of course is to make death the gate of life through salvation history, which is first presented to Adam and Eve in the reriddled form of "mysterious terms" (10.173). Adam's intuitive confidence that this oracular prophecy contains good news (10.1028–46) gives us to know that a riddle is the first object of faith for fallen man. In *Samson Agonistes*, Milton chose as his hero the famous riddler of the Old Testament, whose life has become for him an intolerably painful enigma. Champion of a questionable God, Samson lives in questions. Though God has promised him a deliverer's glory, he is blind, bound, shorn, and scheduled to humiliate himself before the assembly of his enemies. When God reveals his solution, outriddling every voice in the play, the gloating Philistines are (like Satan in *Paradise Regained*) struck "with amaze" (1645), while the Israelites lapse into the dense silence of a sacred amazement.[14] The history of those who covenant with this God is their recurrent participation in the structure of a solemn riddling game: "we oft doubt" (1745), stung and befuddled by the inconsistency between human events and divine promises, but things are "ever best *found* in the close" (1748), when, the solution manifest, we recognize that from the beginning "All is best" (1745).

The riddling of *Paradise Regained* arises from Satan's confusion over

the nature and mission of Christ, but we are wrong to make the answer of Protestant dogmatics into the theme of the poem. It is not his amazement at the three offices of Christ that smites Satan on the pinnacle. Christological approaches to the poem—for that matter, the many attempts to see *Paradise Regained* as exemplifying abstract essences of virtue in order to instruct its readers in the *imitatio christi*—lose touch by the very direction of their discourse with the historical moment of the poem, what could be termed its mythical realism. Unlike the temptation poems of Fletcher and Beaumont, this one contains no allegory. Satan presents Christ with visions, and we are several times directed toward the mechanics of these productions (2.241–244; 2.401–403; 4.40–43; 4.55–56). Although it would be "curious to enquire" (4.42) into precisely how these feats have been managed, Milton's paralipsis informs us that in principle inquiry is possible. The visions are explicable events in a solid world, the last three of them (Parthia, Rome, and Athens) imaging actual places at actual moments. They are not allegorical pageants. Milton has taken care to embed his poem in specific time.

Satan poses his riddle most emphatically after his failure with the Athenian temptation:

> Since neither wealth, nor honor, arms nor arts,
> Kingdom nor Empire pleases thee, nor aught
> By me propos'd in life contemplative,
> Or active, tended on by glory, or fame,
> What dost thou in this World? The Wilderness
> For thee is fittest place; I found thee there,
> And thither will return thee. (4.368–372)

Christ has been proven an anomaly—a man, yet unmoved by the objects men seek; wise, yet neither active nor contemplative. What happens to the candidate for definition who jams the categories in play? Because he defies the classifications of the civilized world, Satan consigns him to the "Wilderness" of indefinite nature. Christ has somehow absorbed the wilderness figure of the masque, the tempter excluded from the order reconstituted at the end, and although Christ will eventually rejoin society, his wilderness wisdom will trouble the hierarchies that pillar social reality in *Comus*. His remedy for the ills of the world will be neither material nor political nor philosophical. Rather, he will fulfill the messianic promises by expressing his Father, dying and rising and departing, being written about by four evange-

lists and interpreted by Paul—in short, by transforming Judaism into Christianity and founding a new culture, unlike any imagined by Satan. This new culture will also whore after false gods, pollute its uniqueness with worldly wisdom, and in the Renaissance will educate its princes in all the accouterments of profane grace that Christ denounces in *Paradise Regained.* What the Creation and the fall are to *Paradise Lost*, Christian culture is to the brief epic. Milton records in "prompted Song" the immediacy of founding events, before they have been fettered by interpretation. The poem stretches from the Baptism, in which the Father proclaims the identity of Jesus, to the declaration on the pinnacle, in which Christ proclaims his own identity. It recaptures the untheologized first Christology in human history, "in secret done" (I.15) prior even to the public mission of Christ. Between the two proclamations, this new Adam, this unclassifiable man, authenticates his novelty.

Satan lurks behind all other gods. To the extent that all cultures save (intermittently) the Hebraic, have united as an outpost of Satan's monarchy, there is a monotheism of evil behind the polytheism of the ancient world.[15] Like Heidegger's *das Man*, the they-self become a god of gods, Satan articulates the wisdom common to everyone. Hence his amazement on the pinnacle.

For his final effort Satan turns to the most delicate question of all in the genesis of Christian culture: what is its relation to Hebraic culture? As he attempted to do with the Father in *Paradise Lost*, he places the Son in an apparent conundrum. I follow Woodhouse in presuming that Satan does not suppose Christ, or any man, able to stand.[16] If Christ casts himself down, he disobeys the law against tempting God and breaks with the people of the Covenant; if he tries to stand, which would require divine assistance, he disobeys the same law. But Christ is the consummate riddler. Once again he bruises this serpentine head. Like George Foreman at that unimaginable moment in Zaire when Ali took the ring, leaving *him* against the ropes, Satan finds the coordinates that anchor his world suddenly reversed. Christ quotes the law that he must break. His initial concession to Hebraic tradition ("it is written") announces a repetition that is simultaneously a transformation.[17] "By matchless Deeds express thy matchless Sire" (I.233), Mary had advised in a speech whose several references to highness have a mocking echo in Satan's challenge ("highest is best") on the "highest Pinnacle." When Christ does *express* the Father, ambiguous at last, he reveals the new meaning that waited unknown in the old law. "Tempt not the Lord thy God; he said and stood." It is

repeated in that Satan has indeed presumed. The law applies and condemns. It is transformed solely by virtue of being "said," right now, by "he," Christ the Lord. The existence of the speaker is the agency of semantic innovation, transforming a law that enjoins man's appropriate behavior toward his God into a declaration of force that enjoins God's truimph over Satan. For all along, in secret, another game has been played out in this encounter: it is not for *das Man* to define the Son, but for the Son to define Satan. He is Satan's God, God of the god of gods, and his life will become the highest ideal of a culture founded on his mediation of the hitherto inconceivable anomaly of being both a man and the master of temptation, proof against sin. The "he" in "he said," able to change the laws of men without changing a word, is the theme of this poem.

The riddle of the Sphinx has been given another answer, wiser than the "man" of Oedipus. To make the action of the passage coherent, Christ has to have been set down in a crouch, evoking the first segment of the classical riddle. If the Greek man was, as the Sphinx added, weakest on four legs, the new man is ready to be strong. "His weakness shall o'ercome Satanic strength" (1.161). When Christ stands, growing up physically and spiritually, ready to begin his mission of salvation, Satan finishes his regressive arc by being cast back to the fall of his infancy. Man as man has been undone in the person of Satan, then rebegun in the person of Christ "As from a second root" (*PL* 3.288): the Cross, destined to replace Christ on the pinnacles of Christian temples, is the third leg in the life of new Adam. The answer to the riddle cannot be the *anthropos* or *theos* Satan expected, for the disjunctive categories of his anthropology are themselves responsible for his confusion. *Theanthropos* is the answer of a higher anthropology.

The centrality of the Greek riddle in the structure, theme, and religious imagination of *Paradise Regained* is confirmed by the rhythms of our experience in reading it. Raphael, Adam's first teacher in *Paradise Lost*, distinguishes between the temporal process of discursive reason and the simultaneous knowing of intuition, both of which are available to unfallen man (5.488–490)—a Renaissance commonplace, especially among neoplatonists, who from Ficino on associated angelic intuition with symbolism.[18] While discursive reason unfolds in search of its conclusion, one step at a time, intuition enjoys the full and immediate presence of knowledge, exactly as one does when experiencing knowingly the "carefully infolded" symbol born in Egypt, according to Henry Reynolds:

> when in the world's youth and capabler estate, those old wise Egyptian priests began to search out the Mysteries of Nature, . . . they devised, to the end to retain among themselves what they had found, . . . certain marks, and characters of things, under which all the precepts of their wisdom were contained . . . And more than thus, they delivered little: or whatever it was, yet always *dissimulanter*, and in Enigmas and mystical riddles, as their following disciples also did. And this proviso of theirs, those images of *Sphinx* they placed before all their Temples, did insinuate; and which they set for admonitions, that high and Mystical matters should by riddles and enigmatical knots be kept inviolate from the profane multitude.[19]

The Sphinx is the first symbol of the first symbols. Thomas Evans allegorized her as learning, whose telos is divinely inspired verse.[20] Answering her riddle, we contract the step-by-step of human time (what moves on four legs, *then* on two, *then* on three?) into one enfolded idea, making the shift from a discursive thinking ruled by the logic of temporal and inferential priority to the all-at-once of intuition. Milton designs his poem about this sort of transition. Found in the close of the brief epic is an "enigmatical knot" abbreviating the whole. The last simile places a dead Sphinx before the temple of God. As his hero transforms Judaism into Christianity, Milton seizes in inspired poetry his own triumphant power to define, Christianizing the traditions of sacred symbolism inaugurated in Egypt and Greece.

The Riddle Transposed

There are few works of literature, and none among the works of Milton, so relentlessly discursive as *Paradise Regained*. "Thought following thought, and step by step led on" (1.192), Christ moves to his rendezvous with the devil. The line epitomizes the many debates, contentions, questions and answers, charges and countercharges in this notoriously *un*symbolic poem. "They also serve who only stand and wait" (Sonnet 19): Christ waits and stands. When he does, speaking ambiguously for the first time, it is as if he triggers a release in the blocked intuition of the poet. The poem passes over into another realm of signification, and some of its careful discursive arrangements become mysteriously reordered there.

Christ was *summa propheta.* The distinction between discursive argument and intuitional symbolism characterizes the two conceptions of

the prophet I isolated in *The Prophetic Milton.*[21] One of the Miltonic prophets teaches, denounces, exposes, taunts, and exhorts (cf. PR 4.356–364); this is the voice of Milton throughout most of his prose and Christ throughout most of this poem. The second prophet mediates through symbol and vision our access to the Godhead, inspired in holy rapture to praise, celebrate, and participate in the unveiling to men of divine mysteries; this is the voice of the invocations in *Paradise Lost* and of the carefully enfolded conclusion to *Paradise Regained.*

Before the angels sing their anthem, Milton sings his own. We have treated the competitive riddling of the climax as a revelation consistent with the entire text, "thought following thought." But as he celebrates the clear victory of his hero, Milton also poses an extraordinary series of enigmas, summarized recently by Edward Tayler.[22] There is, first of all, a latent chiasmus in each of the two similes. Just before Christ stands and Satan falls, the Adversary has lifted Christ into the air, and this action exerts a secondary and puzzling pressure on the reference to Hercules and Antaeus. Similarly, the Oedipean simile could be almost as appropriate were Christ the Sphinx and Satan a failed, devoured Oedipus unable to answer the riddle of who he is. "So Satan fell," we are told at 4.581, "and straight a fiery Globe / Of Angels . . . receiv'd him soft" (581–583). Dunster, writing in 1795, was at a loss to excuse this solecism: "But the grammatical inaccuracy here, I am afraid, cannot be palliated. *Him*, according to the common construction of language, certainly must refer to Satan, the person last mentioned. The intended sense of the passage cannot indeed be misunderstood; but we grieve to find any inaccuracy in a part of the poem so eminently beautiful."[23] We find the *sense* all right, but why does the "common construction of language" urge us to put Satan in the place of Christ? Here is not just "any inaccuracy," and I have trouble believing that the duplicity of the pronoun is not the "intended sense." The discursive thrust of *Paradise Regained* is toward this moment of definitional separation, of theophany and expulsion, but at exactly this time of clarity Milton riddles darkly. Syntax, the medium of discursive thought, is made intuitional when Christ and Satan inhabit for a moment the same *him*. The poem stops, arrested in mystery.

Satan, too, has the power to change meaning solely by virtue of who he is and where he is. The riddle is transposed. What could be the *accuracy* of a pronoun housing both Satan and Christ? How is Satan the theme of this poem?

We can refuse the intuition, stepping on the end. An authorial oversight, no doubt, albeit a disturbing one. Because Dunster

repressed the threat of a seriously double *him* and erected in its place the soothing symptom of authorial oversight (whose true meaning, I would venture, is: "I, Dunster, am committing an oversight"), he failed to notice that the same convergence of Adversary and Messiah is evoked by the angelic hymn. Removing Christ to the flowery valley, they first address "True image of the Father" and then, as if their audience were Satan as well, address fourteen lines to him: "But thou, infernal Serpent!" What the expository "But" works to distinguish, the common logic of address tends to unite. During their first meeting Satan recalls to Christ that he has now and again visited heaven (1.366–377); toward the end Satan compels Christ to dream that he is in hell (4.422–25). Beyond the discursive statement of *Paradise Regained* there is a dimension of symbolism in which its protagonists are strangely identified.

The anarchic potential of the victory passage within the macrocosm of the poem is mirrored, in the microcosm of the passage itself, by the appearance of Oedipus. Since he is not named, this appearance occurs only in the mind of the reader. Yet what reader has not produced his name, and how does his spectral presence affect the internal design of the passage? He is also the double of his protagonist, for like the Sphinx, Oedipus will bring a plague on Thebes and pose a riddle that, once answered, will drive him to self-immolation. Moreover, since *Paradise Regained* explores the meaning of "Son of God," we should be attuned to the familial emphasis in the two similes of combat. In the first we find the triumph of a champion whose strength derives from his father ("*Jove's Alcides*") over a champion whose strength derives from his mother ("Earth's Son *Antaeus*," "Receiving from his mother Earth new strength"). The Sphinx of the second simile continues this pattern, since she descends in Hesiod from the monstrously fertile Echidna, and in this instance has received her power, her riddle, from the Muses.[24] But like the double *him*, the victorious Oedipus seems to uproot male and female, strength and weakness, from the design of the passage and render the whole scheme a puzzlement. Are not mother and father together the undoing of this hero? What do we learn when, given the familial framework in the two similes, we connect Oedipus with Christ, who is "Man by Mother's side" (2.136) and God through his paternity? The last lines of the poem have not forgotten Oedipus:

> hee unobserv'd
> Home to his Mother's house private return'd.[25]

What has the prophetic visionary adumbrated for us?

Such riddles can still smite us with amazement. If I have dramatized them as strikingly as I am able to, risking melodrama, it is to insist on their recognition. I have not conjured them into being—Dunster's was the first annotated edition of *Paradise Regained*. Before reckoning with the startling symbolism at the end of the poem, it would be well to pause over some sober questions about meaning and the posture of interpreters with respect to prophetic literature.

Symbolism Liberates Meaning

In a recent study of cult behavior among tribal societies, Roy Wagner distinguishes prophecy from law:

> Law, as we understand it, is intended *literally*, and its enforcement or interpretation requires an anticipation of the past circumstances and spirit in which the law was written. Prophecy, by contrast, manifests something that law can register only implicitly, if at all: a prefiguring of destiny, projecting the order that is to come. Thus instead of literal acquiescence prophecy demands a commitment to its projected order—a deed or act "in itself" that mirrors or acknowledges this order in some figurative way. . . . For in the act of apprehending and figuratively extending a text, the interpreters assimilate its significance to that of their own creation and shift its semiotic center of gravity from anticipated past to anticipated future.[26]

Because the essence of prophecy consists in anticipating the future, not in reconstructing the past, there is something askew in trying to recover only the intentional meaning of a prophetic utterance. Such speech is turned in the opposite direction, away from the pastness of the historical artifact and toward the futurity we ourselves inhabit only because we continuously generate it: our generative response to a prophetic text *is* its interpretation. Prophecy so conceived resists institutions, customs, dogmatics, and ecclesiastical stability of every sort. In the Bible the encounter between prophecy and legalism is felt most acutely in Christ and the Pharisees. But once the canon of sacred texts became fixed, and once the idea that authentic prophets would continue to appear became a heresy, this encounter reemerged in the field of hermeneutics. These two essential attitudes toward sacred language, the legal and the prophetic, divide Augustine from Pseudo-Dionysius, Erasmus from Luther, Anglican Bishops from Puritan en-

thusiasts. In each case the radical issue is whether the meaning of the Bible can be fixed, or whether instead the Bible forever anticipates a novel future, open to a widening horizon of interpreters who "assimilate its significance to that of their own creation."

The early differences between Latin and Greek Christianity are instructive in this regard. As M.-D. Chenu has noted, Augustine in the *De doctrina christiana* approached the Word as a rhetorician intent on establishing rules of validity.[27] Propositions take precedence over symbols, and moments of "clear" exposition, especially those in Paul, must guide our understanding of puzzling passages. Augustine could value the challenge of biblical symbolism as a psychological ploy, adopted by the Spirit in recognition of the fact that we humans treasure whatever we must labor to achieve, but he never allowed for the possibility that well-interpreted symbolism might bring forth meanings importantly in excess of, much less in contradiction to, the "clear" passages. How do we know when the Bible *is* symbolic and therefore in need of a special interpretive act? "And generally the method consists in this: that whatever appears in the divine Word that does not literally pertain to virtuous behavior or to the truth of faith you must take to be figurative."[28] Augustine declares the reign of the discursive over the symbolic, which is defined as those passages which do not square with the exegetical books of the New Testament until they are interpreted. At such moments the Augustinian reader begins to think symbolically, driving from his mind the meaning apparently before him and all distracting candidates for its hidden good sense. There has in fact been no serious hiding, and the sense is sure to be good: he always begins the act of interpretation in full knowledge of the range of meanings he must eventually "discover" in the text. Biblical symbolism, as a result of this procedure, yields only a semantic duplication of the forthright passages.

Milton availed himself of these and similar principles in his own *De doctrina christiana*. Yet he insisted that conformity to traditional doctrine against the testimony of his inner light would be true heresy, and he considered the anthropomorphic representation of God, which for Augustine was another rhetorical strategy, to have been offered as a dwelling for the human imagination: "Why does our imagination shy away from a notion of God which he himself does not hesitate to promulgate in unambiguous terms?"[29] These positions move him in spirit toward the Greek Fathers, for whom symbols tend to take precedence over the transparent passages. In Pseudo-Dionysius, for example, the *mysterium* of the symbol, bridging the sensible and the intelligible, calls forth a sustained act of contemplation. No more a semantic

redundancy, the symbol occasions spiritual initiation; the exegete does not, as in the Augustinian tradition, discard the symbol once the discursive content has been grasped, or value the symbol solely because of its affective power. It is clung to as a source of inexhaustible meaning. The symbol becomes our guide, as it were, to transcendent knowledge—particularly, as the theology of Dionysius himself would indicate, to the knowledge of angels.[30] Milton's discursive reason ordered and interpreted the Bible in his *De doctrina*. But his imagination, dwelling in its symbols, produced the unpremeditated verse of *Paradise Lost*, our guide to the cosmic scenery and secret purposes of the transcendent world. Half of its books are taken up with the pedagogy of angels.

The prompted song continues in *Paradise Regained*. Are we to read Milton legally, anticipating the past?

Surely the issue is not itself limited to the past. Paul Ricoeur locates the dialectic of law and prophecy in the ways we conceive of metaphor, which is perhaps the common soil of all symbolism.[31] In the substitution theory of metaphor, stemming from Aristotle, the meaning of "lion" in "Richard is a lion" can be stated in one or more regular propositions about Richard: metaphor is in every case a substitute for a nonfigurative utterance. This theory seems adequate to trivial cases such as the workhorse lion that is Richard, but it appears impotent in the presence of significantly novel metaphors. To move from small things to great, would it be possible to capture the meaning of the eschatological vocabulary of the Old Testament prophets—"servant of Yahweh," "Son of Man," "the Kingdom"—by anticipating the past, translating the metaphorical dimension opened in these terms into propositions cleansed of metaphor and attached to their "proper" lexical deployment? Ricoeur replaces the substitution theory with a variety of interaction theory drawn from Benveniste. When the study of semantics is begun at the level of the sentence rather than the word, it becomes clear that metaphor is an "impertinence," a break with established signification that challenges us to seek a new and extralexical pertinence. One cannot crack the meaning of a good metaphor by recourse to values already affixed to either of its terms in that organization of signs so dear to the structuralists. Metaphor signals the victory of *parole* over *langue*, the indeterminacy of novel event over the entropy of synchronic structure. Tradition is decisive for the meaning of dead metaphors. By their very anomaly, good metaphors are prophetic—troublers of stability. Bursting through system, they anticipate the future of an interpreter whose reflections

they occasion. Dwelling in them reflectively, we participate in the genesis of meaning.

Symbols do not contain meaning. They begin it. I would be happy to agree with E. D. Hirsch that "meaning" is what the author meant, limited and indicative, while "significance" is what can be said about meaning, infinite and interrogative, for then I could announce that this book is mostly about significance, and let discussion cease.[32] But the better the poem, the less I am able to find this boundary:

> Cruelty has a Human heart
> And Jealousy a Human Face,
> Terror, the Human Form Divine,
> And Secrecy the Human Dress.
>
> The Human Dress is forged Iron,
> The Human Form, a fiery Forge,
> The Human Face, a Furnace seal'd,
> The Human Heart, its hungry Gorge.

Except in artificial situations, such as the writing of an assigned essay or the preparation of a class, the act of interpretating literature is not motivated by the desire to achieve a bound meaning. We enter into revisionary interpretation because we have been addressed as persons—mystified, exhorted, condemned, flattered, given pleasure. Blake's "A Divine Image" situates human attributes between threatening abstractions evocative of deity and a place of violent manufacturing. Along with the God in whose image we are supposed to be created and, the poem implies, in whose image we create ourselves in any case, the poem *indicts* us. The question of meaning arises inseparably with the question of significance. It can only be the truth of the meaning, our consent to, or vindication from, the indictment, that has consequences.

My natural response is twofold. On the one hand, who speaks to me and with what authority? Recalling that the divine and human forms engraved by Blake seem armored, the muscle blocks welded together like heavy plates, I would want to know much more about the "intention" of this author than its manipulation of language rules in the text of the poem. I would want to know as much as I can about Blake, whether or not such knowledge would by itself permit me to "write off" any aspect of his indictment. On the other hand, the symbolism centers and provokes the widening gyre of a meditation. Here I

provisionally assume that the poem is truly revelatory, if only I can apprehend the description of the world it presupposes. I work through the indictment. Secrecy is our dress. It is known in clothes; clothes reveal that the body is a secret. The face is also "seal'd." Is the face, the one part of the body we do not normally clothe, for that reason changed, prepared, forged and manufactured as if it were a form of clothing? Why is man the creature who reveals his secrecy? To what extent is the Eden myth correct in its assertion that we clothe ourselves out of shame rather than utility? Is all secrecy the result of shame? Is God our shame? What do things manufactured conceal? The work? It might take a lifetime to exfoliate and assess the symbolism of the poem. Rather than closing down and shutting out, symbols open toward a redescription of the world. They provide us with *a particular way into everything*, and this, I think, is the abiding truth of the neoplatonic account of our guidance by them.

On the pinnacle Milton, like Christ, throws open the discursive law. Mythical figures are dislodged from their customary positions in Renaissance culture to be placed in a new and uncertain alignment. We know that within the system of Christian mythography Hercules is a type of Christ, Oedipus (as answerer of the riddle) a type of Christ, Antaeus and the Sphinx types of Satan.[33] Yet typology is barren of meaning in this case. It supplies the abstract principle of similitude—a rule that allows these comparisons to be made with quasi-divine authority, as it would allow many others to be made. We cannot find the meaning of these comparisons by reference to the logic that enables them. Nor can we find their meaning by searching the exegetical tradition. That procedure would make structure reign over event, confusing the commonplace exegesis of symbols with an authentic instance of the symbolic. We must anticipate the future.

Perhaps Plato was right in supposing that a text is inferior to a living voice in its inability to correct a wayward auditor, so that the flawed essence of textuality is to be ever vulnerable to mistaken constructions.[34] But I am not proposing to ignore intentional meaning; even in psychoanalysis one must grasp the intended sense, for the unconscious can only be known as such in relation to the conscious. I am arguing that, as metaphor, symbolism, and prophecy reveal, intentional meaning does not always coincide with determinate meaning, and that meaning is an empty ideal when severed from the validity of meaning, right now, for those who choose to seek it. In *Paradise Regained* the wisdom of "thought following thought" acts to separate Christ and Satan, whereas the mysterious intuition of the climactic passage, only through conflict submissive to this intention, evokes the

possibility of interchanging them; at the same time, the similes draw us into a complex riddle on the theme of sonship and power. Engaging these enigmas in the discursive mode of criticism, I may do violence to Milton's truth. But license is not liberty, and those crying "License!" at the creativity of the critic have sometimes meant to deny him the liberty appropriate to our experience of great literature.

Milton's brief epic portrays how, in the beginning, a world of new meaning rose by the creating logos of one greater man from the chaos of pagan cultures and the legalism of Israel. I will address the symbolism of its conclusion first with respect to the Christian mythology that Milton reimagined, then with respect to the lived myth that authorized the reimagining.

Relation Stands

When the man who began the poem "Son of *Joseph* deem'd" (1.23) deems himself "the Lord thy God," he is imaginatively transmuting the Yahweh faith practiced in the temple beneath him into Christianity and standing between the Father and his worshipers, precisely as he did in *Paradise Lost* when exalted to a position between the Father and angels. The first mediation engendered Satan and unleashed civil war in the land of God. As above, so below: when the Son of God is begotten a second time and sent to intercede between men and God, the important result for Milton, author of the first Gospel epic to take its beginning, middle, and end from the Temptation, is a confrontation between "Sons":

> The Son of God I also am, or was,
> And if I was, I am; relation stands. (4.518–519)

We sense that Satan is comically confused over the proper and metaphorical meanings of "Son of God," but what are these meanings? What is the kinship system?

When God proclaims in *Paradise Lost* that Christ is "By Merit more than Birthright Son of God" (3.308; cf. *PR* 1.166), we tend to hear, and rightly, the note of cosmic liberalism. But the Father is not denying that birthright also and fundamentally entitles Christ to his definitive epithet, for as Milton argued in a heretical passage of his *Christian Doctrine*: "it is as plain as it could possibly be that God voluntarily created or generated or produced the Son before all things: the Son, who was endowed with divine nature and whom, *similarly*, when the time was ripe, God miraculously brought forth in his human

nature from the Virgin Mary" (*CP* VI, 211; my italics). The Son may be called "God," as he calls himself in *Paradise Regained*, because in this case as in the earthly entitlement by which John Milton was the son of John Milton, "the Son receives his name from the Father" (*CP* VI, 260). Christ, then, is the one "Son of God" to receive in addition the unqualified Name-of-God.

Given this subordinationalism—a Son generated in time from the substance of the Father—the difference between Christ as Son of God and Lucifer as Son of God, although of course maintainable, has been considerably weakened in comparison to the orthodox position. But if the heresy lends theological urgency to the conflict between Christ and Lucifer, it renders the doubling on the pinnacle all the more mysterious. Satan insists that "relation stands." When Christ stands to deny the equality of this relation, Milton releases his poetic wiles to suggest that relation *does* stand. If we think of *Paradise Regained* as a demonstration that one of the dangerous consequences of the heresy can be overcome, and the Son of God proven unique, we may shake our heads once again over the *him* suggesting there are two Sons of God. Is there another and secret kinship system?

I will seek for the relation in the recesses of the poem that is the parent of *Paradise Regained*, especially in its depiction of the only father who was never a son. But first, continuing our discussion of symbolism, let us turn to the Satan Crux.

Paradise Lost combines mythopoeic narrative with rational theodicy. Its rhythmic interplay of discursive and symbolic representation is, unlike the prolonged rationality and final symbolic explosion of *Paradise Regained*, constant and various. In many readers the figure of Satan generates a tension between these two poles: his mythopoeic grandeur at the opening of the work opposes his discursive condemnation by the narrator and the heavenly characters. Stanley Fish's solution to this classic dilemma in *Surprised by Sin*, widely adopted by now, has an elegance at once literary and psychological.[35]

In claiming that the tension was deliberate, Fish healed an old division in Milton studies. Provided that our sense of his splendor be corrected repeatedly by the normative declarations of discursive judgment, we may permit the romantic *and* the theological Satan to evolve in us. We are obeying intentional meaning, fulfilling the strategy of the poet, even when our feelings about the mythopoeic Satan contradict this judgment. The psychological elegance of Fish's argument is that the pious reader can entertain potentially rebellious attitudes knowing that, as signs of his fallenness, these attitudes already confirm

the doctrinal content of the poem and therefore have a piety all their own. One price paid for this elegance, however, is the complete subordination of symbol to statement—the sacrifice of metaphor to resurrect unity. Entering the similes of evil with Fish, we discover, in the end, a duplication of discursive meaning. Although the experience of finding this redundancy may have, as Fish claims, great pedagogical force, the overall effect of his reading is to promulgate a tyrannical notion of aesthetic unity at the expense of introducing, without overt recognition, a new and unheard-of flaw in the poem: the alarming idea that its mythopoesis is not generative but repetitive, that its similes, metaphors, and symbols tell us nothing about Christianity that the dogmatizing and sermonizing passages of the poem have not told us already. Fish is the brilliant Augustine of Milton studies, and he has taught us much: there can be little doubt that some of the similes have been designed as didactic redundancies. But what if Fish, and Augustine too, are killing the spirit of the letter in order to preserve the letter of the spirit?

After the Father and Son have voiced their own theodicy in Book 3, deciding upon the course of salvation history, Milton discloses the Father in the mode of symbol: "Dark with excessive bright thy skirts appear" (380). This is the God that prophecy alone can reach. Milton did not require inspiration to write a legalistic theodicy: the invocation to light at the beginning of this book solicits a vision; the poet would "see and tell" (54). What does the vision of "the majesty of darkness" (2.266) tell? What does it mean when the mountain on which the Father's throne rests is said to be pierced by a cave, out of one mouth emitting light and into the other mouth admitting dark (6.4–12)? It is at such moments that Schopenhauer's Jocasta, aroused by our instantaneous response, sends us forth to read about the *via negativa*. "Ah well," says Jocasta, "let us say that God is known to be unknowable—that's what the darkness stands for." Oedipus turns to the *Institutes* of Calvin, leaving the whole question to Tiresias. But Milton wanted to become a Christian Tiresias, and names him in the invocation to light (36). If we give our instantaneous response to the dark God time to achieve a conclusion, we will inevitably find ourselves thinking with Blake and other heresiarchs that brightness invisible produces darkness visible because this God, Milton's God, is the source of evil as well as good. What the discursive argument of the poem denies, the symbol tacitly concedes.

This revelation is entirely unspecific. Placing his mystery in the sanctuary of a symbol, Milton renounces any effort to achieve exacti-

tude about this matter—which is one reason why the revelation is so powerful. We will not find in Milton the labor of Plotinus and Proclus to work out a theology of evil in terms of alterity, or the Gnostic attempt of a Blake to encompass on the surface of myth the divine origins of evil. Milton has, in a way, included them and surpassed them in the vast quiet of his symbol. For what is the task imposed on a religious mind by the darkness of God? It seems to disrupt the moment when man discovers his sinfulness: "If I repent of my being, I accuse God in the same moment in which he accuses me, and the spirit of repentance explodes under the pressure of that paradox."[36] The only way for devotion to survive the revelation that evil originates in God is to assert, in whatever manner, that *to be the* source *of evil is not to* be *evil.* And this, also, the symbol of the dark Father reveals. We see his darkness at a time of celebration, and however imperfect the theodicy of reason may be, the theodicy of faith implicit in the symbol is beyond critique. At the height of his prophetic vision, Milton knows intuitively all that need be known. I will try to unfold his vision, not to indict his God, but to understand its implications for his mythmaking.

The distinguishing feature of Genesis among creation myths lies in its determination to separate the origin of the world, pronounced "good" by the Creator, from the origin of evil. As Ricoeur argues, this separation demands an idea of freedom that philosophy had yet to supply at the time the Adamic myth took shape.[37] Inheriting Christian theology, Milton can fill this lacuna in his myth with the appropriate argument. But his determined pursuit of beginnings drives the argument of *Paradise Lost* so far back behind the myth of Genesis that features of other creation myths become incorporated in his epic. The serpent is no longer a curious presence in the garden: he has become a Christian Titan whose war with God places the existence of discord *before* the Creation; man is not the radical source of evil. Milton tries to make this consequence coherent with his theodicy, since God will pardon man, not Satan, precisely because man succumbs to an anterior evil (3.129–134). Yet everything in Milton that makes him create, in Burden's phrase, a "logical epic"—the need to motivate every action, rationalize every event, answer as best he can all of our discursive questions—will not permit him to let things lie. The enigma of evil gets relocated farther and farther back in cosmic time. Milton represents the moment at which Satan originates in Lucifer, but this moment, which turns on the question of who has created whom, casts our attention to the origin of Lucifer and the origin of the Son. The Father

informs us (5.600–615) that he has created Christ; Abdiel informs us (5.835–845) that Christ has created the angels. So by the same logic that makes Christ the Son of God, Lucifer should be the son of Christ.

But the symbolic Father, dark and light, reaches in mysterious terms to another order of begetting. According to the *De doctrina*, Christ cannot be God entire because "really a God cannot be begotten at all": "For a supreme God is self-existent, but a God who is not self-existent, who did not beget but was begotten, is not a first cause but an effect, and is therefore not a supreme God" (*CP* VI, 211, 263–264). The Miltonic analogue to the effort in Genesis to confine evil to man is the confinement of evil to the second generation of heaven, the angels two removes from the self-existent Father. But once again, the speculations set in motion by this endeavor seem destined to converge with the symbolic disclosure of the Father. A question about "cause" (1.28) initiates the narrative of *Paradise Lost*, and such questions bear the threat of infinite regress—in this case, a regression from Hebraic to Greek and Mesopotamian myths of the origin of evil that must be stopped, as a blink stops sight, by a symbol. The answer he receives, "Th' infernal Serpent" of Genesis, is not sufficient, for he must explore the cause of Satan, who leads us to Chaos and Old Night, and by the end of Book 5 we have learned that it is impossible to think about the origin of Satan without also thinking about the prior origin of the Son. Where can this end? As Milton might have learned from the example of Calvin, a rational theodicy that begins with a moral dualism *and* a strict monotheism must eventually collapse. The evil principle has to originate somewhere: retrace the path of causation stubbornly enough, and there is only one place left for evil to derive from. Calvin did not attempt a rational theodicy.

The God we intuit through symbols should not incite us to reverse the moral polarities of the poem, compiling a lawyer's brief of picky details that justify the ways of Satan. His defiance of God has its own authority, its own grandeur and tragedy, within the plot of the epic; the weight of our intuition concerns the manner in which God is denying Satan. In the large poem, counterparts to the doubling of *Paradise Regained* can be found in the noon-midnight, ascendance-eclipse symbolism explored by Cirillo and Fowler, and in such passages as the apostrophe to sidereal Lucifer, the morning star, during the matins of Adam and Eve (5.166–170).[38] In the order of generation envisioned by these symbols the two principles fused in the Father have been externalized and separated as a Son of Light and a Son of Dark. To the first has been delegated the substance, power, authority, throne, and name.

He is therefore imaged as the creator of the second, and ultimately he will come into his inheritance by turning the power he has received from the Father against his dark twin. The Atonement, in this special sense, represents an internal catharsis of the Godhead. That mercy may reign without occasion to be merciful, in a universe purged of disruption, the dark principle must be destroyed or contained. Then only, "wrath shall be no more" (3.264). So Christ must pay "The rigid satisfaction, death for death" (3.212)—"on mee let thine anger fall" (237). We know he substitutes for Adam, bearing the punishment we deserve. Viewed, however, from the cosmic perspective afforded by the symbolism of *Paradise Lost*, he substitutes for Satan.

This is the Father's terrible riddle: although the meaning of the Sacrifice is the crucifixion of sin, the disarming of death, and the crushing of Satan, the event given this meaning is the wrathful destruction of the wrong Son. Milton could never write with customary strength about the Sacrifice because he felt its intolerable illogic more profoundly than a Crashaw. He reimagines as he can. In his gospel epic he moves the consequence of Atonement, the regaining of paradise, from the crucifixion that murders Christ to the temptation that unmakes Satan. Treating the Father's riddle indirectly, Milton is able to bring meaning back into harmony with event.

Woodhouse said that in *Paradise Regained* Milton proves his Arianism "wholly compatible with the impulse of worship."[39] But the displacement of his worship from the Crucified One to the Tempted One implies a considerable resistance to the riddle of the Father. Might it also imply a solution? Within the full context of his Christian poetry, the momentary doubling on the pinnacle comes like a sudden convulsion in which Milton remembers something he knows about this myth but cannot articulate. What God has put asunder, Milton joins. Relation stands. As I will try to show in some detail toward the end of this chapter, there will be a mysterious substitution. Although the future of his hero is not subject to his wish, the poet somewhere knows that at the hour when the wrath falls on a forsaken Son, charged with the sins of the world, the Son is Satan. The angels receive *him*, preparing *him* for the Cross that will one day stand on the pinnacles of temples as the emblem of a culture.

Oedipus and Sacred Oedipus

In "The Theme of the Three Caskets" Freud shows the recurrence in myth, folklore, and art of a grouping of three women. Their true faces are worn by the Fates. But often, in wishful disguises, they are posed

before a man who must choose one of them in the context of love or affection. Havoc may ensue, as it does in *King Lear*, when the man fails to choose the third woman. Unlike Bassanio in *The Merchant of Venice*, Lear is blinded by the obvious, and his refusal to see through appearances represents, in Freud's view, a disavowal of fate. To all appearances, that is, the old King acknowledges his death in the act of dividing his kingdom, but his unreadiness for the promised end is betrayed in his denial of the cold and reserved Cordelia. Despite her many transformations—Aphrodite for Paris, Cinderella for the Prince, Cordelia for Lear—the third woman preserves in some fashion her true character. She is the figure of Death, the inexorable one, the woman who, as Milton wrote in one of his most haunting phrases, "slits the thin-spun life." This theme also calls forth the best in Freud. At the end of the essay he returns to *King Lear* for his finest moment as a literary critic, which is, not surprisingly, the finest moment in psychoanalytic criticism:

> Lear is an old man. . . . But Lear is not only an old man: he is a dying man. In this way the extraordinary premise of the division of his inheritance loses all its strangeness. But the doomed man is not willing to renounce the love of women; he insists on hearing how much he is loved. Let us now recall the moving final scene, one of the culminating points of tragedy in modern drama. Lear carries Cordelia's dead body on to the stage. Cordelia is Death. If we reverse the situation it becomes intelligible and familiar to us. She is the Death-goddess who, like the Valkyrie in German mythology, carries away the dead hero from the battlefield. Eternal wisdom, clothed in the primaeval myth, bids the old man renounce love, choose death and make friends with the necessity of dying.
>
> The dramatist brings us nearer to the ancient theme by representing the man who makes the choice between the three sisters as aged and dying. The regressive revision which he has thus applied to the myth, distorted as it was by wishful transformation, allows us enough glimpses of its original meaning to enable us perhaps to reach as well a superficial allegorical interpretation of the three female figures in the theme. We might argue that what is represented here are the three inevitable relations that a man has with a woman—the woman who bears him, the woman who is his mate and the woman who destroys him; or that they are the three forms taken by the figure of the mother in the course of a man's life—the mother herself, the beloved one who

> is chosen after her pattern, and lastly the Mother Earth who receives him once more. But it is in vain that an old man yearns for the love of woman as he had it first from his mother; the third of the Fates alone, the silent Goddess of Death, will take him into her arms. (*S.E.* XII, 301)

These two paragraphs are worthy to be set next to any statements that have been elicited by the plays of Shakespeare. There was a side to Freud, this man whose thought, like his therapy, starves illusion, that resembled the implacability of natural law, and it led him directly to the heart of *Lear.*

If *Hamlet* is the great tragedy of the absent father, whose hero finds God, *Lear* is the great tragedy of the absent mother, whose hero finds Nature. It is unbearable. Once he has denied Cordelia, Lear drops into a disillusionment without bottom. When he might finally be able to say "This is the worst," he is dead and unable to speak. Everything maternal in the world save for Death, the mother as natural law, divested of every illusion, fails him. For all the rant about the hell of the vagina and the curse of procreation, the old King conspicuously lacks a Queen. He searches the faces of his daughters for the face of his missing wife. Inside him, meanwhile, another mother presses toward his heart. As he says, during one of those onslaughts of psychosomatic distress scattered throughout the play,

> O, how this mother swells up toward my heart!
> Hysterica passio, down, thou climbing sorrow,
> Thy element's below!—Where is this daughter? (II.iv.54–56)

Before he will weep, "this heart / Shall break into a hundred thousand flaws" (II.iv.282). By the final scene the "sorrow" of "this mother" climbs to his heart. In one of the reversals Freud points to, Mother Earth bearing dead children, the weight of all sorrow and all flaw, becomes Lear bearing Cordelia. "Break heart!"—Kent wishes him death. "Vex not his ghost. Oh, let him pass!" The "break" and the "pass" are imaginatively accomplished by the "hysterica passio" residual from the first and deepest mother. She gave him life once, and she does not repeat herself. It does not matter how many times we say "never," for the third mother is already inside of us, the gift of the first. Like Satan, Lear also lives the riddle of the Sphinx backward: the

last mother is the rediscovery in total disillusionment of what the first has wrought.

Eve became the figure of Death when she "knew not eating Death" (9.792). Thereafter she is doomed to have the children of Satan, who can only sire Death, by oral impregnation. The vacillations in fallen Adam's attitude toward her follow the cleavage in the two common etymologies given for her name, *hevia* (serpent) and *eva* (life):

> Out of my sight, thou Serpent, that name best
> Befits thee with him leagu'd, thyself as false
> And hateful; nothing wants, but that thy shape,
> Like his, and color Serpentine may show
> Thy inward fraud, to warn all Creatures from thee
> Henceforth; lest that too heav'nly form, pretended
> To hellish falsehood, snare them. But for thee
> I had persisted happy. (10.867–874)

> peace return'd
> Home to my Breast, and to my memory
> His promise, that thy Seed shall bruise our Foe;
> Which then not minded in dismay, yet now
> Assures me that the bitterness of death
> Is past, and we shall live. Whence Hail to thee,
> *Eve* rightly call'd, Mother of all Mankind,
> Mother of all things living, since by thee
> Man is to live, and all things live for Man. (11.153–161)

In the first passage, Adam foreshadows his children in feeling that he has inherited death and woe from Eve, that fate is the consequence of his love for Eve. But in the second he senses that "Man is to live" through Eve, and Christ will perfect this refiguring or "rightly calling" of Eve. *Paradise Regained* is a happy repetition of Book 9 of *Paradise Lost*: now Eve stays at home, Adam goes forth to be tempted, and comes home having redeemed the meaning of Eve. The Christian mother is life, Cordelia with the promise of a new futurity. Through Christ, peace returns "Home" to the breast of man, not the wandering womb that overtakes the heart of Lear. How does this renovation establish itself in the symbolism of the brief epic?

Let us return to the pinnacle, bearing with us Freud's evocation of the three mothers in the life of a man—the mother who gives life and nourishment; the mother regained in mature erotic love; and the Earth

who receives us when, in the words of Michael to Adam, "like ripe Fruit thou drop / Into my Mother's lap" (11.535–536). For Antaeus the first mother will become the third, as in a pagan Pietà. For Oedipus, returned to his father's house, the first mother will become the second. It is the work of Christ on earth to redeem tragic man by transforming the third mother into the first:

> he unobserv'd
> Home to his Mother's house private returned.

Like Antaeus after the bout with Hercules, like Oedipus after answering the riddle, Christ in the end returns to his first mother. As we will see, the Miltonic psychology of redemption demands that there be nothing compulsive or fatal about this attachment; and, unlike the Laius who made his deserted son Oedipus ("swell-foot"), Christ's Father will not permit his Son "to dash" his "foot against a stone." Still, two factors orient the closure of the poem toward this tragic fate: the typological identities of Christ and Mary, and its symmetry with the closure of the parent poem. Is this not the home of second Adam and second Eve, toward which our first parents began wandering, providence their guide, at the end of *Paradise Lost*? But while Adam was the mother of Eve (called his "Daughter" at 9.291), new Eve is the mother of new Adam, and the old husband has now become a son. The conventional typology Milton exploits has retained, latent in its symbolism, the idea of their oedipal marriage. As dramatized in *Paradise Regained*, the work of redemption is the sacred complex by means of which this marriage, and much of what it means in psychic life, is transcended through its latent repetition.

The double simile unmistakably urges separation from the first mother. Antaeus is defeated because his strength, depending wholly on this attachment, is also his weakness; Oedipus became, in the words of Neville's Seneca, "the most unhappiest wretch that ever sun did see," because he too delivered his manliness to the first mother; and we might recall here that Death and "th' incestuous Mother" (10.602) are inseparable companions in *Paradise Lost*. Severing this primal attachment is, in the life of every man, the task of forming a self—provisionally completed by the institution of the superego. The savior of *Paradise Regained* makes this severance absolute. He follows the call of his Father into the wilderness, leaving mother and home behind. As a prelude to temptation he fasts, asserting his freedom from the primordial gift of the mother. The meaning of these actions becomes

explicit in the first temptation: as the Word is more nourishing than bread, so the Father replaces the mother as provider of strength.

By "separation" from the mother I do not mean simple distance. Milton reveals a psychological aspect to the theological contention that Mary supplies the humanity of Christ. The hero has assimilated his mother, grieved and identified. She is an internal presence, contributing from within to his work of salvation, and when Christ descends into himself to set his life before him (as Mary does in the matching soliloquy at 2.63–104, calming her pure breast in the absence of her son), the words of his mother appear inside these "deep thoughts" (1.190). Given the context we are evolving now, the speech of Mary still resounding in the mind of her son may be appreciated as one of Milton's shrewdest triumphs. Christ remembers the day she took him "apart" to teach the lesson of apartness:

> High are thy thoughts
> O Son, but nourish them and let them soar
> To what height sacred virtue and true worth
> Can raise them, though above example high;
> By matchless Deeds express thy matchless Sire.
> For know, thou art no Son of mortal man;
> Though men esteem thee low of Parentage,
> Thy Father is th'Eternal King, who rules
> All Heaven and Earth, Angels and Sons of men.
> A messenger from God foretold thy birth
> Conceiv'd in me a Virgin; he foretold
> Thou shouldst be great and sit on *David's* Throne,
> And of thy Kingdom there should be no end. (1.229–241)

He for God only, she for God in him: before the voice of the Father proclaims his lineage to the world at the baptism, second Eve makes her private annunciation. This mother removes from the life of her heroic son any possibility of duplicating the tragedy of Oedipus, which was predicated on his ignorance of his progenitors; when Christ returns home, he knows who he is. She gives him the answer, in a sense, to the riddle of the Sphinx, which Freud interpreted as symbolic of the earliest riddle whose unfolding cannot be extricated from the process of psychic organization: where do babies come from?[40] Christ has the facts of his life. On the pinnacle he will do no more than what Mary has urged. She initiates his separation from her, a mother who does not tempt or retard, but rather propels her son to the paternal

identification that resolves the oedipus complex and organizes his autonomy. At the beginning of her beautifully direct and complete disclosure, she tells him to "nourish" his thoughts (she herself being part of that nourishment, since Christ is thinking this speech), and at the end she points him toward the symbolism of the Bible ("*David's* Throne," "thy Kingdom") he will prefer to material nourishment in the first temptation.

Heir to the oedipus complex, the superego is the universal solution to the tragedy of Oedipus. Yet in preventing this catastrophe, the superego brings its own agony in the either/or of ambition and veneration, rivalry and submission, being oneself and obeying another. When Satan parodies the speech of Mary on the pinnacle, he assumes that Christ's attempt to be "highest" *must entail a conflict with his Father.* But in a speech that gives Christ the wherewithal to rise above Antaeus who could not live separated from his mother, Hercules who did the bidding of women, the Sphinx who must be conquered with the knowledge of humanity, and Oedipus who actualized the marriage latent in the symbolism of second Adam and second Eve, Mary also prepares Christ to outriddle the fallacious Satan. The wisest line in her speech, "By matchless Deeds express thy matchless Sire," the one Christ fulfills on the pinnacle, promises that this son can strive to be *matchless* without competing with his father. The Christian superego, it is suggested, will allow a certain relief from the disjunctions that burden the first superego. Pursuing the highest ambition, "above example high," the Son of God remains, indeed becomes, the image of his "matchless Sire."

"Milton keeps reminding us of Christ's hunger," Watkins writes, "which is his unifying symbol, modulated from food to truth to glory."[41] Satan can be termed "Insatiable of glory" (3.148), and Christ can term himself "hung'ring more to do my Father's will" (2.259): both hunger. By his extension of this vocabulary Milton evolves in metaphor a concept not dissimilar from Freud's libido, which also has its beginnings in hunger. Through the ordeal of his temptation, the Son of God destroys and regenerates the hunger that is Death, Son of Satan.

In the wilderness Christ relives symbolically the entire history of the libido, regaining for the prospect of choice what all the children of Eve undergo as a process. There are two temptations of food. In the first Satan asks Christ to provide nourishment for himself and others in want; in the second Satan would provide food for Christ. The first answers to the narcissism of the undifferentiated infant, for whom the

mother is a part of himself and food can be summoned, as if "By Miracle" (1.337), through the agency of his magical cry. The luxurious banquet, by contrast, acknowledges that nourishment depends on an external source. It is suggestive that Christ first *feels* hunger after the initial temptation (2.244), for the two temptations correspond to a fateful metamorphosis in the history of the libido. At the beginning of life libido is need—biological, appeasable, finite. Soon, however, need yields to the transgressions of desire, psychological and restless; instead of satisfying a mere emptiness, nourishment comes to *mean* something, and as the sign of a desire, becomes caught up in the early dialectic of self and other. Thereafter libido is an energy of infinite "wanting," our involuntary response to the absent and the lost, proceeding from object to object with the excess characteristic of wish—the force in us that will not welcome reality.[42] Far down this line we find the exotic Ovidian banquet conjured by Satan, every dish a sign of wealth, privilege, and cultural refinement, so different from the bread of the first temptation.

For Milton the great symbol of this break in the life of the libido is the fall. Blameless midday hunger took flight into desire, and narcissism was no longer innocent. Doubtless the fasting of *Paradise Regained*, the long debates about the proper objects of desire, are meant to purify that first transgression:

> But now I feel I hunger, which declares
> Nature hath need of what she asks; yet God
> Can satisfy that need some other way,
> Though hunger still remain. (2.252–255)

Fasting reestablishes the difference between need and desire. Anything beyond the constant pressure of hunger is a superfluity—and this entire dimension of excess, the sphere of desire, is referred to God. All the temptations represent a continuation of Christ's separation from Mary. As the source of his humanity, she is the source of his hunger. Taking "some other way," Christ repudiates the long chain of worldly desires that issues from the first mother and ends only in death. God may not give what nature asks, but with hunger in abeyance, the way is clear for God to satisfy in "some other way," creating desire anew. "Me hung'ring more to do my Father's will." The will of the Father *is* the desire of the Son.

But the summary movement from need to desire, the oedipal mother who succeeds the nourishing mother, has been removed from

the chain by the tempter himself. Unlike the Lady of *Comus*, Christ will not be made to declare the sage and serious doctrine of virginity, even though his own life will become its primary sanction for Christian culture. When Belial proposes to tempt Christ with women, Satan seems to be tempted himself, and he reacts with an indignation oddly resembling that of his opponent in the wilderness. This greater man will not, like "*Adam* first of men," fall by "Wife's allurement" (2.133–134):

> for Beauty stands
> In th'admiration only of weak minds
> Led captive; cease to admire, and all her Plumes
> Fall flat and shrink into a trivial toy,
> At every sudden slighting quite abasht:
> Therefore with manlier objects we must try
> His constancy, with such as have more show
> Of worth, of honor, glory, and popular praise;
> Rocks whereon greatest men have oftest wreck'd. (2.220–228)

If there is a coherent figure in the opening lines of this passage (a peacock with tail first spread, then closed, as Dunster suggested?), it soon suffers a puzzling seachange in which the word "shrink" signals both a diminishment of what "stands" within the weak beholder of beauty and a metamorphosis of the plumed object beheld, now "a trivial toy."[43] This likeness of a failed erection leads Satan instantly to "manlier objects," the "Rocks" of his new strategy—a metaphor that seems to have leapt out of the "hard stones" of the first temptation (1.343). Ironically, what Satan says of women is true of everything he offers Christ, who is not destined to dash himself on a stone. Christ is rather the "solid rock" (4.18) on which Satan wrecks, "a rock / Of Adamant" (4.533–534), "firm" (4.534), a proven metal. The triumph is Christ standing on the pinnacle, "highest."

The important thing to remember about "phallic symbols" is that the phallus, as its name proclaims, is itself a symbol—the "signifier of desire," in Lacan's elegant phrase.[44] The tempter displays impressive objects. Is *that* the signified of your desire? Yet Satan has laid down in his own account of diminished female beauty the motif that runs through all of Christ's refusals until, on the pinnacle, the tempter himself falls down in amazement: the Son of God only desires to be the Son of God. The mockery and belittlement of Satan's impressive ob-

jects by the unmoved Christ betrays an almost physical conception of desire. When something appears desirable to us, we have lost to this something in the external world a power in us. "Weak minds" find female beauty strong; it is actually their own expelled strength that binds them. This sense of desire as an investment of mental power answers in Christian terms to the doctrine of *cupiditas*, and in psychoanalytic terms to the cathectic energy of the libido. One can feel the might of Christ grow as his desire is withheld again and again from the objects of the world Satan parades before him:

> Extol not Riches then, the toil of Fools,
> The wise man's cumbrance if not snare, more apt
> To slacken Virtue and abate her edge. (2.453–455)

> Conquerors, who leave behind
> Nothing but ruin wheresoe'er they rove . . .
> Till Conqueror Death discover them scarce men,
> Rolling in brutish vices, and deform'd,
> Violent or shameful death their due reward. (3.78–88)

In the end it is as if all the strength of the world has flowed into him as a consequence of his very rejection of the world. The libido of the sons of men, dispersed like the body of Osiris among the objects desire deifies, is gathered and consecrated in the virgin wholeness of Christ. In the sacred oedipus complex, castration is in part this chosen withdrawal from the profane world, which no longer signifies desire, no longer calls forth power and energy from their interior fortress. This attempt to reclaim desire for volition presents us with a mature derivative of the idealized virginity, the narcissistic form of castration, found in *Comus*. By disassembling the phallus as symbolic bridge between desire and its external objects, Christ gains an interior and spiritual potency—a oneness with the paternal will.

The repressed temptation returns in the vision of Athens, "Mother of Arts" nestled "in her sweet recess" (4.240–242). Satan speaks of women as "such toys" (2.177) and, in rebuking Belial, as "a trivial toy." When he rejects Athenian wisdom, Christ speaks of shallow men reading books as "collecting toys," "Children gathering pebbles on the shore" (4.328–330), and soon thereafter of "swelling epithets thick laid / As varnish on a Harlot's cheek" (343–344). Why are toys a symbol of the despicable, of what desire has vacated? These allusions

to toys in the vicinity of women and books may remind us that learning begins in playing. Successors to the body of the mother, toys are the second objects of knowledge, and books themselves emerge from pebbles and whatnots, the array of playthings. The fact that Christ petulantly claims, "When I was yet a child, no childish play / To me was pleasing, all my mind was set / Serious to learn and know" (1.201–203) alerts us to the possibility that our author associated danger with the learning that is playing. Milton probably began to learn under the aegis of women: son of a charitable mother, he also had an older sister. Perhaps all boys in the full career of their learning move from the maternal matrix of early playing to the more paternal realm of formal education, where the superego oversees their exercises, but that transition was abrupt and the two genders distinctly marked in the Renaissance.

He thanked his father for providing him with this male wisdom in both poetry and prose. Milton scholars know, better than anyone, that he learned. But what was his experience with those books? Why does Satan annul the erotic temptation and conclude the expansion of desire with old teachers and old texts wherein Christ discerns whorish epithets, things "past shame"? When we quote the famous passages on trial and temptation in *Areopagitica*—"He that can apprehend and consider vice with all her baits and seeming pleasures, and yet abstain, and yet distinguish, and yet prefer that which is truly better, he is the true warfaring Christian"—we sometimes forget that their local context is a defense of the reading of tainted books as the safest way for a good man to learn about vice. I suspect that classical literature, which does not blush at "seeming pleasures," was like vernacular romance a play world for Milton, where conflicts of the home took root in culture. That he acquired culturally erotic ways we know from the Ovidian Elegy 5, which equates erection with poetic power, and particularly from Elegy 7, which he printed in 1645 with a retraction dated 1630: the "wantonness" of his youth has been reformed by "the shady Academy" and "its Socratic streams," for "thenceforward my breast has been rigid under a thick case of ice, of which the boy himself fears the frost for his arrows, and Venus herself is afraid of my Diomedean strength" (p. 16). It is the posture of the Lady of Christ's, the Lady, and Christ, all of them strong through the exercise of repudiation. But as we gather from "Il Penseroso," reading is the social life of a solitary man, and it was precisely "the solitariness of man" (*CP* II, 246) that the divine institution of marriage was intended to assuage. Milton could be the master of luxurious books, bringing to them, in Christ's

words, "A spirit and judgment equal or superior" (4.324), but nothing in the play world of culture had prepared him for Mary Powell.

After his impetuous marriage, Milton expounded the sage and serious doctrine of divorce (not for "gross and vulgar apprehensions") with an evident bitterness over the consequences of his chaste and literary youth. Those who sport with Amaryllis in the shade "prove most successful in their matches, because their wild affections . . . have been as so many divorces to teach their experience" (*CP* II, 249–250). The women that Milton knew from books were fantasy women: harlots to be repulsed, virgins to be guarded, "Fairy Damsels" (2.359) like those decorating Satan's banquet. In his maturity Milton had reason enough to denounce a moral life lived by the turning of pages.

The autobiographical sources of the Athenian temptation reach beyond the treacherous instruction Milton had received concerning the second mother in a man's life to the meaning he ascribed to *scientia* itself. The genetic sequence that leads from the body of the mother to toys to books curled back on itself for Milton; books unveiled and exhibited the maternal body of the world. *Ad Patrem* contains the most interesting of Milton's several accounts of his education. Written in the tongue to which his father had given him access, it is our sole record of a disagreement between them. The father, we infer, has expressed some dissatisfaction over the son's dalliance with poetry at the expense of more practical pursuits. As Milton thanks him for his education, he offers high praise unconsciously as well, for the father has given him a mother:

> I do not mention a father's usual generosities, for greater things have a claim on me. It was at your expense, dear father, after I had got the mastery of the language of Romulus and the graces of Latin, and acquired the lofty speech of the magniloquent Greeks, which is fit for the lips of Jove himself, that you persuaded me to add the flowers which France boasts and the eloquence which the modern Italian pours from his degenerate mouth . . . and the mysteries uttered by the Palestinian prophet. And finally, all that heaven contains and earth, our mother, beneath the sky, and the air that flows between earth and heaven, and whatever the waters and the trembling surface of the sea cover, your kindness gives me the means to know, if I care for the knowledge that your kindness offers. From the opening cloud science appears and, naked, she bends her face to my kisses,

> unless I should wish to run away or unless I should find her enjoyment irksome. . . . What greater gift could come from a father, or from Jove himself if he had given everything, with the single exception of heaven. (p. 85)

The gift of male culture might have proven "irksome," inciting a "wish to run away," were it not for the process of erotic sublimation that managed to seduce an oedipal mother (the *publica mater* of "Naturam Non Pati Senium," unwrinkled and ever fertile) in the lessons of the schoolroom. Knowledge, Milton writes in the last Prolusion, is our conquest of Mother Nature:

> almost nothing can happen without warning or by accident to a man who is in possession of the stronghold of wisdom. Truly he will seem to have the stars under his control and domination, land and sea at his command, and the winds and storms submissive to his will. Mother Nature herself has surrendered to him. It is as if some god had abdicated the government of the world and committed its justice, laws, and administration to him as ruler. (p. 625)

Not unlike the abdicating Father of Book 3 of *Paradise Lost*, Milton's father had relinquished the privileged hold over knowledge that all fathers enjoy for a time, transferring the dominion of Mother Nature to a son who would one day design the entire universe, including the incomparable account of Creation in Book 7 of *Paradise Lost*. But the elder Milton had also opened the mysteries of the prophet, and that was the downfall of his paternity. As Milton reveals early in *Ad Patrem*, divine poetry "preserves some spark of Promethean fire and is the unrivalled glory of the heaven-born human mind and an evidence of our ethereal origin and celestial descent" (p. 83). In the religious form of the family romance, the prophetic poet is a son of God.

How can a son assert himself against a revered father? Divide him. "Insofar as I am a poet, I am not your son; my origins are divine, my authority is God, and it is with my God that the struggle we are here engaged in must be adjudicated. Are you not holding now, in your hands, my *Promethean fire*?" Symbolized in *Paradise Regained* in the difference between "son of Joseph" and "Son of God," this splitting of the imago of the father constitutes the major psychological strategy of Milton's life and work.

His father first, then all the derivatives of his father in earthly

authority—teachers, bishops, kings, parliaments, theologians—could be deposed and abused by a rebellious son, while Milton at the same time remained the obedient son of his divine father. This is the generative core of the strong poet Harold Bloom has discerned for us, the unintimidated poet who read his precursors with a "judgment equal or superior," bowing to none of them, not even to Moses and the Evangelists. The two attitudes characteristic of his authorship, celebration and belligerence, derive from the psychological formula of the divided father. Milton's great lesson is that obedience is freedom—and so it is, if one is obedient to what lies outside the world and if through obedience one can do what is otherwise forbidden. His ability to pursue rebellious courses as the injunction of obedience chartered his apartness from social confinements. Guilt could always be overcome. He was not wrong to mature slowly, to desire divorce, to entertain the fantasies that interested him in polygamy, to oppose the church and comply with the killing of the King. When the spurious *Eikon Basilike* was causing some of its readers to regret that execution, Parliament could have found no man in all of Europe whose temperament was better suited to defend their nation against self-recrimination. Tillyard condensed a good deal of insight in remarking that if Milton had been Adam, he would have eaten the apple and commenced to write a pamphlet justifying himself.[45] This man came to the Spirit of Protestantism with a tremendous oedipal defiance.

In *Comus* the split is obscured by the fact that earthly and heavenly authorities promote the same law. Only when the epilogue moves us toward a representation of transcendent destiny do we feel the aggressive desires locked prematurely in the obedience of the Lady. In *Paradise Regained,* written after the sacred complex had been articulated, profane and sacred law oppose each other in lethal combat.

The evolution of paternal defiance into spiritual power is close to being the manifest plot of the brief epic. In exchange for his gifts and his advice, the tempter wants to be worshipped as the Son's god. He hopes to exact from Christ a single acknowledgment: that *Satan is Christ's father.* Has he not always wished to be God? He first appears as an old man. He offers to feed Christ, then to plan his career and tender sound advice, to secure him wealth and power and statecraft, to hurry him and educate him, finally to define his name—and the one thing he expressly withholds from the Son is women. Who played this role in the life of Milton? Somewhere at the ground of the imaginative act that created this Satan (not the Satan of *Paradise Lost*) was the imago of John Milton, Sr., fixed in the unconscious of a troubled oedipal son. In

place of the mother he had offered a cultural world to some degree symbolic of, but therefore alienated from, the genuine desire of the son. In this world, it is true, the son had found sublime means for restoring an archaic self-esteem. But he had also become entangled and belated there, weakening his eyes to learn every lesson but the one that matters most to a man—how to be happy with a woman. In *Paradise Regained* the best teacher takes revenge on the impositions of the Renaissance Name-of-the-Father.

Satan was not Milton's only idea of his father. We are speaking of a split, and if one current was murderous enough to produce a vision of the devil, the other must have lent its veneration to the Miltonic God. As we read in *Ad Patrem,*

> Now, since it is my lot to have been born a poet, why does it seem strange to you that we, who are so closely united by blood, should pursue sister arts and kindred interests? Phoebus himself, wishing to part himself between us two, gave some gifts to me and others to my father; and, father and son, we share the possession of the divided god. (p. 84)

The father had some renown as an organist and composer. It was music, the paternal half of the divided god, "Married to immortal verse" that ravished the "meeting soul" of L'Allegro, and the "pealing Organ" in concert with the anthems of the choir that dissolved the embodied soul of Il Penseroso. Blind but not deaf, Milton again unified this divided god in the operatic heaven of *Paradise Lost.*

Yet Milton cherished at the center of his strength a blasphemous profanation of the father internalized at the resolution of the first oedipal complex. Freud repeatedly noted the "immortality" or conservatism of the superego: the father passes his values to the son, whereupon the son, having embodied them, passes these values to his son (*S.E.* XXII, 67, 178–179; XXIII, 206). The dimension of futurity opened to the ego by the oedipal resolution, as we said earlier, is readily enthralled to the vanities and timidities of social life. From the perspective of the sacred complex unfolded in *Paradise Regained,* the substitutional propulsion of worldly desires under the guidance of the superego is a bondage—a closed arc of hungers that stretches from mother to fatal mother, and those who live solely within this arc, blind mouths from cradle to grave. John Smith maintained that "the only way to *unite* man firmly to himself is by uniting him to God, and

establishing in him a firm amity and agreement with the First and Primitive Being": religion "*restores a Good man to a just power and dominion over himself and his own Will, enables him to overcome himself, his own Self-will and Passions, and to command himself & all his Powers for God.*"[46] This "First and Primitive Being," for Smith and the other members of the Cambridge school an innate idea of God associated with innate principles of virtue, becomes in our psychoanalysis of faith a return to the primitive oedipal father, unspoiled by the ways of tradition, in whose place the ego searches for God. Denouncing Satan, Christ denounces the earthly father as guide to earthly desire. When God dethrones this father at the resolution of the sacred complex, the new Adam can overcome the old antimony of veneration and ambition. Hostile rivalry toward the father, forbidden by the ego-ideal of the first complex, is recovered in the superiority of the interior Spirit to all other "forcers of conscience" (the Westminister Assembly, Milton wrote in "On the New Forcers of Conscience," would "force our Consciences that Christ set free").

The doubling on the pinnacle now appears in a new light. In the name of righteousness, a new ownership of the will, the sacred superego licenses an attack on its rival, the profane superego; but insofar as the complex is being repeated, this sacred obedience duplicates the rivalrous disobedience of the first complex. The ego of new Adam has repossessed the evil energy of old Adam. The structure—temptation, renunciation, obedience—remains the same, yet the sacred complex subverts the profane one.

I have not given much content to the religious superego. Were we to find there some culture-bound list of Christian duties, I would feel that my efforts to discern psychological advent in religious symbolism had been in vain: the new superego would be the old one glorified, a mental space upon which laws can be inscribed. But *Paradise Regained* is notable for the absence of such prescriptions. Christ delivers maxims, yet they are mostly negative ones; we do not even hear the golden rule. He shows how not to become ensnared by the limited vision of the world. The major function of reason in the poem is to refute commonplace conceptions of a successful lifetime, placing our trust in the "inward Oracle" and its "Spirit of Truth" (1.462–463). The poem is a struggle between guides, one of whom is palpable and manifest as the way of the world, whereas the other is secret, interior, waited upon. Guided in this second way, Christ on the pinnacle gains unique selfhood at the moment he claims his Father's name.

Milton attended a college named for him. But this Christ has no need for the "Socratic streams" of the "shady Academy" that had protected the young Milton from wantonness:

> To sage Philosophy next lend thine eare,
> From Heaven descended to the low-rooft house
> Of Socrates, see there his Tenement,
> Whom well inspired the Oracle pronounc'd
> Wisest of men; from whose mouth issu'd forth
> Mellifluous streams that water'd all the schools
> Of Academics old and new. (4.272–278)

Satan speaks beautifully here. Socrates, like Christ an oral man, no writer, "from whose mouth issu'd forth" the virtuous discourse that became Plato's Idea of the Good, Aristotle's magnanimity, the Stoa's *sui potestas*—a great fiat of mind whose reissue in scriptural form has been guarded and systematized by the Western Academy . . . as then in Greece, so later in England. The alternative wisdom that illuminates Christ comes not only from the horizontal stream of history "in our native tongue" (333), but instantaneously from above:

> he who receives
> Light from above, from the fountain of light,
> No other doctrine needs, though granted true. (288–290)

But in what sense is this light a doctrine? It is not handed down. It is not taught in an Academy. To quote Smith again: "Divine Truth is better understood, as it unfolds itself in the purity of mens hearts and lives, then in all those subtil Niceties into which curious Wits may lay it forth. And therefore our Savior, who is the great Master of it, would not, while he was here on earth, draw it up into any *System* or *Body*, nor would his Disciples after him; He would not lay it out to us in any Canons or *Articles* of *Belief*, not being indeed so careful to stock and enrich the World with Opinions and Notions, as with true Piety, and a Godlike pattern of purity, as the best way to thrive in all spiritual understanding. His main scope was to promote an *Holy life*, as the best and most compendious way to a *right Belief*. He hangs all true acquaintance with Divinity upon the doing God's will, *If any man will doe his will, he shall know of the doctrine, whether it be of God* (John 7.17)."[47] Like More and Cudworth, Smith sought to restrain the potential anarchy of the inward light by arguing that innate principles

of virtue would become universally self-evident once the innate idea of God had exfoliated. I doubt whether Milton would have assented to this restraint. From *Areopagitica* to the preface of the *Christian Doctrine* it is clear that the inward light does not tell all men the same thing. But this description of what Christ brought into the world, with its subordination of specific belief to holy life and reason to will, is compatible with the religious vision of *Paradise Regained*. The only paraphrase for the doctrine of "Light from above" is the matchless life of a unique man. The question at issue in the poem is not the ethical "what did Christ tell us to do?" but the existential "who is Christ?" Except in the medium of symbolism, the answer eludes our saying. Christ refuses to be contained by the rational and psychological categories of the world. More akin to oracle than to argument, his unclassifiable doctrine founds a culture whose ideal is forever outside of cultural systems. The interior light makes possible an iconoclastic purgation of the profane superego in which the laws and traditions, the "ways" of a society, reside. This revolutionary culture of free individuals anticipates the future, and its native art is prompted, unpremeditated.

As early as *Comus* Milton knew that reason is choice, but the full implications of this idea are not apparent in his work until the last books of *Paradise Lost*.[48] The poet has celebrated the right reason of prelapsarian man and mourned its failure. In the choosing of the final lines, where the world lies ready to be entered by the guided deliberation of Adam and Eve, Milton foreshadows the starker Protestantism of *Samson Agonistes* and *Paradise Regained*, where choices are not made by a rational faculty invested with universal principles. The old role of right reason is collapsed into the choosing itself, an intuitive act performed in response to the urgings of God. As intellectual intuition transforms the rational pursuit of knowledge into immediate possession, so volitional intuition transforms the rational pursuit of virtue, invariably involving the application of law to circumstance, into the immediate presence of a prompting will. In both cases the ego simply consents to what is there. Reason now becomes the guardian of readiness, keeping the hero from imprudent action during his agon of waiting. One aspect of this guardianship is the management of theodicial doubts, the "patience to prevent/That murmur" (Sonnet XIX) which might inspire action in the absence of divine will. When rousing motions come to Christ on the pinnacle and to Samson after he has refused to be humiliated before Dagon, there is no negotiation with them, no trace of *that* sort of agonizing: "Master's commands

come with a power resistless/To such as owe them absolute subjection" (*SA* 1404–1405). In the two poems generally assumed to have been written after *Paradise Lost*, Saurat concluded, Milton "frees himself from dogma; all he keeps of it is God-Destiny."[49] Obedient to an inward oracle, the ego no longer seeks virtue in struggling to free itself, through a rationally motivated ethics, from the superego. The religious life has accomplished a transvaluation, or recapture at a higher level, of the voluntarism of the primitive superego.

There are obvious risks in this surrender. How many of us, told that writing a book would mean going blind, would proceed to write the book? The otherness within us, the eye that watches and the voice that calls, can become concrete to the point of hallucination in pathological states. It can indeed command irresistible forces. Born of conflict, the superego generates conflict by its presence: the rational ego that would be a law unto itself must always encounter a dispensation of guilt whose habits of accusing and esteeming the ego remain to a large degree incorrigibly irrational. But in the sacred complex Milton came to grips with these conflicts, working through them as well as repeating them. Theodicy itself provides the most lucid example. We have noted that the rough draft of a theodicy undertaken by the oedipal child will confront him with the dark irony that the father, bearer of the law forbidding intercourse with the mother, is precisely the one who disobeys his own law: the desire to be the father is both the model for evil and the model for good. Here we encounter the religious transposition of this embryonic theodicy. God is dark and God is light, for what is it to be evil in *Paradise Lost*? To be evil is to be God, a law unto oneself. In philosophy and symbol the theodicial argument of *Paradise Lost* expansively reworks an outrage intrinsic to the history of conscience.

The superego celebrated in *Paradise Regained* and *Samson Agonistes* has made the prerational otherness at its origins sacred and suprarational. In its passage to the sacred, the superego acquires what I would call a feminine aspect—not in the vulgar sense that women are intuitive, which is a lesser way to think, but in the sense that intuition in all people, men and women, is a form of knowing or being rooted in the early matrix of sudden attunements and shared promptings.

Waller has suggested that the voluntarism of Luther and Calvin encouraged an anxious and fragmented experience of time, since if every moment expresses the will of an unknowable God toward the individual, every next moment can be discontinuous with the present one.[50] Such anxiety proceeds from the assumption that one is set

against time, experiencing the dramaturgy of a God external to the self. In the voluntarist world of his last poems, presided over by a riddling God, Milton shows how this anxiety may be conquered by recovering, in an oedipal structure, basic trust. Christ knows that Satan has been given "permission from above" (1.496) to impose his temptations. But the Son's interior attunement to the otherness of the Father allows him to become the master of circumstance, undefined by circumstance. The foreseeing God of *Paradise Lost* prepared well for the ordeals of fallen man:

> And I will place within them as a guide
> My Umpire *Conscience*, whom if they will hear,
> Light after light well us'd they shall attain,
> And to the end persisting, safe arrive. (3.195–197)

The soul is not solitary. Trusting the God who speaks and illuminates in the felt alterity of conscience, we are led through riddling discontinuities to moments of clarity, light after light, until we come to the end. The father as superego, the father as culture, the father as God—it is through the third father in the life of Renaissance man, less a lawgiver than a guide, that the last mother can regain the power of the first. The oedipus complex is made mortal in order that the ego can be made immortal.

Immortality

Until now my analysis of Miltonic Christianity can claim a certain typicality. Except for details such as his association between *scientia* and the mother's body—not so rare at that, maybe, in a culture conscious of a relationship between *mater* and *materia*, whose icon of Natura sometimes poured from her breast the Milky Way—the religious psychology inferred from *Paradise Regained* has key elements in common with Renaissance Protestantism, especially leftward groups for whom the authority of the church did not blunt a sharp separation between earthly legality and divine impulse.[51] But as we move into the heart of faith, the promise of redemption, there are reasons to suppose that Milton would exemplify a psychology of one. His imagination was not drawn with the usual fervor to the Crucifixion. He was also a mortalist, convinced that body and soul lie down together in the body of Mother Earth to await the quickening voice of the Judgment.[52] Yet Milton contended that his heretical view of death

would not be different, in the experiencing, from the orthodox guess. And however distant Milton kept his art from the lush agonies and weeping wounds so familiar in Renaissance devotional literature, his poems obviously do promise an afterlife. The Father identifies the desert trial as propaedeutic to Christ's conquest of "Sin and Death the two grand foes" (1.159). Christianity has always thought itself the master of these opponents: how are they overcome psychologically? I would like to think that Milton, maker of dense and passionate myths, reveals as well as one man can the resources of his religion.

We are near the limits of a psychoanalysis of religious experience. Where do babies come from? Psychoanalysis from the beginning has equipped itself to understand Oedipus and his riddle. When Milton compares him to Christ on the pinnacle, he occasions us to reflect that Christianity has not only changed the answer, but also the question. Is there life after death? This is the riddle, pointed toward the end rather than the origin of man, Christianity has tracked through myth and symbol, and in these terms, presumed to answer. But the promise of salvation from death cannot sustain itself against the evidence of the world, our sense of the inexorability of natural law, without some fundamental authentication in the psychic life. Dubious though the promise be, it is no simple wish. I cannot follow Hans Loewald, for example, in referring belief in eternal life to the "timelessness" of the unconscious, for as I understand the unconscious, timelessness is a *character* of its mentation, not a content.[53] Freud posited that death may be conceived as narcissistic reunification with the first mother, a notion connected with his "Nirvana principle": because the consolidation of the ego entailed the loss, through separation, of the early mother, the ego may register its destruction as a return to that boundless primary object (*S.E.* XVIII, 37–38). But this theory must be modified to account for the Christian equation of death with guilt. If the fantasy of dissolving back into the pre-ego mother supplies most of the psychic force behind the otherwise incredible notion of a death that bestows immortality, how is Protestantism able to contact this fantasy through its theology of radical evil? It is once again a matter of finding the narcissistic along oedipal pathways.

Freud held throughout his late reformulations of psychoanalysis that feared death—not the one we know as we know a fact, but the one we live toward as our last possibility—acquires its power from the ego's long oppression under the superego (*S.E.* XIII, 57–59; XX, 129–130). The boy gains a conscience by accepting castration, which is the edge between two deaths, propelling him from the first to the third mother. It bars reunification with the earliest love. At the other ter-

minal of life, the death lived toward under the new order of desire instituted by the superego seems an unspeakable injustice, for that is what death means psychoanalytically—the final withdrawal of esteem, the superego retiring from the ego. Like Christ, the great archetype of the unjust death, we die when forsaken by the progenitor to whom we have sworn obedience. Only then will the superego be finished disillusioning us about the first mother. "But it is in vain that an old man yearns for the love of woman as he had it from his mother; the third of the Fates alone, the silent Goddess of Death, will take him into her arms."

Milton also knew that death is the image of our hatefulness and the truth of our guilt. Guiltiness, the first of the four degrees of death defined in the *Christian Doctrine*, "although it is a thing imputed to us by God, is nevertheless a sort of partial death or prelude to death in us, by which we are fettered to condemnation and punishment as by some actual bond" (*CP* VI, 393). Because he had completed to his satisfaction a rational theodicy, Milton could believe that the punishment was just. We are satanic at the instant of death. But what motivates this ultimate desertion of the ego? The eating of the apple? In the archaic world of superego formation, the fantasy crime that would turn the superego into the executioner of the ego is the parricidal enjoyment of the mother. In the large epic similarly, Death comes into the world as the *first oedipal son*; after ravishing his mother, he nearly reenacts the crimes of Oedipus by murdering his father (2.681–726). Satan preinterprets the consequences of the fall he will later devise in conceiving from his brain this dreadful family. Through the Christian centuries Eve will become, like Sin, a cursed mother. Men will recoil from the womb that gave them life. The unappeasable hungers of the son of Satan will become our dark heritage, unconsciously if not openly, as "th'incestuous Mother" informs her charge:

> Thou therefore on these Herbs, and Fruits, and Flow'rs
> Feed first, on each Beast next, and Fish, and Fowl,
> No homely morsels, and whatever thing
> The Scythe of Time mows down, devour unspar'd,
> Till I in Man residing through the Race,
> His thoughts, his looks, words, actions all infect,
> And season him thy last and sweetest prey. (10.603–609)

Because Mother Sin is in our bones, Death claims us. Theodicy has discovered the guilt of natural man: having consented through Adam to the satanic disfigurement, he hurries toward a just death; the same

argument that justifies God justifies the execution of the ego. Exactly for this reason, the sacred complex must reshape the profane superego. An old drama is staged again, and yet once more the world is lost to immediate desire in order to be regained in the futural mode of hope.

As the new Adam transforms the oedipus complex, he inaugurates a new interiority in the symbolism of sacred history. The Godhead is in the mind. God will exact the penalty of justice without abandoning the ego, for if the sacred superego is the Father, the sacred ego must be the likeness of the Son. The *him* at the end of *Paradise Regained*, inhabited by Satan and Christ, is the pronoun of the Christian ego. Theodicy has learned that Satan is the ego at the moment of its just death. Through Christ, death regains the stamp of injustice erased by the work of theodicy. Although it satisfies a legal requirement, his death is not deserved:

> Though now to Death I yield, and am his due
> All that of me can die, yet that debt paid,
> Thou wilt not leave me in the loathsome grave
> His prey, nor suffer my unspotted Soul
> For ever with corruption there to dwell;
> But I shall rise Victorious. (3.245–250)

The resurrected Christ is the ego retrieved in the name of love from the moment of its unjust and pretheodicial death—the man for whom the earth is truly a mother. This exemption demands the mysterious replacement that Milton intimates on the pinnacle. Christ at the end must somehow become Satan in order that the Satan toward which we live as a last possibility may somehow become Christ. Then, through identity with the greater man whose life is an undoing of the life of Death, a complete victory over the crimes of the oedipus complex, there can be psychological authority in the hope that we will be guided safely through the moment of our corruption, returned like Christ to the home of a sinless mother. The expulsion of Satan from *him*, which we must accomplish in order to catch the sense and return home, repeats the triumphant institution of the sacred complex on the pinnacle and adumbrates the triumphant end of this complex in the good death of the saved:

> hee unobserv'd
> Home to his Mother's house private return'd.

Freed from the conscience that begins in external observation, *hee* is "unobserv'd." Lastly *hee* is "private," the possessor of himself, emptied of father, Satan, and death.

Miltonic faith proceeds from a splitting of the original father. God must emerge within a conscience memorializing that father whose final word is death. In this respect the unfolding of Christian psychology is the microcosm of sacred history. As the Father, containing like *him* both light and dark, imaginatively generates the two great antagonists of cosmic time, so the psychic history by which the ego becomes identified with his good Son issues from the rupture of a unitary father into two manifestations, sacred and profane. The Father teaches salvation. He produces duality, but as he inserts tragic oppositions into his universe, he reveals through sacred history the strategy of their resolution. He also teaches creativity, which is a lesson in part opposed to resolution. Everything starts with a rending, a break in the order of the system, for without the split there could be no novelty whatsoever.

The Volume of 1671

After the first readers finished *Paradise Regained*, they turned to its *epithon*—the work "added" according to the title page, *Samson Agonistes*.[56] Moving backward in biblical time, they moved forward in the course of heroic lifetime. Young or old, the hero comes to his triumph in being led to a temple by his enemy and by God. Taken together Christ and Samson form the pattern of a guided life, its riddles solved by the attunement of intuition to divine will.

Among the many reverberations that echo between Milton's last companion poems is the ongoing sanctification of the myth of Oedipus. Compared to Oedipus in the last simile on the pinnacle, Christ comes home to his mother in the last line. Turning the page, we find a Christian tragedy whose opening words, imitating *Oedipus at Colonus*, are those of a blind man following a guiding hand: it is as if *Samson Agonistes* were taking place inside the last simile of *Paradise Regained*, extending that simile into the future. At the end of his life the Hebraic Oedipus receives, through simile, a rebirth from the "ashy womb" (1703) of the Phoenix. Then Manoa consigns the story of his son, along with his body, to the culture of the future.

Buried at the base of this volume and the culture it anticipates is the figure of Oedipus. The poet has appropriated, converted, and transformed him, changing both his riddle and his answer. But like

the sin in our bones, he abides. Milton revealed in this latent tribute the fate out of which was born the prophetic power within himself and his myth.

• 4 •

THE WAY TO STRENGTH FROM WEAKNESS

A reader of Milton must be always upon alert; he is surrounded with sense; it rises in every line, every word is to the purpose; there are no lazy intervals. All has been considered, and demands, and merits observation. . . . What if it [*Paradise Lost*] were a composition entirely new, and not reducible under any known denomination?

—Jonathan Richardson

"I SAW HIM STRONG," strong enough, had he wished, to ruin the sacred truths, Andrew Marvell wrote of the author of this composition entirely new. Probably my fantasy sprang into being over some page in that haunted pool of light where we see, imagine, speak, and listen all at once, so accustomed to this marvel that we forget its affinities with trance and hallucination. I have lost the origin. But it pleases me to imagine someone sufficiently alert to be surrounded with sense as he fumbles through the bookstalls of London in the late 1660s and opens on a lucky whim the first edition of Milton's epic, which was, according to a bookseller in one of Richardson's anecdotes, lying around like waste paper.[1] The surge and thunder of the great poem breaks upon him. With the invocation to Book I still ringing in his mind, he thinks to himself, "The ways of God to men? Mortals in their prayers usually try to justify their ways to God, and the process by which this truly happens, our theologians call justification. No middle flight? Things unattempted? Everyone knows he went down with his cause. Milton is a blind and gouty, defeated old man. *Where did he get the strength*?" At which point I appear, a time-traveling professor from the future, to hand him the present chapter.

Preparations of the Phoenix

The descent of total blindness on John Milton in 1652 almost coincided with the deaths of Mary Powell Milton and their infant son, John. Psychological regression was inevitable in this state of physical

dependency. Milton had to be tended, had to be guided—"Deceivable," as his rueful Samson proclaims, "in most things as a child" (942). As he relearned self-sufficiency Milton must have experienced anew the needs, desires, affects, and especially the frustrations of his childhood, since this second self-sufficiency could never be as perfect as the first. I think he also relived the preparatory years between Cambridge and his marriage. He was wifeless once more. His decision to complete the *Defensio* at the cost of his sight reasserted the commitment to duty that had brought him home from Italy. Considering that his light was spent, he again faced the question of how to use his talent. "They also serve who only stand and wait": the climactic patience of Sonnet XIX, a poem associated in numerous ways with Sonnet VII and the letter to the unknown friend, announces a second moratorium. Again he would wait, affirming nothing but his right to do so.

If, as it seems likely, his duties were reduced along with his salary in 1655, he again entered a period of relative leisure, able to pursue his own designs apart from the clamor of public life. So the blind man, remembering seeds planted by a youth whose bucolic leisure had been the study of all knowledge, proceeded to reap his prosaic possibilities. The educator and the classicist compiled a Latin thesaurus; the statesman wrote *The History of Britain*, published Ralegh's *Cabinet-Council*, and offered his advice to a faltering nation in the pre-Restoration pamphlets; the theologian began his *De doctrina*. Like the poet, the theologian was long choosing and beginning late. This compendium appears to be a mature resumption of the "Index Theologicus" in his Commonplace Book and the lost "perfect System of Divinity" mentioned by Edward Phillips as having occupied him during the 1640s (*EL*, 61). If Milton started on the treatise in 1656, as the evidence suggests, he was simultaneously commencing with a second marriage.[2] We do not know whether this was another impulsive union. Its timing might imply that, his theology taking shape, Milton had managed to renew his lifelong intimacy with the future even as the nation devolved toward the old servility of kingship. When Katherine Woodcock Milton died in February 1658, he soon thereafter (on the testimony given to Aubrey by Phillips) began *Paradise Lost*.[3] From 1658 to 1660 he seems to have worked on the epic and the treatise concurrently, turning to the latter, perhaps, when his poetic vein stopped flowing around the vernal equinox (*EL*, 13). The second hiatus of waiting in his life, punctuated by marriage and bereavement, moved toward fulfillment along the line leading from

treatise to epic. He kept his covenant. Like the Phoenix of *Samson Agonistes*, who "lay erewhile a Holocaust" until she

> From out her ashy womb now teem'd,
> Revives, reflourishes, then vigorous most
> When most unactive deem'd, (1703–1705)

Milton lived himself again, producing in his maturity a youth of ideal accomplishment. With the first draft of *Paradise Lost* complete or at least well under way in 1663, he married yet again.

Milton tinkered with the fair copy of the treatise, probably made in 1660, throughout his remaining years, and in the preface referred to this manuscript as "my dearest and best possession" (*CP* VI, 121). A somewhat alarming affection: why should the treatise be found between the claim to divine favor in the *Second Defense* of 1654 and the evidence of abundant recompense in the published epic of 1667? Theologically, Milton had no pressing need to copy out orthodox doctrine and moral maxims from the comparable books of Ames and Wolleb. What mattered in this regard was the achievement of a systematic theology incorporating his peculiar doctrines: the proximity of God to matter, the Creation *ex materia*, the co-mortality of body and soul, and the partial deconstruction of the Trinity. To justify the ways of God, Milton had first to delineate the deity on trial and the ways to be adjudicated. Like other religious poets, he submitted his desire to the legalities of his faith, teaching obedience and reconciliation. Yet this champion of "Eternal Providence" claimed the freedom to impose some measure of his own disposition on the doctrines concerning God's ways.

The Phoenix is a secretive bird, "vigorous most" when she seems a pile of inert ashes, light spent. There are mighty passions running in the veins of the stony logic of the *Christian Doctrine*. I will try to demonstrate in this and the following chapter that its heresies join rational to psychological coherence, staking out areas of mutual concession between the inheritance of Christian tradition and the psychic labors that had to be completed in order for Milton to leave the tradition his own epic legacy.

In nature as in time, the treatise stands midway between the partisan biblicism of his controversial prose and the articulated cosmos of *Paradise Lost*. But why, heresies aside, did Milton's doctrines have to stop at this way station? Why did he bind himself in lifeless Latin, before, then while, he reimagined the Bible in a composition entirely

new? The treatise was also Milton's grip on a guiding hand. He had lived through a slow disintegration of the phenomenal world whose constancy, rarely in question and therefore rarely the object of reflection, sustains our most jarring hours with a bass note of invariance. At some point an infant "learns" once and for all to distinguish the irruption of flux experienced when his eyes move across a stable visual field from the contrary but phenomenologically similar experience when moving objects appear to a stationary gaze; he masters the instantaneous calculus of orientation that can discriminate the course of the object from the course of the subject when both are moving simultaneously. This is the *sense* of the world, given to us at so deep a stratum that philosophy has never been successful at abstracting our intelligence from its reliance on the world's intelligibility:

the Creator Great
His constellations set,
And the well-balanc't world on hinges hung,
And cast the dark foundations deep,
And bid the welt'ring waves their oozy channel keep.
("On the Morning of Christ's Nativity," 120–124)

Even water, the protean element, has been given its regularity. In Milton this well-balanced world flew apart at one of its deepest foundations, escaping from the channels of binding form.

He experienced visual distortions, some of them severe, in the decade preceding total blindness. At first an iris surrounded the lamp, then the left part of the left eye failed, which "removed from my sight everything on that side" (*CP* IV, 869). Then, should he close his right eye, objects in front became suddenly smaller. Toward the end, "everything which I distinguished when I myself was still seemed to swim, now to the right, now to the left. Certain permanent vapors seem to have settled upon my entire forehead and temples, which press and oppress my eyes with a sort of sleepy heaviness, especially from mealtime to evening." These dazed afternoons (so like the intoxication he despised) remind him of Phineus in the *Argonautica*, who "thought the earth / Whirling beneath his feet" and sank "Speechless at length, into a feeble sleep." When he tried to sleep during the departure of his sight, his closed eyes beheld a brightness from which "colors proportionately darker would burst forth with violence and a sort of crash from within." When these seething disorientations both within and without settled to a "pure black, marked as if with extinguished or

ashy light, and as if interwoven with it," Milton might have wondered if he had not undergone in reverse the Creation of Genesis, victim of an insane Logos whose negative *fiat lux*, striking at the reliability of "Nature's works," had returned him to the "void and formless infinite" (3.49, 12).

Metaphors such as "perceptual nutriment" come naturally to modern psychologists, and they came just as naturally to their Renaissance precursors at the banquet of sense. In concluding the letter from which I have been quoting, Milton connected physical sight with bread, evoking the same passage in Deuteronomy that sees Christ through his first temptation in the wilderness: "But if, as it is written, man shall not live by bread alone, but by every word that proceedeth out of the mouth of God, why should one not likewise find comfort in believing that he cannot see by the eyes alone, but by the guidance and wisdom of God. Indeed while He himself looks out for me and provides for me, which He does, and takes me as if by the hand and leads me throughout life, surely, since it has pleased him, I shall be pleased to grant my eyes a holiday" (870). Reading the Bible, which Calvin call the "spectacles" of Christian perception, was the act that enabled all of Milton's achievements. Scattered among those bright leaves he found the moral, political, theological, and literary attitudes endorsed by God. It was not a passive finding. Our freedom from congealed tradition lay in interpretation (*CP* VI 122–124). True Christians had to search the Word, because those who failed to hear or see the potent book could only fall back on the words of men, and if they heard or saw but failed to reinterpret, on the words of men about the Word—the sources of idolatry, bondage, and true heresy (hearsay). Given the compulsory abstinence of his blindness, Milton could search for an undiluted guidance, like the fasting Christ of *Paradise Regained*. He took the hand and took food at the hand, steadying and nourishing his spirit with the oracles of God.

The traditional metaphor of the book of nature that crowns the catalogue of loss in the prologue to Book 3 of *Paradise Lost* must have come to this poet with a special rightness. A diligent reader since childhood, his acuity and his ambition had matured largely in the universe of a library, where all books were read in the light imparted by the sun of scripture. In the 1640s he had located his theme in Genesis. Christianity would permit him to surpass Homer and Vergil; Protestantism would be his initial advantage over Dante, Ariosto, and Tasso. Blind, he shelved the greatest book where it could not be lost; the treatise we full-sighted readers peruse was to its author a taking in and

filling up, an inhalation of the Spirit. Milton proudly emphasized the fact that here, as nowhere else in his work, he sacrificed his words to the expressions of God: "I . . . have striven to cram my pages even to overflowing, with quotations drawn from all parts of the Bible and to leave as little space as possible for my own words" (*CP* VI, 122). Accurately described in this passage, the treatise contains approximately 7,000 proof texts held within a spare frame of organizational categories—a stupendous undertaking for a blind theologian. It is one thing to know a book by heart; he who can cross-reference any one statement in a book with every other statement knows the equivalent of many books. Milton was surrounding himself with sense. He *claims* the Bible through incessant quotation, placing "all parts" of this book, the atoms of its verses, in a structure comparable to the *ordo cognitionis* of biblical hermeneutics and the proliferating dichotomies of a Ramist *ars memorativa*.[4] The purpose of the treatise was to "assist my faith or my memory" (120): "It was, furthermore, my greatest comfort that . . . I had laid up provision for the future in that I should not thenceforth be unprepared or hesitant when I needed to give account of my beliefs" (121). The treatise was indeed a "possession" of his "best and dearest" book. In the mythical language of *Areopagitica*, Milton endeavored to "re-member" the dispersed body of Truth, closing up piece with piece in the act of internalizing a "wholesome body of divinitie." As he assembled this figure within him, he began devoting his talent to *Paradise Lost*, where his own words, absorbing and transforming the words of the Bible, bring Truth to life. The obedient labor of interpretation, mitigated by the formulation of heterodoxies, preceded the imaginative freedom of poetry, mitigated by the contribution of divine agencies.

We know that blindness for Milton was a sign of artistic power and spiritual favor, linking him to God in proportion as it distanced him from men. Yet naturally enough, we are still disposed to think of his affliction as an obstacle overcome. *Paradise Lost* was accomplished despite a handicap, and therein lies the heroism of its author. So far as it goes, this view of the poem is indisputable. With my psychological preoccupations I am struck by the mere persistence of his literary ambitions. The common defense against loss is to devalue the object: I never loved the person who does not love me; I did not want what I have been denied. Satan indulges this weakness, debasing through word and deed everything he has been excluded from; he speaks contemptuously to the sun. The sightless narrator, however, praises light of every kind. Milton might have been expected to find reasons for denigrating sight and two of its great privileges, reading and writing;

respectable arguments depreciating physical seeing were everywhere available in the Renaissance. A motive of this kind may well have entered into Christ's condemnation of classical learning in favor of "Light from above." Yet, accommodating himself to all the new inconveniences, Milton remained an author, creating something visible for the experience of a sighted audience that he could never join again. Although prophetic seeing was superior to physical seeing, Milton did not return with vengeful fervor to the repudiation of material existence expressed in *Comus*. The theology of his blindness elevates, to the point of heresy, the dignity of the physical world. On the surface of it, these are victories to be admired, bearings kept despite his loss. But this is truly the surface of it.

"The leaves should all be black whereon I write," he said prophetically in his uncompleted poem on the Passion, foreshadowing the "universal blanc" of the invocation to light. I am convinced that progress in our understanding of Milton's creative revival depends on inverting the lingering attitude that blindness was an obstacle overcome. Why could Milton compose his masterwork *only* when blind? Why should blindness have made him decisive? Viewing his career in prospect, the movement from scholarly retirement to political controversy had by no means solved his difficulties with the "one talent." The conception of the deathless poem remained bound to paralyzing conflicts of superego formation. In a sense, he had chosen against his poetic future; for almost two decades the fear of exposure continued to outmaneuver the fear of belatedness. Milton tried to reinclude the abandoned desire for poetic immortality through his aesthetics of the true poem, which makes heroic deeds in the public arena a precondition for writing in the major genres, and he might well have gone on forging, right to the end, strategies of honorable delay. The plot of his life makes better sense if we assume, however we then construe the implications, that *he did not want to begin*. I think he welcomed his public involvement more than he cared to admit. Glancing through the long list of suggestions for the plot of a tragic drama that Milton recorded in the early 1640s, one observes that the subtlest way to postpone unwanted choice is to multiply alternatives. After the subject had been determined, they kept on multiplying. In what genre? How organized? We have four outlines for a tragedy on the fall, and Phillips saw Satan's address to the sun in a fuller version of one of them (*EL* 72–73).

Resolution formed in the dark. Internal and external factors alike suggest that *Paradise Lost* was inaugurated by, not despite, his loss. The poem actually does begin in the realm of severance, the "darkness

visible" of hell, which the narrator himself renders analogous to "What in me is dark" and "what is low" (1.23–24). External to the work, we know that Milton held a climatic theory of national character that associated the cold of the north with brute strength and the warmth of the south with the cultivation of the arts.[5] In the epic itself, where a drama of finishing replaces the lifelong drama of beginning, "cold / Climate" (9.44–45) is among the things that may leave the poet "Deprest" (46), his intentions unfulfilled. The epic repeatedly calls attention to the benefits of the sun's penetrating and generative warmth. Milton, however, had his light in ashes: the inspiration for a poem recording the sad loss of perpetual summer came to him with the solar decline of the autumn equinox and departed in the spring (*EL*, 13). Unable to believe this testimony, Richardson uncharacteristically declared, like a Bentley of biography, that Milton composed from spring to autumn.[6] But one must have a mind of winter. The second word in the title of the poem cancels the first, and the strength of its author coalesced mysteriously in "absence, darkness, death; things which are not."[7] The Christ of *Paradise Regained* rejects the ways of the world, but throughout much of the work the specific objects of his rejection are *visions* of the ways of the world. It was through blindness that Milton, imitating his hero, became worldless in order to reconceive in poetry the sense of the world.

"There is a certain road which leads through weakness, as the apostle teaches, to the greatest strength," Milton wrote in the *Second Defense* (*CP* IV, 589), alluding to II Cor. 12: "when I am weak; then I am strong." During the 1650s he several times wrote the following declaration in autograph albums: "My strength is made perfect in weakness."[8] The secret of this renovation dwells in the magnificent and difficult verses of the invocation to light. Here is Milton on the threshold of his fulfillment, about to take his poem "Above th'*Aonian Mount*" to the court of the Father. As he calls upon the light, psychic patterns expressed elsewhere in his work fall together in a remarkable clarification. The invocation gives us pause. It is one of the moments in *Paradise Lost* when the hermeneutic circle of the reading experience, that continuous readjustment in which we move back and forth from the meaning of the part to our projection of the whole, seems to stop wheeling, and part and whole, arrayed before us, offer themselves to our reflection.[9] The whole of the epic is in several ways a part of this part.

Analyzing the symbolism of rising and falling, darkness and light, Jackson Cope concluded that Milton is "the micro-cosmic mirror of his own argument."[10] But this mirroring must not be understood to

imply that Milton and his blindness, as representations in the poem, have become rootless operators in the semic network of an aesthetic object, meanings only, while themselves remaining what they were outside this network. There is a contrary flow from the poem into the life of the poet. Blindness is saturated with symbols, such that we can no longer think of it as the condition or empirical datum implied in our "despite blindness": the "blindness" in "What did blindness mean to Milton?" cannot finally be distinguished from the significance of *Paradise Lost*. Composition is composure. Representing his act of creation, Milton in the invocation shows us that this project is simultaneous with, radically inseparable from, his effort to interpret his blindness with symbols. Blindness is a sign, but a broken sign, like a word as sensible object only, a pure *signans* whose fulfilling *signatum* Milton searches for. In his invocation we find this broken sign situated at the crucial junction between the intentional meaning of the poem and an unintended sense struck in the unconscious of the poet. I would not wish to say, as some psychoanalytic interpreters might, that poetic creation for Milton was a denial of his blindness—a compensatory structure that became "relatively autonomous," or in effect, successfully compensatory.[11] The compensation of *Paradise Lost* arose from a centering acceptance of his blindness as a sign in need of meaning and a meaning in need of interpretation, both of which Milton created by first submitting to the significance of this sign already created in his unconscious.

The invocation can be divided into three units: 1–22, where the light addressed filters down from the essence of God to the *Fiat* of Genesis to the "sovran vital Lamp" of the sun; 22–40, where the acknowledgment of blindness is followed by an assertion of the continuity of his pursuits, as if the poet himself were saying "despite blindness" ("Yet not the more"); 40–55, where a second and more detailed acknowledgment of loss is followed by a commanding renewal of the address to light in the first unit. Before proceeding with the analysis, it is necessary to consider the contiguous passages that frame this moment. Evoking a speculative style often taken by intellectual historians to be constitutive of Renaissance thought, they direct us to the size and value of what is being achieved in the centerpiece.

Vast Imagination

My every step
might be a fall. I am a prisoner
shuffling through a time that feels like dream,

taking no note of mornings or of sunsets.
It is night. I am alone. In verse like this,
I must create my insipid universe.

—Jorges Luis Borges, "The Blind Man"

An insipid universe of second-hand words issues from this tired but moving parody of the invocation to light. The first man to shuffle through "a time that feels like dream" flanked his prologue on either side with enormous visions, forming a triptych whose subject is the ability of imagination to make time and space astonishing to reason. These visions belong in a Renaissance tradition that, from Nicolas to Bruno, introduced into philosophical discourse an element of wonder, a new Plotinian inebriation, by rethinking the ancient macrocosm-microcosm correspondence as the coincidence of the maximum with the minimum, the atom with the universe, time's right now with eternity's forever now, the dislocation of an infinitesimally shrinking sphere with the everywhere of an infinitely expanding sphere. Subjecting concepts to a muscular rhythm of *complicatio* (enfolding) and *explicatio* (unfolding), these figures sought to heal a new *tméma* or incision dividing absolute from perspectival knowledge; the pressure of ideas such as infinity exploded the inherited notion of the whole as the sum or container of its parts.[12] In philosophy proper the culmination of this trend occurs in the metaphysics of Descartes. Leaving the object of thought unspecified, his *cogito* condenses into one brief formula the affirmation of subjective unity given in each discrete thought. Infinity, the first of Descartes' innate ideas, confronts the ego with a concept that could not have been produced from within its existence: "the mark of the workman."[13] Our fundamental certainties derive from a double *complicatio*: the *cogito* is the one enfolded in all ideas; infinity is the all in one idea. Through Descartes primarily, the problematic of this tradition, along with its affinity for wonder, passed into the idealist systems of Kant, Fichte, Schelling, Schopenhauer, and Hegel, and can still be observed in latecomers such as Josiah Royce, whose famous map of England confounds the purpose of a map, which is to represent a stable terrain, by mapping itself. This ever-shrinking hole in the fabric of determinate knowledge would have delighted Cardinal Nicolas.

Florentine Neoplatonism belongs to this tradition only tangentially. Although their philosophy betrays a strong cosmological interest, Ficino and Pico do not thematize infinity. For this reason the appropriation of the Cusan tradition by English writers in the seven-

teenth century produced an extremist version of the Florentine aesthetic of *concordia discors*; Henry More's trope of the stars as drops spilled from the pen of God, dipped in light, as it swept across the page of the universe, is light-years of sensibility from the *festina lente* of the Italians. It was not long before the advent of wonder in literature exposed a defect in conventional notions of aesthetic unity—harmony among the parts, concordant arrangement of discordant parts, the shapely presentation of an action—corrected somewhat by the eventual revival of Longinus, who encouraged one to treat parts *as* wholes.[14] The ascendancy of Milton is closely associated with this revival.

Nicolson argued that the new wonder in seventeenth-century literature engendered a conflict between an "aesthetics of content," with its preference for the bounded density of the microcosm, and an "aesthetics of aspiration," with its enthusiastic strife toward uncircumscribed immensity. Milton was the pivotal figure, exemplifying both attitudes. Nicolson isolated the "two persistent aspects of Milton's personality, one satisfied with proportion and limitation, the other revelling in the luxuriant and the unrestrained."[15] Even when Milton revels in imagination's mastery of the world, he does so in a manner fundamentally alien to the tradition he draws from. Sir Thomas Browne captures the attitude characteristic of literary purveyors of wonder, whether in the tiny or the great, as he guiltlessly indulges in his favorite subject: "The world that I regard is my selfe, it is the microcosme of mine own frame, that I cast mine eye on; for the other, I use it but like my Globe, and turne it round sometimes for my recreation."[16] Easiness, the promise of inexhaustible wonders ready-to-hand, is the regnant tone of this tradition—the "easie Philosopher" of "Upon Appleton House" for whom the woods, like everything else, are an occasion to translate the world's wit into wit's worlds.[17] But Milton sang, in Marvell's phrase, with *gravity* and ease. His play with the all and the one has little of the casual impudence that makes the earth into a globe idly spun. We find extravagance in this poet, even nonchalance, but whatever is "easy" in his unpremeditated verse seems crossed by images of effort and exertion.

On one end the invocation to light is bounded by the last segment of Satan's journey. He makes a splendid figure in departing from the court of Chaos and Old Night:

He ceas'd; and *Satan* stay'd not to reply,
But glad that now his Sea should find a shore,

With fresh alacrity and force renew'd
Springs upward like a Pyramid of fire
Into the wild expanse, and through the shock
Of fighting Elements, on all sides round
Environ'd wins his way. (2.1010–1016)

Assuming that his wings are spread open, then rapidly closed, the apex of this "Pyramid of fire" points downward. It is possible that Milton remembers the diagrams of Robert Fludd, who often represents the designs uniting macrocosm and human microcosm by triangles, sometimes triangles of fire, as in his diagram of the relationship between sensuality and intellect.[18] Here we see two interlocking triangles superimposed on the trunk of a man. One apex is located at the head, the other (the triangle of fire) has its apex in the penis; their legs intersect in the sphere of the heart, where reason and passion blend. The shape made by Satan springing upward answers in Fludd to the distribution of intellectual energy from its zero-point of sexual desire to its uncontested reign in pure idea. Perhaps Milton knew without benefit of romantic commentary that his apostate angel embodies an "alacrity and force" indistinguishable from the soul's vitality.

This is speculative. The pyramid of fire bears another and I would say *certain* allusion, ultimately consonant with the fiery triangle of Fludd. Satan is leaving "The Womb of nature and perhaps her Grave" (911), where "embryon Atoms" (900) fray each other in perpetual strife. In this context Milton would surely have recalled the *locus classicus* of atomic energy in the *Timaeus*. Like the Miltonic God, the Platonic Demiurge shapes preexistent matter, and immediately after describing the chaotic states of this substance Plato begins expounding his theory of the ultimate constituents of created matter. Each of the four elements has a shape constructed by the manipulation of equilateral triangles: the shape of fire's atom is, precisely as in Milton's simile, the tetrahedron or pyramid, one of the precedents that led Christian physicists to analyze light and vision by means of triangular structures.[19] Patterns suggest themselves. The atom of fire probably has its place in the profuse and interlocking schemes of zodaical, vegetable, and mineral symbolism in *Paradise Lost*. But the Milton who reveled in the unrestrained also had a hand in this simile, whose pith depends on the ideology of Satan's insurrection:

who saw
When this creation was? remember'st thou
Thy making, while the Maker gave thee being?

> We know no time when we were not as now;
> Know none before us, self-begot, self-rais'd
> By our own quick'ning power, when fatal course
> Had circl'd his full Orb, the birth mature
> Of this our native Heav'n, Ethereal Sons. (5.856–863)

Launching himself from Chaos, self-raised by his own quickening power, Satan is nothing less than his own idea of himself. Light, one would suppose, has served as the metaphor for so many absolutes in part because, as fire or as the sun, light appears to be its own origin. "He ceas'd"—Chaos ceased—and Satan renews his journey in an image of self-creation, tracing the shape of hell's element.

Defying Chaos, this journey culminates in the first sight of heaven and of our universe. Within ken of the "sacred influence / Of light," between chaos and cosmos, Satan glides on the air

> at leisure to behold
> Far off th' Empyreal Heav'n, extended wide
> In circuit, undetermin'd square or round,
> With Opal Tow'rs and Battlements adorn'd
> Of living Sapphire, once his native Seat;
> And fast by hanging in a golden Chain
> This pendant world, in bigness as a Star
> Of smallest Magnitude close by the Moon.
> Thither full fraught with mischievous revenge,
> Accurst, and in a cursed hour he hies. (1046–1055)

The eighteenth-century exclamations of the Richardsons seem wonderfully apt:

> What a vast imagination! What an idea of distance, the distance from Hell to where Satan now is! He is as it were at his journey's end, and yet so remote as that the new creation,that immense heavens wherein are placed the fixed stars—this vast globe, to which our earth is but as a point, an atom—appears but as the smallest star.[20]

Adam can imagine "this Earth a spot, a grain, / An Atom, with the Firmament compar'd" (8.17–18)—a comparison that comes in its fashion to all who behold the nighttime sky. But Milton has wrought a marvel from this humbling proportion, showing us a star of smallest magnitude, an undifferentiated point of light, that eliminates alto-

gether the difference between earth and firmament. Plotinus opposed the definition of beauty as a harmony among parts by reminding us of "a single star at night," a partless, unanalyzable whole.[21] Beautiful in both ways, Milton's smallest star contains all stars, places, and points, each and every part of our world. Satan springing upward as an atom of fire to view at his leisure the universe squeezed into an atom of light! The passage puts one in mind of Bruno's reply to those cramped minds who suppose that we cannot imagine infinity: closer students of imagination realize that, always conceiving a beyond by virtue of which, and only by virtue of which, the finite is finite, we cannot imagine finitude.[22] As the devil abandons this tremendous vista, Milton leaves Satan to renew his own journey in a sudden blaze of light, fire's sublimation.

On the other side of the invocation Milton enters heaven in a passage that modulates from the vision of God to a participation in God's vision:

> Now had th'Almighty Father from above,
> From the pure Empyrean where he sits
> High Thron'd above all highth, bent down his eye,
> His own works and their works at once to view:
> About him all the Sanctities of Heaven
> Stood thick as Stars, and from his sight receiv'd
> Beatitude past utterance; on his right
> The radiant image of his Glory sat,
> His only Son; on Earth he first beheld
> Our two first Parents, yet the only two
> Of mankind, in the happy Garden plac't,
> Reaping immortal fruits of joy and love,
> Uninterrupted joy, unrivall'd love
> In blissful solitude; he then survey'd
> Hell and the Gulf between, and *Satan* there
> Coasting the wall of Heav'n on this side Night
> In the dun Air sublime, and ready now
> To stoop with wearied wings, and willing feet
> On the bare outside of this World, that seem'd
> Firm land imbosom'd without Firmament,
> Uncertain which, in Ocean or in Air.
> Him God beholding from his prospect high,
> Wherein past, present, future he beholds,
> Thus to his only Son foreseeing spake. (3.56–79)

God sees everything "at once," the whole of space-time in a single leisurely glance composed of two moments of attention: first on earth, garden, and man; then on "Hell and the Gulf between, and *Satan*." Stupendous as it is, the look of Satan *positions* him. But the triangle implied by the two foci of God's surveying glance will not locate him, for he looks from "above all highth," encompassing a totality whose coordinates cannot, therefore, determine him. Rosalie Colie spoke of the diminishment of Satan in the enlarged perspective of God.[23] It seems to me, however, that Milton is careful not to specify whether God in bending "down his eye" sees his objects small or large. His vantage does not have a point. We also sense that the alternatives of human focus—vista or detail, whole or part—do not apply to his vision, which is able to grasp precisely what man and Satan are doing, while simultaneously retaining the horizon of context.

Surveying space from a height outside of space, foreseeing time from a now outside of time, this vision dramatizes without recourse to theological vocabulary one of the ways in which God bridges the ontological gulf between himself and his Creation. Phenomenologists of perception have noted that, while the other senses require some friction between the object and ourselves, for the viewer there can be a neutralization of this engagement. In order for us to hear, the object must *do* something, and in the modes of the other senses the subject generally appropriates the object in a relationship of practical involvement. But the seen object "lets me be as I let it be."[24] When I see a deer standing in the distance, its being is revealed to me without being disturbed by me. This is one reason why vision has been the traditional theological metaphor for the presence of God in the world he has made. The creatures in the foreseeing gaze of the God of *Paradise Lost*, even Satan, are left free to execute their own designs.

Having seen that it is not good, God turns his eye from the world to announce his awareness of the fall and to proclaim that his gift of freedom is irrevocable. In the vision that precedes his theodicial explication, we have been presented with the activity of this letting be: foreseeing is the mode of foreknowing, presence without intervention. The gaze of Satan astonishes us in the object seen, whereas the gaze of God astonishes us primarily in the way he sees; yet this seeing casts us back with new understanding to the eyes that behold our universe as a smallest star. Satan's glance at the apple of the world is prelude to an imposition. He is sea in search of a shore, a breaking wave that will still be dashing itself against the Christ of *Paradise Regained*. He will traverse the distance between himself and the object seen, entering it.

When he emerges from the universe, he will discover that a bridge has been built "following the track / Of *Satan*, to the selsame place where hee / First lighted from his Wing, and landed safe / From out of *Chaos* to the outside bare / Of this round World" (10.314–318). This bridge materializes, as it were, his line of sight at the end of Book 2—indeed, all the sightings that guided his flight from hell. To tempt is etymologically to taste. Seeing that ends in taste, that does not preserve the otherness of the object but incorporates the object into the being of the subject, will make the world into the spoils of Death.

Between these two visions we have an invocation to light by a blind poet of the seventeenth century who wanders through time and space, containing in his apparent finitude the vastly imaginative reductions of all to one, history to now, on either side of him. Where does he belong in the history of the aesthetic reorientation that fully acknowledged his powers? The sublime of Schopenhauer provides a useful description of one aspect of Milton's triptych—his exaltation, not just in the grandeur of Creation, but in the grandeur of its *createdness* within his poetry. For this philosopher the experience of the sublime separates in a single moment the two poles of our nature. The sublime presence dwarfs or endangers man as a creature of will, bound to a fragile body and engaged in creaturely pursuits; it overwhelms our narcissism by manifesting "the insignificance and dependence of ourselves as individuals."[25] Pleasure arises from our capacity to detach ourselves from this tense apprehension, achieving before the work of art or nature "the immediate consciousness that all these worlds exist only in our representation": "The vastness of the world, which previously disturbed our peace of mind, now rests within us; our dependence on it is now annulled by its dependence on us."[26] Schopenhauer's sublime is not a reconstituted narcissism. It is the experience, simultaneously, of omnipotence and its failure—the containment of anxiety.

Milton produces the baroque extreme of this effect by revealing that the creator of these visions of an enormous outside sees only a dark inside. His visual deprivation invests the world-encompassing sights of the poet with an element of personal exaltation; able to compress our "Orbicular world" (10.381) into a star, he seems almost contemptuously unconfined by the opacity of his dry "Orbs." If I am right in supposing that it reveals to us the experiential origins of the sublime glance at the end of Book 2, a biographical detail allows us to take the measure of his triumph. Milton told Philaris that "upon the eyes turning" the mist surrounding him admitted "a minute quantity of light

as if through a crack" (*CP* IV, 87). But the mind is its own place. There was, somewhere in the reaches of imagination, a perspective from which that chink of light was the entire universe, and Milton found it.

The wonder of vast imagination remains secondary, however, to the sublime encounter that takes place in the invocation itself. Here Milton effects a transition, within his poem and within himself, from the seeing of Satan to the seeing of God. The first has been accomplished. Now the blind poet comes before a light necessary for the continuation of his work. His address subverts the element of proud independence in his poetic vision by granting, though in no simple way, his dependency on this light. The vitality of poetic will precedes the power of the representation, and for Milton the transaction by which representation comes about exceeds in sublimity, is prior in every way, to the representation achieved. The light he addresses and requires is grander than either of the flanking visions. Milton's is the profoundest sight, the black (w)hole enfolding the entire illumination of *Paradise Lost.*

First Light

Philosophy was born and nourished by poetry in the infancy of knowledge, and with it all those sciences it has guided toward perfection; we may thus expect them, on completion, to flow back like so many individual streams into the universal ocean of poetry from which they took their source.

—Schelling, *System of Transcendental Idealism*

Hail holy Light, offspring of Heav'n first-born,
Or of th'Eternal Coeternal beam
May I express thee unblam'd? since God is Light,
And never but in unapproached Light
Dwelt from eternity, dwelt then in thee,
Bright effluence of bright essence increate.
Or hears't thou rather pure Ethereal stream,
Whose Fountain who shall tell? before the Sun,
Before the Heavens thou wert, and at the voice
Of God, as with a Mantle didst invest
The rising world of waters dark and deep,
Won from the void and formless infinite.

Thee I revisit now with bolder wing,
Escap't the *Stygian* Pool, though long detain'd
In that obscure sojourn, while in my flight
Though utter and through middle darkness borne
With other notes than to th'*Orphean* Lyre
I sung of *Chaos* and *Eternal Night,*
Taught by the heav'nly Muse to venture down
The dark descent, and up to reascend,
Though hard and rare: thee I revisit safe,
And feel thy sovran vital Lamp. (3.1–22)

Milton traces the Christian-Platonic history of a sunbeam. Light appears first in the invisible world of heaven, flowing out ("effluence") from the "bright essence" of the deity, then "before the Sun" in the first fiat of the visible Creation, then finally, as the sustainer of visibility to all but the unrevisited blind, in the "sovran vital Lamp" of the sun. The downward course of light from the divine essence has met the upward journey of the poem, also joining three stations, from the "*Stygian* Pool" of hell to "*Chaos* and *Eternal Night*" to Satan's view of our pendant world. Milton's cosmic itinerary suggests that if the poet would make order out of chaos, he must first create Chaos, which even the light of Genesis required as a receptacle. That much he has done. Utter and middle darkness are within the compass of the work; Satan, the finest characterization of the poem, is brilliantly underway. His epic is itself a "rising world" now prepared for light.

To enter the psychodynamics of this creation we must come to grips with the challenging language of symbolic light and heavy abstraction in the opening eight lines. The constraining light of conscience in the Ludlow masque became the light of guidance and intuition for the transformed conscience of *Paradise Regained.* At the gate of this passage shines a greater and more mysterious light; before determining the psychological positions mobilized in the poet, we must allow his symbol to "Shine inward" (52). I will broach an interpretation of this first light, the light somehow contiguous with God, by meditating on the challenge posed. Perhaps the difficulty of everything lies ultimately in the simplicity of its being. But these opening lines activate ontology, surrounding us with the sense of poetry itself. Milton has found a territory that, shared with the theologian and the philosopher, belongs to the poet as his primordial ground.

This symbolism is unique in the invocations. Milton generally prefers anthropomorphic accommodations for the deity, and later in

Book 3 light will be contained within familial metaphors when the poet hymns the Godhead of an invisibly bright Father and a visibly bright Son. There is no mistaking the terrestriality of heaven and humanity of God in this poem. Its portrait of God is so defiantly anthropomorphic that Milton seems at times to be rebuking other Christian poets for their recourse to "purer" symbols, as if they found scandal or embarrassment in the central accommodations of scripture. God-as-father or God-as-king evoke a known schema by means of which the strangeness of "God" becomes as-if-familiar. By initiating religious emotions in relationships preshaped within our ordinary experience, and preserving for prayer and revelation the sender/receiver framework of human communication, these metaphors make the otherness of God accessible. They tell man where to begin, which can only be where he is.

The other invocations follow this path. In Book 1 (Heavenly Muse) and Book 7 (Urania) Milton addresses personifications well established in the history of his culture. He claims to know where the metaphor leaves off and things divine assert their namelessness: "The meaning, not the Name I call" (7.5). But in Book 3 personification appears nakedly in the very act of addressing a light and questioning this light concerning its lineage. God-as-father retains an irreducible strangeness *because* we know what a father is, and know how we are related to a father. But what is light? What schemata inform this symbolism? For the first and only time in his late poetry Milton proposes and explores a philosopher's metaphor for deity, one in which the symbol is nearly as elusive as the symbolized.

In the first speech Adam utters, when he is still a pagan, he asks the "fair Light" of the sun and "thou enlight'n'd Earth" to bring him knowledge of his creator (8.273–274). Prior to revelation, light is nature's accommodation for the Godhead. Milton's first invocation, reciting the history of inspiration from the mountains of Jehovah to the streams of Sion, tells how the argument of the poem reached its author. Now he appears to abandon his initial emphasis on a revelation mediated through the history of the "chosen Seed" (1.8). Opening his invocation to light with one of the handful of philosophically ambitious passages in the epic, Milton adopts a figure for things divine given to us outside of this history and outside of cultural categories such as father and son: every newborn opens its eyes on a world bathed in light. The poet would have understood that the natural revelation of light had been developed within the limits of unaided reason by classical thought. From the Gospel of John onward, that great

assimilation by which a small mystery cult became a world religion took place in the language of light; it is virtually axiomatic in the history of Christian theology that wherever the symbolism of light flourishes, the religion of Jerusalem is being informed with the metaphysics of Athens. The philosophical density of the opening lines repeats this confluence of traditions, yoking the light of revelation to the light of nature.

Light is the theological metaphor beloved of Christian Platonists especially. For what is light? Is light a thing? Ficino supposed that light, being self-identical and incapable of division, was "almost an embryonic consciousness."[27] Milton's Samson is similarly perplexed over the thingness of light:

> Since light so necessary is to life,
> And almost life itself, if it be true
> That light is in the Soul,
> She all in every part. (90–93)

The bond between mind and light was assumed even in the scientific tradition, where students of optics observed that because light alone of nature's works operates in straight lines and regular angles of refraction of reflection, light transports the ideality of geometry to the order of nature.[28] Does light have a body, then? Do we see light? It had long been understood that we do not—and this, I think, is the scheme evoked by the symbolism of the opening passage. Because the light addressed may be "coeternal" with God, it is before all created objects, distinct from them. At the same time, this light is distinct from its source. It is not the God of light, but the light of God, in which he dwells. Pure light, light between source and object: this is the unapproached light that, in nature as in heaven, no eye can see. Had Adam dallied with the sun he would have realized, as Milton did, that the sun we see is not light: "We cannot imagine light without some source of light, but we do not therefore think that a source of light is the same thing as light" (*CP* VI, 312). Although we see all things by virtue of light, light qua light extinguishes itself in performing this office.[29] Light "as with a Mantle didst invest / The rising world of waters dark and deep"—a mantle that eventually found its way into the new wardrobe of the Emperor, for the world is being clad in nakedness. *Lux* is an invisible idea of the surface of *lumen*, the one dispersed in the many. Its fate is the opposite of that of a sign: while our awareness of the fleshliness of signs is obliterated when the

intention of an utterance is fulfilled in meaning, the singular spirituality of light is obliterated in the fleshliness of the visible. Adam refers the multiplicity of the "enlight'n'd Earth" to the sun's "fair Light," which might be taken as a sensory emblem for the way in which all objects seen refer us to unseen and unimaginable light, the signified of "the Book of knowledge fair" (47). As the referent of the natural poetry of vision, light *is* in the soul. It falls between the sensible and the intelligible, where Plato located the event of beauty.[30]

Milton's lovely phrase for God, "bright essence," epitomizes the invasion of his religion by the light of pagan philosophy. The sometime antagonist has set his torch in the sanctuary: God is an *essence. Paradise Lost*, the epic of Genesis, is a supreme example of this same confluence of cultures in the realm of poetry. Rather than briefly evoking theology as another and different result of this confluence, Milton is reabsorbing religion and philosophy into the matrix of poetry—a poetry that seeks to be philosophical and religious, justifying the ways of God to man. The act of naming God, first of all, represents the fundamental poetry of religious devotion. Adam knows the name appropriate to every creature, but not the name of the creator:

> O by what Name, for thou above all these,
> Above mankind, or aught than mankind higher,
> Surpassest far my naming, how may I
> Adore thee, Author of this Universe. (8.357–360)

This lacuna introduces a dynamics of creative metaphor into the system of the first language, and in the next chapter we will see how the aspiration of words toward God mirrors the physical and spiritual ascent planned for prelapsarian man. Devotion is essentially poetry—hence the parallel between the unpremeditated verse of *Paradise Lost* and the "Unmeditated" prayers of Adam and Eve (5.149).

If naming God is the poem of the religious life, the name chosen in the invocation to Book 3, light, is the poem of the philosophical life. Critics who bring to the passage a metaphysics of light patched together from the history of philosophical theology give the impression that the habitual use of light in describing mind was either merely figurative or (in the case of Ramus, say, or the Cambridge Platonists) unfortunately literalist—in any case, that these figures could be avoided, that someone could translate these metaphors into a language of analytical purity or provide an adequate account of mind without them. Put out that light, however, and you cannot relume philosophy

as anything beyond symbolic logic; that too may require light, as Wittgenstein's early theory of the *Bild* or "Picture" suggests.[31] Derrida's superb essay on the "white mythology"—the hidden mythology—of metaphors in philosophical discourse, preeminently variations on the "heliotrope," argues that light chases away demons of illogic from the "clear and distinct ideas" of Descartes as well as the Good of Plato.[32] This old sun also rises on the phenomenology of Husserl, whose transcendental ego, unable to have itself as an object, can map its unity solely through the "beam" of intentionality cast on the multiplicate and shifting *cogitatum*. It shines on the thought of Heidegger, who interprets *aletheia* (truth) as "unconcealedness," a clearing demarcated by the "ontological difference" wherein, just as things seen conceal light, the profusion of unconcealed beings conceals Being.[33] With respect to its indispensability at least, the trope of light may be termed a philosophical archetype.

Milton, as he strives to name the divine through the language of light, is reclaiming the sense in which religion and philosophy are poetry, flowing back, in Schelling's words, "like so many individual streams into the universal ocean of poetry from which they took their source." Light must stand high among the thoughts "that voluntary move / Harmonious numbers" (37–38).

Aside from the noteworthy concession of "bright essence," philosophy is the partner denied its traditional contribution in Milton's recovery of the matrix of poetry. He avoids any hint of necessity in following the emanation of light from heaven to the eye of the flesh. Even in the crowded metaphysics of his beginning he projects an Old Testament sense of awe and deep restraint. "Hail holy Light": through apostrophe he comes into the presence of light, yet preserves intact the zone of its sacrality. This light is set apart from all other lights. It is hallowed, "unapproached." If the holy light is the unoriginated "coeternal beam" of the "eternal" God, then it has never, in the state addressed at first, been touched at all, inasmuch as touch implies a breach of separation. Language itself must touch this light befittingly, with an accuracy of reference indistinguishable from the righteousness of the speaker. In its initial purity between source and object light is already a provision; this "excessive light" supplies the dwelling of God, just as later it will clothe the dark waters and, as the sun, dispense warmth and vitality. Considering "a common nature that unites and tyes the scattered and divided individuals into one species," Thomas Browne discovered "the opinion of Plato, and . . . Hermeticall Philosophers" confirmed by the spread of holy light:

> However, I am sure there is a common Spirit that plays within us, yet makes no part of us, and that is the Spirit of God, the fire and scintillation of that noble and mighty Essence, which is the life and radicall heat of spirits, and those essences that know not the vertue of the Sunne, a fire quite contrary to the fire of Hell: This is that gentle heate that brooded on the waters, and in six dayes hatched the world; this is that irradiation that dispels the mists of Hell, the clouds of horrour, feare, sorrow, despaire; and preserves the region of the mind in serenity.[34]

The pneumatology of the epic, to be discussed in Chapter 5, may well have been influenced by Hermeticism, and may well be related to the light of the invocation. But "the fire and scintillation of that noble and mighty Essence" fails to gloss "Bright effluence of bright essence increate" to the extent that Milton, unlike Browne, cares above all about the createdness of his light. Is light the first "offspring of Heav'n"? Is light filial or copaternal? Milton is alive to the question of generation. The God who is light seems to inform the light addressed with his paternal power; Milton's favored accommodation of God-as-father dwells inside, so to speak, the philosophical metaphor of light—a containment to be turned inside out later in Book 3. In a manner most uncharacteristic of a tendentious poet who would "see and tell / Of things invisible to mortal sight" (54–55), Milton permits the light to keep the secret of its own beginnings: "Or hear'st thou rather pure Ethereal stream, / Whose Fountain who shall tell?" By raising the question, then turning from it as from an impious curiosity, Milton has stoked the fires of a long debate over the divine entity symbolized by his holy light.

The major positions on this question are all open to serious objections. Maurice Kelley, identifying the "holy Light" with the light of Genesis I.3, would have us believe that this passage is "simply an invocation to light in the physical sense, which is turned in the concluding lines into a plea for metaphysical light—for insight and inspiration."[35] But "physical sense" is treacherous, all the more so when thought "simply." This reading ignores the three-stage procession of the light. The "Heav'n" of the opening lines is an eternal realm where Milton tries out hypotheses about the relation between light and the God of light; with "before the Sun," he moves from the question of what is before light to what we know, from the account in Genesis, light to have been before; the "Celestial light" at the end of the invocation, requested by a man who lives in a world divided by the sun

into day and night, circles back to the "holy Light" in the "Heav'n" of the opening lines, which is outside the bounds of Creation proper—the light to be found in the presence of God, a place of "things invisible to mortal sight" where Milton intends, through visionary rapture, to bring his poem. William Hunter maintains that the light addressed at the beginning is, like all the figures invoked in *Paradise Lost*, the Son.[36] Yet this light is expressly differentiated from the Heavenly Muse of the first invocation, who "taught" the poet to master the infernal realms of the first two books. Hunter's thesis must also get around the alternatives of the opening lines, since the *Christian Doctrine* is emphatic on the createdness of the Son; he is the firstborn offspring of God without the softening of an "or." Must we suppose that Milton changed his mind on the heresy most thoroughly argued in his treatise, or else retreat to some version of Lewis' idea (the one fleshed out, for example, by Patrides) that Milton diluted or concealed his heterodoxy in order to appeal to the broad spectrum of Christian readers, naming these multitudes "fit audience . . . though few" with his customary ineptitude?[37] Or worse still, must we suppose that Milton wrote "who shall tell?" when he could have told in order to evade censorship—that his voice was not, after all, "unchang'd" by dangerous circumstances? This would seem to be a case where interpretation has Milton doing penance for his heresies at the price of his integrity. The third and oldest interpretation, revived periodically, equates the light addressed with the Holy Spirit.[38] This reading demands subtle dialectics or staunch faith to explain away Milton's flat statement in the *Christian Doctrine* that the Spirit, if there is a Spirit, "cannot be a God nor an object of invocation" (*CP* VI, 295).

The best from all three of these interpretations will remain, minus their problems, if we take the holy light to symbolize what becomes of orthodoxy's Holy Spirit after Milton has disassembled the triune Godhead and erected in its place his own dyad of uncreated Father and created Son.

Traditional analogies for the Trinity articulate the *homoousia* of theology into two entities joined by a relationship: the Holy Spirit. The metaphors of the light and the stream endorsed by the Council of Nicaea and adopted by Milton present the Spirit as the motion, separable only by artificial analysis, joining the fountain to the river or the source of light to the object illuminated; Spirit is the flow or energy betwixt source and end, potency and realization. In Augustine's psychologizing of the Trinity, Spirit is imaged in the loving will whose volitional consent drives the projects of the intellect

(Son), through their realization in action, back into the memory (Father), which has in turn supplied the intellect with its principles of understanding. Another familiar trope has the Father as lover, the Son as beloved, and the Spirit as the love between them, sometimes represented as a kiss in which the *spiritus* passes back and forth from mouth to mouth.[39] Breath is life, the constant exchange between the not-me of exteriority and our interior being; in the interstice between realms, neither this nor that, we find the third person of the Trinity, a bird in flight between here and there or a bird on the waters between Chaos and Creation. *Betweenness* permeates the conception of the Spirit.

Much of what Victor Turner has argued about the meaning of liminality in social process applies to the volatile role played by the Spirit in the history of Christianity.[40] During the liminal phase in rites of passage, between the "no longer" of one recognized position in the social structure and the "not yet" of another and different position, neophytes experience a direct and leveling *communitas* relatively free from the obligations of deference or responsibility prescribed in the structure itself. They are often exposed to a symbolism that elicits creative interpretations, thus complementing the inertia of social structures with a seminary of novel possibilities. It is at least a partial truth that the Passion served Christianity as a charter myth for its priestly class. Standing in the place of Christ—physically so, if there were a Crucifix on the altar behind him—a representative of the ecclesiastical hierarchy doled out the food of the Son. But the gifts of God brought by the Spirit were never so easily contained within a structural mediation. We find the Spirit associated with conversion, idiosyncratic vision, holy madness, and of course reform. It was in the name of a *communitas* of the Spirit that Luther and Calvin asserted the spiritual parity of priest and congregation.

As a theologian asking that his work be assessed "as God's spirit shall direct you" (*CP* VII, 124), Milton listed the many senses of "Spirit" in the Bible, interpreting the brooding Spirit of Genesis 1.2 as "the Son, through whom . . . the Father created all things," rather than merely "the power and virtue of God" (282). Yet a page later he summarized the foregoing interpretations as follows: "Unquestionably, all these passages and many others of the same type in the Old Testament, were understood to refer to the virtue and power of God the Father" (283). There is no contradiction, since the Miltonic Son is the created God who manifests the power and virtue of the Father from whose substance he has been made. For Milton, the space

between Father and Son occupied by the orthodox Spirit is a space first of creation, then of communication. The generative act by which the "unapproached" Father produced his visible manifestation in the Son is twice repeated in *Paradise Lost*, figured both times as a transfusion of light from prior to secondary, source to object:

> and on his Son with Rays direct
> Shone full; hee all his Father full exprest
> Ineffably into his face receiv'd,
> And thus the filial Godhead answering spake. (6.719–722)

> So spake the Father, and unfolding bright
> Toward the right hand his Glory, on the Son
> Blaz'd forth unclouded Deity; he full
> Resplendent all his Father manifest
> Express'd, and thus divinely answer'd mild. (10.63–67)

In Milton's theology, conventional metaphors for the Holy Spirit are truer than the Godhead they were meant to accommodate. The light between is not an entity, but a medium of power. The Son is the similitude of the source of this light, illuminated by this light, but at the beginning of Book 3 this light is conceived at a time when, "from Eternity," the Son is yet to be. Milton hails the divine prototype of a vitality felt in his art and his temperament alike.

"Truth is compar'd in Scripture to a streaming fountain," he wrote in *Areopagitica*: "if her waters flow not in a perpetuall progression, they sick'n into a muddy pool of conformity and tradition" (*CP* II, 543). Unlike Dante, this poet revels in acts of creation, the quickening of a "rising world" rather than the entropic arrest of a finished one. The pilgrim of the *Commedia* traverses, step by step, a clarified cosmos pivoting on a chained Lucifer, its liminality confined to the tidy dynamics of Purgatory, whereas the angels and devils of *Paradise Lost* go about their purposes in a space always being disturbed (Pandemonium erected, Hell explored and settled, Chaos clashing, Heaven's landscape torn by warfare, the world created, the bridge from hell to the world unfurling, the bent axis of the fallen earth, Adam and Eve walking through the definitive liminality of mankind, between Eden and the world, and all those similes making size itself dynamic). A golden chain attaches the universe to heaven, but in the center of this fixed expanse the happy couple celebrate "ceaseless change" (5.183). Whether angelic, demonic, or human, the figures of

the poem continually happen on the previously unknown, discovering. The means of locomotion through the unrolling horizons of this teeming space is flight, suspension between origin and end, and the master of this sublimity also flies, composing his poem "now." I remarked of *Paradise Regained* that Milton tries to represent the oncoming immediacy of sacred history before its sense has been decided by the traditions of men. Imaginatively, the internal drama of composition places the epic itself beyond interpretation. For how does one regard a work coming into existence now? We receive it rather than determine it, checking our urge to interpret as its possibilities become increasingly actual. Interpretation presupposes an acccomplished text. We will interpret *Paradise Lost* in the light of this necessary assumption. Yet any passage we fix upon, save the last, will forever proclaim that its author is in flight. In the light of Book 3, which becomes the light of poetic inspiration, everything is pushed back into possibility—a universal prolepsis, the promise of everything.

The evening and the morning made the day. Characteristically, Milton in his blindness misses "the sweet *approach* of Ev'n or Morn." He surprises the first light as pure indeterminate potency between creator and creature. It is God as he spills over, about to entify; in the language of the fourteenth-century nominalists, it would be termed the *potentia absoluta* prior to the *potentia ordinatus*.[41] God famously proclaims in another philosophical passage:

> Boundless the Deep, because I am who fill
> Infinitude, nor vacuous the space.
> Though I uncircumscrib'd myself retire,
> And put not forth my goodness, which is free
> To act or not, Necessity and Chance
> Approach not mee, and what I will is Fate. (7.168–173)

God associates his spatial infinity with his "free" or boundless will. The light of the invocation solves in metaphors the paradox of a deity both uncircumscribed and retired. As "offspring" or "beam" or "stream" light radiates outward from a "bright essence," rendering this essence endless *and* distinct. Did God create the medium or vehicle of his creative will? "Whose fountain who shall tell?" The open question of the createdness of light is not a theological uncertainty, but a theological proposition. Milton addresses the will of God "put forth" in a form, light, symbolizing creative intent. In escaping the disjunction of created/uncreated, this light eludes the categories that

necessarily define other creatures. It is neither God nor other than God—a penumbra of free and indefinite possibility around the circumference of the divine essence. The ipseity of clear circumscription in the Greek idea of essence has yielded to the limitless freedom of a volitional God.[42] He inspires not self-emptying contemplation but the flight of an artistic task: "to see *and* tell." His light could only be stilled in a symbol, since a pure stream of coming-into-being, named as a concept, would have already been given a being. Confining the light to an hypostasis, orthodoxy loses its spirit. Poetry alone can name the concept that is not a concept, surrounding the essence of God with a brightness that dispels the limitations of "essence," watchword of philosophical theology.

There is not a detail of *Paradise Lost*—not a river, a god, a simile, a word—that is not somehow the progeny of this light. Christ is first: "on thee / Imprest th' effulgence of his Glory abides" (3.387–388). At its first termination, when effluence becomes effulgence and beam becomes impression, the light is the "Divine Similitude" impressed on the Son. Released by the Son, this light will be angels ("Progeny of Light" at 5.600, "Sons of Light" at 5.160) and, thrown like a mantle over the waters, the cosmos we both are and inhabit. This open "I will" prior to "Fate" becomes, for its creatures, the object of ethics as well as worship, for the creative intent of God is known to its creations as law—the "Necessity" of existing in a certain way in a certain circumstance, and the "Fate" of having a creator who wills. Christ's pronouncement on glory in *Paradise Regained* contains the principle of all true action: "I seek not mine, but his / Who sent me, and thereby witness *whence I am*" (3.106–107). For Milton the full truth about our existence lies in two crucial addenda to the *cogito*. I think, therefore I am; I am, therefore I was created; I was created, therefore I am religious. Refusing these corollaries, Satan can only generate, in his theism of the self, a parody of being. And like all parodies, he would not exist were it not for the very same source he exists to deny. "But were I myself the author of my being," Descartes said, sobering himself after the hyperbolic doubt of the *cogito*, "I should doubt nothing and should desire nothing, and finally no perfection would be lacking to me; for I should have bestowed on myself every perfection of which I possessed any idea and should thus be God."[43] This is the core of self-defeat in Satan. The absolute denial of God issues inevitably in the absolute imitation of God, so that even as he creates himself, Satan creates himself in the image of God—which is a parody of how God created him. He too bears witness whence he is.

The orthodox trinity invites diagram and analogy, as in the triple

verbs and three conceits of Donne's "Batter my heart, three-personed God." The Miltonic Godhead coheres, yet not by the inherent design of shared essence, and not by shared substance alone. His Godhead is better captured in drama than in diagram. We might say that the contemplative *epistrophe* or "turning back" of the Plotinian Nous in the second "moment" of its generation, constituting itself as a unity in the image of the One, has a Miltonic counterpart in the obedience of the created Son to the paternal will he thereafter expresses.[44] But here contemplation and self-knowledge are absorbed in a fundamentally ethical gesture: what goes out as paternal will returns as free submission in Milton's voluntarist theogony. Because the Son might not have been, he also owes for the gift of life. Stella Revard has written perceptively on this theme, arguing that the Miltonic Christ is a moral exemplum in his heavenly as well as his earthly existence.[45] Her insistence that we "listen to the subtle, yet still lovely, music of Milton's Trinity" is particularly welcome in an area of Milton studies given over, in the main, to taxonomy (Arian? Semi-Arian? Subordinationalist? Dogma or Drama?).

The distinct essences of Father and Son, no longer merged in the Holy Spirit, achieve correlation through a union of wills whose earthly manifestation we studied in the previous chapter. God's desire becomes Christ's law. If it is not to be felt as law, but embraced as shared desire, there must be a complete "subordination"—an emptying, almost, of the Son, that he may be filled with his creator's intent:

> this I my Glory account,
> My exaltation, and my whole delight,
> That thou in me well pleas'd, declar'st thy will
> Fulfill'd, which to fulfill is all my bliss. (6.726–729)

In this respect, the usual descriptions of the Incarnation—the "putting off" of Christ's heavenly power, for example—apply to this Christ from the beginning. Without developing the idea as yet, we can see that Milton's "Begotten Son" (3.384) is the divine model for all the strengths that come from weakness, including the poem created from the selflessness of inspiration. He is sent forth to create the world dressed in the armor of the Father, riding the chariot of the Father, accompanied by the Father's "overshadowing Spirit and might" (7.165), bearing tools forged by the Father—"all his Father in him shone" (7.196). Swearing never to "disjoin" the praise of the Son from "thy Father's praise," Milton invents a doxology of subordination:

Thee next they sang of all Creation first,
Begotten Son, Divine Similitude,
In whose conspicuous count'nance, without cloud
Made visible, th'Almighty Father shines,
Whom else no Creature can behold; on thee
Impresst th'effulgence of his Glory abides,
Tranfus'd on thee his ample Spirit rests.
Hee Heav'n of Heavens and all the Powers therein
By thee created, and by thee threw down
Th'aspiring Dominations: thou that day
Thy Father's dreadful Thunder didst not spare. (3.383–393)

For nearly ten lines Christ is praised without once appearing as the subject of a verb. On or by the Son, it is the Father who acts; his invisible light shines on the countenance of the Son. When, "thou" being allowed to supplant "thee," the Son acts, his strength derives from "Thy Father's dreadful Thunder." Milton creates the strange effect of praise withheld in the giving, a foreground constantly consumed in a background. Yet this undisjoined praise expresses the complex unity of his double or disjoined Godhead; Christ, having willed his submission, joining what the Father put asunder, is authentically the cause of his praise. He is something "more" than an ectype: "By Merit more than Birthright Son of God" (3.309). In his cosmic kingship Christ will "assume" his "Merits" (318–319). If the Father has glory, the Son alone has merit, which results from voluntary obedience and results in—power.

Before the otherness of Christ, cosmos, or creature, before the order of obedience or disobedience instituted in this otherness, before the Father was a God of law, he was a God of poetry. Milton, ready at last, invokes the pure power of making. In a gesture of sublimity equal to any in the history of our poetry, he summons from God the divine analogue of the beforeness that he has exhibited throughout his entire poetic career. For as the medium of creative will, the light is also time, traditionally represented as the spilling over into motion of eternity's *nunc stans*—more accurately, it is the not-yet of time, time's arrow threaded in the bent bow of its origin. "Now had th'Almighty Father from above": after the invocation is complete, its light requested and presumably, since the poem continues, granted, the first word of the first line is the strong trochee of an eternal Now whose coiled energy is ejected into the rhythm of the poem, reinitiating both the "now" of narrative and the "now" of composition. "Milton's meaning is re-

markably freed of the burden of anteriority," Harold Bloom has written, "but only because Milton himself is already one with the future, which he introjects."[46] He is thinking of the many strategies by which the poet transforms his belatedness in the poetic tradition into an imaginative earliness. The wish, however, is grander still. As the holy light is absorbed and distributed in the body of the epic, Milton introjects, not just a specific tradition, but futurity itself. "Necessity and Chance" do not approach this unapproached light. It is not beholden to contingencies. It will make actual whatever pleases God, and God's desire lacks want. Considered in the opening lines without regard to any creation except its own, *light is the energy of wish-fulfillment unconditioned by a prior moment of satisfaction.*

This is the power of which poets dream. If Milton has this light within him as he illustrates and justifies its ways in a particular future, he would have access to the indefinite potential of an absolute origin, freed of the burden of anteriority in a radical sense: his would be the creative will that founds rather than repeats. "May I *express* thee unblam'd?" May I wield this power to manifest the unseen, speak the unheard-of?

First Sight: Sexuality, Ambition, and Revelation

If my interpretation of the light convinces, we must revise our earlier assertion that the generation of the Son is the first event in the chronology of the epic. Milton has looked into the generation of the power that generated Christ. At the risk of tiring my reader with a catalogue of the obvious, I wish to reacquaint us briefly with the large impulse our poet indulged in Book 3 when probing the origin of origins.

From start to finish, Milton is the poet of the *arche*. Many Renaissance voices equated stature with earliness, reaching for the sources of value and achievement, but Milton turned back with distinctive intensity. The Old Testament supplied the true cultural forms corrupted by Athens and Rome; primitive Christianity supplied the true ecclesiastical forms defaced by the Roman Church. He wrote the early history of his nation. As a poet he delighted in the topos of bestowing lineage on mythical figures. In "L'Allegro" he dismisses the usual genealogy of Mirth in favor of a "sager" one of his own invention—an urge to reconceive that he later imposed on the Christian Godhead. "Lycidas" begins at the "sacred well," then twice returns, coaxing from its fountainhead the genre whose limits it tests and ex-

tends. Even the Miltonic style is marked by its exploitation of etymology. His very language perpetuates a linguistic past, giving new vigor to the genesis of words in a present whose corruption may be measured in part by its forgetful drift away from these earlier and revelatory definitions.

Paradise Lost is the poem of genesis par excellence. Almost all its events are the first such events, patterns into which subsequent history will fall. Milton designs the poem so as to catch up his readers in an excited search for priority. Since epics begin *in medias res*, we know when we awaken into the argument with Satan's rousing of Beelzebub that this adventuresome poet, shunning the middle flight, intends to advance his plot further back into the recesses of extrabiblical history; the fulfillment of this anticipation arrives with a formal closure, like a rhyme in action, when Satan—"he of the first, / If not the first Arch-Angel" (5.659–660)—awakens Beelzebub to begin his rebellion (5.671–672). Sin, Adam, Eve, and the serpent recount the story of their birth. Uriel tells Satan, Raphael tells Adam, and Urania tells Milton what transpired on the "Birthday of heaven and earth" (7.256). The personifications addressed in the first three invocations have all enjoyed proximity to the source; the last invocation, a soliloquy, alludes to a time when "first this Subject for Heroic Song / Pleas'd me long choosing, and beginning late" (9.25–26)—the first inkling, the genesis of *Paradise Lost* itself. Like other professionals of the time, Renaissance physicians played the game of discovering the first practitioner of their trade. God, they were happy to realize, invented the arts of surgery and anesthesiology by putting Adam to sleep in order to extract a rib without offending his patient.[47] But Milton's imperturbable Adam witnesses the operation:

> Mine eyes he clos'd, but op'n left the Cell
> Of Fancy my internal sight, by which
> Abstract as in a trance methought I saw,
> Though sleeping, where I lay, and saw the shape
> Still glorious before whom awake I stood;
> Who stooping op'n'd my left side, and took
> From thence a Rib, with cordial spirits warm,
> And Life-blood streaming fresh; wide was the wound,
> But suddenly with flesh fill'd up and heal'd:
> The Rib he form'd and fashion'd with his hands;
> Under his forming hands a Creature grew. (8.460–470)

One senses that God's promise to give Adam "Thy wish, exactly to thy hearts desire" would have been nicked by a small but irrepressible

disappointment had not Adam's internal sight been as open as his left side.

When Milton calls on the light in his most comprehensive and personal attempt to see the beginning, he acknowledges the possibility of a hostile component in his wish to scrutinize first things. He may not be "unblam'd." The chance of inciting divine wrath seems to be carried over into the first and famously irritable speech of the Father, whose defensive questions anticipate attack:

> For Man will heark'n to his glozing lies,
> And easily transgress the sole Command,
> Sole pledge of his obedience: So will fall
> Hee and his faithless Progeny: whose fault?
> Whose but his own? ingrate, he had of mee
> All he could have. (3.93–98)

For the sake of opened eyes, among other things, man fell, and the "Ingrate" sin of discontent with "All he could have" cannot be entirely missing from the blind poet who seeks an extraordinary favor. Seeing does not always let be: it aspires. In his brilliant pages on the sensuality of Milton's verse, Watkins noted the phallic aggression implied in the realized physicality with which eyes are *rolled*, *bent*, *cast*, *thrown*, *darted*, and *sharpened*.[48] Vision's orgasm, we might say, is the bursting into flame or illumination—the sudden blaze of "Lycidas," the Promethean fire of "Ad Patrem," the renewed force of Satan as a pyramid of fire, the intensified brightness of Christ before his missions, the Phoenix of Samson's violent triumph—that provides Milton's choice trope for ambitious action. One chief embodiment of this metaphor is the "potent" sun (4.673) of *Paradise Lost*, "like the God / Of this new World" (4.33–34). It is the ambition of God as fait accompli, and once the poem reaches our pendant world, we are continually apprised of its whereabouts. What reader of *Paradise Lost* could assent on his pulses to Raphael's dry dictum that "Bright infers not Excellence" (8.91)?

"We boast our light; but if we look not wisely on the Sun itself, it smites us into darkness" (*CP* II, 550). Restrained sight, associated with the distance between creator and created, appears on various levels of Milton's universe. Angels fold their heads into their wings when the unapproachable Father shows his beams through a cloud, which cannot be said to signify worship or humility, caused as it is by the sheer dazzle of his "excessive bright" (3.380). The stars of our world "keep distance due" from the "Lordly eye" of the sun (3.578),

and as Satan says, reading the signature of the creator with a rebel's insinuation, "Hide thir diminisht heads," like the angels, in the daylight glory of the sun (4.35). This gesture becomes an expression of shame when fallen Adam desires to conceal himself from God or angel in woods "impenetrable / To Star or Sun-light" (9.1086–1087). As in *Comus*, the eye of conscience and ultimately of God is the prime instrument of judgment; the "Chariot of Paternal Deity" mounted by a warfaring Christ is "set with Eyes, with Eyes the Wheels / Of Beryl, and careering Fires between" (6.755–756). Must not seeing be the prime instrument of crime? Eve sees her fall in a dream, and at the moment of actualization her proleptic imagination leaps off from the platform of riveted vision: "Fixt on the Fruit she gaz'd, which to behold / Might tempt alone" (9.735–736).

Everything points to a guilty seeing, a *blamed* seeing to be uncovered somewhere in the archaeology of Milton's ambition. Like the symbolism of crime and judgment in *Paradise Lost*, the suggested link in Chapter 2 between the spending of his hoarded talent and exposure to a punitive gaze exemplifies Freud's "talionic law," an eye for an eye, governing the fantasy worlds of childhood. Bergler has maintained that in the creation of literature, an undertaking charged with the dynamics of narcissism, the scopophilia of the author is transformed into exhibitionism. Having imagined the hidden sins, solitary deeds, or private interiors of their characters (or themselves), writers deliver this privileged vision to their audience; he who saw is then beheld.[49] Freud accented this *verkehrung ins gegenteil* or "turning around upon the self" in a theory of fantasy based on elementary transformations of grammar. For example, the two postures of seeing and being seen represent two versions of the same libidinal assertion: since the subject in a voyeuristic fantasy becomes the object in an exhibitionistic fantasy, exhibitionism is scopophilia in the passive voice (*S.E.* XIV, 109–140; cf. XVII, 175–204). In Milton the desire to write poetry was attached genetically to an intrusive sexual looking, a gaze that would take in, possess, even steal, while the text he postponed was equated in his unconscious with the discovery or confession of this trespass. His ambition hardly swerved at all from its first public expression at the Vacation Exercise of Cambridge University in 1628:

Yet I had rather, if I were to choose,
Thy service in some graver subject use,
Such as may make thee search thy coffers round,
Before thou clothe my fancy in fit sound:
Such where the deep transported mind may soar

Above the wheeling poles, and at Heav'n's door
Look in, and see each blissful Deity . . . .
Then sing of secret things that came to pass
When Beldam Nature in her cradle was. (29–35; 45–46)

Almost forty years later the poet would also "see and tell / Of things invisible to mortal sight"—of heaven, creation, first things.

And now the fruit falls. It is not difficult to give a psychoanalytic name to those naked thoughts that knocked loudly to be born from the mind of a nineteen-year-old undergraduate and knock still in the invocation to light. *Paradise Lost* stages the primal scene over and over again.

Later in the invocation the poet will compare himself to "the wakeful Bird" that "Sings darkling, and in shadiest Covert hid.' The nightingale serenades Adam and Eve while they perform the mysterious rites of love in a "shadier Bower / More sacred and sequester'd" (4.705–706) than the bowers of classical myth:

it was a place
Chos'n by the sovran Planter, when he fram'd
All things to man's delightful use; the roof
Of thickest *covert* was *inwoven shade*. (4.690–693; my italics)

This is the place God set aside for the primal scene, and Milton, alone in shadiest covert, continues the unpremeditated song born there in paradise. *His poetic act transpires inside a blind consciousness equated figuratively with the very place our Grand Parents lay together in the beginning*. The fantasy theater of the primal scene: this is where Milton as a poet is at.

As I have intended to demonstrate in my partial catalogue of origins and fragmentary discussion of seeing, the epic affords its readers a prolonged observation of generative acts. We overhear the formulation of the great stratagems of heaven and hell, then witeness their execution. We stand watch as Adam and Eve revel in their inmost bower:

nor turn'd I ween
Adam from his fair Spouse, nor Eve the rites
Mysterious of connubial Love refus'd. (4.71–743)

The striving to see and the discovery of seeing's limits, perceptually and morally, intertwine throughout. God cannot be seen as an object, nor are the stars, so far from earthly sight, "to be scann'd" (8.74) Yet

"Say first," the epic proper begins, "for Heav'n *hides nothing* from thy view": through the assistance of the Heavenly Muse, Milton would enjoy in spiritual form the magnified vision of the telescope that fascinates him in his similes; he compares his scrutiny of Satan to Galileo observing the surfaces of the sidereal king and queen, the male sun (3.588–590) and female moon (1.286–291). Urania, a muse baptized by previous Christian poets, has not lost her old association with astronomy in the work of this poet, who in his "ever-during dark" views celestial bodies. Seeing is the scope of his desire.

Identifying the centrality of the primal scene in Milton's imagination would allow one, in an extended analysis, to place apparently unrelated or unimportantly related features of the epic in a new structural alignment. But what sort of procedure is this? We generally think of explanation as a process by which the unknown we question is referred to the known we understand, and in this case, given the ordinary understanding of the primal scene, we might better proceed as if the epic interpreted psychoanalysis. It would be a barren exercise to forfeit the meaning of this poem to a falsely completed and reified conception of the primal scene—the Big Scene, as Nabokov called it, in the Freudian melodrama. Such a conception does in fact predominate in the psychoanalytic literature. Milton is in my opinion *the* poet of this fantasy. We must try to learn from him the poetry of the primal scene rather than teach the poet our commonplaces.

The invocation addresses the absolute *complicatio* unfolded in all creation, that it might be unfolded anew in the creation of *Paradise Lost*. Does the primal scene motivate this address? Does the invocation express yearning for the primal scene, acknowledging the forces that led to its repression in the complementary signs of the invisibility of pure light and the blindness of the poet? Positive answers will usually assume that the fantasy, memory, fantasized memory, or remembered fantasy of parental intercourse was exiled to the unconscious, persisted there like a material object lodged in a niche, and managed to achieve a monument to itself, an external reiteration of its internal immortality, in the poem we read. The fantasy yearned for the work; the ego of the poet did no more than consent to desire his desire: "the artist will yield positive content to that which, in the unconscious, is mere dispossession."[50]

This notion of art bespeaks a narrow formalism concerning both the psyche and psychoanalysis. Against this view, I contend that the act of creating this passage, in its fullness, is the *completion* of a psycho-

genetic sequence initiated by the primal scene, which is not a psychoanalytic concept to be limited to distinct symptoms in a clinical setting, but a fundamental event that gathers together the great themes of psychic development. Stating his monism in the sphere of metaphysics, Milton argued that the superior principle always contains the inferior one; the formal cause always coexists with the material cause.[51] In the invocation, similarly, the telos of a life, the moment toward which Milton has lived, reveals its necessary continuity with a specific sensory image. It is an epiphany of the imagination, bringing all of desire to light.

In Freud the primal scene is intimately connected with the *nachträglichkeit* or "deferred action." This term denotes a special relation between two events, sometimes widely separated.[52] Let us say that a boy in his second year hears or sees or (the room is dark) thinks he might have seen his parents pressed together, doing something. Perhaps none of the three makes anything much of it; perhaps the child is discovered and humiliated, gruffly sent away. At a later time, typically under the pressure of the oedipal dilemma, a second event triggers a "revision" of the first as a scene of intolerable anxiety—and the primal scene is repressed. It is as if a traumatic future enters the world of the child, but defers announcing itself until the child is ready all at once to experience and disavow this entrance. Todorov has noticed the arresting likeness between the nachträglichkeit and biblical typology: the first scene awaits its traumatic elucidation in the revised scene, just as the dark figures of the Old Testament are plucked out of time or "saved" for an unveiling in the New Testament that both manifests and overcomes them.[53] This parallel reminds us that the action deferred in the nachträglichkeit is an advent of meaning. We have some warrant to expect that the evolution of significance produced by the deferral of the primal scene may not be concluded with the particular form of the scene elected for repression—with the letter, as it were, of psychic history. If repression immortalizes the first scene, it also makes room for metaphorical substitution. An element of invariance will join the substitute to the repressed original, but the substitute will necessarily possess an element of novelty as well. We could speak, then, of a "deferred sublimation" implicit in the acquisition of the primal scene.

The scenario of the nachträglichkeit must not mislead us into thinking of two events only. The first scene, if not imaginary to begin with, is revised and varied in the imagination of the boy, and this work of revisionary interpretation coincides with the vector of maturation. A past flows into the scene. If we suppose that the boy acquires

the primal scene when he is mourning a damaged oneness with his mother, we can see immediately that the primal scene is the genetic successor to the image of the absent object about which the differentiation of the ego occurs. The image of a nourishing union with the mother, denying her absence, is replaced by a new and triangulated image of sexual union with the mother, libidinizing her absence. This new image concedes and fills in the absence of the mother; when she is not in the presence of the boy, she is absent-with-another. The physical union at the center of the primal scene may permit the boy to defer his grief over the impaired narcissistic union, which helps us to understand the fusion of separation anxiety and castration anxiety discovered in *Comus*. The primal scene is the site at which the issues of ego differentiation become translated into the issues of the oedipus complex. The entire complex unfolds within its structure. Literally and conceptually, the primal scene *includes* the complex.

Knowledge awaits the child in the beckoning dimensions of the image. Imagining himself in the position occupied by his father, the boy solidifies his oedipal love. He articulates the (at first) confused representation of sexual union. He locates the phallus in the scene, finds the vagina there, and interprets the meaning of the difference. But the scene is inescapably triangular. ("Three," a friend remarked on hearing my ideas about the primal scene, "is a complicated number: *there is always someone watching*.") As the child actively projects himself into the role of the father, the father occupies the slot he has vacated—watcher, discoverer, and in this position, punisher. This is one reason why the role of seer might become, as it was for Milton, so near in association to being caught and condemned. In evading his gaze, the child may also place himself in the position of the mother, who then becomes the observing rival: all of the currents of the complex are realized in imaginative permutations of this scene in psychic reality. Inhabited from its three-cornered perspectives, the scene becomes a space. It is the first *space of sexual desire*, and as such, destined for repression.

The meaning deferred in the original scene is the mourned passing of the narcissistic ego translated into the love, jealousy, and terror of the oedipus complex. Trauma comes at last as an announcement of the child's readiness for oedipal resolution—the moment he knows that he cannot survive in this space, and simultaneously, because he inhabits it, that there is no way out of this space. Thus entrapped, the first ego of sexual desire commits the "suicide" of repression. But although the contents of the scene undergo repression, its structures remain in the

ego. Insofar as the repression of the space of sexual desire either presupposes or engenders the superego, the ego under the superego is the successor to the imagining son who actively injected himself into the scene, situated hereafter as an agent observed. A new intrapsychic structure captures and fixes the guilty son. Yet, just as the severed attachment to the mother encourages the child in the realistic hope for her restitution in another, so the banishment of the primal scene, with its terrifying "no exit," makes possible its sublimation in a *space of ambition*. The evidence of his career suggests that the ambition of Milton retained the original positioning of the scene, opposite to the correlation of ego and superego: the poet who would look in is the successor to the imagining son who strove to see.

Psychoanalysts regularly link the primal scene with curiosity, which Christianity, long before, had regularly linked with the fall. Milton tells us the myth of the origins of evil. Adam and Eve awaken to find their "Eyes how op'n'd," their self-regard spoiled in the presence of "this new comer, Shame," who sits on their genitals, reproaching sight (9.1097–1098). To tell this story the poet must be able to see innocence innocently. One of Satan's tragedies is his inability to recall his innocence without these memories becoming infected by his crimes: when he imagines himself in heaven, he invariably begins to imagine the recapitulation of his rebellion (4.93–102). Milton first shows us our unfallen paradise and the pure embrace of our naked parents through the eyes of Satan so that we may look beyond him at what we, too, have lost and must thereafter covet. Past his gaze, in the exuberance of the poetry, a more primordial sight discovers paradise. When we only look at, not through Satan, we find a seeing disturbed by the desire not to see. He is torn between the devouring gaze of "unespi'd" voyeurism (4.399) and the pain of his exclusion from the joy beheld—a hateful seige of contraries magnificently rendered in the flesh when he both turns aside from, yet eyes askance, the bodies of man and woman "Imparadis't in one another's arms / The happier *Eden*" (4.506–507). But Milton, hailing the light by virtue of which everything is and everything is seen, would be "unblam'd," purged of the guilt of looking. At the first sight of the human body in the poem, he proclaims that the "show" of shame hides true purity, just as clothes hide our genitals:

> Nor those mysterious parts were then conceal'd,
> Then was not guilty shame: dishonest shame
> Of Nature's works, honor dishonorable,

Sin-bred, how have ye troubl'd all mankind
With shows instead, mere shows of seeming pure,
And banisht from man's life his happiest life,
Simplicity and spotless innocence.
So pass'd they naked on, nor shunn'd the sight
Of God or Angel, for they thought no ill. (4.312–320)

The untroubled vision that purifies the world was there in the beginning, before the triggering of the nachträglichkeit. How can we, as sin-bred creatures accused from within, recover the lost innocence of the primal scene?

The poet is our guide. Milton shows through symbols that the way back does not lie in an undoing of the repression. Satan stands between us and the vision of innocence in order to assert the justice of the repression. The way back to the goodness of the light lies in blindness, in this respect a symbol for the prototypical hysteria of repression: "I do not see." In the *Second Defense* Milton insisted that blind men are "almost sacred, nor do these shadows around us seem to have been created so much by the dullness of our eyes as by the shade of angels' wings"(*CP* IV, 590). This metaphor recurs in *Paradise Lost* as the forced blindness of the angels, compelled by the "excessive bright" of the Father to cover their eyes with their wings. The gestures connect guiltlessness with an inability to see—a celestial preenactment of the oedipal repression.

More than guilt is deferred in the primal scene. Through repression the child may acquire something akin to the "neutralization" of vision I spoke of in discussing the way God sees. "I do not see" may also come to mean "I let the marriage of my parents be; I will not intrude on the space they inhabit." The evolved significance of the primal scene lies in this direction.

Laplanche and Pontalis have noted that the three "primal fantasies" of classical psychoanalysis comprise a rudimentary etiology.[54] They are myths for a tribe of one: the seduction of a child by an adult concerns the origins of sexuality; castration the origins of death; the primal scene the origins of the subject. So considered, the image of parental intercourse possesses two contrary aspects. It foretells the end of narcissism in the knowledge of contingency and createdness: the imaginer is as nothing before the fully revealed meaning of his image—there in the image, he *was not*. This view of the primal scene belongs to the psychogenesis of the sublime. An awesome revelation of one's nothingness must somehow be countered. And if all one is

stems from a contingent event during which one was as nothing, nowhere to be found, has not the imaginer of this event triumphed over his initial nothingness in the act of producing and beholding its image? He who was not now sees all. The fantasy announces in complex tones the mystery of being. It points the way to gratitude and to debt. The full meaning of the primal scene is acquired *through* the repression of the scene as space of sexual desire. The father will no longer be seen. As the superego, he sees. *Ecce homo*! There in the light stands a son. He will model himself on his progenitors, undertaking a psychological continuation of the act that gave him being; this obedience to the superego recognizes, more than a merely cognitive knowledge could, the createdness of his ego. As Thomas Weiskel has also argued, submission to an internal father resolves our first encounter with the sublime.[55] The subject, having discovered his nothingness, steadies himself in acting as one created—as one who was once nothing.

Milton winds back the clock until there is only God and light and all things are as nothing: he hopes, in his blindness, to be revisited by this light. We are now able to appreciate again and more thoroughly why the complex is the precondition for his religious life. God dethrones the father as creator. From the viewpoint of the sacred, the first superego is itself a nachträglichkeit, deferring its revision in God. Religion expands into a sacred history the knowledge of createdness embedded in the primal scene. Transposed to religion, the primal scene becomes a *space of revelation*: "In the beginning God created the heaven and the earth." Access to the sacred space also demands repression of a sort from the creator of *Paradise Lost*. The child's turning away from the primal scene, relinquishing his intrusive vision and letting the marriage be, Milton reenacts as a philosophical and religious discretion: his "unapproached" essence being unique among things that are, the invisible God is not an object and is not to be known as an object. As the child found ambition in repressing the space of sexual desire, Milton finds revelation in "repressing" the space of ambition: his will be an inspired poem, manifesting the workmanship of a divine author. Coming back to the goodness of the light in the beginning through a long acknowledgment of guilt and finitude, he regains the innocence of the guiltless child who saw or thought he saw his mother and father pressed together, doing something.

There must be many reasons for the custom of recording births and deaths in a family bible. One of them probably derives from the psychogenetic process by which this book comes to mean, to be felt as authentic, to provide a sense of uplifted or transcended life. Intuitively

one inscribes, in the sacred text as on a tombstone in hallowed ground, the facts of birth and death: for the evolved significance of createdness, entering psychic life through the delayed trauma of the primal scene, gives rise to sacred texts and hallowed grounds. When Milton recorded his creation next to the beginning of Genesis, he left a concise analysis of his creative urge. "John Milton was born": these plain words, when understood in their full resonance, abbreviate the psychic history that bore him from the primal scene, through repression, to an immense ambition that, "long choosing," enveloped the space of revelation cleared in Genesis. He would write the primal scene of everything and everybody—a total genesis. Moses had seen, but given his position in history, had not seen the entirety; Milton summoned a light prior to the Mosaic first light. Without losing sight of his creaturehood, he attempted a poetry of extended revelation that repeats and revises Genesis. *Paradise Lost* is the *nachträglichkeit* or "revisiting" of Creation.

Men come and go, the limits of their lives inscribed on bibles and tombstones. Milton wanted to inscribe a religious innovation on the sacred space of his own life. "May I express thee unblam'd?" May I imagine you, create you, before whom all things are as nothing? May I master in this way the religious sublime?" At the level of the sacred complex, this self-conceived ambition effects, like the space of ambition in the first complex, a return of the repressed. Milton would know origins, and he would handle the potent light of the father as his artistic power to represent these origins. But the volition sacrificed in inspired poetry provides one means of sanctifying these potentially blameworthy liberties. In the sacred complex the superego must consent to the ego's revelation.

Let us revisit Satan:

> who saw
> When this creation was? remember'st thou
> Thy making, while the Maker gave thee being?

Not I, no. Satan is not exactly lying. He is the empiricist of createdness. Adam begins his autobiography by pointing to a narrative difficulty more severe than the translation of heaven to earth Raphael has just accomplished: "For Man to tell how human Life began / Is hard; for who himself beginning knew?" Everyone's story, like an epic, begins in the middle of things, out of the sense of absent source. But as a result of this blindness, the first man quests for his scene of origin:

Thou Sun, said I, fair Light,
And thou enlight'n'd Earth, so fresh and gay,
Ye Hills and Dales, ye Rivers, Woods, and Plains
And ye that live and move, fair Creatures, tell,
Tell, if ye saw, how came I thus, how here?
Not of myself; by some great Maker then. (8.273–278)

The visible world is innately categorized as "here before I was here." Sight itself directs Adam to the beginning that cannot be known as things experienced can be known. We human beings are so constituted that we imagine our creation, and the acquisition of this image is a crucial turning point in the organization of our psychic reality. Satan refuses to produce the image. Defying its evolved significance, he is left with the entanglement of narcissism and oedipal rebellion that in a more hidden fashion wrought a stasis in the youth of Milton. He casts himself into the human primal scene of Eden to sow enmity between father and mother, collapsing the spaces of desire, ambition, and revelation into a psychic battlefield for his resistance to the Godhead.

The Milton of *Paradise Lost* departs from Satan. The devil's *invidia* at the elevation of the Son can be understood once again as a recoil from the ego-ideal; he hates, not only the privileges of the Son, but the meaning of the Son, unable to bear a created order in which the sole way to merit lies in reflecting the glory of God. Heinz Kohut has isolated two narcissistic positions. In one the "grandiose self" of the child contains in its absolute omnipotence the imago of the parent. In the other the child attaches himself to the imago of the idealized parent, great in being contained within this superior greatness.[56] Satan, forever bearing the cross of his wronged grandiosity, could not abide the second sort of narcissism, for that is the posture with a future. Destined to be redefined by the primal scene, finitude's next teacher, the narcissism of the contained ego prefigures submission to the otherness of the superego. Doubtless the poet would, like Satan, make himself, for when thoughts are voluntarily moving harmonious numbers, when the words of *Paradise Lost* are emerging in the consciousness of its maker, the self creating cannot be distinguished from the self created: at such moments, Milton was nothing less than the meaning of *Paradise Lost*. But the poet seeks for this self-quickening power in origins that bespeak the otherness of God, moving through the superego to recover the contained narcissism fulfilled by the superego. There are several good explanations for why Milton begins with Satan,

brightening his character with surprising nobility, only to degrade and humiliate him in the course of the epic, leaving him a snake with ashes in his mouth. We are made to undergo a moral reorientation, according to Fish; we are experiencing, according to Tayler, a progressive unveiling of an essence there from the beginning.[57] I would add that Satan represents the origin, the original original sin, *Milton* turns away from.

Tantalus and Orpheus

thee I revisit safe,
And feel thy sovran vital Lamp; but thou
Revisit'st not these eyes, that roll in vain
To find thy piercing ray, and find no dawn;
So thick a drop serene hath quencht thir Orbs,
Or dim suffusion veil'd. Yet not the more
Cease I to wander where the Muses haunt
Clear Spring, or shady Grove, or Sunny Hill,
Smit with the love of sacred Song; but chief
Thee *Sion* and the flow'ry Brooks beneath
That wash thy hallow'd feet, and warbling flow,
Nightly I visit: nor sometimes forget
Those other two equall'd with me in Fate,
So were I equall'd with them in renown,
Blind *Thamyris* and blind *Maeonides*,
And *Tiresias* and *Phineus* Prophets old.
Then feed on thoughts, that voluntary move
Harmonious numbers; as the wakeful Bird
Sings darkling, and in shadiest Covert hid
Tunes her nocturnal Note. (21–40)

The blindness of three of the poets and prophets listed was a punishment from the gods. Are they in this sense "equall'd with [him] in Fate"?

Mauss proposed in his classic *Essai sur le don* that the fundamental obligations sustaining social groups arose about the gift: to give, to receive, and to repay.[58] The metaphors defining the poet's blindness imply a violation of these principles. He revisits but, like the victim of a social slight, is not revisited; he has returned to light, but later in the invocation, "not to me returns" the pageantry of the seasons. When we deal with each other, we look into each other's eyes. An exchange of

visual images of vision accompanies the various forms of social exchange, and Milton, "from the cheerful ways of men / Cut off" (46–47), seems intuitively to posit a connection between his blindness and his solitary, uncompensated endurance. Etymology condenses the breach of obligation with the affliction itself, since he dutifully "revisits" in the English meaning, yet in the Latin meaning is not "seeing again."

This muted evocation of insult is not unconscious. Milton gave his eyes to forward liberty, and in the *Second Defense* he began his first catalogue of blind seers with the principle of recompense:

> Or shall I recall those ancient bards and wise men of the most distant past, whose misfortune the gods, it is said, recompensed with far more potent gifts, and whom men treated with such respect that they preferred to blame the very gods than to impute their blindness to them as a crime? The tradition about the seer Tiresias is well known. (*CP* IV, 584)

Those who failed to understand the divine repayment of "far more potent gifts," it would seem, blamed the gods. If we assume that one does not justify a God who has not first been indicted, why might the creator of *Paradise Lost* blame God? An outrage deeper than unfulfilled obligation surfaces in the invocation. When light first touches Milton, he greets its presence with eyes that "roll in vain." His vain quest to see a light he only feels belongs to the motif of the taunting God so prevalent in *Paradise Lost*. Milton is, in this instance surely, of the devil's party.

Tantalization—the inverse of nourishing, the gift displayed but never given—is the recurrent image of torment in *Paradise Lost*. Devils try to drink from the "tempting stream" of Lethe, yet

> of itself the water flies
> All taste of living wight, as once it fled
> The lip of *Tantalus*. (2.612–614)

This is the prototype of a cruel mnemonics. The nearness of Lethe makes the wish to be guiltless unforgettable: unable to reach its object, desire repeats itself forever. A similar reminder torments the spirits in the Paradise of Fools, where Saint Peter "seems / To wait them with his Keys, and now at foot / Of Heav'n's ascent they lift thir Feet, when lo / A violent cross wind from either Coast / Blows them

transverse ten thousand Leagues awry" (3.484–488). When Satan flies past the gate of heaven,

> The Stairs were then let down, whether to dare
> The Fiend by easy ascent, or aggravate
> His sad exclusion from the doors of Bliss. (3.523–525)

No meal can appease the son of Satan, endlessly hungry Death. Satan's whole career has been shaped into a tantalization from the beginning of the poem, when God releases him from bondage, we are told, only that he may forge his chains again, "and enrag'd might see" the goodness his malice produces for everyone but himself (1.213–220). In our last view of him, Satan is again tantalized, the moment of his victorious return to hell transformed into humilation. The fruit that becomes ashes in the mouths of snaky demons returns this motif to its classical form—the arousal, then thwarting of appetite. God makes certain that, beyond the frustration inherent in their ambition, the devils will suffer an aggravated frustration.

Milton himself, who created the righteous nastiness of the lowered stairs, refers to the plight of Satan as a "sad exclusion." Maybe he intended to convey through this cruel derision an aspect of his deity that creatures cannot fully savor; maybe, with regard to this theme, one might concede something to Empson and Waldock, for Milton cannot command our sympathies when he makes the creator of the universe behave like a spoiled child. In any case, his control over the portrayal of Satan remains unimpaired. The psychological acumen that counterbalances theological abstraction in this characterization, preventing Satan from dissolving into mere evil, is powerfully evident at the moment when, inhabiting the serpent, Satan invents his origins:

> To satisfy the sharp desire I had
> Of tasting those fair Apples, I resolv'd
> Not to defer; hunger and thirst at once,
> Powerful persuaders, quick'n'd at the scent
> Of that alluring fruit, urg'd me so keen.
> About the mossy Trunk I wound me soon,
> For high from ground the branches would require
> Thy utmost reach or *Adam's*: Round the Tree
> *All other Beasts that saw, with like desire*
> *Longing and envying stood, but could not reach.*
> Amid the tree now got, where plenty hung

Tempting so nigh, to pluck and eat my fill
I spar'd not, for such pleasure till that hour
At Feed or Fountain never had I found. (9.584–597; my italics)

The taunts of God have won their way into the depths of Satan's imagination. This is his scene—to be engaged in self-creation, gorging himself on the desirable, while others unable to reach look on, tantalized. The relation between the other animals and the serpent in this scene is Satan's wishful revision of the relation between himself and the God he imitates.

Tantalization occurs after the self-definition of a true temptation. It is the travesty of free choice. When tempted, one may choose this or that. The tantalized, having made his choice, is forced to remain with *that*. He makes the same wrong decision over and over again, or, like the devils reaching for Lethe, he may unmake the same wrong decision, losing over and over again what he chose once to abandon. It is a reenactment for the purpose of mockery, like a school lesson repeated with no intent to educate, only to humiliate. The God who extends the false invitation of the lowered stairs declares the finality of damnation: "What you have willed is your fate, but still you desire otherwise." Fallen man will sometimes feel tantalized as he awaits the delivery of divine promises. But patience is the antidote to fallen man's sense of tantalization, and God does not tantalize unfallen man.

Milton feels the warmth of the light, but his tantalized eyes, repeating their loss, "roll in vain / To find thy piercing ray, and find no dawn." That tiny chink of light seen "upon the eyes turning" was not always the universe. How often and, though he might have denied it, how bitterly he must have rolled his eyes toward the single piercing ray that never became a dawn!

Yet Milton and Satan, having together escaped detention in hell and flown to the light, diverge. When Satan addresses the sun at the opening of the next book, he concludes with one of the few resolutions available to a tantalized being:

So farewell Hope, and with Hope farewell Fear,
Farewell Remorse: all Good to me is lost;
Evil be thou my Good. (4.108–110)

On lowered stairs of light Milton would enter heaven. He invokes his hope instead of bidding farewell to it.

Nor is he tantalized utterly. Before the wound of blindness, he was

"Smit with the love of sacred Song," and in his blindness, "not the more / Cease I to wander." The nightly visits that follow appear to be the meditations of a man wandering through his memory of books, seeing the "places" that Renaissance mnemotechnicians recommended for the good ordering of recall. In the "Clear Spring" that refreshed the pagan poet and the "flow'ry Brooks" of Sion that healed the blind man in John, Milton avows the persistent vitality, cultural as well as natural, of the "pure ethereal stream" that flowed, shaping, into the dark waters of the abyss. The bible *is* within him, and "chief" within him, his "best and dearest" place. Siloa, moreover, is a "warbling" brook. Quickened into life by its placement in the flow of the invocation, this transfigured cliché joins the waters of Genesis, "at the voice/ Of God" given light and form; the waters by which Christ, light of the world, gave light again to the blind man; and the equivalence of seeing, in the ruling trope of the invocation, to Milton singing or "warbling" his poem. It also looks forward to the singing nightingale fed on such thoughts.

The search for repayment implicit in the failed visitation of light emerges openly in his memory of blind predecessors. Equality of fame would equilibrate equality of fate, and the achievement of this wish once more demands the revisitation of light. The three appearances of the word "So" in the invocation, all of them initiating a line, mark the logic of his recompense. The "So" of the magnitude of affliction ("So thick a drop serene hath quencht thir Orbs"), the "So" of wished compensation ("So were I equall'd with them in renown"), the "So" of actual restitution ("So much the rather thou Celestial Light"): quietly, in the architecture of the passage, Milton is building a nest of deservedness to hold the divinely created light. The thrust of allusion, metaphor, and syntax combines with our primordial sense of obligation to invest the regiving of light with an inevitability inseparable from justice. *God must, or must be blamed*. Milton will not roll in vain. His characteristic poetic effect, prolepsis, is here indistinguishable from desire, and he gives to this prolepsis of desire a theodicial force.

Now memory flows into creation. Unlike the tantalized, Milton has received nourishment from his remembering. "Then feed on thoughts, that voluntary move / Harmonious numbers": the regular iambs of the first line, which does what it means, signal the passage to creation with a poetic version of Austin's performative utterance, while the rapid anapest in the second foot of the next line, stringing "Harmonious" together with "numbers," glides out from the vast variety compressed for this poet in the meaning of "Harmonious."[59] The

thoughts move "voluntary" or of themselves into poetry, just as the eating of food will nourish without a second act of volition. What Hölderlin taught Heidegger to wish, Milton claims to have achieved: "Poetically man dwells."[60] The union of form and matter in the Creation of God also inheres in the creation of poetry, for poetry here is not the shaping of thoughts or the arrest of thoughts, but thinking itself, the stream of consciousness become a warbling brook. It has been noticed that the lines on nightly visits echo "Il Penseroso" in their progression form "shady Grove, or Sunny Hill" to solitary nocturnal pursuits such as religious meditation and prophetic song.[61] In the context of Milton's youthful verse, the nightingale simile is especially rich:

> as the Wakeful Bird
> Sings darkling, and in shadiest Covert hid
> Tunes her nocturnal Note.

The first poem Milton published, the epitaph for Shakespeare, proclaims that the dramatist's "easy numbers flow" to "th' shame of slow-endeavoring art"—already a conventional encomium in the infancy of Shakespeare idolatry, but one foretelling the slow-endeavoring career of its author. In the early "O Nightingale!" Milton vowed that "Whether the Muse, or Love call thee his mate, / Both them I serve, and of their train am I." By the time he was writing *Comus*, these two services had become incompatible, their disjunction a symptom of the conflicts that made the poet belated. But the sexual revelry of Adam and Eve is succeeded by the note of the wakeful bird. "These lull'd by Nightingales imbracing slept" (4.771). Now that Milton, like the nightingale, is again the servant of the Muse and of Love, his harmonious numbers flow with the ease he once attributed to "my Shakespeare."

"Sings darkling" and "nocturnal Note" appear to be redundant, a quality also felt in "shadiest covert hid." As editors note, night was doubled for the blind man. But there are other implications in the "dark dark dark" language of the simile. The composing of inspired poetry, we inferred from the reticent, undivulged Lady of the Ludlow masque, is a hidden act, and therefore a safe one. On a dark night, inside a blind body, thinking poetry that has yet to be written down, continuing a poem unfinished and unawaited, the wakeful consciousness of a poet lacking renown earns the redundance of "darkling," "nocturnal," and "in shadiest Covert hid": much of the

joy in the simile emanates from the singer's knowledge that he is wholly unseen. Without recognizing our presence at his performance, Milton opens to our vision an unprecedented interiority in the poetic act. More profoundly than Satan rising in defiance from the burning lake, this is the genesis of English romanticism—the definitive evocation of what creating might be, though rarely is. Poetry has arrived at one of its great destinies. As the art evolved from an oral, public, and tribal setting into a textual tradition whose participants were unique individuals striving for achievement, the act of composition became the structural descendant of the "singing" of the old bard. In the autotelic joy of his wakeful bird, Milton gives voice to the grand utopian wish that poetry, pushed into a more interior space by the textual tradition, delivered to mankind: that a perfected consciousness would sing, its very being poetic.

In subsequent invocations the solitary and active visitation of memory is replaced by the collaborative and more passive visitation of inspired dreams. Having suddenly acquired a threatening audience and an anxious recollection of Orpheus, the endangered Milton of Book 7 lives in "solitude,"

> yet not alone, while thou
> Visit'st my slumbers Nightly, or when Morn
> Purples the East. (7.28–30)

It does not require an elaborate psychology to understand what dreaming could mean to a man gone blind in his maturity. "Day brought back my Night" concludes the poet of Sonnet XXIII, whose awakening cost him the embrace of his veiled wife. In dreams as in memory, Milton could see once again and enjoy the freedom of movement denied him in the waking world. "Methought I saw," says entranced Adam of the creation of his wife, echoing the traditional English formula for the dream vision. The imaginer and the rememberer would not long forget their blindness; dreams must have been the better relief. His vision a dream vision, the poet lived in the circumstance of the convention. For this reason perhaps the invocations seem to understand the dream as more than an as-if trope for inspired creation. It is presented as an instance of, and prototype for, such creation. I suggested in my previous book that the invocations offer "the paradox of an action both voluntary and involuntary," a creation willed ("voluntary move") and yet, Milton being at times as much an amanuensis as his employees, transcribed ("dictates to me slumb'ring").[62] The Muse herself is "unimplored" at 9.22 and "implored" at 7.38.

Inspiration does not make sense to many people today, even to some who know firsthand the elusive magic collaborating in the production of a rough draft. But witness dreams. In the experiencing they appear to be given to us; we are recipients, as we feel ourselves to be recipients of perception. Yet we also have the knowledge or access to the knowledge—it flickers on the margins of dream consciousness—that we are in truth dreaming. Otherness and mineness are nowhere so entwined in our ordinary experience as in dreaming—a phenomenological observation consistent with the psychoanalytic theory of dreams, which posits that the dreamer is being re-presented with "places" in his memory from which he has become estranged. Now and then I hear that this theory has been made obsolete by the discovery of REM patterns and dream-associated brain waves in the fetus, but are the signs of physiology rigid designators, as attached to their referents as "Socrates" to Socrates? Surely the existence of a biological "prestructure," a split within the inchoate organism, is not prima facie inhospitable to the psychoanalytic idea of the unconscious, since the properly psychological process of filling and ordering this structure must still occur early in infancy, keyed to the rhythm of feeding. Given the initial lack of differentiation among mental regions, agreed upon by all schools of psychology—a "blooming buzzing confusion" in the still timely words of William James—we may yet assent to Freud's "*Dreaming is a piece of infantile mental life that has been superceded.*"[63] The first superceded piece, I assume with Freud, is the wish-fulfilling hallucination of the absent mother. The simultaneity of mineness and otherness in the dream preserves, every night, the indistinct boundaries of our old relation to this archaic figure—and it is to the ambiguity of the boundary that Milton is attuned. "All" in the dream must not be "mine" (9.46). He is in "solitude, yet not alone" in dreams. The otherness felt in dreams has become the inspirer, source of poetic power.

Some strands of my argument come together here. Blindness encouraged regression. Milton entered a second moratorium parallel to the retirement at Horton and Hammersmith broken by the death of his mother. The artistic wish for deathless renown he would now fulfill derived from a narcissistic attachment to his mother prior to the oedipal disillusionment. In the Muse of *Paradise Lost* Milton resuscitated an Eden lost in the maze of the primal scene:

> my Celestial Patroness, who deigns
> Her nightly visitation unimplor'd,
> And dictates to me slumb'ring, or inspires

Easy my unpremeditated Verse:
Since first this Subject for Heroic Song
Pleas'd me long choosing, and beginning late. (9.21–26)

The oneiric patroness visits each night, nourishing dreams with her presence. She is also there in the "Easy" moments when thinking is poetry. Her gifts can scarcely be distinguished from the "voluntary" motions of the thoughts the poet feeds on. She teaches, she knows. "Thou from the first wast present"—present from the "first" instant this song was conceived. She raises and supports, lifts the poet on wings, guides him securely in his flight to high heaven, and when this flight is completed, will "with like safety guided down / Return me to my Native Element" (7.15–16). She is the preoedipal mother who carried us through the horizons of a burgeoning space that was, as dreams and inspired poems still are, ours and not-ours.

Milton split the imago of the preoedipal mother exactly as he did the oedipal father. His lifelong fascination with the legend of Orpheus is the clearest indicator of this transformation, for the overwhelming fact about the Thracian bard, so obvious that we have tended not to see, is that his mother is his Muse. Milton's fullest treatments of the myth are the residue in symbolism of his attempts to harmonize the conflictual interplay of mother, love, and poetry.

The descent to hell in search of Eurydice appears in both the companion poems, climactically in "L'Allegro." If the happy goddess can supply music that would have "won the ear / Of *Pluto*, to have quite set free / His half-regain'd *Eurydice*" (148–150), Milton means to live with Mirth. This future, where shepherds sport with shepherdesses and Hymen would "oft appear" (126), projects the compatibility of sexual pleasure and poetry. Orpheus used his mother's power to secure a woman of his own from a jealous god; Mirth would perfect the rescue. In the Orpheus passage of "Il Penseroso," however, the bard is called forth to sing the same old notes that "made Hell grant what Love did seek" (108), a concession that did not "*quite* set free" the captive lady. Leaving Eurydice with Pluto, the literary traditions of virgin melancholy number Chaucerian and other romances, the genre of chastity, and widen at last to the anthems of the church and the prophetic strain of the hermitage. It was probably not incidental for the young Milton, certainly not for the later Milton, that Pluto returned the poet's woman on the condition that he would not look back at her. In the end, Milton had become a blind and perfected Orpheus, singing "other notes" to win his desire from hell. Having

"Escap't the Stygian Pool," the Christian Orpheus of Book 3 wishes to look only towards the light.

If the young poet imagining a union between poetry and sexuality would require the full recovery of Eurydice, transcending the failure of Orphean love, the poet who chose to live a virgin had eventually to confront and transcend the death of Orpheus:

> What could the Muse herself that *Orpheus* bore,
> The Muse herself, for her enchanting son
> Whom Universal nature did lament,
> When by the rout that made the hideous roar,
> His gory visage down the steam was sent,
> Down the swift *Hebrus* to the *Lesbian* shore?
>   Alas! What boots it with uncessant care
> To tend the homely slighted Shepherd's trade,
> And strictly meditate the thankless Muse?
> Were it not better done as others use,
> To sport with *Amaryllis* in the shade,
> Or with the tangles of *Neaera's* hair?

Orpheus was the Antaeus of poetry. His strength descended from his mother, and when caught in the alien, unmaternal element of "the hideous roar," he died. Milton's repeated "the Muse herself" locates the horror of the myth, not primarily in the dismemberment of Orpheus, but in the impotence of the "thankless" Calliope, Muse of epic, the matriarch of his ambition who demands the sacrifice of erotic pleasure but cannot prevent death. As Parker emphasizes, the author's mother had died just months before the composition of "Lycidas."[64] The failure of the Muse to forestall the "blind Fury" is in effect the collapse of the defensive mythology of *Comus*: omnipotence lingering from the narcissistic mother is a frail guardian for those "that wander in that perilous flood." Perhaps the future taught Milton that a lethal gift had also been inserted into his beginning. I have discussed his brief remarks about the charity of his mother and the timing of her death; we know little else about her. But one detail recorded by Aubrey seems important in the context of muses and poets, mothers and sons: "his father read without spectacles at 84," whereas "his mother had weak eyes, and used spectacles presently after she was thirty years old" (*EL*, 4–5). Aubrey bothered to collect this information, obviously, because he was interested in knowing whether the poet was predisposed to blindness and, if so, by which side of the family. The contrast

between weak maternal and strong paternal sight being so pronounced, it is difficult to believe that Milton did not make the simple deduction that his mother had transmitted to him the possibility of going blind.

Let us assume, with more evidence forthcoming in the next chapter, that blindness was in one important way a maternal inheritance descending from the line of Eve. From the same source, however, came inspiration, intuition, and the originally secure expansions of the narcissistic ego. Milton could resolve the tension of this ambivalence within the symbolism of the sacred complex, where the strong-sighted father has been supplanted by the omnipresent vision of God. From the limitless strength of the Father he could solicit a "Celestial Patroness" free from the taint of earthly mothers—a mother who will not disillusion:

> But drive far off the barbarous dissonance
> Of *Bacchus* and his Revellers, the Race
> Of that wild Rout that tore the *Thracian* Bard
> In *Rhodope*, where Woods and Rocks had Ears
> To rapture, till the savage clamor drown'd
> Both Harp and Voice; nor could the Muse defend
> Her Son. So fail not thou, who thee implores:
> For thou art Heavn'ly, shee an empty dream. (7.32–39)

The poet would be the son of a Muse able to regain for his art the animism of a primitive ego whose life, like the song of Orpheus, called the world to life. Poetry began in the Rhodope of childhood ("where Woods and Rocks had Ears / To rapture") organized by her attentions. But she—and this favor, unlike her visits, is implored—must also be able to "defend / Her Son" from a malign audience, "dangers compast round" (7.27). She will be able to perform this service because of what she is and what she is not. She is not, like Calliope, "an empty dream," a dream all his. She does not tantalize, bringing a vain dream to eyes that roll in vain, condemned even when dreaming to see their own insides. She is "Heavn'ly," bearing gifts dispensed through the good will of the invincible Father.

This is half the representation of the creative act in *Paradise Lost*. To my psychoanalytic eyes it has both beauty and truth. Milton is trying to secure, through symbols of great psychological as well as cultural authority, a power of innovation that these symbols represent as intimate otherness. The new is the not-self: radically interior, the act of

poetic creation need not be, for Milton must not be, solitary, for the radical interior of the differentiated ego retains its archaic partner. Milton's entire psychic orientation stands behind his belief that the "bright essence" of the superego generates this power, this otherness. But it is as if the structure of the superego, when stretched out in the cosmic dilations of religious symbolism, loosens to reveal its full contents. A Muse was hidden in the strict taskmaster. The mother, too, inhabits this psychic agency, and she is more than another voice to obey. Insofar as the child completes his mourning for the preoedipal mother in forming a superego, the felt alterity of a vigilant and watchful superego preserves by internalization her early being for the child—her love and her ministrations. She is not the dominant pole of the first superego. But through obedience to the paternal law, the boy supplies himself with the possibility of reencountering, in a future of lessened severity reached through this obedience, maternal intuition. It is in this sense no merely symbolic truth that only the Father can dispense the presence of the Muse. Finding this sense, Milton ended the antagonism between the oedipal father and the narcissistic ego whose symbol was once the paralysis of *Comus*.

Illumination

Thus with the Year
Seasons return, but not to me returns
Day, or the sweet approach of Ev'n or Morn,
Or sight of vernal bloom, or Summer's Rose,
Or flocks, or herds, or human face divine;
But cloud instead, and ever-during dark
Surrounds me, from the cheerful ways of men
Cut off, and for the Book of knowledge fair
Presented with a Universal blanc
Of Nature's works to me expung'd and ras'd,
And wisdom at one entrance quite shut out.
So much the rather thou Celestial Light
Shine inward, and the mind through all her powers
Irradiate, there plant eyes, all mist from thence
Purge and disperse, that I may see and tell
Of things invisible to mortal sight. (40–55)

Men have always told time by means of light. Milton feels, in his "ever-during dark," severed from the temporal rhythm that vivifies

the order of nature. The repeated nocturnal action ("Nightly," "nor sometimes forget," "Then feed on thoughts") of the previous lines is absorbed into the "Thus" of an elapsed year, whose returning seasons elicit Milton's catalogue of loss. Without experience of "Day" or the transitions from dark to light and light to dark, he has lost the differences of the daily round. Via the seasonal progression from vernal to summer flowers, his catalogue moves up the chain of being to animals, then to the sight missed most—"human face divine," the visible evidence of flesh and spirit, emotion and reason reconciled. Milton is cut off from the *ways of men* in this cheerful dance of time. We should hear an echo of the famous "ways of God" in the opening invocation, for someone partaking of "the cheerful ways of men" would obviously feel less urgency about justifying the ways of God.

The several definitions of what the blind man does see—a night, a veil of dim suffusion, a cloud, an ever-during dark—culminate in the splendid metaphor of the book of knowledge become a blank book. Scholars have certified the antiquity of this trope, often citing Curtius, and developed its Renaissance variations (the emblem book of nature, for instance).[65] But once again Milton has freshly imagined tradition. This is the "Book of knowledge fair," implying that there are other books of knowledge, but not "fair" ones: he is implicitly contrasting the visual poverty of print with the opulence of God's moveable typeology. Vergil and Moses could be read to him, their visual marks returned to the music of the living voice, yet who could speak a flower or recite the approaching evening? Milton sees a "universal blanc," an unprinted or white (*blanc*) page. The book has not been taken away, but "expung'd and ras'd"—erased, expurgated, *censored*. If I seem to be pressing here, it is because the man who now beholds his blindness as an expunged book once wrote: "as good almost kill a Man as kill a good Book; who kills a Man kills a reasonable creature, Gods Image; but hee who destroyes a good Booke, kills reason it selfe, kills the Image of God, as it were in the eye" (*CP* II, 492). The blank book of *Paradise Lost* obliterates the "sight" of the "human face divine," image of the image of God. Milton is, moreover, *presented* with this book. All the possibilities of this verb exert pressure here: confronted with this book, as when we say an object is "presented" to us; given this book, as a gift, by another; and temporally constituted or "presented" by this book, as one caught in the "ever-during" presence of a flattened present. Does a blank book have meaning? Why should this trope lead immediately to the influx of celestial light, returning the invocation to its initial symbol?

The book is there, presented and blank. "Nature's works" have been unmade, but the book that held them remains. Might one dream on it? Imagine on it? Write on it? Robert Fludd began his diagrams of the Creation with an absolutely black plate, representing by the deprivation of vision the infinite abyss confronting the creator.[66] Milton's "universal blanc" is an aftermath able to receive a new beginning—a space of undetermined possibility, and as such, the human and artistic counterpart of the invocation's first light. Pausing in his narrative, Milton demonstrates that he, like the deity symbolized at the beginning of the address, has been wound back from "grateful / Vicissitude" to a *nunc stans* ("ever-during dark") of time and will, ready to undergo the venture of creation; this release of shaping energy actually occurs in the first line following the invocation, where Milton breaks his pause by pouring the overspill of God's "Now" into the illuminated panorama of all Creation. Thomas Browne gathered his divinity from the book of nature and the book of scriptures.[67] The light of God, in Milton's universe, required a "void and formless infinite" to generate the former; the empty book of his blindness becomes a corresponding substratum on which the God of revelation may regenerate the latter. "Presented with a universal blanc"—a paradise lost, without flowers, flocks, herds, or faces, ready for the illumination of *Paradise Lost*. One must be as nothing to be as one created.

When illuminated, the negative symbol heals and enables. It calls for light. The equivocal gift of the blank book belongs among the several voids we have drawn from the inventory of psychoanalysis. The void of absence provokes the first image, precursor of the linguistic sign. The void in the wake of repression provokes substitution, the possibility of sublimation. The void of castration provokes the internalized otherness of the superego. (Others might be added, such as the "dream screen" of Bertram Lewin, which he defined as the primitive incorporation by the drowsy, feeding infant of the expanse of the breast, upon which dream images come to be projected.)[68] Each of these voids appears between loss and creativity. Through the symbol of the "universal blanc" Milton becomes conscious of the void he is conscious of, enacting in the awareness of his poetry the sort of transvaluation provoked by absences in the course of psychic history. His God of light, whose nature is to "fill infinitude," abhorring voids, becomes a supreme symbol of the developmental power initiated by void. If the light shining inward in the last lines of the invocation stands for the composition of *Paradise Lost*, then the epic was

composed on blindness, and the actual book we are reading, which was once a blank book, becomes a palpable metaphor for the creation of the poem. The text we read, Milton invites us to imagine, was printed on blindness.

The fecundity of the void may be discerned in another and unspoken trope that nearly surfaces in the *fiat lux* of the concluding lines. It has silently dominated the presentation of blindness throughout the invocation. Milton's orbs are quenched because, in *gutta serena*, tumors prevent the animal spirits from entering the cavities of the eyes. Whether seeing was thought to be the result of extramission or intramission, spirits had to be present in the eyeballs. Milton was blind, then, because something inside could not get out. But the metaphors of the invocation invert this state of affairs: something outside cannot get in. Failed activity resides out there in the world, where light could revisit him but does not, and things seen might return to him but do not; he searches for a "piercing ray." It is appropriate for him to say that wisdom is "at one entrance quite shut out," but it would have been just as appropriate, in view of the physiology of the time, for him to say that wisdom is at one exit quite shut in. This passivity directs us back to the woman in Milton.

Blindness was the punishment Oedipus imposed on himself in his delayed deference to the prerogatives of Laius. Tiresias, who knew the crime of Oedipus, was blinded either because he saw a naked goddess, in which case the punishment befit the crime in making the crime thereafter impossible, or because he revealed the superior sexual pleasure discovered when living as a woman, in which case the punishment *was* the crime. Blindness means castration, whatever castration may mean, and given that the anxiety of the male child comes to seem "realistic" partly through his apprehensive etiology of the vagina, to be castrated may mean to be a woman. The emasculation of blindness must have been particularly harrowing for a poet whose vision was intrusive and ambitious as well as erotic—all the more so if he assumed that weak eyes had come to him through his mother. Although I am uncomfortable enough in the area of triumphal forms to doubt, for example, whether the short lines of "Lycidas" repeat the dismemberment of Orpheus, I do think that the rupture of enjambment between "men" and "cut off" in the invocation to light reduces to one the strokes of a paternal vengeance that, beginning with the headaches of a young student, smote and smote until, light denied, it smote no more.

But there is a way to strength through weakness. Accepting the unconscious meaning of his affliction, Milton invites the celestial light to enter his "shadiest covert." He will be the vessel of God's light, and the ever-during dark that "surrounds" him, as if delimited by compasses, will present an inner space within which the creative force of the deity can realize itself once again. As the femininity of the wakeful bird tuning her note metamorphoses into "the mind through all her powers," the blank book almost expressly becomes a desiring womb. Milton asks God, not simply to undo, but to fulfill—to undo by fulfilling—the psychological meaning of blindness and cast his bright seed into this dark enclosure. "There plant eyes"—a figure that links artistic creation with the mating of seeds and sun at the ground of the chain of animate being, metaphorically equating darkened mind with earth, the dark element represented in Book 7 as the "great Mother" from whose womb creatures emerge (281). Freud alerted us to the fact that our jokes may be as revelatory as our formal autobiographies. One of the anecdotes from the anonymous biographer about the composition of *Paradise Lost* captures the victory through acquiescence we have deduced from the poem itself. Having awakened with a passage to be transcribed, Milton would grumble about the house when his amanuensis was late, saying "*he wanted to be milked*" (*EL*, 33). The joke condenses both of his representations of the creative act. To the extent that the male poet is the child and the female Muse his mother, the poem is food: "Then feed on thoughts, that voluntary move / Harmonious numbers." To the extent that in the invocation to Book 3 the poet is female and the light of inspiration is male, the text becomes his child, maturing as it is nourished.

Birth without the medium of woman occurs twice in the epic. Sin emerges full-blown from the head of Satan, whom she addresses as "my Author" (3.864). Since Adam had the idea of Eve, demanding the fulfillment of his "wish" against the playful strictures of God, Eve also materializes a mental pregnancy; like Sin, she addresses her male mother as "my Author" (4.635). Inviting God to enter him and inseminate him with the power to actualize his idea, Milton clearly intends his creation to follow the second model. Differences aside, Satan, Adam, and Milton require the assistance of femininity to fulfill their creative urge. Is not the Father, albeit darkly, the ultimate authority for them all?

Sexuality in *Paradise Lost* descends from the bisexual angels to the clear-cut male and female of the created universe, "Which two great

Sexes animate the World" (8.151). During the astronomy lesson Raphael locates these principles in the male sun that radiates light and the female moon that reflects light (8.148–152), an enskied sexuality emphasized in his narrative of Creation, when Christ makes "first the Sun" (7.354) as a male and kingly "Fountain" (364), then with light borrowed from this luminary produces the queenly moon (375–386). Sun and moon, monarchs of day and night, compose a sidereal emblem with two fields of reference, "downwards" to the subordination of Eve to Adam and "upwards" to the subordination of Son to Father. With respect to redemption, the parallels between Christ and Eve have often been mentioned.[69] After the fall Eve is the first to achieve self-accusation; she offers to perform the redemptive gesture of returning "to the place of judgment" and praying that "The sentence from thy head remov'd may light / On me" (10.934–935). Yet the likeness extends beyond sacrificial love to the heavenly existence of the Son. Both these beings come second, both are formed of the substance of their fathers, and both have a distinct essence whose moral ideal is voluntary submission to their mothering fathers. They are also alike in the toils of motherhood: executors both, they embody the image of another. As a moon reflecting the invisible sun of the Father, Christ manifests the paternal will, multiplying the divine image he received from his Father. In history the Son appears to us as the known deity, "Father of his Family" (10.216) whose "Humiliation shall exalt" our fallen manhood (3.311–314). But Milton, extending the vision of man, would have us know by mysterious analogy that the Son is, to the Father, woman and womb. It is this Son, the feminine Son, the first *lumen* to receive the divine *lux*, who stands as the prototype of strength that comes from weakness, exemplifying the movement from self-emptying obedience to self-fulfilling exaltation Milton would imitate in the blank book that becomes *Paradise Lost* and the ever-during dark of blindness that becomes a womb full of futurity.

It may seem puerile blasphemy to propose that Milton's Son is somehow his Father's Eve, receiver and reflector of his phallus of light, but this is one of the radical fantasies varied in *Paradise Lost*: the father inseminating the son; the son as mother, materializer. Several times the poem aggressively incites us to imagine the Godhead in this way. God himself, after all, raises the issue of his sexual life, or at least of his marital status. When Adam persists in his desire for a mate, the deity replies:

> What think'st thou then of mee, and this my State,
> Seem I to thee sufficiently possest

> Of happiness, or not? who am alone
> From all Eternity, for none I know
> Second to mee or like, equal much less. (8.403–407)

Critics argue over the identity of this divine speaker, mounting elaborate theories to explain how, if the speaker is the Son, he could justly say these words, and why, if the speaker is the Father, as most readers have assumed throughout the centuries, his apparent monotheism contradicts the two-person Godhead, not to mention the intimacy between these persons, found elsewhere in the poem.[70] Before we begin rationalizing the puzzling speech, it has prompted a more immediate response—the very one that incites rationalization. "But no," we say, assuming the speaker is the Father and having looked in at heaven through the eyes of our visionary poet, "you have the Son, your divine similitude and sole complacence." At the prompting of God, we place Christ in the position of Eve. The allegory of Satan, Sin, and Death, read as a parody of the Godhead, also encourages us to entertain a relationship between Christ and the woman. In my view, this relationship supplies a key to one aspect of the allegory.

Refusing his source, turning away from the Father and Son analogous to the earthly father and mother, Satan both loses and looses the femininity of obedience displayed by the bisexual angels and prototypically exemplified by Christ. Sin becomes the horrid disfigurement of the woman that all religious heroes, open to the will of the Father, possess inside them in Milton's late poetry. The harmonious arrangement of the Father, feminine submission, and filial ambition in the psychology of the angels splinters in the head of Satan. With the expulsion of Sin and conception of Death, he externalizes his mental strife, creating the family as a league of hatred. As human pathology reveals the normal, so the decomposition of Satan reveals the angelic psyche. The bottom half of the woman degenerates to an animal. Death attacks her womb, then hell hounds gnaw her entrails. The objects Satan recoils from—his obedience, his source, the Son who is his source—are symbolized by the tormented womb of the woman once a part of him.

The psychological scheme that renders the Miltonic Godhead as-if-familiar arises from the intertwining of positive and negative oedipal currents at the moment of superego formation. The Son is in the image of the father, yet this act of obedience is in the image of the mother. Femininity symbolizes the something "more" in Christ, the epistrophe of voluntary love that makes him exceed a mere echo of his

creator. For Christ, as for angels and men, femininity constitutes the sphere of freedom. Although sometimes credited with a dull and orthodox view of the superego, Ernest Jones could not understand why fear or respect were sufficient to motivate the internalization of the father: "How can we conceive of the same institution as being both an object that presents itself to the id to be loved instead of the parents and as an active force criticizing the ego? If the superego arises from incorporating the abandoned love object, how comes it that in fact it is more often derived from the parent of the same sex?"[71] I do not have the clinical experience necessary to assess his suggestion that "the superego usually arises from identification with the father where the initial hostile rivalry has been replaced by homosexual love," but if so, this would be a paradoxical homosexuality in which the ego, offering itself as a love object to the alliance of superego and id, gains love from this internal father by imitating its heterosexuality. A settlement along these lines, a knot of oedipal resolution tied with paternal *and* maternal identifications, seems implicit in Milton's heterodox Godhead.

Let us suppose that the child, unable to occupy the position of the father in the primal scene, tentatively projects himself into the role of the mother. He has lost her as a sexual object, but retains her as a state of the ego—as an identification, which is how the process of mourning concludes. It is, moreover, an identification that lays the groundwork for the *acceptance* of castration that ideally brings the complex to an end. He also discovers in the maternal role what could be termed the "Holy Spirit" of the parental relationship, a love that passes, not only from father to mother as in the dominant oedipal position, but also from mother to father. When the scene is repressed, these explorations in the negative complex help to shape the correlation of ego to superego, contributing a warmth to the cold structure of disclosed guilt, observing father and observed son. A psychological wedding resolves the complex: to some degree at least the ego is to the superego as the oedipal mother to the oedipal father. Shattered, transformed into psychic structure, the physical union at the center of the primal scene is in the fullness of time reconstituted inside the child. His internal world with its apportionings of dependence and independence, law and desire, has become the living remnant of the marriage that gave him being. This ideal is the psychic coherence of the Miltonic Godhead. Here on earth it is one way we create ourselves by recreating our creators.

The poet sings "in shadiest covert hid," his mind a place symbolic of the bower where nightingales celebrated the union of Adam and Eve.

In the song arising here we find the entire transformational history of this space in psychic life. The composing of the song demands and receives a cooperation among the various relationships or states of the ego that have achieved organization in this space, moving back from the Father and the father-as-superego to contact the preoedipal mother, and with the poet's ego in the role of oedipal mother, moving back to contact the potency of the oedipal father. Milton was indeed in "solitude; yet not alone" when he sang in the shadiest covert of the paradise within. Inside him was an internalized marriage whose resources the epic taps and celebrates, filling the letter of the primal scene with completions of the spirit.

The opening invocation prefigures the strategies of Milton's creation, anchoring them, as so often, in the Creation:

> And chiefly Thou O Spirit, that dost prefer
> Before all Temples th'upright heart and pure,
> Instruct me, for thou know'st; Thou from the first
> Wast present, and with mighty wings outspread
> Dove-like satst brooding on the vast Abyss
> *And mad'st it pregnant: What in me is dark*
> *Illumine*, what is low raise and support. (17–23; my italics)

Pregnancy and illumination are already contiguous. In Book 3, Milton isolates the male side of this "brooding" *and* "impregnating" Spirit to fill with inward light the void left by the emasculation of blindness, interpreting his blindness, through symbols, as the mark of a femininity "deeper" than castration. Creation by the Godhead is the religious heir of the primal scene. To be made the true poem of this Godhead, Milton places himself in two of the slots made available within the sacred scene: the "shadiest covert" of his darkened mind becomes the "vast Abyss," the apeiron, the substratum, the matrix, the earth, the womb, while his poetic voice becomes the feminine Son or Logos, directing a light not his own. Later, in Book 7, he solicits the "brooding" of the first invocation's bisexual Spirit, replacing the "empty dream" of Calliope with the maternal guardianship of a Heavenly Muse. Blindness enables. The stasis of *Comus* derived from the fusion of the two fundamental anxieties of man in psychoanalysis: separation from the mother and castration by the father. Their opposites are reunion with the mother and the giving of a phallus by the father, and these are precisely the symbols of power that nourish the easy flow of the creation of *Paradise Lost*.

The Fate of the Debt—and Guilt

Blindness was Milton's opportunity. Sacred texts of his faith promised a restorative future: the last would go first, the meek inherit, the lame be made whole. These religious formulae for bringing strength out of weakness were congruent—as they must be, if they are to hold good on this earth—with psychological formulae extending into the unconscious. Accompanied by regression, blindness gave him a new claim on the image of the early mother; he found the evidence of her presence in dreams that permitted him to see and to wander, becoming strong in the weakness of the child. Frustrating the desire to look and everything that desire meant, blindness doubled back on the voids of absence, repression, castration; he centered on his fate, becoming strong in the weakness of the woman. But the harmonies discerned in his poetry cannot alone account for the authority of his resolve. Why such confidence after a lifetime that could be described, at best, as partially evasive?

The answer turns on his indebtedness. When Satan expelled the woman in himself, he exited from the psychology of the debt; it was there Milton found his majesty. For decades he had wished to illuminate the secrets of genesis in a work of art, but the poet who published divine secrets might render himself subject to exposure, like the blind loquacious Samson. Angels fold their wings. God is light, but a blinding light, the essence of his ways taboo to created eyes. Milton could solicit prophetic vision, command the light with godlike authority, and accomplish his life's ambition because, blind as he composed, *he had already suffered the punishment* men dread for such ventures. He did not come to the light with criminal intent—far from it; the psychic transaction that brought him *Paradise Lost* is probably as close as any religious poet has ever come to a non-satanic self-creation. He appeared before the court of God, I think, with a stilled apprehension. For if the result was guiltworthy, he had paid the penalty before committing the crime. Has not the Son, before he stands in the place of guilty men and suffers the stroke of his Father's justice, been given his throne? To balance the books, then, Milton demanded his wish, assembling about him symbols of his desire and his deservedness. It is an inversion, a "turning around upon the other," of his old psychology of debt that lends to the invocations, especially in Book 3, their sublime proleptic sweep. Milton went blind without daring to see—now he could risk looking in. He had suffered the agony of Oedipus—now he could risk the liberty. Indebted for the gift of a talent, Milton repaid by collecting a debt from God.

Whether profane or sacred, the superego remains to some degree a dispensation of guilt. My argument here assumes that Milton, not altogether unconsciously, received his blindness as a punishment. At first, certainly, it was a noble sacrifice for liberty, but as that liberty failed, the meaning of his sacrifice became newly mysterious to him. "They also serve who only stand and wait." Although Sonnet XIX cannot be securely dated, it reveals that Milton was able to strike a posture of utter indecision toward the meaning of his spent light: the sense of the condition is unknown; patiently he awaits an interpretation. Every one of the symbols interpreting blindness in the invocation to light, when situated vis-à-vis the creative acts of the Godhead, informs us that ever-during dark is the strict precondition for the accomplishment of the poet's task. This private theodicy would have no persuasive heft unless the deed of the poet were commensurate with the fate he has suffered—unless the poem sanctioned by God were equivalent, an eye for an eye, to the blindness sanctioned by God. The inversion of the debt presupposes that his affliction is deserved as well as deserving, and that in turn presupposes his guilt. Confined for his sins in a "moving Grave" (102), Samson rises from an "ashy womb," "From under ashes into sudden flame." Confined for his sins in the "darkness visible" of hell, Satan rises with fiery virtue to master Chaos and conquer earth. The guilty earth itself, in the end, "shall burn, and from her ashes spring / New Heav'n and Earth" (3.334–335). Milton, too, lived to behold contrary meanings in the eclipse of light.

But the public revelations of this poet betray a meager appetite for self-accusation. The glaring difference between the Augustinian and the Miltonic confessions is that Milton never once acknowledges inexcusable moral fault. One would suppose that theodicies are undertaken to preserve the spirit of repentance. But when, in public, was he the least bit penitent? In the *Second Defense* he explicitly denies that blindness is penal:

> For my part, I call upon Thee, my God, who knowest my inmost mind and all my thoughts, to witness that (although I have repeatedly examined myself on this point as earnestly as I could, and have searched all the corners of my life) I am conscious of nothing, or of no deed, either recent or remote, whose wickedness could justly occasion or invite upon me this supreme misfortune. (*CP* IV, 587)

Then let those who slander the judgments of God cease to speak evil and invent empty tales about me. Let them be sure that I feel

> neither regret nor shame for my lot, that I stand unmoved and steady in my resolution, that I neither discern nor endure the anger of God, that in fact I know and recognize in the most momentous affairs his fatherly mercy and kindness towards me, and especially in this fact, that with his consolation strengthening my spirit I bow to his divine will, dwelling more often on what he has bestowed on me than on what he has denied. Finally, let them rest assured that I would not exchange the consciousness of my achievement for any deed of theirs, be it ever so righteous, nor would I be deprived of the recollection of my deeds, ever a source of gratitude and repose. (*CP* IV, 589)

Paradise Lost reveals that Milton had not finished examining himself "on this point." The truth is that in his way he did feel regret and shame for his lot, did discern and endure the anger of God, and for this reason enjoyed "his fatherly mercy and kindness." The first gesture of theodicy, as God demonstrates, is the condemnation of "Ingrate" man. Though encoded in a peculiar language reconciling metaphor and fact, self-accusation occurs exactly where one would expect it to occur in the epic: noontime in the garden.

• 5 •

"ONE FIRST MATTER ALL": SPIRIT AS ENERGY

MILTON never conquered the last infirmity of his noble mind. "So were I equall'd with them in renown," he writes of blind predecessors in the invocation to light, pricked as in his youth by the spur of fame:

> I began thus farre to assent both to them and divers of my friends here at home, and not lesse to an inward prompting which now grew daily upon me, that by labour and intent study (which I take to be my portion in this life) joyn'd with the strong propensity of nature, I might perhaps leave something so written to aftertimes, as they should not willingly let it die. These thoughts at once possest me, and these other. That if I were certain to write as men buy Leases, for three lives and downward, there ought no regard be sooner had, then to Gods glory by the honour and instruction of my country. (*CP* I, 810)

Preparing his claim on the attention of aftertimes, the scrivener's son remembered that when three men purchased a lease, the lease was good "downward" until the partner who was, at the purchase, endowed with the most future owned everything by himself. What does it mean "to write as men buy Leases"? Note the triads: friends abroad, friends at home, inward prompting; labor, intent study, strong propensity of nature; God's glory, honor of my country, instruction of my country. Twice the greatest of the three is the last to be announced—inward prompting, strong propensity of nature. Then, resolution having "possest" him, God's glory emerges as the first-named and future-happy partner in the lease on fame, imaginatively subsuming inward prompting and strong propensity of nature, just as "honour" reaches back to the regard of friends and "instruction" to "labour and intent study." The passage shows that an unconquerable desire for fame, rooted in those grounds of the self closest to God, in the strength of one's nature and the call of one's conscience, can be converted from an infirmity to a motive for righteousness provided that the glory sought remains subordinate to God's glory.[1] It is as

well, like so many of Milton's pronouncements before the completion of *Paradise Lost*, oracular. Writing as men buy leases, for three lives and downward, Milton produced three major poems, or, to divide his literary creations in another way, three books: *Poems* (1645, 1673), *Paradise Lost* (1667, 1674), and *Paradise Regained . . . To which is added Samson Agonistes* (1671).

The ambition to gain a lease on new life in aftertimes led Milton to search famous works for whatever was constant or perennial about their fame. Thus he would glorify God while doing honor to his native language and instructing his countrymen—the invariant essence of enduring reputation: "That what the greatest and choycest wits of *Athens, Rome,* or modern *Italy*, and those Hebrews of old did for their country, I in my proportion with this over and above of being a Christian, might doe for mine" (*CP* I, 812). But as we know from the poems themselves, Milton also searched monumental literature for lacunae, "Things unattempted yet in Prose or Rhyme" and deeds "unrecorded left through many an Age." The attunement of his ambition to what famous literature lacked went beyond an eye for worthy but untreated subjects. In a deeper sense, the history of fame could offer him a partial guidance only, because fame in essence is not historical. It is, like life, a thin-spun thread, essentially now and in the future: the once famous has fame no more. "I might perhaps leave something so written to aftertimes, as they should not willingly let it die." The syntax makes "to" two, doing double duty with "leave (to)" and "written (to)." Writing to aftertimes—and all the more so when contemporary England proved itself dishonorable and incorrigible—Milton would have extrapolated from the ideals and dissatisfactions of seventeenth-century literary culture those principles which, consistent with inward prompting, strong propensity of nature, and God's glory, might be expected to thrive in the culture of aftertimes. It would be self-defeating to write for the monks who had spun out the thread of fame for Vergil and Ovid.

The literary expectations of readers, Protestant readers in particular, were changing dramatically during the seventeenth century. Nursing his poetic ambition through the two decades of the Puritan experiment, Milton could not fail to have noticed that the rhetorical culture of Renaissance humanism, so deferential to the status of literature, was being displaced in his own lifetime because its revered texts neglected "solid," "useful," and "profitable" knowledge of "things." Grandeur of invention might no longer be capital enough in the changing economy of literary fame. Sidney could glory in the aloof fictionality of literature, its "second nature" answerable only to a lofty and abstract conception of virtue, but after Baconianism had worked its way into the reforming

spirit of Milton's own political and religious allies, the old defense might have looked more like an exposure. Milton himself would not have questioned the desire for high imaginative literature to strive, decorum permitting, to be accurate about the world. The thingness or substantiality of this world was, in his theology, no small part of the glory of God. He favored a marriage between the old humanism and the new empiricism, not a divorce. Although he was critical of the universities on partially Baconian grounds, his *Of Education* was unique among the many publications commissioned by Samuel Hartlib in its refusal to sever Baconian reforms from the standard humanist curriculum, and he expressed special warmth for authors who, like Celsus, combined reliable science with literary values.[2] An epic on Genesis afforded him a grand opportunity to introduce new harmony where others were beginning to perceive an inevitable disparity.

We are familiar with the critique of heroism resulting from conversion—or as Milton would see it, the reversion—of classical epic to biblical materials in *Paradise Lost*. Yet the ancient epic was not a univocal form. If Homer and Vergil had celebrated the history of heroic action, godless Lucretius had celebrated the "natural history" of scientific truth, his "great argument" being the sufficiency of matter to have produced, and to sustain in all its variety, the universe. Might not he, too, be converted? The subject of *Paradise Lost* invited a rapprochement between the two precedents, an elegant new variation on the plasticity of epic form, for in Christian culture the Book of Genesis had from the beginning accommodated both these urges, serving as the foundation of salvation history and the foundation of empirical knowledge. Whereas the pagan epic had concerned either the cultural world or the natural world, Genesis had always been understood as the reconciliation of narrative and encyclopedia, history and hexaemeron. So the Bible according to Milton would welcome, within limits, scientific speculation: the various attempts to establish a credible science in *Paradise Lost* reveal a poet who had imposed on himself a double and peculiarly modern obligation, responsible to the nature of things as well as to the meaning of Christian history. In the magic folds of his epic container we find political, legal, military, and pedagogical oration, drama, hymn, panegyric, love lyric, elegy, allegory, dream allegory, Ovidian metamorphosis, pastoral, prayer, sermon, exegesis, philosophical symposium, and the education of a prince—all of these kinds of speech act, and more, fitted to the narrative of mythical events. We also find, on the side of scientific explanation, an elaborate representation of the ubiquity of matter, a treatise on atoms, a dialogue on world systems, georgic, the opposite of georgic in the satanic inventions of mining,

gunpowder, and weaponry, and dietary advice ranging from a cosmic physiology to some notes toward a vegetarian cookbook. Milton's joining of the Homeric-Vergilian tradition with the Lucretian tradition is among the most sustained of the many generic mergers in the poem. It is better managed, perhaps, than the shuffle of cetology with heroic quest in *Moby Dick*.

Galileo complained to Kepler about the blind humanism of those who preferred books to the direct observation of nature: "Really, as some have shut their ears, these have shut their eyes toward the light of truth. This is an awful thing, but it does not astonish me. This sort of person thinks that philosophy is a book like the Aeneid or Odyssey and that one has not to search for truth in the world of nature."[3] Sturdy science was Milton's bulwark against the dissatisfaction overtaking classical poets and, in some quarters, holy writ itself. By inserting Galileo into the epic world of Homer and Vergil, he could at the same stroke outdo the past and protect himself from obsolescence in the future.

Fame was not the sole motive. Other wishes, just as intimate, converged on the Lucretian half of his mission. The poem is usually read with heroism at the center and science at the periphery. But Milton achieved such integration in *Paradise Lost* that we can transpose this model without losing either the moral design of the epic or the character of its author. Philosophical and theological commitments spurred him in the same direction as fame, and beneath this complex of motives we can discern the psychological drama between the poet and God whose successful resolution is the very condition for the existence of the poem, the illumination of "what is dark" and support of "what is low" without which Milton could not have justified the ways of God to men.

Earth and Sky

The sidereal vistas of *Paradise Lost*, though triumphantly responsive to the emotional tone of Renaissance astronomy, bear signs of trouble and confusion when judged by standards appropriate to Lucretian epic. The universe is unmistakably terracentric; we look through the telescope of Galileo (the only contemporary mentioned in the epic) at the universe of Ptolemy. One supposes that Milton, preparing the boldest of his three late signatures on the lease of fame, correctly sensed Galileo to have been the most future-happy man he ever met, of whom all Europe still talks from side to side. Yet there were, above and beyond the conflicts within astronomy, conflicts about it. As fame dug in the spur, something else pulled back on the reins.

When Adam initiates the dialogue on world systems by expressing his bafflement over the inexplicable *copia* of the firmament (8.15–38), he is appropriating the bedtime curiosity of Eve in Book 4: "But wherefore all night long shine these, for whom / This glorious sight, when sleep hath shut all eyes" (658–659)? Satan's dream-given answer to her query betrays the dark context impinging on such innocent questions. "Why sleep'st thou *Eve*?"

> now reigns
> Full Orb'd the Moon, and with more pleasing light
> Shadowy sets off the face of things; in vain,
> If none regard; Heav'n wakes with all his eyes,
> Whom to behold but thee, Nature's desire,
> In whose sight all things joy, with ravishment
> Attracted by thy beauty still to gaze. (5.41–47)

In a gesture repeating his arousal of Beelzebub, Satan calls Eve forth into the night, the matriarchal half of day ruled by the "Full Orb'd" moon, and explains the abundance of stars as a vast field of sparkling eyes yearning to observe her sovereign beauty. Given that the question Satan answers here, when revived before an angel in Book 8, yields an admonishment to earthly sight, it is possible to recognize in this portion of the dream a moral and psychological indictment of the lure of astronomy—the science practiced at night, its implicit assumption being that the heavens shine "in vain, / If none regard." The stargazer is gazed upon. Behind the desire to see the heavens may stand a prideful exhibitionism, an urge to know that, mastering everything, would install itself as the sole object of wonder in the universe. Yet Milton wanted to know, and insofar as he chose a particular star map, presumed to know. Compounding empirical uncertainty with an attitude near to hypocrisy, Milton—his poetic design behind him, nodding "yes"—has Raphael answer Adam's straightforward curiosity about the heavens with a moral assault on the very idea of astronomy.

These brief remarks about the firmament in *Paradise Lost* alert us to some of the pressures arising from Milton's effort to accommodate diverse kinds of seventeenth-century truth. The contradiction between the professed unknowability of the world system and the tacit choice of a specific system remains in the poem as evidence of his acceptance of, and discomfort with, the demands imposed on him by the Lucretian side of his intentions. The disparity between astronomy as presumption, a prideful effort to see what God has placed at a distance from man, and

the complicity with astronomy implied in the allusions to Galileo and his telescope reflects a real ambivalence over the question of what limits there are, or should be, to human knowledge. It is a symptom of conflicts nearer to the moral core of the poem—about the limits placed on wonder, the extent to which the fall is irreversible, and the ultimate rationality of divine commandment.

Necessary as a stage for the poem's action, sidereal order has no important relation to Milton's theology. But other positions in *De doctrina christiana* are science-related, and some of them might be considered science-dependent. The systematic theologian deduced from scripture a heretical monism of matter and spirit, and if this doctrine were to be fleshed out in an epic at once Vergilian and Lucretian, Milton would require convincing scientific support. His portrayal of matter, though motivated by theological and exegetical reasoning, had to meet the standards of empirical discourse that threaten his presentation of astronomy—clarity, consistency, plausibility. The science of the sky was a plaything by comparison.

Raphael establishes this priority. "Heav'n is for thee too high / To know what passes there; be lowly wise" (8.172–173). Concluding the dialogue on world systems, this injunction to "be lowly wise" returns Adam to the beginning of their symposium, to Book 5 and the questions about nourishment. This "lowly" inquiry rose from a point of hospitality to information about the nature and health of a body, a definition of the celestial world in terms of biology, not astronomy, and from thence to a revelation of how the heavens, spatially dividing earth from God, might be properly traversed by man. Unlike Adam's astronomical questions, these were answered without equivocation, and the answers were conspicuously plentiful, giving far more than Adam requested, not less. After his uncomfortable entanglement with the science of the sky, Milton has his angel point back to the more important delineation of the first science to be founded on this earth—medicine, a "wise" discipline fit to human proportions. "Think only what concerns thee and thy being." Since food is the daily renewal of the gift of life, and since God's one commandment announces an exception to the causal bond between eating and living, Adam has every reason to know what it means, in *Paradise Lost*, to be an organism.

Like astronomy, medicine was an embattled discipline in the seventeenth century, but its disputes bore immediately on life and death, religious belief, educational theory, and social allegiance generally. The deeper one reads in the prose of the century, the more one realizes that the gradual transformation of alchemy into chemistry, rather than

the shift from astrology to astronomy, was responsible for the stormiest passions in the climate of the new empiricism. While the heliocentric theory precipitated a largely metaphysical crisis, supplying what is probably the most striking example ever of the appearance/reality dichotomy, medicine *mattered*. The struggle between Galenists and Paracelsians was thoroughly involved in the social turmoils of Renaissance Europe. In England, Puritan educational reformers almost unanimously preferred the new medicine, developed to a large extent by scholar-gypsies proudly unaffiliated with established academies, to the well-funded Galenism of university physicians; during the Interregnum a class conflict between the College of Physicians and the apothecaries of London could crystallize in an intellectual disagreement over the relative efficacy of traditional herbal medicines and the new chemical potions of Paracelsus and Van Helmont.[4] In his utopian fragment "Macaria," Gabriel Plattes continued the Rosicrucian tradition by projecting an ideal England in which priests would be doctors, trained in the marvels of medical alchemy; his example may serve to remind us that in the utopian narratives so popular during this period hidden societies are regularly associated with "hidden" or occult sciences, and that an abiding feature of Renaissance utopianism is the representation of an ideal state of the flesh, regaining by medical art the long life enjoyed by the patriarchs of Israel.[5] Another of Hartlib's associates, Noah Biggs, alluded favorably to Milton in an attack on Galenism. While writing *Paradise Lost*, Milton numbered among his friends Henry Oldenburg and Theodore Haak, both aligned with the Paracelsian movement. But the important acquaintance in this regard was Nathan Paget, who offered his cousin as Milton's third wife and was accepted; this physician possessed a library remarkable for its holdings in chemical philosophy.[6] Scholars and annotators, however, have generally denied that Milton had serious knowledge of medicine, even a lively amateur interest in the issues dividing this science.

"How much it means," he wrote in Prolusion VII, "to get an insight into the delicate anatomy of the human body and into its medical treatment."[7] It did not mean all that much, according to Svendsen, who judged that "the range of information and depth is narrow and small": "his medical allusions are traditional, even old-fashioned, familiar and acceptable to a mid-seventeenth-century reader if not to a late seventeenth-century scientist. . . .This is a literary and moralized kind of nature philosophy, not experimental or speculative."[8] Svendsen garnered his own medical lore primarily from the encyclopedias of Bartholomew, Batman, and La Primaudaye. Outmoded as they were

published, these superficial epitomes are probably adequate to the range of learning displayed in Milton's prose, where medical analogies, if numerous, belong to the world of Celsus rather than Paracelsus. There is nothing in *Paradise Lost* that would lead us to suppose its author an expert layman capable of assessing the work of "a late seventeenth-century scientist." But this is a high standard: would Abraham Cowley, author of "To Dr. Scarborough," "Ode Upon Dr. Harvey," "To the Royal Society," and *A Proposition for the Advancement of Experimental Philosophy*, belong in this category? There *are* indications in *Paradise Lost* that the blind poet, possibly through discussions with men like Paget, brought himself up-to-date on certain medical questions.[9]

Arguably the Christian Lucretius could not have written his epic without a physiology of some sort. For what was death? What happened to the human body in its quick transition from immortal to mortal state? What were the concrete effects of that "mortal taste"? In the seventeenth century the Old Testament, like pagan mythology in the hands of Bacon and Henry Reynolds, was receiving a highly rationalized exegesis. Theologians such as John Weemse, whose works have been said (imprudently, I admit) to contain more parallels with *Paradise Lost* than those of any other Protestant theologian in our language, understood that a physiological event of some kind must have accompanied the fall and endeavored to hammer out in medical terms a description of this event respectful of orthodox theology.[10] Were he not committed to the inseparability of matter and spirit, body and soul, Milton might have finessed the problem by treating death as a moral condition only. But he held his several materialist heresies in esteem. It was their prestige, their pride of place at the beginning of man's education, that made Raphael point back from the vagaries of astronomy to the science of the earth, that low wisdom framing questions about life and health—and in a fallen world, about death and disease.

I am not in a position to say whether the monist arguments of the *De doctrina* constitute distinguished exegesis. As philosophy, they are not sophisticated. What impresses me most about them is the rough determination of the thought, its willingness to risk unpromising consequences for the sake of one tremendous loyalty: *the two worlds of orthodox Christian idealism must be joined at the cost of materializing the soul.* Speculating about the origins of matter, Milton constructs a clumsy scenario involving its emergence *ex deo* and its status when, awaiting the informing power of God, it stands outside the coordinates of creation. Whenever God makes, however, he infuses matter with a sub-

stantial vitality. This vitalism—the prejudice Milton begins with and the truth toward which his awkward prehistory of matter leads—is impervious to miracle. God may or may not be able to "annihilate" his creations (Milton wavers on this point; see *CP* VI, 310, 419), but he cannot separate the vital form from the material receptacle once creation has occurred. Forms "are themselves material" (*CP* VI, 308). Souls cannot exist apart from bodies, because soul and body are not, within the context of any conceivable creation, including heaven, two entities. For example: God could not produce an orthodox death, raising a bodiless soul from a material corpse, even if he wished to do so. In that they are truly creations, and therefore distinct from God, all possible worlds have as their *arche* the permanent cancellation of a dualism between mind and matter.[11]

This spiritual materialism suggests that Milton approached the two sides of his epic with a deep presumption of unity. Medicine and moral theology would seem to be, on his own account, aspects of the same subject. It is also important to reflect on the distance between these ideas and the orthodox ones. Milton has subverted the fundamental "picture" of the world by which most of his contemporaries navigated. When they looked at two trees they were disposed by centuries of metaphysics to a particular construction of the percept. What makes the trees different? Their matter, their substance. And what makes them alike? Their form, their essence. But for Milton they are alike in stemming from "One first matter," and the form joined inseparably with this matter is the principle of individuality—a profoundly heterodox view of things shared by Giordano Bruno: "That which is common has the function of matter; that which is individual and brings about the distinction, has the function of form."[12] The degree of Milton's heterodoxy gives an urgency to the cosmological or Lucretian coherence of *Paradise Lost*. More than creating a world, he was demonstrating one.

And more than demonstrating one. If astronomy for Milton was a science of intellectual curiosity, medicine was a science of personal crisis. The twinned themes of body and soul in *Paradise Lost* afford a glimpse at what "mortal taste" meant to our author in the most private world of all.

Victim of a Crudity

Every reader of Milton, coming to the "blind fury" of "Lycidas" or the speeches on nighttime vice in *Comus*, must be struck by the irony

of these early associations—so innocent, so impersonally traditional!—between blindness and evil. A similar irony should arise when we read in *Of Education* about the value of being able to "manage a crudity"— a spell of indigestion—"which he who can wisely and timely doe, is . . . a great Physician to himself" (*CP* II, 392–393), since Milton certainly believed that digestive failure was the efficient cause of his blindness.

Consenting to describe the course of his symptoms to Leonard Philaris, he began with this statement: "It was ten years, I think, more or less, since I noticed my sight becoming weak and growing dim, *and at the same time* my spleen and all my viscera burdened and shaken with flatulence" (*CP* IV, 869; my italics). A few sentences later he says that in the present, "permanent vapors seem to have settled upon my entire forehead and temples, which press and oppress my eyes with a sort of sleepy heaviness, *especially from mealtime to evening*" (my italics). Early biographers, sharing that odd interest people used to have in what great men eat, took note of his temperate diet (*EL*, 29, 36, 194, 207, 279), but Aubrey recorded that, nonetheless, he often used the laxative "manna" (*EL*, 8)—not an uncommon practice in the seventeenth century with its heavy fare of meats and pastries. The connection between improper digestion and loss of sight in his letter to Philaris must not be mistaken for a quirk of personal fastidiousness. It rather indicates Milton's formal knowledge of what *should be* connected in his case history. The Englishman Walter Bailey summarized the advice of many physicians, Fernel and Riolan among them: "As generally in the preservation of health: so especially to continue the sight, it is convenient that the body be obedient, and do his office of evacuation accordingly."[13] Revealing to Philaris his chronically disobedient digestion, Milton was organizing his symptoms to accord with the assumptions of Renaissance ophthalmology. By this time (September 1654) he already knew the diagnosis.

"So thick a drop serene hath quencht thir Orbs, / Or dim suffusion veil'd" (3.25–26). We infer from this passage in *Paradise Lost* and from the description of the normal appearance of his blind eyes ("without a cloud") in the *Second Defense* (*CP* IV, 583) that Milton believed himself a victim of *gutta serena* or *suffusio obscura*, in which a congealed, "thick" humour obstructed the optic nerves, leaving the "orbs" (L. *orbitas*, "eyesocket") "quencht" or dry from the absence of vitreous spirits. The immediate cause of the obstruction, in this as in all cases of cataractic blindness, was the failure of the body to evacuate "vapors" produced during the digestion of food. Collecting in the troughs of the eyes, these unpurged vapors hardened into opacity. Richard Bani-

ster, the foremost English ophthalmologist of the seventeenth century, lists the "outward causes" of *gutta serena* as follows: "The forerunners and (as it were) outward causes of this disease, are much rawnesse, or ill digestion, drinking of pure wine, great heat of the Sunne, or cold on the head, continuall reading, bathes after meate, vomiting, immoderate company with women, holding in of the breath, as we see in Trumpetors: for these things fill the head with vapours."[14] From this catalogue we can confidently select those windy behaviors Milton believed to have resulted in his blindness. Digestive vapors, the inner winds of an unnatural flatulence, had condensed in his optic nerves, a process hastened by "continual reading"—specifically, as we are told in the *Second Defense* (*CP* IV, 587–588), the preparation of his first assault on Salmasius.

Seventeenth-century physicians did not enjoy the aura of sanctity that settled on their posterity. The medical charlatan was a stock character of Renaissance drama and skeptical ridicule of physicians a common motive in Renaissance satire. Thomas Browne noted in the opening of *Religio Medici* that the "generall scandall of my profession" made the very title of his book a paradox, alluding, probably, to a proverb quoted by his follower Joseph Glanvill: *ubi tres medici, duo athei* (show me three physicians, and I'll show you two atheists).[15] Yet these commonplace attitudes are nowhere heard in the poetry of Milton, and rarely in his prose. He liked physicians and appears to have respected their science.

Charles Diodati, the great friend of his youth and the son of an eminent physician, went abroad to study medicine. In midlife we find Milton contributing to the projects of Samuel Hartlib, champion of utopian medicine. As blindness closed in, he seems to have consulted many physicians. The account of Edward Phillips implies that family and friends tried unsuccessfully to dissuade him: "It is most certainly known that his sight, what with his continual study, his being subject to headache, and his perpetual tampering with physic to preserve it, had been decaying for above a dozen years before" (*EL*, 76).[16] There is good reason to doubt whether the damaging effect of "physic" ever became "most certainly known" to Milton himself. The anonymous biographer also cast this information in an indefinite construction: "And the issues and setons, made use of to save or retrieve that [the sight of his left eye], were thought, by drawing away the spirits which should have supplied the optic vessels, to have hastened the loss of the other" (*EL*, 28). Yet by whom was this "thought"? Although we have poems, pamphlets, and letters in which Milton discussed his

blindness, he never once blamed the affliction on "tampering with physic." I suspect that the word "tampering" in Phillips expresses the long-standing exasperation of relatives forced to deal with a stubborn patriarch, and trying to enjoy in this instance the belated triumph of "we told you so." When, after he had lived in total blindness for two years, young Philaris urged him to record his symptoms for the French physician François Thevenin, Milton did so willingly, if not hopefully. It must have been tempting to blame the worsening of his disease on the interventions of his physicians, as many Renaissance patients did, but Milton seems to have forsworn the familiar defensive drama ending in the accusation of betrayal. Paget was among the closest friends of his old age, and given the doctor's keen interest in poetry, he may have represented for Milton a partial restoration of the lost Diodati.[17] Samuel Barrow, physician-in-ordinary to Charles II, donated a friendly poem to the second edition of *Paradise Lost*. As Milton did not disparage the light that had deserted him, so apparently he did not indict the profession that had failed him.

Is it not reasonable to assume that references to digestion in his subsequent poetry would exude something of the emotional intensity and intellectual gravity that characterize his late references to darkness, light, and vision? To confine myself to two obvious examples, the medical theory of catharsis as digestive purgation in *Samson Agonistes* and the banquet scene added to the sources for *Paradise Regained* take on new resonance as the artistic decisions of a man whose food had solidified into an impenetrable barrier between his mind and the world. But the effects of Renaissance diagnostics cannot be limited to gross correlations between Milton's illness and his works: in the explanations given for illness by all cultures—unappeased ancestors, the evil eye, germs—personal and public mythologies commingle and perpetuate each other. Psychologically, the Renaissance understanding of *gutta serena* would tend to release fantasies of having been poisoned, undone by a dependency on food and those who provide it, and such fantasies have their roots in the primordial expression of love between mother and child. Milton chose to write of the great myth of the evil meal before he became blind, but not perhaps before the simultaneous appearance of impaired sight and regular flatulence. Dalila is "a pois'nous bosom snake" (763). Her love-test, demanding that a husband show his devotion by entrusting her with the truth of his strength, repeats the pattern laid down by Mother Eve: what hair is for Samson, the potency derived from good nourishment is for Adam and other men. Did Milton feel, however irrationally, that Mary Powell, the wife in charge of feeding him during the years of rapidly deteriorating sight,

was guilty of a culinary lapse comparable to that of Eve in Book 9? Although blindness became for him a sign of prophetic mission, he might well have considered in his bleakest moods that his youthful association between pure diet and invulnerable virtue (*Comus*) or poetic ambition (Elegy 6) had proven tragically accurate. We know that his third marriage was based on a contract: should Elizabeth Minshull, cousin to a trusted physician, prepare food to his instructions, she would inherit his estate.[18]

It is probable that Mary Powell, like Dalila, rehearsed an ancient pattern, inasmuch as Milton's first nourisher (psychologically, if not actually), Sarah Milton, suffered from weak eyesight. The feeling that mothers give us the balm of life and simultaneously the poisons of disease and death is not uncommon in our mortal species. The art of Milton, which is uncommon, sets in motion the ambivalences surrounding women, food, trust, love, and death in a remarkably direct and undisguised manner, a manner, it should be said, true to the manifest import of the myth he is narrating, then orders these emotions in the lofty structure of a "great Argument" comprehending the symbolism of his religion. We have traced the redemption or refiguring of woman in *Paradise Regained* and in the poetic act as it is internally portrayed in *Paradise Lost.* Until the fatality of Book 9, the argumentative structure of the large epic tends to deflect our rage and regret over the loss of paradise from their customary objects—Adam and Eve, Eve primarily. Negative emotions light instead on Satan, the "cause" of our loss located in Book 1, or hover before God himself, suspended in the presence of an unimaginable object made a possible one by the sheer daring of theodicy itself: is *he* to blame for the entanglement of life in death? But as the theodicy gradually succeeds in establishing God's innocence of responsibility for the fall, Eve in particular stands accused, and the last three books of the poem seek to make peace with the ambivalent gift of life and death she bequeaths to her children. Ultimately Eve is to be cherished and celebrated. My psychological speculations about Sarah Milton and Mary Powell have literary consequences, in that the achievement of this composure would be nothing without the force, and centrality, and dexterity, of the ambivalence.

Comparing poetry to a chess game between the poet and experience, J. V. Cunningham asserts that

> The nature and quality of the game the two of them will play will depend not only on his own skill but also on the adroitness and resourcefulness of his opposing experience. If experience is

> simple, unseasoned, without depth, the poet may have an easy time of it but the game suffers; it lacks depth and interest. And if the opponent is dishonest, if he throws away a queen to extricate the poet from a difficult spot, the game will be shoddy and not worth playing. There are poems that are praised for their order, simplicity, and charm that so far as I can see represent a victory over nothing at all. We must distinguish between the simplicity of a Capablanca against Nimzovitch, of a master of chess against a master of chess, and the simplicity with which one defeats a friend when he negligently forgets to castle.[19]

In Milton, experience always castles:

> Meanwhile at Table *Eve*
> Minister'd naked, and thir flowing cups
> With pleasant liquors crown'd: O innocence
> Deserving Paradise! if ever, then,
> Then had the Sons of God excuse to have been
> Enamour'd at that sight; but in those hearts
> Love unlibidinous reign'd, nor jealousy
> Was understood, the injur'd Lover's Hell. (5.443–450)

This is a choice example of the "recognition, implicit in the expression of every experience, of other kinds of experience which are possible" that Eliot erroneously denied to Milton. Christopher Ricks, supplying a more Miltonic description of this quality, speaks of the "invoking of what is then deliberately excluded"—more Miltonic, not just in the word "invoking," but also in associating this kind of poetic effect with the mastery of conflict.[20] The obvious motive for the passage is to inform us that the nakedness of the hostess caused no libidinal complications in Eden. Yet, experience being a stiff opponent, this cannot get said without a strong evocation of its opposite. If ever angels, like the classical gods, had lusted for earthly women, it should have been . . . "then, / Then," the narrator repeats, his own excitement palpable. The libidinal power of Eve at this moment would have been, futhermore, *excuse*! Perhaps the ingenious treachery of sexual relations for this poet, the source of male vulnerability, can be epitomized by the fact that women, whether trustworthy like Elizabeth Minshull and the Eve of this passage, or poisonous like Mary Powell and the Eve of 9.983–999 ("For never did thy Beauty . . . so inflame my sense, / With ardor to enjoy thee, fairer now / Than ever"), were

never so attractive as when they "minister'd" at table. When they are most dangerous to men, women are most alluring. It was over the dinner table, according to the testimony of the maid, that a buoyant Milton, savoring the plate before him, exclaimed, "God have mercy, Betty, I see thou wilt perform according to thy promise in providing me such dishes as I think fit whilst I live, and when I die thou knowest that I have left thee all." The remark suggests that the match between the poet and his myth was immediate and visceral as well as contemplative and parabolic. Even the oath seems significant—God's mercy seen in a woman who "wilt perform according to [her] promise."

Milton was still tampering with physic in the design of his epic. The first education of Adam attempts to define, and make plausible, a conception of the original physiology of the human body. On the basis of this conception, Raphael projects a destiny for unfallen man answering to everything in Milton that would wish away physical affliction, blindness especially, without wishing away the body. Raphael is an authority suited by tradition to expand heavenward the idea of health, for this angel had long been the favorite of physicians. He delivered the prescriptions of the *Book of Noah*, an early Jewish treatise on medicine. Ficino invoked this precedent when defending the medical efficacy of grace, "because the soul is dependent on God, and the body on the soul. Do not the Hebrews consider that the Archangel Raphael practiced this art?"[21] His expertise was familiar enough for Cotton Mather to have included a chapter entitled "Raphael, Or, Notable Cures from the INVISIBLE WORLD" in *The Angel of Bethesda.* Entering the visible world of *Paradise Lost*, he begins his lessons by revealing that the chain of being is a food chain. Then he demonstrates: "and to taste / Think not I shall be nice" (5.432–433). To understand this crudity is to understand a good deal about *Paradise Lost.*

"Knowledge is as Food" I: The Wonderful Alchemist Within

One must begin with the shape of Milton's "great Argument." In the summary theodicy of Book 3, God first explains that he "made" Adam and Eve "just and right, / Sufficient to have stood, though free to fall" (98–99). Sufficient to have fallen, though free to stand; sufficient to have stood or fallen, though free to stand or fall: vary it as we will, freedom in the choice between two realizable alternatives is the initial premise of this theodicy. Although the freedom of a temporal creature

is oriented necessarily toward the future, since the past cannot be affected by will and therefore cannot be free, freedom must come to us from the past, inherent in our nature. God "made" man free, and because of this beginning, man is justly condemned and punished for his fault.

By the close of his speech God has redirected his theodicy to the question of mercy, distinguishing between the rebel angels, who will receive justice, and errant man, who will receive justice *and* mercy:

> The first sort by thir own suggestion fell,
> Self-tempted, self-deprav'd: Man falls deceiv'd
> By th' other first: Man therefore shall find grace,
> The other none: in Mercy and Justice both,
> Through Heav'n and Earth, so shall my glory excel,
> But Mercy first and last shall brightest shine. (129–134)

With regard to the event of the fall, the fact that it happened, the justification of God requires freedom. Once there is a fall, however, we freely fallen humans may find God just in executing the sentence for disobedience, but merciless: as argued in Chapter 3, justice for Adam and Eve coincides with the success of Satan in conquering an empire for Sin and Death. Nothing in justice alone compels us to love its executor. Theodicy must temper the result of its first argument in order to protect the impulse to love and worship God. Lest he align himself with the rebellion of justly doomed Satan, man must again have something to lose. So mercy is the complement to freedom. As freedom lies in nature, mercy lies in grace. We are "made" free, but "shall find" grace. Mercy will be gradually bestowed. It is a promise or covenant inextricable from the redemptive history contrived in the ensuing conversation between Father and Son.

In sum, two theodicial arguments. The first is significantly philosophical: because man is "self-knowing" (7.510), he will discover freedom in the act of knowing himself. The second is specifically Christian, a renovation in faith proceeding from "the Mercie-seat above" (11.2). Freedom—the gift of the beginning, grounded in our nature, a state we already enjoy—exonerates God from the charge of being responsible for the event of the fall. Mercy—the gift of the end, complete only in the hereafter, a promise we live toward in history—exonerates God from the charge of exemplifying a cold justice consistent with the reign of Satan. The two educations of Adam, by Raphael in

Books 5–8 and by Michael in Books 11–12, amplify and complicate this large division in Milton's theodicy. Thus, God instructs Raphael to advise Adam "of his happy state, / Happiness in his power left free to will, / Left to his own free Will, his Will though free, / Yet mutable" (5.234–237), while Michael is to "reveal / To *Adam* what shall come in future days, / As I shall thee enlighten, intermix / My Cov'nant in the woman's seed renew'd" (11.113–116)—in other words, to manifest the merciful history concealed in the "mysterious terms" (10.173) of judgment.

In the myth of Genesis equals cannot visit unfallen men. Society is restricted to the relationship between mates, and as a result, large areas of our experience appear to be absent from our myth of its perfection. Milton remedied this defect. By having Raphael call on Adam and Eve, he introduced the themes of hospitality, the alien, and the social life generally into his vision of bliss. This narrative and thematic decision may remind us that Milton worked in foreign affairs for the Commonwealth, just as the relationship struck between Adam and Eve and their visitor may remind us of Milton's other profession as a tutor. The socializing beings are, as they had to be, unequal. But Milton went to great lengths to stress the being these different beings share. One of the finest strokes in the poem lies in showing us how the ontological relationship between angels and men is filled out in the biology and psychology of unfallen man, at once widening the scope of his bliss and establishing the meaning of his fall. The angel reveals a lot about our happy state that, in many versions of the myth, bliss is bliss only by remaining ignorant of.

Amplifying the happiness in freedom and the freedom in happiness, man and angel range through various subjects over their long vegetable lunch. Do angels eat—and if so, how? The cosmological material elicited by this question leads to the history of Satan's rebellion and the victory of the Son. Adam's curiosity about the origin of his world prompts the sequel to this victory in the story of the Son's Creation in Book 7; his curiosity about the heavens prompts the dialogue on astronomy in Book 8. Finally, his own birth narrative turns the conversation to love, and the symposium ends with a question reminiscent of the initial one. Do angels have sex—and if so how? One suspects that the sequence is not a desultory one, although to my knowledge no one has convincingly defined its rationale. While I do not wish to banish other possibilities, the eutaxy made available by psychoanalysis is particularly elegant in this instance. Once we realize that the framing questions suggest maturation, the intervening

material sorts itself out in accord with the beginning and end. Food and digestion; a rebellion against the Father taking, as we will see, an anal shape; the triumphant installation of Christ as the ego-ideal; the reward of paternal and creative power given to this ideally obedient Son; the limits of ambition; genital love: the progress of this intellectual afternoon reproduces in broad outline the developmental history of the male child's psychological concerns.[22] The first education of man evokes the lived education of all men.

There is, however, an oddity befitting paradise. Whereas the theme of this history in the maturation of fallen men is the gradual submergence of the yearning for omnipotence, Adam's instructor gives rational endorsement and rhetorical encouragement to a fantasy commonly expressive of this yearning.

Adam keeps discovering the superiority of the angel body to his own. Presumably his senses inform him that, in the narrator's words, "what redounds, transpires / Through Spirits with ease" (5.438–439) in the process of digestion. During the narrative of heavenly warfare, he learns that, if mutilated, the angel body heals itself "by native vigor" (6.436). Matter would not be matter unless it were somehow cohesive, but as matter approaches God in *Paradise Lost* its composition becomes "more refin'd, more spiritous, and pure" (5.475)—and this elasticity accounts for the great advantages of the angel body in eating and battling. As angelic corporeality allows for the pleasure of eating, it avoids the embarrassment of evacuation (an embarrassment, I infer, since evacuation is the one everyday function of the body whose Edenic exercise Milton has not seen fit to tell us about directly), and as it permits genuine warfare, it precludes genuine mutilation:

> Yet soon he heal'd; for Spirits that live throughout
> Vital in every part, not as frail man
> In Entrails, Heart or Head, Liver or Reins,
> Cannot but by annihilating die;
> Nor in thir liquid texture mortal wound
> Receive, no more than can the fluid Air:
> All Heart they live, all Head, all Eye, all Ear,
> All Intellect, all Sense, and as they please,
> They Limb themselves, and color, shape or size
> Assume, as likes them best, condense or rare. (6.344–353)

More perfectly than the man who can manage a crudity in *Of Education*, angels are physicians and surgeons unto themselves. The healer,

external to the patient in a fallen world, is incorporate within them. Let us set aside for a moment the monism of the author: why such an emphasis on the angel body in the discourse of this teacher to this pupil? Adam, "frail man" even in paradise, is clearly (note his "sudden mind" at 5.452–460) being seeded with a wish.

To show that there could be many different gifts of the Spirit without the Spirit being disunified, St. Paul appealed to the articulation of the body:

> For to one is given by Spirit the word of wisdom; to another the word of knowledge by the same Spirit; to another faith by the same Spirit; to another the gifts of healing by the same Spirit. . . . But all these worketh that one and the selfsame Spirit, dividing to every man severally as he will. For as the body is one, and hath many members, and all the members of that one body, being many, are one body: so also is Christ. . . . If the foot shall say, Because I am not the hand, I am not of the body; is it therefore not of the body? And if the ear shall say, Because I am not the eye, I am not of the body; is it therefore not of the body? If the whole body were an eye, where were the hearing? If the whole were hearing, where were the smelling? But now hath God set the members every one of them in the body, as it hath pleased him. (1 Cor. 12:8–18)

"And if they were all one member," he continues in the next verse, "where were the body?" Among the angels, Milton answers. Satan healed soon, but the bodies of the apostate angels coarsen as the result of their disobedience (6.661). The loyal angels, on the other hand, "under one Head more near / United" (5.830–831) and possessing all the gifts of the Spirit given to Christ (5.844–845: "all honor to him done / Returns our own"), enjoy completely the ability to refashion at will the relationship between the whole of the body and its parts. It is as if, in Milton's imagination, a change in the spiritual situation Paul defines required a corresponding adjustment in the actuality of the trope he chooses to define it by: one who has every kind of Spirit will have every kind of body. And what do frail men wish for if they wish for this? Freedom as positive power, freedom to do "as they please," "as likes them best."

Angels are the object of wish because their bodies are wholly at the command of wish. Attachment to the human body complicates even prelapsarian freedom with a realm of involuntary process. But the

will of an angel lays hold of its entire being, and substance is at one with spirit. The angel body has not been *organized*, fixed into an arrangement of parts, each of which performs a designated function, nor has this body been *engendered*:

> For Spirits when they please
> Can either Sex assume, or both; so soft
> And uncompounded is thir Essence pure,
> Not ti'ed or manacl'd with joint or limb,
> Nor founded on the brittle strength of bones,
> Like cumbrous flesh; but in what shape they choose
> Dilated or condens't, bright or obscure,
> Can execute their aery purposes,
> And works of love or enmity fulfil. (1.423–431)

All heart, all head, all eye, all ear they live, and since every part can become any part, all genital "or both"—no permutation is inadmissible. I assume that no one fails to perceive the envy of the narrator in this passage. When Raphael takes up the question of angelic sexuality, his diction, echoing the narrator's "cumbrous flesh," invites a similar response from Adam. For he does not merely say, as some critics give the impression in their haste to clothe the divinization of the flesh at the end of Book 8 with the shameful catastrophe of Book 9, that angels also do this thing. He says they do this thing "In eminence" because they are not burdened, like humans, with the gravity of "restrain'd conveyance."

Reflecting the natural rhythm of the day in his shift from a noontime concern with what the great eat to an evening concern with how the great make love, Adam imagines two ways in which angels might couple:

> Love not the heav'nly Spirits, and how thir Love
> Express they, by looks only, or do they mix
> Irradiance, virtual or immediate touch? (8.615–617)

Love in the sexual imagination of Adam is *touch*, whether "virtual" as in looking or "immediate" as in the contiguity of boundaries. Raphael expands his horizons. Angels mix totally, exempt from the law that two bodies cannot inhabit the same space, which binds only dense and "exclusive" corporeality:

Whatever pure thou in the body enjoy'st
(And pure thou wert created) we enjoy
In eminence, and obstacle find none
Of membrane, joint, or limb, exclusive bars:
Easier than Air with Air, if Spirits embrace,
Total they mix, Union of Pure with Pure
Desiring; nor restrain'd conveyance need
As Flesh to mix with Flesh, or Soul with Soul. (8.622–629)

I take the sense of the last lines to be "Easier than air with air, if Spirits embrace, total they mix, a union of pure desiring with pure desiring." "Desiring," that is, is not a verb, but an emotional act become a substantive, a piece of word-smithing that imitates the angelic unity of mind and thing: "desiring angel" is technically redundant, since a desiring angel *is* pure desiring. This desiring is pointedly superior to our own "restrain'd conveyance." Both touch and sight are always of surfaces. To put this as clearly as Raphael does, we take joy in the friction of surfaces, and in intercourse flesh is the wall or limit or "obstacle" into which our desire repeatedly thuds. But angels, freed from "exclusive bars," become utterly and not superficially one. "Easier than Air with Air," they can will to be each other, for matter is what they have in common and its forms, what they have distinctly, are subject to volition.

Influential Miltonists have been reluctant to indulge this lavish fantasy. A. J. A. Waldock, reacting like a Renaissance Christian to the taint of Muhammedanism, calls Miltonic angelology "rather nonsensical," mutters snidely about "the excursus into the mysteries of angelic copulation (for which there was no real excuse at all)," and contrasts such indelicacies with the "supreme literary tact" of Dante, who excelled at fantasies of torture rather than pleasure.[23] On what basis, the criterion of sense having been invoked, is one view of angels more nonsensical than another? When he tried to imagine himself having lunch with an angel, Waldock apparently did not suppose that Adam's question would occur to him, or that a *real* angel would answer it—"there was no real excuse at all." Other critics, while admiring the passage, refuse to enter into the fantasy, to give themselves over to its concretion, but shroud it in transcendent mists. C. S. Lewis, who thought himself particularly skilled at detecting joy, cautioned that "these angelic fusions, since angels are corporeal, are not without pleasure: but we must not imagine it after the pattern of our

own specialized and rebellious senses." (In other words, don't imagine it, after having conceded with characteristic magnanimity that it is "not without pleasure," when in fact it is without anything else.) Even Joseph Summers, writing about the reverence for sexuality in Milton and placing him with Blake and Lawrence, concludes that "This 'eminence' of joy, this miraculous total 'mixing' without loss of identity, is beyond human capacities and therefore beyond even the desirable for man."[24] But human capacities do not determine the desirable. It is often the other way around; we desire to fly, for example, and attribute this power to gods and angels until we eventually acquire the capacity for ourselves. In any case, as we will see, this eminent joy is *not* outside human capacities in the unfallen world, which is why the rhetoric of the angel is calculated to solicit desire. Milton's angels fulfill a recurrent and impossible dream of earthly lovers. They desire and simultaneously possess: their "Union" of desiring and desiring would have been an empty solecism in the *Symposium* of Plato, where desire always presupposes lack. Remembering the fundamental purpose of marriage in Milton's view, we may say that angelic sexuality offers an absolute cure for the "defects" (8.419) of solitude.

As the myth of the fall is commonly related and understood, a punishment is specified for disobedience, but no reward established for obedience beyond the perpetuation of the status quo. Milton elaborates the motivational structure of prefallen virtue by instituting the reward normally thought to be superfluous. Free will, he seems to have reasoned, demands an ideal object, unattained but not unattainable, and this object is no less than its own perfection. Raphael invites Adam to desire the compliant body of an angel because it is his good destiny to acquire one:

> And from these corporal nutriments perhaps
> Your bodies may at last turn all to spirit,
> Improv'd by tract of time, and wing'd ascend
> Ethereal, as wee, or may at choice
> Here or in Heav'nly Paradises dwell;
> If ye be found obedient, and retain
> Unalterably firm his love entire
> Whose progeny you are. Meanwhile enjoy
> Your fill what happiness this happy state
> Can comprehend, incapable of more. (5.496–505)

Here Raphael, obeying the instructions of God, informs man of his "happy state." If Adam and Eve enjoy their "fill" of current happiness,

as beings "incapable of more," they will by natural process be "Improv'd," and thereby become capable of more. To gain they must act as if denied. If they were to fly from earth to heaven, for example, they could presumably determine the celestial design, extending the reach of "lowly" wisdom. Milton erects a symmetry between the two halves of his theodicy: like the promise of mercy given to fallen man, the freedom given to unfallen man is meant to unfold gradually from a beginning toward a perfect end. Nourished on the proper nutriments, the human body will slowly become an angel body, "wing'd" for ascent and "at choice" able to subsist on the airy food of heaven.

Any Christian writing of this myth would undergo an inner drama of mourning in creating a world commensurate to his wish, the paradise, then suffering its loss, learning in the most intimate way the truism that in this life such a world is available only to the imagination: "Sad task" (9.13). But as has been said many times from many perspectives, Milton could not have selected a more auspicious discipline for his talent. From the beginning his poetry activates and endeavors to resolve conflicts between wish and necessity—"Yet can I not persuade me thou art dead" in the elegy on his niece; the sudden "no" and severe "must" of "wisest Fate" in the Nativity Ode, curtailing the celebration of bliss; the complex bargaining for diverse wishes in the companion poems and in *Comus*; the fond dreams and false surmises that must be corrected before the consolation can be delivered in "Lycidas." In *Paradise Lost* the good destiny of the body intensifies the wish that must be sacrificed, and does so in a highly personal manner. For there could be no better fortification against *gutta serena* than an angel body. The diminished plasticity of human flesh entails its fragility: because my arm is just my arm, my eyes my eyes, they may be taken from me. The anatomical precondition for blindness, as Samson laments, is the fact that sight has been "To such a tender ball as th'eye confin'd" (94). No vapors could solidify in the optic nerves of vaporous beings who cannot be said to possess optic nerves, only substance temporarily imbued with that form, and at will become "condense or rare." They expel "what redounds . . . with ease" and live "all Eye." Perfect surgeons and physicians unto themselves, angels stand protected against dismemberment in all the forms human beings experience it, from the archaic threats and wounds evoked by the maturational design of Adam's education to the blindness of the poet. A body turned "all to spirit" would expunge frailty from our psychology as well as our physicality. The object of desire implanted during this education materializes the invulnerability, potency, and consummate wholeness of the ideal ego against which all our losses strike.

The placement of this body in *Paradise Lost* is far more artful poetically and psychologically than the placement in *Comus* of a similar wish for fleshly purity. Here the pure body has, besides the full authority of Christian myth, a powerfully dialectical setting. From the moment the angel appears and offers its corporeality to Adam and Eve as a possibility they might achieve, we understand that the possibility is destined to be lost and understand, therefore, that the desire of the poet (and of ourselves, to the degree we share his yearning) must constantly know itself against the truth. Paradise is a stern teacher. What is it to know oneself as fallen? To know that I was born in pain, labor hard, and will die. But what is it to know all that? It is at least to know myself as a creature in conflict with necessity, a creature, that is, who wishes. In order to desire the unclotted body of *Comus*, one must imaginatively forfeit the sexual life, which is a high price to pay, even in imagination, and in the end will probably teach you nothing. When we engage the angel body of *Paradise Lost*, however, we enter into the war between desire and history whose complex etiology is the subject of the myth of the fall. By the end of this chapter I hope to have shown that the full acknowldgment of a wish and the full mastery of its disillusionment empowers the creation of the epic. For the time being, we can discern that two parallel dramas on the theme of "the better fortitude / Of Patience and Heroic Martyrdom" (9.31–32) are proceeding on the same stage. As Adam discovers his happy state, and then in his second education prepares for a world negating his first happiness, so Milton relives through artistic invention his own agon with necessity, and particularly with blindness, necessity's epitome.

The irony of the temptation is that Satan offers a reward that God has already offered and, it would seem, already bestowed in the making of man. Transforming an awkward detail of the Bible into his strongest and most empirical argument, Milton's serpent claims to have eaten and become able to speak, rising a notch in the chain of being; a man among animals, he promises Eve that she will become, by "Proportional ascent" (9.936), an angel among men. Her last unfallen thought is "what hinders then / To reach, and feed at once both Body and Mind?" (9.778–779) Failed temptations always end this way. What hinders? What is stopping me? Only the law, but the law does not *stop* me. What hinders then to reach? Nothing. It is exquisitely appropriate that Eve conclude her unfallen life with a question whose answer is "nothing," the word tolling at the silent moment of the fall, and exquisitely appropriate that she reduce a spiritual possibility to a physical one, a higher order of hindrance to a lower one, since that is

precisely the reverse of the process by which bodies are fed, and men become angels. On every other tree man has the food that will "feed at once both Body and Mind." Yet the true route to the angels is a slow natural process, while the false way claims to be instantaneous, "at once," a magical transformation. The moral task set for man was to remain obedient, but also to be patient in awaiting the full dilation of freedom. It was not wrong to aspire. It was wrong to prefer the lying magic of Satan to the authentic science of God.

But this science is, as Waldock claims, "rather nonsensical." His remark has the benefit of prompting us to reflect on angelology. Whether angels were conceived of as princes, courtiers, couriers, soldiers, servants, spies, philosophers, magicians, or physicians, the rule governing speculation about these elusive creatures was to make their existence consonant with encyclopedic knowledge about nature. It was a branch of theology close to science, and its governing rule—that the fantastic be not impossible—resembles the mode of thought required today in the genre of science fiction.[25] Thus medieval angels were assigned to drive the Ptolemaic spheres, making the sky into a vast palimpsest of Western religions, with Hebraic-Christian angels to guide the Greco-Roman myths enshrined in the planets and constellations, and a Babylonian zodiac to house everyone. In the Renaissance, Cartesian conceptions of space and matter inspired Henry More, dubbed the "Angel of Christ's" by his classmates, to reconceive angelology in ways that Marjorie Nicolson and C. S. Lewis found compatible with *Paradise Lost*.[26] Occupying the border between nature and divinity, angelology is a fine example of the huge labor of cross-referencing that unified Renaissance knowledge, not different in kind from medieval thought but more intricate to the degree that the natural and human sciences were beginning to discover their autonomy and the correspondences linking part to whole were necessarily more finely textured and more evidently improvised. Still, if one sought out and consented to the correct authors, the seal of the whole could be observed on each part, and each entry in the cognitive encyclopedia could bear the inscription "see every other entry."

It facilitated art.[27] An intimate and, I would think for many readers today, patent fantasy surfaced in Milton's representation of the angel body. It is easy for us to ridicule these exhibitions of the unconscious, which is one reason why primitive fantasies rarely surface with so little distortion as this one in the literature of the last two centuries, save in those genres of the fantastic, including science fiction, set aside for them. Science has come to confer belief, to define the really credible,

more stringently than ever before, and its devastating rebukes to fantasy have tended to make the artist guarded, involuted, knowingly artful in putting his wishes into play. But the somewhat mythical empiricism of the Renaissance, teeming with magic and metaphor, beckoned to wish. Insofar as artistic creation involves the public elaboration of unconscious fantasy, the assumptions about man and the universe sanctioned by Renaissance science assisted this emergence. When Milton invested the citizens of heaven with spiritual bodies, the three masters of his ego—the superego, the id, and reality—could approve for their own different reasons the same representational act. So there is, then, a bit of science to legitimize Milton's nonsense and implant one of the commanding tropes of *Paradise Lost* in the field of reality. Exquisitely appropriate once again, it concerns sublimation. The desire for an immortality of fame (roughly, the stake of the ego), the internal logic of the great argument justifying the Father (roughly, the stake of the superego), and the psychological drama of the poem's creation (roughly, when imagining paradise, the stake of the id) are all vitally interested in the real way we digest food.

Angels eat. Milton testily disowns "the common gloss / of Theologians" (5.435–436) when Raphael falls to his fruit with enthusiasm. As in a Platonic dialogue, the polite questions of the host serve the pedagogical aims of the wise guest, enabling Raphael to expound the alimentary structure of a universe wherein all things, planets as well as angels and men, require nourishment from a lower order of substance. The angel insists that his eating is neither miraculous nor exceptional:

> Wonder not then, what God for you saw good
> If I refuse not, but convert, as you,
> To proper substance; time may come when men
> With Angels may participate, and find
> No inconvenient Diet, nor too light Fare. (5.491–495)

A few lines before, the narrator anticipates this dismissal of "wonder." Like Adam, readers are not to be astonished:

> So down they sat,
> And to thir viands fell, nor seemingly
> The Angel, nor in mist, the common gloss
> Of Theologians, but with keen dispatch
> Of real hunger, and concoctive heat

To transubstantiate; what redounds, transpires
Through Spirits with ease; *nor wonder*; if by fire
Of sooty coal the Empiric Alchemist
Can turn, or holds it possible to turn
Metals of drossiest Ore to perfet Gold
As from the Mine. (433–443; my italics)

Obviously Milton was skeptical about the claims of alchemy.[28] The "true wonder" here is the inner alchemy of the digestive tract, where transmutation is a daily occurrence—a "wonder" more reliable than the wonder of the analogue introduced to authenticate it. To appreciate this claim one must think as a child: you eat; you are replenished. But you are not what you eat, which is dead, insensate, and does not think. What you eat becomes you, and for a monist that means a transubstantiation from almost pure matter to almost pure spirit. Whether or not the alchemist turns "drossiest Ore to perfet Gold," the inner alchemist of the body does. "Wonder not then, what God for you saw good / If I refuse not, but convert, as you." Eve need only have remembered this statement when listening to the serpent's tale of magical transformation, which also begins "Wonder not" (9.532) and becomes credible to her in stages marked by progressive amazement—"Not unamaz'd" (552) at "such wonder" (556), "Yet more amaz'd" (614), and at last in simile "th'amaz'd Night-wanderer" (640) who will return to Adam with a tale "wonderful to hear" (862), her very first words, "Hast thou not wonder'd, Adam, at my stay?" (856), imitating the opening "Wonder not" of the serpent in their effort simultaneously to evoke and capture the capacity to wonder. The positive side of the law against eating one fruit is the injunction to eat every other fruit, and in this feast lies a transubstantiation, a true wonder, a credible alchemy.

Raphael can "convert, as you, / To proper substance." There is no generic interruption between the hunger of man and the "real hunger" of the angel, and in keeping with this principle, the physiology of the passage would not have seemed wonderful to contemporary readers. Raphael eats precisely as they supposed themselves to eat, breaking down his food in the manner favored by conservative medical thought. Galen presented "concoctive heat" as the active agent in digestion, and compared the stomach to a cauldron heated by the "burning hearths" of adjacent viscera. Food was cooked, literally "boiled" during the first of the six stages of digestion—the separation of chyle from solid excrement. Well into the Renaissance the stomach was referred to as "the

Microcosmes Kitchin."[29] Milton's analogy of the alchemist begins by expanding this orthodoxy; both digestion and alchemy achieve their effects by "fire." But this fire transports us from Galenism to its major rival in Renaissance medicine. Robert West termed the simile a "casual analogy" intended to make the heresy of angelic eating plausible.[30] But if so, why would he question the success of the alchemist in achieving transmutation? To the contrary, the "Empiric Alchemist" shows Milton looking for plausibility in the best speculations of Hermetic physicians.

While Reuchlin and Fernel, among the Galenists, admitted discomfort with heat as the sole agent in digestion, the foremost challenge to this belief came from Paracelsus and his followers.[31] For Paracelsus the normal activities of the body were each governed by a specific *archeus*, and not, as in Galenic thought, ruled by the general balance of the humours. He named the *director digestionis* an "inner alchemist" because it was, like the "outer stomach" of alchemy proper, capable of dissolving substances into an earlier state of existence from which genuine transmutation was possible. The metamorphosis of food was, for this physician as for Milton, a grand religious mystery. An alchemist might labor for years at his weird work, uncertain of success, but the inner alchemist of a healthy body produced transmutation with quotidian regularity. This is close to the spirit of Milton's simile. The body of man or angel does not heat "by fire / Of sooty coal," but rather contains in a sovereign state the "vital virtue . . . and vital warmth" (7.236) infused into matter by the Spirit of God, the life-heat whose sidereal castle is the "Arch-chemic Sun" (3.609). Paracelsus thought as a child: food must revert, through the agency of the inner alchemist, to the "one first matter" from which all substance derives, and then be reconstituted as the various substances operating under the various *archeae* of the body. So the kitchen became a laboratory.

His disciples spoke with pride of this new beginning on the subject of digestion. Oswald Croll—the most influential of the early propagandists—praised "the archeus of man, or internal alchemist, born within and implanted by God."[32] Some Galenists appropriated the doctrine, allowing it, as Milton does, to absorb the old idea of concoction. Thus Timothy Bright in his *Treatise of Melancholy*: "so fareth it with nourishments, whose divers parts are layd open by so manifold concoctions, and cleansings, and strainings, as are continually without intermission practized of nature in everie mans bodie: noe gold finer more busie at the mine, or artificiall chymist halfe so industrious in his laboratorie, as this naturall chymist is in such preparation of al nourishment: be it meat, or drink, or what sort soever."[33]

In England, Charles Webster maintains against the views of Kocher and Debus, the medicine of Paracelsus was "largely assimilated by 1600," in part because Paracelsus drew so much from late medieval alchemy in which Englishmen figured prominently.[34] The influence grew in the seventeenth century, then became overwhelming after 1640. William Harvey, in most respects a staunch Aristotelian, informed his anatomy students at Oxford that the "liver, spleen, stomach" acted "as alchemists by means of different furnaces and heats."[35] Within the Paracelsian tradition itself, the suggestive metaphor of the inner alchemist ultimately issued in a radical breakthrough—one that England heard a great deal about during the Interregnum. The great J. B. Van Helmont, whose *Ortus Medicinae* was among the flood of Paracelsian texts translated during this period, recognized that the process of cooking left the structure of food unchanged.[36] As the new metaphor indicated, digestion was a true transmutation beyond the power of mere heat, of fire like the fire of sooty coal, to affect. In a real chemical laboratory, acid alone can precipitate a dissolution. So Van Helmont reasoned that organic acids ("ferments") transmuted ingested matter in a specific chemical reaction known as normal digestion; in all likelihood he had also identified the reagent as hydrochloric acid ("Spirit of Salt") and understood the process of neutralization. Van Helmont, whose own treatises invariably began with a vitriolic attack on current medical theory, was brandished like a weapon against university physicians by the reformers gathered around Hartlib.[37] The enemy took note; one of the questions disputed at Oxford in 1651 was "whether concoction in the stomach was caused by acid ferment or by heat."[38] Paget owned the complete works of Van Helmont in several editions.

Milton starts his account of the natural marvel of digestion with the "concoctive heat" of Galen, but moves quickly to the "transubstantiation" and "fire" of the inner alchemist. The passage sets Galen against Paracelsus, just as the dialogue on astronomy sets Ptolemy against Copernicus—yet here the balance of conviction tips toward the new science, and other passages tacitly confirm the choice:

> O *Adam*, one Almighty is, from whom
> All things proceed, and up to him return,
> If not deprav'd from good, created all
> Such to perfection, one first matter all. (5.469–472)

> As new wak't from soundest sleep
> Soft on the flow'ry herb I found me laid . . . .

Straight toward Heav'n my wond'ring Eyes I turn'd,
And gaz'd a while the ample Sky, till rais'd
By quick instinctive motion up I sprung,
As thitherward endeavoring, and upright
Stood on my feet . . . .
Myself I then perus'd, and Limb by Limb
Survey'd, and sometimes went, and sometimes ran
With supple joints, as lively vigor led. (8.253–269)

Now Morn her rosy steps in th' Eastern Clime
Advancing, sow'd the Earth with Orient Pearl,
When *Adam* wak't, so custom'd, for his sleep
Was Aery light, from pure digestion bred,
And temperate vapors bland. (5.1–5)

How can immortal man "return" to God? He need only feed the "quick instinctive motion" and "lively vigor" there from the beginning. The physiological impulses of the body, like the "wond'ring Eyes" of Adam, are aimed "Straight toward Heav'n." The phenomenon we now and again experience of awaking from sleep, our daily surrender to the involuntary, refreshed and invigorated, is in Eden a small spiritualization "from pure digestion bred." We have paid due attention to the beauty and pleasure of Milton's Eden, but not to its abounding healthiness.

"Man's nourishment," Raphael tells Adam, is "by gradual scale sublim'd" (5.483). The alchemical language of the phrase implies that the agent of ascent works in the interior heats of the body. Naturally and emblematically, the sublimating flesh is tuned to rise with the "Arch-chemic Sun." Digestive alchemy is the inborn dynamic of spiritual aspiration, the science of God's summons to man, pointing his way back to the source of life. The quickening touch of "our Nourisher" has designed our physiology, not only to elevate the thingness of food, but also to melt this too solid flesh into the fluid substantiality of angels: the means of ontological promotion were infused in the dust of Adam. Raphael relates "Great things, and full of wonder in our ears" (7.70; cf. 8.11). Yet while an angel's eating may be a "wonder" in the sense of an occasion for awe, it is not a miracle disrupting natural law. Milton denied that the Eucharist occasioned a miracle of any sort (*CP* VI, 554–560). It would appear from *Paradise Lost* that the consequence of detaching marvel from one ritual meal was to charge with new energy the cliché that God is our nourisher

and make all eating a creator-centered sacrament of transubstantiation. In Eden at least, digestion is the process by which involuntary process seeks to relinquish itself to will.

Medical alchemy supplies a context for Milton's notion of unfallen freedom. But what of its association with angels? What of the education that invites Adam to desire angelhood? The crucial role of "wonder" in the life of man, from the moment Adam opens his eyes, through the wonders and not-wonders of Books 5–8, to the fallacious wonder of the temptation in Book 9, guides us to a fuller intellectual context, consistent with the philosophical currents promoting medical alchemy in the Renaissance, for Milton's designing of the first education.

Aristotle proposed in the *Metaphysics* that wonder created philosophy. Milton could well have had the passage in mind when introducing philosophy into Eden in Books 5–8:

> it is because of wondering that men began to philosophize and do now. First, they wondered at the *difficulties* close at hand; then, advancing little by little, they discussed *difficulties* also about greater matters, for example, about the changing attributes of the Moon and of the Sun and of the stars, and about the generation of the universe. Now a man who is perplexed and wonders considers himself ignorant (whence a lover of myth, too, is in a sense a philosopher, for a myth is composed of wonders), so if indeed they philosophized in order to avoid ignorance, it is evident that they pursued science in order to understand and not in order to use it for something else. This is confirmed by what happened; for it was when almost all the necessities of life were supplied, both for comfort and *activity*, that such thinking began to be sought. Clearly, then, we do not seek this science for any other need; but just as a man is said to be free if he exists for his own sake and not for the sake of somebody else, so this alone of all the sciences is free, for only this science exists for its own sake.
>
> Accordingly, the possession of this science might justly be regarded as not befitting man; for human nature is servile in many ways, and so, as Simonides says, "God alone should have this prerogative . . . " If, then, there is something in what the poets say and Deity is by nature jealous, he would most probably be so in this case, and all men of intellectual eminence would be unfortunate.[39]

Aristotle proceeded to defend the position that God is not jealous, and in any case would have no right to be, since God as cause and prin-

ciple is the ultimate object of philosophical thought. The jealousy of a God who withholds powerful and worthwhile secrets from man is in *Paradise Lost* a serpentine theme, appearing emphatically in the final words of Eve's tempter (9.727–731). But as we know from Adam's birth narrative, wonder in Milton's universe creates religion first. A "self-knowing" man possesses the free nature at his origins, where he also finds an originator far different from the self-contemplating god of Aristotle. Adam's God wills, and Adam's reverence is open to his will: a conscience born of gratitude precedes the orientation of wonder toward matters "close at hand," such as whether the angel eats, and seemingly "greater matters" like "the stars" and "the generation of the universe." This religion, in which the disposition of free will takes priority over free speculation, cannot be squared entirely with Aristotle's concept of the autonomy of the philosophical mind. Indeed, the difference between the wonder that founds religion and the wonder that founds philosophy constitutes the major conflict of unfallen life.

For Milton agreed with Aristotle about the excess intrinsic to *scientia*. We do not desire to know *in order to*: utility is not the motive for intellectual strife, which stirs when "all the necessities of life" have been secured. Wonder, the state of not knowing but wanting to know, is an intellectual appetite, as Nicolas of Cusa, alluding to the passage in Aristotle, made explicit: "just as a preceding unpleasant sensation in the opening of the stomach stimulates the nature for being restored, so wonder stimulates the desire-for-knowing which is naturally bestowed upon all human beings."[40] Precisely because "Knowledge is as food," Milton has Raphael say, knowledge "needs no less / Her Temperance over Appetite" (7.126–127). Wonder must be curbed by a rule of temperance. In order to discipline this appetite Milton revalued those concerns that Aristotle had placed prior to philosophy—the useful and wonderless knowledge focused on "the necessities of life."

When he accepts the ban on sidereal curiosity, Adam maintains that wonder is exceeded by utility. The presence of these sentiments in a poem so forthrightly speculative as *Paradise Lost*—"a lover of myth, too, is in a sense a philosopher, for a myth is composed of wonders," and in two senses a philosopher if, lacking "supreme literary tact," he urges the truth of his wonders—reveals that the conflict we are discussing cannot be limited to unfallen life:

> How fully hast thou satisfi'd me, pure
> Intelligence of Heav'n, Angel serene,

And freed from intricacies, taught to live
The easiest way, nor with perplexing thoughts
To interrupt the sweet of Life, from which
God hath bid dwell far off all anxious cares,
And not molest us, unless we ourselves
Seek them with wand'ring thoughts, and notions vain.
But apt the Mind or Fancy is to rove
Uncheckt, and of her roving is no end;
Till warn'd, or by experience taught, she learn
That not to know at large of things remote
From use, obscure and subtle, but to know
That which before us lies in daily life,
Is the prime Wisdom; what is more, is fume,
Or emptiness, or fond impertinence,
And renders us in things that most concern
Unpractic'd, unprepar'd, and still to seek.
Therefore from this high pitch let us descend
A lower flight, and speak of things at hand
Useful. (8.180–200)

This from the man who "with no middle flight intends to soar"! But there can be no doubt that Adam shows himself responsive to one thrust of Raphael's instructions. Mind or fancy is *apt* to rove: the very word seems eager to produce *appetite* and *apple*. In reply to questions about the heavens, Raphael simply multiplied the questions. Now temperate Adam is "fully . . . satisfi'd" with his ignorance, wonder-full, because he has been "by experience taught" from Raphael's refusal to satisfy him the better lesson of contentment. Unchecked wonder disturbs "the sweet of life." The proper sphere of knowledge is domestic, practical, useful, at hand. Three interlocking metaphors characterize the pursuit of interminable knowledge: the maze opposite to home and finitude ("intricacies," "perplexing," "wand'ring," "no end"); the vaporous opposite to the solid and nourishing ("notions vain," "fume," "emptiness"); the flight opposite to "That which before us lies," immediacy and contentment ("dwell far off," "seek them," "rove / Uncheckt," "to know at large of things remote," "high pitch"). It is hardly surprising that these are the three metaphors that blossom during the fall in Book 9. Adam should enjoy his sweet "fill" of happiness. "O yet happiest if ye seek / No happier state, and know to know no more" (4.774–475).

Paradise Lost as a whole constantly reminds us that it was written

during one of the few periods, maybe the only period, of British history when Platonism could be said to be the dominant philosophical tradition; but this passage seems to look forward in English thought all the way to the current reverence for the instrumentalism of Wittgenstein, the most English of the philosophical Germans. That is probably to dignify it overmuch. Here the elevation of the verse seems at odds with the constriction of the thought, an effect that Milton may have intended, since he is being doctrinal and delivering the law to the inherently lawless roving of "Mind or Fancy." It does not work for me, however, and I wonder if I am alone. "Nor with perplexing thoughts / To interrupt the sweet of life"—nobly said, but a vulgarity is a vulgarity. One can plead context. Ptolemy or Copernicus, one inhabited world or many inhabited worlds—it does not matter. Yet surely Milton knew that he was writing a persistently etiological poem wherein each of the episodes unfolds toward universality. Were wonder to be checked by law alone, by the authority of God's commandment, the resistance provoked by the passage would be more profound. Obedience must eventually recognize its dependence on the will of God. It is Milton's attempt to ground intellectual temperance, like temperate eating, in reason's discovery of its own health, thus deferring the encounter with voluntarism, that leads him to give a mandate to ignorance. It is all the more striking in view of this mandate that Milton should confront Adam and Eve with the possibility that they might grow wings and fly to heaven. Why disturb the sweet of life with the alluring prospect of acquiring an angel body?

Something of the ambition we associate with Renaissance man survives in Milton's Eden despite his obvious suspicion of "this intellectual being, / Those thoughts that wander through Eternity" (2.147–148). As was suggested in the last chapter, wonder is the attitude particularly prominent in Renaissance neoplatonism—a tradition that listened with approval to alchemy and other occult arts. Pico della Mirandola, true to his name, saturated the opening of his *Oration on the Dignity of Man* with wonder: "Most venerable fathers, I have read in the records of the Arabians that Abdul the Saracen, on being asked what thing on, so to speak, the world's stage, he viewed as most greatly worthy of wonder, answered that he viewed nothing more wonderful than man. And Mercury's 'a great wonder, Asclepius, is man!' agrees with that opinion . . . . Why should we not wonder more at the angels themselves and at the very blessed heavenly choirs?"[41] To settle this question, Pico tells the philosophical creation myth that has often been thought to summarize the Renaissance vi-

sion of humanity. When he came to fashion Adam, God discovered that he had already filled every slot in the chain of being. So he made man without *quidditas*: a man has by nature no fixed nature, enjoying instead an unchained being free to act toward his own self-determination rather than, like other creatures, condemned to act out an essence inserted into him from the beginning. Our likeness to uncircumscribed deity makes us the envy of the angels themselves. "Let a certain holy ambition invade the mind," Pico urges: "let us compete with the angels in dignity and glory. When we have willed it, we shall not be at all below them."[42] The supreme exercise of freedom lies in the emulation of angels, and Pico performs this exercise with holy delirium.

Milton's subdued adaptation of these themes comprises a fascinating index to whole dimensions of his art and temperament. Pico offers, first of all, a blatantly Pelagian view of man, and Milton repositions such issues, properly enough, in the life of unfallen man. He extends to both the natural and supernatural domains the Piconian idea of an existence prior to essence. Originally all being is unchained. The neoplatonic aspiration to return to the source, gaining thereby the nature of those closest to the source, appears in Milton as a characteristic of matter in all its forms:

> O *Adam*, one Almighty is, from whom
> All things proceed, and up to him return,
> If not deprav'd from good, created all
> Such to perfection, one first matter all,
> Indu'd with various forms, various degrees
> Of substance, and in things that live, of life;
> But more refin'd, more spiritous, and pure,
> As nearer to him plac't or nearer tending. (5.469–476)

Matter is "plac't or nearer tending," but note that if all things "up to him return," they must be "plac't" *and* "tending." Under the influence of celestial heats, fruit sublimates the elements, and under the influence of vital heats, digestion sublimates fruit. Angels, toward which our physiology strives, have not been fixed in the ontology of Milton, who was ever grateful for vicissitude. They rather enjoy as no other creatures the form-receiving potential of the "one first matter" able to become "all." Pure volition, they are pure possibility—creators of the microcosmic universe of themselves. They may therefore be called, with more propriety than is conventional, "gods."

Milton desired this kingdom of volition. To turn to his life, con-

trasting the unfallen physiology with the treacherous digestion that left him blind, allows us to appreciate in a personal way a yearning manifest in the poem. Milton probably felt that all men would desire this kingdom, although as his harsh words against "Hyprocrites" in the nuptial hymn imply (4.744), he was wise enough to know that all men would not acknowledge this kind of desire. The first education of man does more than warn him. It frees his wish for happiness. To have every pleasure of the body, but none of its liabilities; to move, but not to be burdened; to be substantial, but not to be fixed; to fly and to merge as one with another; to be free of the anxiety of being wounded, free of clot and stain and mark; and above all to be unimpeded will, such that, if you preferred the friction of surfaces, for example, as a perfection of design not to be improved upon, nothing would stop you from doing that thing, or any other thing, as you do them now. Milton liberated the fantasies of mankind in this happy incarnation of his ideal ego. Is there an item in the vocabulary surrounding "feeling good" that would not apply in eminence to the state of angelhood? Satan corrupts the fantasy when he invites Eve to desire this state, not to feel good in eminence, the counterpart of being good in eminence, but to be worshipped as an object of desire. This monist angelology provides a welcome ballast to the wholly mental eminence found in the spirit worlds descending from Pseudo-Dionysius, Pico's among them.

But there, in being the object of desire, was the rub. Adam never responds to the rhetorical invitations of Raphael. His desire for angelhood remains unpronounced. Angelhood is not, and must not be, the known object of desire:

> God hath bid dwell far off all anxious cares,
> And not molest us, unless we ourselves
> Seek them with wand'ring thoughts, and notions vain.

No more than God, Milton could not allow Adam to *will* to be an angel. He is to will to be the man he now is, in the happy state he now is, and to will temperate limits for his appetites. Milton created for himself poetic problems of considerable difficulty: there were to be angels, and angels happier than man in ways than men can understand, but no hint of angel worship and no hint that man has been endowed with a mean lot short of his desire. He triumphed with a supreme literary tact—and more. *Milton's solution to the problem of how an unfallen man could desire angelhood*, which of course he would, *while preserving*

this desire from the contamination of becoming a motive, which of course he must, *was to invent in Renaissance terms the unconscious mind.*

Body and mind are one in kind. An innocent vital process, an energics of the spirit "desires" and "emulates" angelhood. This Renaissance unconscious, like the modern one, is between categories, at once involuntary and directed, a physiological process and an anthropomorphic performance—in short, an inner alchemist. Freud wrote: "Thus the content of a repressed image or idea can make its way into consciousness on condition that it is *negated.* Negation is a way of taking cognizance of what is repressed; indeed it is already a lifting of the repression, though not, of course, an acceptance of what is repressed" (*S.E.* XIX, 235–236). And thus it is in Milton. The negations of "nor wonder" and "wonder not" dispose us toward the continuity between human and angelic digestion:

> Wonder not then, what God for you saw good
> If I refuse not, but convert, as you,
> To proper substance; time may come when men
> With Angels may participate, and find
> No inconvenient Diet, nor too light Fare:
> And from these corporal nutriments perhaps
> Your bodies may at last turn all to spirit. (5.491–497)

No wonder? But the hunger of our unfallen digestion *is* an unchecked appetite for wonder. Insofar as the moral life is to discipline this appetite, permitting it to rove unchecked only in the realm of the involuntary, the moral life may be summarized as the maintenance in consciousness of the "not" negating "wonder"—"a lifting of the repression, though not, of course, an acceptance of what is repressed," as Freud wittily observed. To wonder not is to entrust our wishes to our nourisher. Had Eve really not wondered at the serpent who began "Wonder not," the unconscious would be conscious, and fantasy truth. What a wonder, Asclepius, man would have been.

The angel does not say as much. A conditional construction, "time may come . . . perhaps," appeals to the heats in us that desire this time and this destiny. But for Milton wish-fulfillment in our happy state would not have been hallucinatory (see God's statement of our original destiny at 7.154–161). This is an unfallen unconscious whose dreams are coming true, the power to fulfill having been given, in Eden, to the energy of the wish itself.

"Knowledge is as food" II: Poetry is as Digestion

During my analysis of satanic self-creation I tried to summarize in Cartesian epigrams Milton's view of how human existence tends innately to unpack from itself religious existence: I think, therefore I am; I am, therefore I was created; I was created, therefore I am religious. I also suggested that the absence of a name for God, by introducing a need for metaphor and dislocating words from their given designation in a semantic search for God, establishes in unfallen language a poetics continuous with religious devotion. Returning to the nature of poetry from another angle, I would like to place a third Miltonic addendum beneath the *cogito*: I am religious, therefore I am poetic.

At the outset of his fine defense of Milton's style, Christopher Ricks concedes a few passages in *Paradise Lost* to his opponents, among them the "aridities" of angelic digestion.[43] I sympathize to some degree. The Lucretian episodes we are interpreting seem at first glance arcane and dated, defended against ridicule by odd bits of discredited science, or simply boring in their distance from charged human event, like an "Aire and Angels" turned inside out in the manner of Benlowes or More, making love into a labored metaphor for pneumatology. But the second half of Book 5 has special authority in *Paradise Lost*. It contains the axioms of reality. That this reality was carefully formulated, weighed in the poet's mind for two decades, everyone knows. But it is not widely appreciated that the success of the epic depends to a large extent on the kind of poetic effect that is the best defense of the ontological moments in Book 5 . This effect is offered, not imposed. If, however, we do not take the offer, the study of style will probably remain what it is in Ricks: the appreciation of local successes, the journey of the critical soul through bright fragments.

Paradise Lost is, after all, a poem of considerable length. In the memories of its nonprofessional readers such a poem may exist as an indefinite swirl. Who spoke second at the counsel of Book 2? What happened during the second day of the civil war in heaven? When did Eve narrate her birth? Perhaps all has not been lost. Perhaps the reader has experienced some inarticulable improvement of sensibility. But one must insist that, if poetry survives by occasioning a revelatory redescription of the world, the poem as indefinite swirl is dead, a lost time, nor has the world been revealed. Of course the threat of this diffusion will be proportionate to the intricacy of the narrative, graver

with respect to the *Faerie Queene* than with respect to *Paradise Lost*, and that is a point I mean to emphasize.

Milton's poem coheres for us, or has the possibility of becoming coherent for us, because he was able to root the entirety with logical and emotional force in certain privileged passages approximately as long as the bundles of "Ten, Twenty, or Thirty Verses at a time" (*EL*, 73) in which it was composed. He was, as it were, able to assimilate the work of years to the work of some few visionary days, allowing his readers to treat themselves to an effect I call the "enfolded sublime." We know from the concept of the hermeneutic circle that interpretation involves the constant adjustment of part to whole and whole to part. But one trouble with fashionable talk about the hermeneutic circle is that this process shapes all experience and thus all reading, the cereal box you gaze at over breakfast and the late Heidegger that puts you to sleep: ubiquity invites reiteration. The enfolded sublime, on the other hand, is a high aesthetic achievement. It is the arrest of the hermeneutic circle, as if the poem were saying, "Stop here, I will occupy you." One problem with fashionable exhibitions of readers' responses is that the readers have not responded to this gesture, or else they would not read as they do, lost in the woods of endless self-correction and forever in quest of the poem denied them by their very procedure: process cannot yield solution. The enfolded sublime is not absolute closure. Yet by virtue of the enfolded sublime we have the poem, and having it, can proceed to ask our many questions.

The miniature of this effect is common enough in the doll's house of the lyric. During the seventeenth century poets such as Herrick and Lovelace made an aesthetic principle of "much in little," although it seems to me that these poets often give us much little and little much. In the long poem, however, the enfolded sublime is a rare effect available only from the greatest artists. Milton was its master. Spenser was not, as Milton told us:

> Not sedulous by Nature to indite
> Wars, hitherto the only Argument
> Heroic deem'd, chief maistry to dissect
> With long and tedious havoc fabl'd Knights
> In Battles feign'd; the better fortitude
> Of Patience and Heroic Martyrdom
> Unsung; or to describe Races and Games,

Or tilting Furniture, emblazon'd Shields,
Impreses quaint, Caparisons and Steeds;
Bases and tinsel Trappings, gorgeous Knights
At Joust and Tournament; then marshall'd Feast
Serv'd up in Hall with Sewers, and Seneschals. (9.27–38)

Surely this should be added to what is reputedly the small number of humorous passages in the epic. It is perceptive humor. Commentators have noted only two of the three indictments of romance, leaving out the perceptively humorous one. This form glorifies the lesser heroism of war; its heroes and its battles are "feign'd" rather than historical. Yet Milton also indicts the genre for being "long and tedious," then condenses hundreds of poems and expanses of story into one catalogue of a day in the life of romance.

It is just this command of essence over detail that romance lacks—headless narratives stuffed with minute descriptions of the surfaces of objects. Although he assumed the obligation inherent in allegory of mastering with idea the proliferation of "tilting Furniture," Spenser was unable to arrest the spin of narrative in still points of enfolded sublimity. He tried to supply these well-wrought contractions, without which a long poem becomes long and tedious, destined to remain for most of its readers a complex impression rather than an event of meaning—Redcrosse surveying the world in Book I, Colin's interrupted vision in Book VI, the bowers in betweeen. But these moments, to make a long story short, are not able to ingest the bulk of the poem with sufficient elegance. Too much redounds. The hypertrophied and repetitious details in Spenser invite a labor of cross-referencing not unlike the order to be found in the Renaissance encyclopedia of synthetic knowledge. The poem, like the sense of the world that produced it, delays comprehension by multiplying allusion: "to understand this, see that," "to understand that, see this other." Meanings in literature are not of a piece; they have characteristic "shapes" or modes of presentation. The best student of Spenser, James Nohrnberg, accurately conveys the meaning of the *Faerie Queene* as a texture of proliferating analogies no mind could survey at a glance.[44] A single analogy is potentially infinite; we arrest the interaction when the two terms cease to reveal anything importantly new about each other. Lacking the enfolded sublime, which gives authoritative instructions to guide us through the maze of unfolding analogies, the parts of Spenser's poem weave themselves into each other with infinite

stitches, and the only satisfactory alternative to a vague impression becomes the retention of the poem in complete detail—a work of art that, as a burden to memory, would have delighted the Ireneo Funes of Borges. But if Spenser is the Renaissance reproduced, Milton is the Renaissance understood.

The enfolded sublime is not available to a first glance. We realize this effect after the linear reading is finished, and even then it must be won by a reflection determined to have the poem, to make it occur as a thought. The moments selected can enfold no more than we are able to unfold. Since the entirety settles about these passages as if the poem had designated them its chosen reductions, they permit us a sense of provisional completion—an order subject to correction and elaboration, but of a higher level than the one we correct and elaborate in the linear process of reading: those who argue for the primacy of "interpretive decisions" in literary studies often transform a genuine philosophical issue into a specious *weltanschauung* by claiming that all such decisions are equally remote from external sanction; for it is not philosophy, but rather the trivial security of being irrefutable, to argue that every argument is decided by power or persuasion. Readers will differ, of course, over the ultimate sense of the poem arrayed about these moments, which drives everyone concerned back to the linear process of reading, and they will differ as well over the moments to be privileged. In fact, competing interpretations of *Paradise Lost* and other long works usually have their point of departure in a difference over the selection of the sublime passage. Practitioners of deconstruction typically enter the text at a moment traditional readings have thought insignificant, believing it to be subsumed in the controlling revelation of more famous passages, then labor to elevate this victim of oppression into a new sort of sublimity—I would be tempted to call it "unenfolded"—in which what is meant *and* what is repressed, or what cannot achieve direct expression so long as the meant is meant, are simultaneously focused. Since I am by no means opposed to this technique (what admirer of Freud could be?), I want to explain why I am endeavoring to merge psychoanalytic interpretation with an enjoyment, traditional in literary study, of the order enfolded in central passages. My first reason is simply that what is not meant in Milton is most eloquently not meant at crucial moments—when, for example, the Lady is left paralyzed after successfully repelling her tempter. Secondly, in the three late poems the psychological, philosophical, and theological "not meant" is in many ways, though not in every way,

consistent with the manifest statement of the poem, and since this remarkable consistency seems to me the crux of Milton's achievement, I am trying to define it. Finally and obviously, the sublime coherence of these poems places us before the wisdom that can be acquired from the symbols of a major religion, many of which, the more so as Milton presented them, are held in common with another major religion.

Paradise Lost contains several examples of this effect. The poem coheres in different intonations about different centralities. While they prompt us to complete our sense of the entirety, the great enfolding moments offered to a reader of late Milton are perhaps unique in also representing a genesis: out of these passages the entirety, including, more or less explicitly, the poetic act, can be seen to have been derived. I have organized my discussion of the epic about some of them. In the triptych of sublime visions at the transition from Book 2 to Book 3, Milton occupies the center, where he approaches the act originating everything in the act of creating his poem. I should explain that there is no contradiction between excellence in the enfolded sublime and my repeated stress on process in Milton, for I am speaking of the way in which the long poem presents its meaning or comes into coherence, whatever that meaning and that coherence be. At the center of the triptych is unmistakably a poet of temptation, conflict, and crisis: there he is, like a hyphen between antagonists, his poetic versions of Satan and God massed on either side of him, eyeing each other's works. Here the three dimensions of the sacred in the poem—the demonic, the human, and the divine—are contiguous, and Milton's deployment of the symbols found in these regions has a full epitome in the invocation to light. In the next chapter I will turn to the enfoldment at the conclusion of the epic—where the human once again is central, but the resolution of conflicting forces appears to have shifted problematically from the personal resolution of Book 3. In this chapter, seeking the acknowledgment of guilt required by the psychological balance of the invocation to light, I am tracing a spiral about the ontological speech beginning "O *Adam*, one almighty is." Already the promise of enfoldment has been heard. A sublime phrase introduces ontology—the "might" compressed between "al" and "is," the philosophical collision of "one" and "all," and the whole discourse prefaced by the gentle exclamation "O," which images visually the ideal circle of matter's return that the speech will proceed to explain scientifically and theologically. A few lines hence Raphael reshuffles these immensities in defining creation as unfolded from "one first matter all." One. First. Matter. All. These four words alone, fully heard,

comprehend a great poem with much to say. To have them before us in a meaningful syntactic sequence is to be made momentarily giddy with the plasticity of huge meaning, able to present itself to us with harrowing abstraction, as when Satan at the end of Book 2 enjoys a vision of our universe contracted into an atom of light ("O . . . "). The enfolded sublime seems to have been especially available, and especially sought for, at the end of the Renaissance. Descartes creates this effect, as do Kepler, Galileo, and Newton in their elegant reductions of copious motion to single formulae. On the pinnacle of *Paradise Regained*, in the double chorus on the triumph of Samson, and in a handful of masterful passages masterfully placed in the design of *Paradise Lost*, Milton was its poet.

I return to my widening gyre. Students of the Miltonic style cannot afford to discard the tabletalk of Book 5, for there are clear indications that Milton intended to root poetry in the natural vitality of Eden; the "Holy rapture" of the "unmeditated" prayers at the beginning of Book 5 (147–149) has often been linked with the "unpremeditated Verse" (9.24) of the epic itself. We might be inclined to regard the element of spontaneity joining a digestive physiology and holy poetry as a casual association, despite a passage such as

> Then feed on thoughts, that voluntary move
> Harmonious numbers;

but Book 5 has been designed to provide a logical development of this metaphor. Beginning as a medical ontologist, Raphael then becomes a "Divine / Historian" (8.6–7) in the epic mode, whose high matter will leave the ears of Adam "full of wonder" (7.70) and his intellect eager to "magnify" (7.97) the works of God. Understood systematically in the context of the axioms of reality, poetry continues at a higher echelon the refinements of the inner alchemist, making the relatively physical relatively spiritual. The possibility of this connection can be found in the ontological discourse itself. Thus "flow'rs and thir fruit"—terms for poetry and poetic invention familiar to Renaissance scholars, as for example in Sonnet VII and the invocation or funeral lament in "Lycidas"—are "by gradual scale sublim'd" (482–483). The next item in the first education of man awakens this sleeping significance. Poetry belongs to the movement of "all things" that "up to him return." In the passage cited at the opening of this chapter Milton took note of his "strong propensity of nature" to inspire a nation and glorify the Lord. The art of such a poet—and

Raphael is one—provides a channel in conscious life to express and refine the somatic appetite for wonder. Poetry is as digestion.

When Raphael, about to narrate the wars of heaven, defines the semantic strategy he has adopted to perform this task—presumably Milton's confession of his own strategy—he does not expect Adam to forget the previous lesson about nourishment. Comprehension of heaven, which Raphael gives to man, builds upon prior comprehension of the food which man gives to Raphael. Both activities strike an ontological relationship between "spiritual" and "corporal":

> yet for thy good
> This is dispens't, and what surmounts the reach
> Of human sense, I shall delineate so,
> By lik'ning spiritual to corporal forms,
> As may express them best, though what if Earth
> Be but the shadow of Heav'n, and things therein
> Each to other like more than on Earth is thought? (570–576)

This much-debated passage on accommodated speech begins by likening the forthcoming epic, the first to be heard on earth, to the nourishment also "dispens't" for the "good" of man. At first it would appear that much of heaven is an intellectual fruit beyond "the reach" of our senses; if so, flimsy similes must be thrown across the void separating two worlds distinct in essence. Then, through the example of objects above having shadows below, "the reach / Of human sense" seems to be capable of a true grasp, and the formerly earthbound similes delineate the actual relations between the two worlds. Madsen says that "Milton is using 'shadow' here not in its Platonic or Neoplatonic sense but in its familiar Christian sense of 'foreshadowing' or 'adumbration,' and . . . the symbolism of *Paradise Lost* is typological rather than Platonic."[45] The epic teems with typology, but the shadow of these lines has (with Miltonic adjustments) a neoplatonic cast.

Insofar as Raphael addresses the dualists of Christian theology, his questions about what "on Earth is thought" can only be audible to the reader of the poem. But before we think of scholastic philosophers, we must take account of the dramatic occasion: when has anyone "on earth" ever "thought" his world excessively distinct from heaven? It was just a moment ago, when Adam feared that the fruits of Eden would be "unsavoury food perhaps" (401) for an angel. He was told, as Raphael now reminds us, that human food "may compare with Heaven" (432), and mistook truth for politeness:

Inhabitant with God, now know I well
Thy favor, in this honor done to Man,
Under whose lowly roof thou hast voutsaf't
To enter, and these earthly fruits to taste,
Food not of Angels, yet accepted so,
As that more willingly thou couldst not seem
At Heav'n's high feasts to have fed: *yet what compare?*
(461–467; my italics)

The ontological discourse was the second and completer reply to this incredulity. In context, Raphael's "though what if . . . more than on Earth is thought?" is a professorial allusion to Adam's "yet what compare?" and the answer he received. The angel's question is clearly rhetorical, inviting Adam to reach for a truth he has already reached, since Raphael has just asserted that "things" in the two worlds come from "one first matter all" and are indeed "Each to other like, more than on Earth is thought."

The early pages of the *Christian Doctrine* propose an unusually positive theory of the truth-value of biblical metaphors for God. "Indeed he [God] has brought himself down to our level expressly to prevent our being carried beyond the reach of human comprehension" (*CP* VI, 133–134).[46] So guided, the exegete deduced from scripture his monist heresies. But if fallen education begins in the presence of the Bible, making hermeneutic procedures prior to ontology, Adam is educated in the order of things. He has an example from the world, an angel that really eats, to make the good sense of divine metaphor a logical deduction. Within the epic, then, monist ontology underwrites an unconventionally generous view of God's accommodation to man. As spiritual is to corporeal in the nature of things, so spiritual is to corporeal in the names of things: the universe being as it is, meanings borne on earthly terms can reach the lofty signification of heaven. Normally in Christian thought the fact that we can speak metaphorically of divine matters is a sign of our deficiency. God knows no metaphor. Here, however, words delineate: speech about heaven is not language in the act of defeating itself, straining hopelessly to capture the unnameable, but language realizing its potential to name everything. Raphael conveys these consequences of ontological monism through his two uses of the concept of likeness. "Lik'ning" refers to our usual notion of simile as a comparison with no stipulations outside its own framework, whereas "Each to other like" in his rhetorical question posits an ontological characteristic—similarity—inherent in the two elements of this linguistic similitude. Man

and angel will not understand the same meaning, yet their different meanings will be "like" to the same extent that their different physiologies, rational powers, and worlds are "like"—"Differing but in degree, of kind the same" (490). Because Raphael can stoop to assimilate the food of Adam's world, Adam can rise to assimilate the meaning of Raphael's world. A heretical conception of the spirituality of matter issues in an unorthodox and optimistic conception of the materiality of spiritual discourse.

Seventeenth-century philosophers such as Ralph Cudworth were already attacking Hobbes and Descartes on the grounds that their efforts to speak of minds and things in a single vocabulary involved, in Ryle's phrase, a "category mistake," leading to anthropomorphism with respect to matter and misplaced concreteness with respect to mind.[47] In using the word *res* to name both body and mind, Descartes alerted suspicious readers to the need to reexamine the foundations of the dualism he continually asserted. But it is important to recognize that Descartes committed the category mistake unwittingly: his intention was to distinguish between things in space and minds in time, not to confuse them. Milton, like Hobbes in this regard, committed the category mistake knowingly at the level of ontology; and if he also assumed—as he almost certainly did, it being an assumption so common in his times that it was hardly even noticed—that ontology would correspond to semantics in the event of truth, then there was to Milton's way of thinking no problem with category errors in language. Matter being the only category, the only "kind," those cases in which the vocabulary of things migrates into the description of mental life, and vice versa, evidence the power of language to mirror the world faithfully. Although his hope was to elevate matter rather than debase spirit, Milton proceeded with all the determination of a Hobbes to construct this all-purpose language, producing what we now call the Miltonic style. A graded continuum in the meanings of nourishment was his model case, and the meal of Book 5 the foundation of his poetry, struck deep in reality.

Words and rhythms rooted in physiology at the material end of the monist continuum strive, like food itself, toward a higher state of application:

> But Knowledge is as food, and needs no less
> Her Temperance over Appetite, to know
> In measure what the mind may well contain,
> Oppresses else with Surfeit, and soon turns
> Wisdom to Folly, as Nourishment to Wind. (7.126–130)

When man and angel speak over their meal, they enact this trope, including its moral lesson. Once again Milton was aided in his semantic-ontological demonstration by the fact that many organizing structures in Renaissance medicine had arisen from projections of large and obvious processes, such as digestion, onto other activities of the organism, such as thought. In this passage, the "as" transforming healthy physiology into good knowledge is the signal of a scientific analogy, not a literary conceit. Renaissance neuroanatomists still accepted the general theory of the three cerebral cavities established by their medieval predecessors.[48] The anterior cell of the brain, soft and moist, housed the *sensus communis* where the sensorium was formed from the input of the sensory organs and varied by imagination; the middle ventricle, warmest of the three, housed reason, which organized information passed back from the anterior cell into categories (the most general being similarity/dissimilarity), performing its acts of judgment; the third and final chamber, cool and dry, was the warehouse of information, the place of memory storage, admirably suited for retention by the firm case of the cerebellum. The underlying logic of this scheme is digestive, as Walter Pagel has maintained.[49] What imagination gathered, reason separated and memory assimilated. The human brain collected, digested, and incorporated the material of the external world. Raphael is truly witty. "Knowledge is as food" is a proposition: unwanted data, the useless or the forgettable, was thought to escape in the form of vaporous spirits from the cranial sutures at the back of the head, and was actually called, in medical Latin, *flatus*.[50] In the Renaissance the word "digest" meant "to order," as it sometimes does today. But what we now think of as the literal and abstract senses of the word had yet to be clearly differentiated even in scientific thought. Informed by metaphors understood as propositions, Renaissance medicine was an invaluable resource for someone presenting a monist universe in which concretion had to seem omnipresent.

The narrative of heavenly warfare is a good food offered to the "reach / Of human sense," its similes are to be understood on the basis of digestive similarities between heaven and earth, and the narrative itself shapes Satan's defeat in the "Intestine War" (6.259) as the evacuation from heaven of "what redounds": "but the evil soon / Driv'n back redounded as a flood on those / From whom it sprung, impossible to mix / With Blessedness" (7.56–59). The fates of food—eaten, assimilated or rejected—serve as the major "corporal" delineation in bringing all heaven before the sense of Adam. Satan turns wisdom to folly, and Milton captures his ironic contribution to

his own defeat in a perfect and perfectly primitive accommodation. The rebel leader parodies Raphael's ontological speech in his hymn to the vitality of the underground, where things "dark and crude" strive for "ambient light" (6.472–483). Acting on his advice, the devils have "concocted and adusted" the "blackest grain" to charge their cannon. Its discharge condenses violent voidings—belching, vomiting, defecating:

> Immediate in a flame,
> But soon obscur'd with smoke, all Heav'n appear'd,
> From those deep-throated Engines belcht, whose roar
> Embowell'd with outrageous noise the Air,
> And all her entrails tore, disgorging foul
> Thir devilish glut. (6.584–589)

The volley has defeat written all over it. There is no more graphic and familiar instance of aggression against a nourisher coincident with aggression against the self than the rejection of food. On the third day a crystal wall "Roll'd inward, and a spacious Gap disclos'd / Into the wasteful Deep" (861–862); while "Disburd'n'd Heav'n rejoiced," Christ drove the rebel angels headlong into this "mural breach" (878–879). We have seen how food and appetite are extended to the temperate knowledge and wonder of the moral life in subsequent segments of Adam's education. In the end, as Raphael describes angelic copulation, nourishment is again the corporeality made spiritual. For when in human experience do we enjoy the object of our desire within us and mix totally? The sexuality of the angels conflates eating and pregnancy—a fusion already announced in the key word "fruit" (see Raphael's compliment to Eve and her meal at 5.388–391).

Begun concretely in the eating of a meal, and abstractly in the axioms of ontology, the progressive dilation of nourishment through Books 5–8 instructs us in the ultimate good sense of a great complex of metaphors spread into every corner of *Paradise Lost*. One branch of this complex is Milton's variation on the old theme of the scatological devil; as indicated in the previous paragraph, nourishment takes our senses to hell as well as to heaven. Inversions and perversions of alchemical digestion, Michael Lieb has shown, characterize Satan throughout the poem.[51] His negative alchemy turns sweetness into stench, vital life into the food of Death, all places into hell. His, too, is an alimentary universe, but one in which the purer feeds the grosser. Opposing him, the judgmental alchemy of God separates the saved

from the damned, what can be assimilated and improved at the top of the universe from what must be evacuated and disowned at the bottom. After changing the fruit in the mouths of devils into ashes, God enunciates in digestive terms the irony of salvation history:

> I suffer them to enter and possess
> A place so heav'nly, and conniving seem
> To gratify my scornful Enemies,
> That laugh . . .
> And know not that I call'd and drew them thither
> My Hell-hounds, to lick up the draff and filth
> Which man's polluting Sin with taint hath shed
> On what was pure, till cramm'd and gorg'd, nigh burst
> With suckt and glutted offal, at one sling
> Of thy victorious Arm, well-pleasing Son,
> Both *Sin*, and *Death*, and yawning *Grave* at last
> Through *Chaos* hurl'd, obstruct the mouth of Hell
> For ever, and seal up his ravenous Jaws. (10.623–637)

In hell reside the unknowing housekeepers of God, busy with the chore of containing their own mess. The topos of the scatological devil is normally a simple strategy for heaping abuse on Satan, and it serves this function in Milton. But the resonances of digestion are so intricate and so orderly that this abusive rhetoric becomes embedded in a full portrait of Satan, and Satan embedded in a fuller portrait of the three dimensions of the sacred. As with "Knowledge is as food," the metaphor comes true.

Watkins in particular appreciated the poetic results of Milton's monism:

> We cannot overstress a fundamental truth about Milton which we find endlessly proliferated in his work. At his most creative, he accepts the whole range from the physical, specifically the senses, to the ultimate Divine as *absolutely unbroken*. This glad acceptance means that he is free to speak of any order of being (extending to inanimate matter) in identical sensuous terms as the great common denominator. . . . Few poets . . . have come so close to making what are ordinarily abstract concepts thus tangible.[52]

Although Ricks does not cite Watkins, and makes nothing of monism, his observations time and again exemplify this statement.

His remark about the "Tree / Of Prohibition" at 9.644–645 is typical of many of the local effects described in his book: "what begins as a moving and ancient metaphor (lead us not into temptation) crystallizes with terrifying literalness."[53] These and other critics have taught us that Milton's ability to unite what Ricks nicely terms "incompatible greatnesses" appears with special clarity in his exploitation of the physical and moral senses of words such as *glory*, *amaze*, *oppression*, *height*, *place*, *supplant*, *end*, *wander*, *alienated*, *distance*, and *distaste*. He regularly gives us the impression that there is nothing alogical, nothing merely incidental about the polysemy of words: a good Miltonic word not only means in the realms of action and mentality, but meaningfully relates these realms.

Examples being many and familiar, I will limit myself to two. When Christ takes the field in the heavenly war, he commands the mountains to return to their bases:

> This saw his hapless Foes, but stood obdur'd. . . .
> In heav'nly Spirits could such perverseness dwell?
> But to convince the proud what Signs avail,
> Or Wonders move th'obdurate to relent?
> They hard'n'd more. . . . (6.785–791)

It is one thing to moralize about pride. It is monist poetry to offer the physical hardening of these "obdurate" wills, "gross by sinning grown" (661), as an ontological way of showing that it takes enormous concentration, not light carelessness, to deny the obvious. Milton halves the word "obdurate" and unfolds it as an idea of stubbornness and an occurrence—a prerational manifestation, a hardening landed before our senses. We are reading of the first situation that demanded the notion of "obdurate," and the poem recovers the birth of this idea in the *rightness* of a metaphor, as if hardening had been given to our senses in order that we might signify and understand pride. Milton incites us to think through the basis of an idea glossed over in moralizing talk about the "hardening of the will." The questions in the passage have exactly this effect: such pride is a mystery, not something well-understood. Freud was in fact anticipated by the poets, and much of what he said about the defense of denial, including the trope of defense, is available in contemplating this passage. Lastly, the first monist effect in the poem is the "fruit" dangling at the end of the opening line, both victual and consequence. The twinned epics of nature and moral action unfold from the torque of this word, which is varied constantly throughout the poem.

Monism also concerns the nature of the mimesis in *Paradise Lost*—what gets represented, what the poet cares about from episode to episode. As Watkins suggested, it provides a psychological encouragement before the fact, and a philosophical justification after the fact, for the wide childlike imagination of our poet. When approaching the transcendent, Milton does not set aside the categories of familiar experience; he insists on their application. The divine and demonic worlds are not concealed behind walls of paradox. Like a curious child, Milton is interested in how hidden things work (political counsels, the warfare and loving and marking of time in heaven, digestion), how things came about (imagine a catalogue of everything), and what things feel like (falling nine days, being hit by a mountain, having a rib extracted, changing into a serpent). Intellectual motives converging with psychological advantages in the creative process, he retained for his poetry strong vestiges of the tone and topography of our earliest researches into the world.

Milton resumed and renegotiated his oedipus complex in the religious symbolism of his late poetry. The first complex remains in the poetry just as, in the best symbols, we often discern an implicit history, a prior experience "requesting" new meaning. The metaphor "time is a river," for example, propels us into a reflection about the riverness of time, but whatever we may gain from this reflection will be indebted to the phenomenal experience the metaphor both preserves and requests transcendence of—the small vertigo when the flow of the river, different and yet the same, presences the flow of time, different and yet the same. As time is a river, as God is the Father, Miltonic Christianity is the oedipus complex. The principle guiding Milton's psychic life beyond the profane complex, whose settlement kept him a virgin for exactly half of his sixty-six years, to the sacred complex of his final poems, stands at the heart of his ontology in *Paradise Lost*. The entire poem is enfolded about it; the lower strives toward the higher, the earthly toward the heavenly, the unconscious sublimation toward poetic sublimation, without abandoning the concretion of the lower, earthly, and unconscious. The vitality of the body aspires and, in an unbroken continuum, passes over its task and its energy to the delineation of true wonders in religious poetry. This central nexus in the poem is an attempt to bring into awareness the creative process on the assumption that reason does not initiate the desire for meaning—once again, a romantic theme emerging from Renaissance ideas. Developing his ontology poetically, Milton came closest to understanding the psychogenesis of his cultural achievement.

To prepare for the fall so loudly anticipated by the material of this

chapter, I will present as clearly as I can a psychoanalytic view of this material. In the midst of this much emphasis on food there are surely archaic elements. The earliest sublimity of being at one with the mother, of recognizing no distinction between nourishment and nourisher, is discernible in Milton's conception of an inner alchemist that sublimates nourishment heavenward toward its nourisher, seeking to merge with "one first matter." This unconscious desire for the restitution of narcissistic love divulges itself in fantasies such as angelic copulation. Yet even here it is an oedipal ego—vulnerable in a solid body, subject to law, threatened with death for disobedience—before whom this fantasy is situated. Suffering in fact the fantastic tragedy of the oedipal child, Milton had been wounded, and his blindness was an epitome of many disappointments. Digestion had struck against his eyes. We have already noted the narcissistic meaning of this event as the consequence of impure or poisoned nourishment from a weak-eyed mother and an embittered wife: stated bluntly in terms of the myth, Eve poisoned Adam. In his poem, although he made it Adam's first response, Milton rejected this narcissistic interpretation for an oedipal one. At the later and more "refined" level of the oedipus complex, blindness was unconsciously construed as punishment, an interpretation supported by the many representations of intrusive and aggressive seeing in the poem. What of digestion at this stage of unconscious meaning? The phallus is the target of dismemberment because it is the organ of the sexual urge, yet the urgency of the oedipus complex is more than sexual. Its phallus is the heir of a long attachment to the mother, and in this sense represents the earlier bond between child and mother: appetite loved her first; the penis symbolizes hunger. With the accuracy of justice, the father shot a dart at Milton's eyes from the organic system that loved, perhaps in all of us will always love, the early mother. Stated bluntly in terms of the myth, Adam fell because he could not sever himself from Eve.

But the lower aspires. When enmeshed in religious symbols, the oedipus complex turns upon itself, submitting the ego to divine authority outside the traditions of men. In order to enter the sacred complex one must dethrone the profane superego by deriving a spiritual sonship immediately from God. Adam, to whom the philosophical, moral, and psychological education of Books 5–8 is addressed, gives the imagination of the poet an entrance into this second oedipal alignment, since he is the only man besides Jesus whose father is God. The "universal blanc" of Milton's blindness requested new meaning and new authorization from the resources of his psyche. As

we have stressed, once castration has been acknowledged in the first complex, obedience can also be motivated by the positive expectation of regaining in the future a semblance of the lost wish. This structure holds good, mutatis mutandis, in the second complex. How might Milton gain prophetic vision by identifying his blindness with the catastrophe of Adam? How might the logic of the sacred complex permit him to fulfill, in the name of justice, his poetic wish?

An Extra Spirit, the Pathology of the Fall, and the Two Cures of Michael

The volcanoes of hell, the mining of Mammon, the dogs gnawing at the entrails of Sin, the dart of Death, the explosive belly of the cannon, the penetration of Eden, the poisonous dream poured into the ear of Eve, the possession of the serpent—throughout Milton associates Satan with violence to inner parts. The most lacerating of these assaults is of course the evil meal of Book 9. But we cannot fully appreciate this violation without proceeding farther into Milton's physiological postulates. The friendly office of the inner alchemist, who guides our food across the boundary of traditional dualism, presupposes a substratum of organic matter stretched without break from dense body to rarified thought. To construct this ontological bridge from earth to heaven, a task parodied in the "Causey" (10.415) built by Sin and Death, Milton turned again to the speculations of hermetic neoplatonists.

Galen's doctrine of the three bodily spirits was an incarnation in physiology of the Platonic soul.[54] The liver produced "natural spirits" in the venous blood; the arterial chamber of the heart generated "vital spirits"; and the percolation of the blood through an intricate network of arteries, the so-called (and nonexistent) *rete mirabilis*, refined the "animal spirits" of the brain: life, passion, and thought were interludes in the upward journey of food through the bloodstream. Christian physicians inherited the problem of unifying this materialist hypothesis, developed in the matrix of Greek and Stoic pneumatology, with the doctrine of the immaterial soul. The usual solution was that the rational soul, being above the indignity of location, is entirely present in each part of the body; animal spirits circulating through the chambers of the brain function as but "the chariot of the soul."[55] In Christian medicine, Galenic pneumatology survived by tightening man's double notion of himself, holding together a physiology (a theory of digestion) with a psychology (a theory of mental faculties

modeled on digestion). Whenever orthodox theology or philosophy put the spirits in their place, the place was material. As Pierre Charron stated, "the Soule is in the bodie as the forme in the matter . . . and there is no mean or middle that doth unite and knit them together: for betwixt the matter and the forme, there is no middle, according to all Philosophie."[56]

But when Donne spoke in a poem of the subtle knot of mind and body woven by these blood-begotten entities, he was closer to the spirit of spirits. Two analogies suggest themselves for the role of the spirits in Renaissance thought. The first is the appeal to *mana* in primitive mentality as defined by early anthropologists such as Durkheim, Mauss, and Lévy-Bruhl.[57] *Mana* in their version of the primitive mind is an impersonal force suffused throughout the world that collects about powerful objects and powerful people. Its arrivals and withdrawals explain to men why things happen as they do—why that magic worked, why that man got sick, why that object is not to be touched by that other object. The spirits, similarly, were more than explanatory; they embodied the *principle* of causal explanation. Thus, when the chariot of God transports itself in *Paradise Lost*, this legerdemain does not merely happen, as something beyond the reach of explanation. It occurs as the effect of a cause, for the chariot is "Itself instinct with Spirit" (6.752). Spirits permeated Renaissance science, philosophy, theology, and poetry. Almost everything in the repertory of human behavior—sensations, acts of mind, love and lust, sickness and health, death—was open to causal definition by describing events in the invisible (you cannot check the evidence) but material (it is there nonetheless) paraworld of the spirits. The second analogy is with the "economic point of view" in Freud, which reduces mentality to the freeing, binding, and discharging of psychic energy. Accurate or not, Ricoeur contends, this energics made one of the vocabularies of psychoanalysis into a defensible "semantics of force" committed to representing desire *as* the pressure of a physical system.[58] Pneumatology equipped medieval and Renaissance man with his own rhetoric of force. When vital spirits cling to the image of a lady in the heart of a stricken lover or desert the heart of a quivering coward, we confront a reduction similar to Freud's cathectic energy. Motive becomes quantifiable, and as the immediate corollary of this reckoning, compulsive, a physiological tyranny to which lovers and cowards submit.

The explanatory power of pneumata received its grandest expansion in the work of Descartes. Spirits that could materialize the soul had

another use: they could deanimate the body. In the medical spirits Descartes found a physical apparatus able to explain the vital motions of the body as an autonomous performance. Flesh was nothing but geometry in motion, spirits behaving spiritually, obedient to the laws of impact and hydraulic pressure. The horseman was out of the chariot. "Itself instinct with Spirit," the chariot had become a self-sufficient automaton not unlike those tricky creations sometimes placed in Renaissance gardens to move or sing by means of pressures fed through hidden pipes. We can detect a similar confidence in the scientific purity of pneumatic explanation in Bacon, Telesio, and other figures of the scientific Renaissance. Newton himself would look to them in search of a physical medium for the force of gravity.[59]

In another tradition of Renaissance thought, mind was being fusedwith body, not subtracted from it. The God of the neoplatonists, if pure idea, had spilled over into progressively more material emanations. Philosophers in this tradition, with their emotional interest in the ascent of contemplation and their intellectual interest in proving rationally the immortality of the soul, sought to define the precise gradations of mind's articulation with matter: earth and heaven might learn to communicate by means of substance. If infused with life, spirits that served in the soulless universe of Cartesian mechanism could serve with equal prominence in the vitalist universe of hermetic neoplatonism. Agrippa and Ficino invoked the medical spirits to explain the ability of imagination to operate at a distance. In his *De triplica vita* (a title suggesting the Galenic tripartition of *spiritus*) Ficino argued that commerce between the magician and celestial inhabitants implied a kinship between astral matter and cerebral spirits. Pneumatology of this kind sanctioned an aggrandizement of the will; inspired by a Pelagian nostalgia for the astral powers given Adam in the *Corpus Hermeticum*, pious magi endeavored to reclaim the original dominion of man. By the time Ficino's magic reached Campanella, D. P. Walker has shown, similarity had become identity.[60] Spirits, souls, and angels were consubstantial. *Spiritus* had come to supply the "middle term" denied by orthodoxy, and all traffic between earth and heaven could be rerouted accordingly.

Alexander Ross, the furious watchdog of convention in seventeenth-century England, acknowledged that ordinary language invited such excesses. Although the single word "spirit" denoted bodily substances, souls, and angels, souls were not medical spirits, and neither souls nor medical spirits were angelic. Weemse also cautioned against accepting

the false lure of polysemy.[61] But Milton saw truth liberated where the dualists saw language abused:

> one first matter all,
> Indu'd with various forms, various degrees
> Of substance, and in things that live, of life;
> But more refin'd, more spiritous, and pure,
> As nearer to him plac't or nearer tending
> Each in thir several active Spheres assign'd,
> Till body up to spirit work, in bounds
> Proportion'd to each kind. So from the root
> Springs lighter the green stalk, from thence the leaves
> More aery, last the bright consummate flow'r
> Spirits odorous breathes: flow'rs and thir fruit
> Man's nourishment, by gradual scale sublim'd
> To vital spirits aspire, to animal,
> To intellectual, give both life and sense,
> Fancy and understanding, whence the Soul
> Reason receives, and reason is her being,
> Discursive, or Intuitive; discourse
> Is oftest yours, the latter most is ours,
> Differing but in degree, of kind the same. (5.472–490)

Everything is in transition. The root metamorphoses into stalk, stalk into leaves, and leaves into flower, whereupon the plant offers its being to man, "bright" only to one who sees and "odorous" only to one who smells. The bright essence at the summit of vegetable aspiration then becomes the root of human aspiration. If we assume that in the phrase "Man's nourishment" the "flow'rs and thir fruit" have already been refined to natural spirits, there are again four stages in the human sublimation: natural spirits (root), vital spirits (stalk), animal spirits (leaves), and intellectual spirits (consummate flower). Like the "bright" and "odorous" flower, the consummate refinement of man lifts his "being" into communication with a higher order of creation, the "kind" of heaven. The extended meanings of "spirit," far from treacherous, correspond to ontological process. Milton has ventured the radical equation of cerebral spirits, souls, and angels.

A modification in Galenic pneumatology accompanies this equation. To the three bodily spirits Milton has added a fourth or "intellectual" spirit, complementing the familiar distinction between discursive and

intuitive reason with a heretical physiological distinction. As Babb explains, the poet "assumed—or deduced—the existence of a material intellectual spirit which is the substance of the angels and which man alone among earthly creatures possesses, and thus he bridges the gap between man and angel."[62] Because reason is divided into discursive and intuitive and man produces animal and intellectual spirits "whence the Soul / Reason receives," the natural inference would be that animal spirits operate in the lower sort of knowing, which is temporal and logical, while intellectual spirits operate in the higher sort of knowing, which is simultaneous and nondeliberative. Had man remained obedient, he would have become "all spirit" like the angels—predominantly intellectual spirit, angels being "pure / Intelligential substances" (5.407–408). The unfallen life of man contains numerous instances of intuition's cooperation with "discourse." It is the power by which Adam knows God before finding him ("Whom thou sought'st I am"), dreams realities, and gives voice to the created order in "Unmeditated" prayers. Considered physiologically, which is one of the ways Milton chose to consider it, the fall is a disaster for the intellectual spirit especially.

Satan twice attacks the spirits of man. With the "devilish art" of Book 4, he tries to "taint" the "animal spirits" of Eve by his dream of false ascent (804–805). But "Evil into the mind of God or Man / May come or go, so unapprov'd, and leave / No spot or blame behind" (5.117–119). To think through an evil action in the discursive mode, whether awake or dreaming, is blameless. A taste that puts an end to thought precipitates the fall. At that moment evil reaches inward to bring "spot" *and* "blame," a guilt inseparable from physical stain, to all the spirits. Representations of digestive illness surround the crisis. At each disobedience nature suffers the cramps of a bellyache, "Sighing through all her Works" (783) and trembling "from her entrails" (1000). These macrocosmic symptoms in the body of mother earth herald the arrival of a pathogenic nature, corrupt with flatulence: "Vapor, and Mist, and Exhalation hot, / Corrupt and Pestilent" (10.694–695). But an endogenous source of disease is born simultaneously in the inner weather of man's storm-tossed pneumata. The "dilated Spirits" (9.876) praised by Eve have actually become "Encumber'd" (9.1051).

Unfallen love sponsored the refinements of digestion. As the inner alchemist sublimates nourishment "by gradual scale" until "body up to spirit work," so true love "refines / The thoughts" and "is the

scale / By which to heav'nly Love thou may'st ascend" (8.589–592). The first "operation" attributed to the fruit is the reorientation earthwards of love's alchemy:

> but that false Fruit
> Far other operation first display'd,
> Carnal desire inflaming. (9.1011–1013)

These are the fires in which you burn, akin to hell as the heats of unfallen digestion were akin to heavenly vitality. "There they thir fill of Love and Love's disport / Took largely" (1042–1043). Lust, too, is an intemperate meal. Our disobedient parents move as the chariot wills, subjected to a physiological "operation."

The Pelagian program of Renaissance hermeticism has now been banished to the far side of human evil. "No more" will men enjoy the "Venial discourse" with angels (9.1–5) that occult philosophers hoped to promote through pneumatology. Spirits as vehicles of aspiration, leading the will intuitively toward God and the flesh naturally toward an intellectual body, belong to our prelapsarian condition. The intimate defilement of the fruit founds a new psychology on a new biology:

> Soon as the force of that fallacious Fruit,
> That with exhilarating vapor bland
> About thir spirits had play'd, and inmost powers
> Made err, was now exhal'd, and grosser sleep
> Bred of unkindly fumes, with conscious dreams
> Encumber'd, now had left them, up they rose
> As from unrest, and each the other viewing,
> Soon found thir Eyes how op'n'd, and thir minds
> How dark'n'd. (9.1046–1054)

A "force" has violated "thir spirits" even to the "inmost powers." The meal that Satan promises will transform mankind into angels does transform us—but into the opposite of angelic freedom. "Necessity and Chance," God proclaims, his words appropriate in kind if not in degree to angels, "Approach not me, and what I will is Fate" (7.173–174). At the fall, necessity invades man and his will must expend great energies on adjusting to fate rather than creating it. With the "mortal taste" of that fallacious fruit, pneumatology becomes for the first time in Eden a semantics of reduced volition, a vocabulary

that tells of an "operation" that "Made err," conveying the surrender of mankind to appetites in excess of biological need. If intuition is the transcendence of will in apodictic knowledge, intemperance is the debasement of will in mechanistic desire. From this one choice, choice will forever be conditioned by "dark'n'd" impulse, by illusion and by bitter conflict. This new order of mechanistic intemperance, Michael informs Adam, will cause "Diseases dire" (11.474). Satan has managed to fray the bright thread joining immortal man to immortal angel:

> As with new Wine intoxicated both
> They swim in mirth, and fancy that they feel
> Divinity within them breeding wings
> Wherewith to scorn the Earth: but that false Fruit
> Far other operation first display'd. (1008–1012)

In this brilliant passage we behold the unfallen destiny of man transformed into an illusion of inebriation, the false promise of a mere wish, and worse than that, a defensive wish, for all the while an "operation" weighs them downward in wingless obedience to gravity to copulate on the ground.

One consequence of impaired intuition, the "inmost" of human "powers," is damage to the "self-knowing" constitutive of man:

> Thus they in mutual accusation spent
> The fruitless hours, but neither self-condemning,
> And of thir vain contest appear'd no end. (9.1187–1189)

Milton ends the book of the fall with the insight that judgment, turned away from the recognition of inmost crimes, will tend hereafter to operate more accurately with respect to the other than with respect to the self. Subsequent lapses in intuitive apprehension measure how darkened their minds have become, as Adam and Eve reason inconclusively about the nature of death and think to abort posterity through sexual abstinence or suicide—the desire to die, which will also be expressed in the "Lazar-house" of Book 11, where diseased humanity begs relief from a perversely tardy Death (491–493), being the definitive slavery of the mind to the sovereign imposition of a fallen body in a fallen nature.

Intuition survives in their hopeful suspicions about the "mysterious terms" of divine judgment. But, leaving aside the rare miracle of assumption whose interest for Milton I have treated elsewhere, the patho-

genesis of the fall is irreversible.[63] Adam receives from Michael a full exposition of the new rhetoric of force. As the reader had been led to expect, the natural death in the future of fallen Adam is a perfect inversion of the natural elevation in the future of unfallen Adam, as if the first physiology had been turned over and dropped, producing a headlong fall toward brute matter: "in thy blood will reign / A melancholy damp of cold and dry / To weigh thy Spirits down, and last consume / The Balm of Life" (534–546). Lifeblood itself is subjected to a weight. Whereas Raphael enticed Adam with a physiology of aspiration, body working up to spirit, Michael sobers him with the physiology of fatality, spirit working down to body, that will in time become the object of medical science. The body is now, as it has never been before, the dominion of necessity over wish. Under the reign of leaden melancholy, the downward course of the spirits duplicates as mere process the downward plunge of desire into mechanism observed at the fall. Death is the biological repetition compulsion of our primal crime. Like Satan, we "consume" ourselves in the end.

For centuries Christian theologians had rigorously dissociated the forbidden fruit from the magical potions consumed in Homer and Ovid. Milton adopted this tradition in his treatise on doctrine. Inherently the fruit was no more than a fruit: "It was called the tree of knowledge of good and evil because of what happened afterwards: for since it was tasted, not only do we know evil, but also we do not even know good except through evil" (*CP* VI, 352). But in the epic real symptoms—intoxication, digestive vapors, restless sleep, darkening mind—are predicated of the fruit. At the very moment it is first tasted, Milton is oddly uncertain over its inherent quality: "such delight till then, as seem'd, / In Fruit she never tasted, whether true / Or fancied so" (787–789). Why is he unwilling to say whether the forbidden fruit is really the sweetest one? Sweetest or not, he has given us the material to build a fairly strong case for its having been poison fruit. Because the sentence of death is being carried out by "The Law I gave to Nature" (II.49), Milton might have decided that it would be efficient and forward-looking of God to have put death in the fruit—in which case "fallacious" Satan lied to man about the *particular* "force" of the fruit and not about its forcefulness. Favoring this view would be Milton's reluctance elsewhere in his monist poem to separate the letter from the spirit of his metaphors. The question cannot be decided absolutely, and in any case nothing much hangs on its outcome: however God executes the sentence, man ate and man dies. The fact that there *is* a question tells us more than its answer would.

Milton has made the pathogenic fruit in his Lucretian narrative of the fall concrete enough to belie easy translation into mere metaphor. Before us, in a primary way, are the signs of an illness.

It comes as no surprise that Adam requires the attentions of a physician. The angel ministers to his need:

> but to nobler sights
> *Michael* from *Adam's* eyes the Film remov'd
> Which that false Fruit that promis'd clearer sight
> Had bred; then purg'd with Euphrasy and Rue
> The visual Nerve, for he had much to see;
> And from the Well of Life three drops instill'd.
> So deep the power of these Ingredients pierc'd,
> Ev'n to the inmost seat of mental sight,
> That *Adam* now enforc't to close his eyes,
> Sunk down and all his Spirits became intranst:
> But him the gentle Angel by the hand
> Soon rais'd, and his attention thus recall'd.
> *Adam*, now ope thine eyes. (11.411–423)

Michael removes two obstructions from the head of Adam. There is nothing metaphorical in the presentation of the first cure; the passage opens with the founding of our medical tradition. Like the physicians known to Milton, the angel can lift an incipient tumor and purge the optic nerve with "Euphrasy and Rue," curative herbs used commonly during the Renaissance for disorders of the eye. Fruit as victual "Had bred" fruit as consequence, depositing a film in the eyes of man: in the materialistic narrative coincident with the moral narrative, Adam was developing cataractic blindness from the unpurged vapors of the fruit. Then Michael clears a pathway to "the inmost seat of mental sight." His cure of this second obstruction does indeed leave empirical medicine behind. The angel administers three drops "from the Well of Life"—a wishing well nowhere mentioned in the rest of the poem. By the end of the passage Milton has at last severed medicine from miracle. Here certainly is a "wonder."

But the affliction treated in the second cure is no less physical than the film bred in Adam's eyes. Recalling the "inmost powers" invaded by the force of the fruit, the "inmost seat of mental sight" must be the habitation of the intellectual spirits. Michael renovates that intuitive power manifest in visionary dreams and unmeditated songs. Adam is now being readied for another education, and enabled to see, in the

temporary brilliance of his forthcoming visions, what Michael intends him to see. Reestablishing the frayed link between man and angel, prophecy considered as a physiological event demands a new enlightenment of the intellectual spirits. Immediately "all his Spirits" fall into a holy trance, prepared to be taught the new ways of God.

Milton implies that, beyond death and disease, the material price of the fall was to make the innermost treasure of the spirits of intuition virtually inaccessible by natural means. If we carry a paradise within us, we carry also a paradise lost within us. This fruit of original sin will never be the province of medical science, for nature operating naturally in the body pulls intuition down to deliberation in the inevitable descent of the oppressed spirits. Such a recovery can only be "enforc't" from above by the radically exogenous cure of inspiration: as a countermeasure to the "force" invading our darkened minds at the fall, the three drops touch the spotted parts of the mind as intimately as the fatal fruit. Divine intervention of this kind will be known throughout human history in the Protestant guidance of the "spirit" within. It is felt by an "enforc't" Samson when he suddenly loses his will and gains divine purpose, declaring, "I begin to feel / Some rousing motions in me" (1381–1382). In *Paradise Lost* Milton represents the first of these benign intrusions by the (for once) Trinitarian dosage from the "Well of Life." What can the Well of Life be, given the suppositions of the epic, but the pure vitality of God that has, through the fateful act of creation, merged indissolubly with "one first matter"? Inspiration is a second gift of life for man.

It will take our argument back to the psychological drama of the creative process to note the serious theological repercussions of presenting the fall as a pathogenic event reversing the upward aspiration of *Paradise Lost's* ontology. Milton held in the *Christian Doctrine* that the fruit was forbidden for no other purpose than to manifest our obedience: "it was necessary that one thing at least should be either forbidden or commanded, and above all something which was in itself neither good nor evil" (*CP* VI, 351). But the forbidden fruit of the epic cannot be, as many critics have alleged, the orthodox "arbitrary fiat" defended in the treatise. Perhaps no coherent poem could preserve this doctrine unless it were the author's overwhelming intention to do so, since good art tends by nature to make its details seem inevitable. However that may be, the God of this poem could not have prohibited any other action, such as standing on one's head, slapping a face, or spitting into the wind, without changing altogether the meaning of the fall. Eating embodies ontological and historical rhythms; the

semantic dilation of food to knowledge, appetite to wonder, and temperance to obedience can be traced in much more detail than I have attempted. The fall focuses a great design of unbroken harmony. Seeking forbidden nourishment is not, as given sense by this design, a crime solely because of arbitrary authority: it is *the* trial and *the* crime, the ontologically appropriate crime, its wrongness shot through the texture of Milton's universe. The forbidding is not arbitrary: it is "necessary," as the treatise grants, in order that there be obedience and therefore virtue. That this one fruit was forbidden may well be arbitrary, although its effects would not appear to be those of just any fruit, and Milton himself allows that it might have been the tastiest one. In its ultimate good sense this poem *attacks* the severe voluntarism that some critics believe Milton to have espoused.[64] Milton directs this attack, furthermore, at voluntarism's seemingly impregnable instance: "But of the tree of the knowledge of good and evil, thou shalt not eat of it: for in the day that thou eatest thereof thou shalt surely die" (Gen. 2.17). Huge blocks of prehistory context this law in *Paradise Lost*. The more we learn about Milton's universe, the more we learn that we are "of kind" to be able to know more, and the more we suspect that, if we did, God might be justified apart from any deep concern with his absolute will. Is Milton resisting the God of incomprehensible power in the sacred complex, evading a feared encounter with sheer will? Is he lessening the otherness of the law, which would be a step toward assimilating the superego into the ego?

To an extent, undeniably. Absolute voluntarism, a divine will of power alone, is the theological position given to Satan in the poem. God is not always so riddling and involuted as he was with Samson, and sometimes the inner oracle of *Paradise Regained* will sanction a rational doctrine "granted true." When imagining the institution of the first law, Milton tried to mitigate Satan's view of a fearfully irrational God. (That sort of deity would appear soon enough in the tale of Abraham and Isaac, which Milton interestingly omitted from his selective bible in Books 11 and 12.) In any event, the monist ontology of the epic does not imply that Milton would have preferred reason to be a law unto itself, rendering divine commandment superfluous. Ontological appropriateness is a long way from the suggestion that reason might have *discovered* the law, which is the only real opponent of voluntarism in Christian theology, and a rare one at that—Aquinas on the Decalogue, Kant on the categorical imperative, a few lesser figures in the history of Christian ethics. Understood within the entirety of *Paradise Lost*, the law against eating a particular food is not so much

generated by reason as it is a revelation to reason. Our sense of its appropriateness arises from what the poem has taught us about those elements of our humanity that lie beneath reason, disturbing its false autonomy—the biology and psychology of man, creature of complex appetites. There can be little doubt that Milton felt his way to this appropriateness, not in the empyrean of pure reason, but in the experience of *gutta serena*. Blindness as interpreted by seventeenth-century medicine supplied him with the history of what transpired in the secret recesses of the body when man ate "the Fruit / Of that Forbidden Tree, whose mortal taste / Brought Death into the World, and all our woe."

Andrew Marvell was a reader of acute intuitions.

> the Argument
> Held me a while misdoubting his Intent,
> That he would ruin (for I saw him strong)
> The sacred Truths to Fable and old Song
> (So Sampson grop'd the Temple's Posts in spite)
> The World o'erwhelming to revenge his sight.

Marvell feared, "while misdoubting his Intent," that Milton would collapse the temple of "sacred Truths" into the ruins of "Fable and old Song," revenging himself on the God he ostensibly came forth to justify. In fact, the purgative denial of this very aggression is writ large in the poem Milton solicited from his God. He destroyed the unfallen world with "Death" and "all our woe" in the act of accusing, not the theft of his sight, but the fate of his blindness. In the etiology of his condition, nourishment had turned to wind, a gross vapor had solidified in his optic nerves, obstructing his spirits and darkening his eyes. Blindness to Samson is "a living death": "Myself my Sepulcher, a moving Grave, / Buried" (100–103). Although Milton never expressed this agony in his own voice, he must on occasion have seen death in his blackness, such as the "murmur" recorded in Sonnet XIX, where his "dark world and wide" is a living anticipation of the "death" of the unprofitable servant in the parable of the talents, cast into the "outer darkness" (Matt. 25.30). As he wrote his epic on Genesis, this blindness by digestive failure became the sin tainting all men, a hardened infection whose cause, like the tumors in his eyes, could not be expunged. The taint Adam suffered has been fixed in our solid flesh: through our first Parents, we have all supped full of horrors. The reversal of ontological aspiration, condemning the body to weight and

fixity, makes *gutta serena* the radical instance of an illness everywhere to be found in fallen biology and fallen psychology. Milton recorded his own ongoing struggle with the general pathology in the last words of his last invocation:

> unless an age too late, or cold
> Climate, or Years damp my intended wing
> Deprest; and much they may, if all be mine,
> Not Hers who brings it nightly to my Ear. (9.44–47)

Continuing the tradition begun with Michael's renewal of "the inmost seat of mental sight," the flight of inspiration must contend in a fallen mind with melancholy "damp" and "deprest" spirits. These forces may be held in abeyance for a time, but there is no escape, as there was in *Comus*, from growing "clotted by contagion." The epic poet has already been struck by our virulent common illness. It is not to be dispensed with. On the contrary, his effort is to secure from this weakness, fully acknowledged as weakness, his creative power.

He never solicits the double cure given unimplored to Adam. The theme of the Lucretian epic in *Paradise Lost* pursues in another register the theme of "better fortitude" governing its epic of heroic action—our restricted freedom and our acceptance, without indicting God, of a world no longer to our wish. To be the proponent of this theme, the poet must be a true poem, containing "in himselfe the experience and the practice of all that which is praiseworthy" (*CP* I, 890). The physiological representation of the fall, ending with Adam being cured from incipient cataracts, demonstrates the fit between the life of the poet and his art. More importantly it reveals that Milton has mastered the moral valor he celebrates. To Adam is given the happy cure Milton no longer expects. He will not bargain with God over his blindness, nor will he defend himself against his loss, as he did in the Second Defense (*CP* IV, 583–591), by idealizing the deprivation. The poet deals *with* his blindness.

"So much the rather thou Celestial Light / Shine inward." We left the invocation to light with the conviction that its reversal of the formula of debt, making God the obligatee and Milton the obligator, would not be psychologically coherent unless blindness were a punishment owed for the potential trespass of the poem he had yet to write. And that is precisely what it is. "Blindness must not be considered a judicial punishment." Not a judicial punishment, the sentence specific to a given crime, but a punishment, a sign in his flesh of the com-

munal guilt of original sin, and the means by which Milton could identify himself with Adam, for whom the physiological deterioration making him vulnerable to blindness *was* a judicial punishment. The poem indicts blindness as a fate justly deserved. To justify the ways of God in a personal sphere was to conceive of his condition as unalterable, as punishment, and still as just. Completing this private theodicy in his representation of "Man's First Disobedience," Milton links blindness to the primal crime of Adam, thus indicting himself rather than God, and as the flesh of Adam assumes responsibility for the sentence of *gutta serena*. He can therefore fulfill in the name of justice his desire to see "what surmounts the reach / Of human sense." He has suffered the punishment due for whatever may be unlawful in such ambition, and suffered it *as lawful*, the just imposition of a just God. Having created the wish to be invulnerable in the flesh, indulged the desirability of this wish, and then, without indicting God, sacrificed this wish, Milton can rightfully claim his due: "to see" (as Adam and Michael see in Book 11) "and tell" (as Adam and Michael tell in Book 12) "Of things invisible to mortal sight." *As blindness is equivalent to the curse of the fall, so the writing of the poem, which the guilt of this blindness paid for, must be unconsciously equivalent to the guilty aspiration of the fall.* It was because of its intrusive curiosity that this literary project was "long choosing, and beginning late," and it is evident in the poem that Milton's appetite for wonder exceeds on "bolder wing" the limits set for Adam.

The feeling that one has been punished before one has become worthy of a guilt that would motivate the punishment is intrinsic to our lives. If it were not, no one would bother to ask the murmuring questions that lead to a theodicy. From the beginning, long before we are able to discern power from justice, we have been sentenced to disease and death. The romantic criminal, sensing the arbitrary exaction of these penalties, feels free to disregard justice altogether. But Milton defended the justice of our condition, while in the same breath satisfying his desire to violate mortal limits. Unlike the romantic criminal, whose archetype in our literature is his own portrait of Satan, he was able to avail himself of the redemptive paradoxes of Christianity: to fall is to rise, to be guilty of unfettered desire is the way to attain unfettered desire. If the writing of the poem is unconsciously equivalent to the guilty aspiration of the fall, that aspiration is ironically equivalent, in the manifest meaning of the poem, to the future outlined for unfallen man. The doubling of unfallen and fallen aspiration permits a moral sublimation in the psychology of the aspir-

ing poet. We can infer the achievement of this sublimation from the paradoxes inherent in the formula of the transposed debt. The potentially guilty act of writing the poem is justly owed to Milton, provided that he indict his blindness as a morally just punishment: the potentially guilty act is therefore morally just, for as the poem must be written, Milton will justify the ways of God to all men, including himself.

The same redeeming sublimation can be inferred from the imagery of blindness and sight. *As blindness is punishment, so seeing is forgiveness, and the visionary composition of the epic renders the poet symbolically guiltless.*

> What in me is dark
> Illumine, what is low raise and support;
> That to the highth of this great Argument
> I may assert Eternal Providence,
> And justify the ways of God to men.

The "great Argument" of a private theodicy, requiring Milton to condemn his blindness as the summation of what is "dark" and "low" in mankind, is the psychological plot of the poem's creation, the means by which "I may assert" became "I did assert." At the center we have circled about in discussing the wish evoked in Books 5–8 and its disillusionment in Book 9: in order to claim prophetic vision, and become in that sense guiltless and unfallen, Milton must purge through moral accusation another version of the same wish, demonstrating that his desire not to be blind, guilty, and fallen contradicts the justice of God. This, rather than the defensive stainlessness projected in *Comus*, is the happiest strategy of the oedipal child. To achieve the wish, first slay the wish. Elevating the complex to religion, Milton restaged its sad yet heroic drama of relinquishment, achieving its resolution in a work of art that spent with passion his hoarded talent.

Prophecy for the late Milton was a partial and exceptional relief from the pressures of the fall. For the few and by the few, prophetic missions could not serve the grand hopes of utopian Christianity or the quasi-scientific visions of reconquering through pious magic the dominion of old Adam. But these currents of Renaissance ambition touch his work nonetheless. Milton believed himself to have undergone a privileged recovery from the second disease cured by Michael. His mind irradiated "through all her powers," he repossessed

the intellectual spirits darkened and dissipated by the trauma of our first crime—lodged in us useless—to enjoy once again the watch of inward sight with its gifts of unpremeditated song and true prophetic vision. His renewal of the intuitive powers originally implanted "inmost" in man trails with it the Pelagian aspirations checked at the fall. Heaven is again in communication with earth. Again a man is reasoning with the unpremeditated grace, the poise of voluntary and involuntary, characteristic of intuition. His eyes how darkened, but his mind how opened!

The metaphor deep-rooted in *Paradise Lost* for the poem itself is the angel body. Composing Book 3, its poet has actually visited heaven and "drawn Empyreal Air, / Thy temp'ring" (7.14–15), reclaiming in art the lost destiny of the human body. With the "sense variously drawn out from one Verse into another," its unobstructed lines forsake the "vexation, hindrance, and constraint" of rhyme, stanzaic pattern, gnomic syntax, and other restrained conveyances in the corpus of poetry. Elevating deliberation to the higher state of intuition, its unpremeditated flow joins earth and heaven. It will allow no dismemberment in the nature of things, for that would leave a blind poet as "Cut off" from the ways of God as he is from "the cheerful ways of men." Resting on a substratum of "one first matter," it boasts a language able to name earth and heaven in full semantic propriety. It limbs itself as we please. Its parts are "condense or rare," enfolding the entirety with no loss in local concentration. It varies its body at will, dilating fact into metaphor and reintegrating metaphor into fact. It flies. And to these graces of angelic corporeality—fineness, substantiality, plasticity, flight—we must not fail to add the immortality it strives for, successfully up to now. Fame was not, at last, an infirmity of noble mind, but a mental body of unfallen vitalities.

Milton wished for more than an immortality of fame. Like all Christians, he wished for immortality. But he wanted immortality in a particular and heterodox universe. The orthodox may expect the resurrection of the body, but Milton went much further in stipulating the materiality of the hereafter. Miltonic man, in a sense, *does* live by bread alone; in heaven as on earth, he will eat, and he will exercise the rest of his senses as well. For the divided and distinguished worlds of Christian orthodoxy Milton substituted a graded continuum by which we climb the scale to heaven. There is no otherness beyond the solidity of matter, only degrees of solidity. One inalienable characteristic of this universe is its tangibility, and the tangible to a blind man—the respiration of air, the taste and weight and internal companionship of

food, the coordinates of known objects—must constitute the real more concertedly than it does for the sighted. We are told by the "So much the rather" of the invocation to light that "To see and tell" of the invisible transcendent realm will compensate our poet for his lost perception. "And so," Masson wrote, "his very blindness . . . assisting him in his stupendous task, by having already converted all external space in his own sensations into an infinite globe of circumambient blackness or darkness through which he could dash brilliance at his pleasure, there did come forth a cosmical epic which was without a precedent and remains without a parallel."[65] By creating *en image* the cosmos of *Paradise Lost*, Milton filled the emptiness of his "darkness visible" with masses, colors, and motions. He was insistent about the possible actuality of this compensatory world of images; no echelon of Miltonic being is in principle unimaginable . In the most famous critique of this emphatic imagining, Samuel Johnson directs us to the right point in coming to the wrong conclusion:

> Another inconvenience of Milton's design is, that it requires the description of what cannot be described, the agency of spirits. He saw that immateriality supplied no images, and that he could not show angels acting but by instruments of action; he therefore invested them with form and matter. This, being necessary, was therefore defensible; and he should have secured the consistency of his system, by keeping immateriality out of sight, and enticing his reader to drop it from his thoughts. But he has unhappily perplexed his poetry with his philosophy.[66]

Johnson assumed that Milton believed in the immateriality of the spirit world, but his mistake reveals one of the reasons Milton opposed this belief. On the "universal blanc" of his sightlessness the poet imagined a phenomenal heaven. Where he an orthodox imaginer, he would have assumed, as orthodox Johnson assumed, "that immateriality supplied no images." But if his compensatory images signified an orthodox heaven, whose authentic mode of being was unimaginable because extraphenomenal, then in the process of understanding them their solidity would have to be dissolved back into the very emptiness Milton had contracted to escape from. The sublime creations of an orthodox Milton would, in the end, have replicated the "blanc" they were intended to cancel.

Christianity often projects as the great object of our desire an impalpable and foodless world. But this empty idealist heaven held no

reward for blind Milton. The dream visitation of his veiled wife in Sonnet XXIII appears "such, as yet once more I trust to have / Full sight of her in Heaven without restraint"—and in the universe designed for *Paradise Lost*, full hearing, full smell, full touch, full taste. Milton had experienced the impoverishment of mere mentality. Having dropped into a state resembling pure mind, he came to discover profound reassurance in the abiding presence of material things. Monism in this respect was a guardian philosophy, guaranteeing the steadfastness of this world by locating redemption within it. A mature mind may decide with good reason that "he has unhappily perplexed his poetry with his philosophy," and find amusement in those sad fragments of Renaissance hermeticism that Milton collected together to lend this philosophy its poignantly vulnerable credibility. I reply that heavens, whatever we must accomplish to reach them, exist essentially to be desired. Do not some of our most disorienting fantasies cast us into a fearful combination of blindness and immateriality? Wherever one might go in the monist universe of *Paradise Lost*, he would always be in touch.

• 6 •

THE PLACE OF REST

> By picturing our wishes as fulfilled, dreams are after all leading us into the future. But this future, which the dreamer pictures as the present, has been moulded by his indestructible wish into a perfect likeness of the past.
>
> Sigmund Freud, *The Interpretation of Dreams*

> To the cleavage the *yes* to Freud introduces into the heart of the faith of believers, separating idols from symbols, there corresponds the cleavage the *no* to Freud introduces into the heart of the Freudian reality principle, separating mere resignation to Ananke from the love of Creation.
>
> Paul Ricoeur, *Freud and Philosophy*

MILTON devoted his public career to elucidating "three varieties of liberty without which civilized life is scarcely possible, namely ecclesiastical liberty, domestic or personal liberty, and civil liberty" (*CP* IV, 624). At a minimum, to be civilized is to hold a place in common with others, and Milton taught civilization the supreme but precarious freedom of schism—the liberty to divorce the spouse, to depose and execute the sovereign, to disobey and overthrow the ecclesiastical hierarchy. These were hopeful rebellions, inspired by his yearning for great ideals, but in each case his efforts fell short of restoring the ideal. The three estates of family, nation, and church were sites of disappointment for this man. As Martz has recently emphasized, he was a "poet of exile," dwelling in riven homes.[1] The first wife deserted him, the second died soon after their marriage, and the third, if a good cook, quarreled bitterly with the daughters of the first. The nation, proving itself enslaved in spirit, welcomed another monarchy. Milton was imprisoned for a time in 1660; he wrote in 1666 that the virtue of "devotion to my country almost drove me from my country . . . . One's country is wherever it is well with one."[2] How roomy a place was that when "fall'n on evil days, / . . . and evil tongues; / In darkness and with dangers compast round" (7.25–27)?

He never wrote his nationalist epic on King Arthur, and a digression on contemporary England was excised from his *History of Britain.* As for fellowship in the house of God, we have no evidence that Milton, church-outed by the prelates as a young man, even attended a church during his maturity. After formulating a theory of matrimony, he refused to be married in one, and the Spirit of the invocations "dost prefer / Before all Temples th' upright heart and pure." Milton doubtless enjoyed the friendship of learned and cheerful men. In London in the 1640s he would now and then "keep a Gawdy-day" with "some Young Sparks of his acquaintance" (*EL*, 62). Later on there were similar days, when the blind man resolved "to drench / In mirth, that after no repenting draws" (Sonnet XXI). The letters of his last years are full of fatherly advice for the admiring young. Yet in fundamental respects his search for a good place to reside, happily centered in family, nation, political faction, or congregation, ended in solitude. Like an uncanny emblem of these failures, blindness also excommunicated Milton, severing him from the external world vision best reveals to us and, in the mood of the invocation to light, from the cheerful ways of men involved in their surroundings.

The fact of the matter is virtually allegorical. Milton was born on Bread Street under the sign of the eagle, a bird celebrated for its clear-sighted delivery of the fire of Jove. He inherited this first home from his father. In 1666, when we suppose him to have finished *Paradise Lost*, "By the great Fire which happened in *London* . . . he had a house in *Breadstreet* burnt, which was all the real estate that he had left" (*EL*, 48). Was he moved to add a passage to the still unpublished manuscript of *Paradise Lost*, or did the prophetic poet simply experience one of those apprehensions of inhuman coherence that steal upon us now and again in this riddling life, remembering that he had granted the dispossessed founders of mankind, before they stepped out of his poem, a look back at the lost real estate of their birthright?

> They looking back, all th' Eastern side beheld
> Of Paradise, so late thir happy seat,
> Wav'd over by that flaming Brand, the Gate
> With dreadful Faces throng'd and fiery Arms.

My subject in this final chapter is the symbol life prepared Milton, step by step, to create and comprehend—paradise lost, the once and future home.

In the three late poems Milton deliberately set about to enfold his

Christian vision in the symbol of home. It is found three times in the place of closure. These three homes, moreover, are situated in the beginning, the middle, and the end, like Freud's three mothers—the home we are born into, the home of the grave awaiting us, and, in between, the home we choose. Having refused to go to his first home with Manoa or to his middle home with Dalila, Samson is brought "With silent obsequy and funeral train / Home to his Father's house" (1732–1733), destined in Christian times to become a symbol for the afterlife. Christ returns to his first home, reconstituting the union of mother and son found at the end of the Nativity Ode; having acquired a beginning, he must soon leave the first home to undertake his mission. This leavetaking is the situation of first Adam and first Eve at the close of *Paradise Lost*.

After she eats the fruit, and before he eats the fruit, our first parents reaffirm their wedding vows; but a day later they discover that, alienated from God, they are also estranged from each other. The Expulsion at the end of the poem concludes the reconciliation of man to woman, and mankind to God, begun simultaneously with the contrite tears of Book 10. As the love between man and woman must be reaffirmed in fallen circumstances, so must the love between mankind and God. Michael's lessons amplify the half of Milton's theodicy concerned with postlapsarian man, defining the lot of a creature whose God has tempered justice with mercy: having refused the first gift of life by abusing his freedom, man must accept his existence on altered terms, sent forth from the garden "though sorrowing, yet in peace" (11.118).

Raphael began the first education of Adam with a clear statement of procedure. To narrate celestial events he would avail himself of ontological likenesses between earth and heaven; Adam's understanding of the nature of his own world could be taken for granted. But the task of Michael does not lend itself to a single rule of symbolism, an invitation to ascend from known to unknown on the rungs of true analogies, for the second and prophetic history lesson progressively unveils the "mysterious terms" of judgment through historical time, where each event demands a grasp of future events, ultimately the last event, in order to be interpreted fully. It is true that the rebellion of Satan, to be interpreted fully, requires knowledge of the fall, whose full understanding requires knowledge of all subsequent history. Yet direct concentration on the anticipatory thrust of Satan's defiance would have violated the decorum of freedom. Raphael does not tell our forefather that he "must" or "will" succumb to Satan, only that he

will be tried, guiding him to a knowledge of the context of the forthcoming decision: what it means to rise (become angelic) and to fall (become satanic). Once fallen, Adam enters the hermeneutic circle of typology to be apprised of a different and singularly human consequence. Triumphing in the end over the choice posed in the garden, man will rise through having fallen.

Adam initiated his first education by twice asking whether an angel could be nourished on earthly food. The answer was yes: the terrestrial category holds good throughout the cosmos, and because it does, the immediate experience of man can "delineate" in imagined forms supraterrestrial affairs. Within the second education the structural equivalent to the questions about nourishment is Adam's fear that in leaving Eden he will have left God; like the earlier questions, this fear allows his instructor to state the relation between human knowledge and the transcendent realm. "This most afflicts me," Adam proclaims, "that departing hence / As from his face I shall be hid, depriv'd / His blessed count'nance" (10.315–317). Michael replies that God is omnipresent. But the truism countering this anxiety is not analogous to the "yes" that stilled Adam's polite uncertainty over how to entertain an angel. There has been a break from Edenic knowledge. In order to be present in the fallen world, God has adopted a new mode of presence:

> Yet doubt not but in Valley and in Plain
> God is as here, and will be found alike
> Present, and of his presence many a sign
> Still following thee, still compassing thee round
> With goodness and paternal Love, his Face
> Express, and of his steps the track Divine.
> Which that thou may'st believe, and be confirm'd,
> Ere thou from hence depart, know I am sent. (11.349–356)

God will not show the expression of his face, but express his expression with "many a sign." He will be present in signs. It is the large purpose of the second education to prepare Adam to live in good faith through a history driven by signs and types, shadows and enigmas, wherein his descendants, until the end, will always be engaged with partialities. The Advent is a brief cessation in this long labor of interpretation: just as God hides his face from man after the Expulsion, so Christ departs after his new beginning, his guidance to be sought "only in those written records pure / Though not but by the Spirit

understood" (12.513–514). Man must accept a second existence on mysterious terms, vesting his faith in fallen symbols that do not rise from presence to transcendence in an unbroken ontological continuum, but, fulfilling and anticipating, link partiality with partiality across catastrophes, wars, dispersals, and the deaths of empires. Since the fall, we know good only by knowing evil (*CP* VI, 352), and in the same fashion know God's presence only through volatile signs whose very existence declares his absence.

When Michael explains the meaning of Christ, Adam understands the most mysterious term of the judgment. The biological paradox "seed of the woman" is a redeemer and restorer, the "one greater" (1.4; 12.242), Moses perfected, who will "bring back / Through the world's wilderness long wander'd man / Safe to eternal Paradise of rest" (12.312–314); "then the Earth / Shall all be Paradise, far happier place / Than this of *Eden*, and far happier days" (12.463–465). Adam then articulates the *felix culpa*—according to one critic, "the central paradox of Christian history and of *Paradise Lost*".[3]

> O goodness infinite, goodness immense!
> That all this good of evil shall produce,
> And evil turn to good; more wonderful
> Than that which by creation first brought forth
> Light out of darkness! Full of doubt I stand,
> Whether I should repent me now of sin
> By mee done and occasion'd, or rejoice
> Much more, that much more good thereof shall spring,
> To God more glory, more good will to Men
> From God, and over wrath grace shall abound. (12.469–478)

"More" is what Adam sees, defying the forlorn "No more" that opens Book 9, and the most pointed "more" in this speech is surely the first. One cannot imagine a happier welcome for the entrance into human consciousness of the *felix culpa*: Adam explicitly prefers Michael's education, with its "more wonderful" knowledge of the destiny of fallen man, to Raphael's education, with its knowledge of what "creation first brought forth" and the destiny of unfallen man. Humanity, like the poem itself, is being reoriented from the beginning to the end.

Michael again brings the world to its "great period" (12.467), whereupon Adam, seeing time from the judgment in the garden to the Last Judgment, summarizes his folly ("to obey is best") and formally converts to Christianity: "Taught this by his example whom I now /

Acknowledge my Redeemer ever blest" (12.572–573). At this point, the summit toward which these lessons have been climbing, Michael introduces a new symbol enclosing the entire history of mankind:

> This having learnt, thou hast attain'd the sum
> Of wisdom; hope no higher, though all the Stars
> Thou knew'st by name, and all th' ethereal Powers,
> And secrets of the deep, all Nature's works,
> Or works of God in Heav'n, Air, Earth, or Sea,
> And all the riches of the World enjoy'dst,
> And all the rule, one Empire; only add
> Deeds to thy knowledge, answerable, add Faith,
> Add Virtue, Patience, Temperance, add Love,
> By name to come call'd Charity, the soul
> Of all the rest: then wilt thou not be loath
> To leave this Paradise, but shalt possess
> A paradise within thee, happier far.
> Let us descend now therefore from this top
> Of Speculation. (575–591)

Milton does not want us to think of the loss of paradise as one event, a disaster, and the promise of divine providence as another and compensatory event laid over the first. He has designed them as a single act—the insertion of a symbol into man, uniting his past, present, and future. The "paradise within" abbreviates the exegesis of the mysterious terms, enfolding the meaning of the oracular judgment, now wholly revealed, in a newborn symbol of undarkened clarity. The curse expelling man from paradise has thrust him into a symbol of paradise that spans the cleavage, psychological and temporal, between the garden paradise he will remember and the transcendent paradise he will anticipate.

Milton closes the epic at the moment Adam and Eve acknowledge their loss of paradise as presence and begin to find their way in the new order of presence-in-absence:

> Some natural tears they dropp'd, but wip'd them soon;
> The World was all before them, where to choose
> Thir place of rest, and Providence thir guide:
> They hand in hand with wand'ring steps and slow,
> Through *Eden* took thir solitary way.

Paradise Lost ends at the beginning of the middle. The middle, history, is the age of mysterious terms. As they enter this middle, Adam and Eve must turn their backs on the immediacy of the supernatural. Except for a few chosen individuals, mankind will live, like Adam and Eve at this moment, "solitary," bereft of God or angels. Hereafter faith resides in things unseen; Adam has learned that he must "walk / *As* in his presence" (12.562–563), living with arduous symbols, not with immediate presence. Their first search is for a home, "Thir place of rest," but the world Michael discloses will never be hospitable to the paradise within. We bear a symbol rooted in a lost home and pointed toward a new home, neither of which we will find in this world. The mind is its own place, furnished with the memories and anticipations of absent places. Providence guides and promises. But the fact remains that we are fundamentally homeless in history, which is a time of unstable significance, "not but by the Spirit understood."

If not contrary to the sense of the poem, this reading of the last passage is by design a negative one, highlighting a single side of its twilight mood. I hope that it may serve better than a lengthy exposition to remind us that Milton employed a timely symbolism in two of the titles and three of the closures of his final works. As the dark congeries of terms such as *alienation, estrangement, displacement, unheimlich,* and even the *différence / différance* of *deconstruction* suggests, homelessness is a preeminent metaphor, maybe *the* preeminent metaphor, in modern culture. Abrams demonstrated in his excellent *Natural Supernaturalism* that the Christian myth of oneness, exile, and return—the history Milton contracted into his symbolism of the three paradises—persisted in the great romantic philosophies of Fichte, Schelling, and Hegel, where a self-perfected subject traversed its own interior and secular version of the old salvation history, ending at the ground without which it could not have begun; a self-consciousness propelled into otherness, the philosophical mind returned from the long detour of thought to the System that regathered the other into the play of the same—Hegel's Absolute Mind.[4] The break characteristic of the modern period occurs when the three paradises of this secular myth collapse into the second. There are no islands of bliss on either side of exile. There is no place from which reason departs, as from itself, and no promised land at which reason arrives, as at its own completion, to be again at one with reality through its possession of the System that is reality. From Nietzsche to Derrida, the heroes of modernism, with

the exception of Husserl, have struck against the philosophical afterlife of Christian myth. The second paradise, which is paradise lost, becomes a value to be argued for and to be defended against illusion. Heidegger, the ponderous magician of philosophical symbolism, insisted on our homelessness in *Being and Time*, and in an early lecture on Novalis' "Philosophy is strictly speaking a homesickness," proclaimed that "our homesickness makes us into human beings": how can we be at home in a world we have not chosen, but have been thrown into as visitors who must die?[5] As he answered this question in *Being and Time*, Heidegger could be said to have mounted a theodicy in the absence of God, resisting the illusion of religion while retaining from the theological tradition in which he was educated a contempt for our comfortable repose in this world. If Heidegger was the philosopher of homelessness, Freud was its scientist. Psychic life is "uncanny," *unheimlich* or "un-homely," because conscious desire has been detached from its earliest objects, which are lost to us through repression, and subjected to the (from one point of view) contortions of projection, identification, and the various mechanisms of defense. Milton presents us with a major statement of Christian myth placed historically at the threshold of a cultural revolution characterized by its progressive defection from this myth, and more arresting still, a major statement of Christian myth attuned to precisely the issue that this revolution now revolves about.

He told us something moving and important about the second paradise and the homelessness implicit in our symbol of home. Many readers have felt this. Students of *Paradise Lost*, however they assess the merits of Adam's second education, testify almost unanimously to the satisfaction of its closure. Richardson was hugely unembarrassed about his gratitude:

> O *Milton* thou hast employed all thy vast treasure of wit, learning and ability, . . . all the fire and beauty and sublimity of imagination peculiar to thyself, . . . and together with all these a genius perfectly poetical, if ever any man's was, and that regulated by a most solid judgment. All these thou hast consecrated to produce a poem, more instrumental than any other human composition, to calm and purify the mind, and through the delightful regions of poetry, to exalt and fix it to the mysteries, sublimities and practices of religion; to a state of tranquility and happiness, the uttermost mortality is capable of.[6]

Hyperbole, the scholar's self-aggrandizing enthusiasm for his subject—but not by much. Those who have survived the zaps and jolts of our electric world with enough intellectual determination to read *Paradise Lost* at a single sitting will find that its closure produces a state of tranquility and happiness, the utmost mortality—considered as man aware of his death, Christian or not—is capable of, at least in English poetry. But why? Because of the happy resurrection awaiting at the end of these wandering steps? Because Adam and Eve, suspended between paradises, embody lessons wholly expressed in the second education? I think the closing passage enlivens the spirits of readers because it rectifies a mistaken tendency in the second education, regaining something lost, or in grave danger of being lost, in the Christianity to which Adam converts. The conclusion of *Paradise Lost* is the end of the poem, our place of rest, in no trivial way. Here Milton offered to us in a wonderful artistic moment the nonsectarian wisdom he had harvested from the long work of mastering the sacred complex. It is the conclusion earned by the creation of the poem.

To measure the power of this gift we must be willing to examine the Christianity of the second education from a position informed by the modern defection from Christian myth. When someone wants to sequester the meaning of a text in the enclosed gardens of historicism or aestheticism, he usually assumes that this meaning has value, and hence must be protected from defacement by voracious and antithetical ideas. But a great work of literature is not a national park; it is a ground to be plowed. Until we have risked thinking Milton amid the great arguments that followed his own, we have no way of determining what he still has to say to us, and hence no way of profiting from his accomplishment.

Theodicy and Loss: The Problem of the Happy Fall

It would appear that evaluating the final two books of the epic has replaced explaining Satan's heroism as the dominant "problem" Milton studies exists to ponder with respect to *Paradise Lost*. The results have been extraordinary. There is probably no clearer instance in which the modern Miltonist enjoys an understanding of the poem superior to that of his predeccessors. In 1942 C. S. Lewis could speak casually of the "grave structural flaw" of concluding the epic with "an untransmuted lump of futurity," but today we know for certain, in a manner that Lewis himself would have admired, that "futurity" has been trans-

muted with all the architectural power one expects from this poet: Milton hath quit himself like Milton.[7] Yet, despite the elegance of design, and despite the care given to detail, I think the last books remain problematic. The difficulty stems from Milton's heavily eschatological presentation of the *felix culpa*.

Cultivated properly, the interior paradise is "happier far." Michael's catalogue of the knowledge excelled by the meaning enfolded in this symbol (12.575–591, quoted above) opens with the names of all the stars, recalling the first education, where this knowledge was too high for man in his current state but not outside his destiny. As the catalogue widens to incorporate the entire burden of Raphael's instruction, we learn that the knowledge of Christ's ultimate victory excels every wonder dispensed to unfallen knowledge—"all th' ethereal powers" (knowledge of angels), "secrets of the deep" (knowledge of pre-Creation), "all Nature's works, / or works of God" (knowledge of the Creation, of matter and physiology). The last items in the catalogue ("all the riches of the World," "And all the rule") evoke the unfallen state of man, his occupation of Eden, "In narrow room Nature's whole wealth" (4.207), and his dominion over the earth. Looking forward, they also suggest the offers of Satan during the temptation of Christ; the mountain on which Adam and Michael stand has been compared to "that Hill . . . / Whereon for different cause the Tempter set / Our second *Adam* in the Wilderness, / To show him all Earth's Kingdoms and thir Glory" (II.381–384). Christ, too, we infer, taught the *felix culpa*, choosing the fallen state of man over Satan's travesty of our first heritage: although obediently rather than disobediently, second Adam confirmed the decision of first Adam by giving up the world.

This is a militant *felix culpa*, and the good fortune of the fall appears to rest in the end rather than in the effort of getting there. The concluding words of the angel advise Adam and Eve to be

> sad,
> With cause for evils past, yet much more cheer'd
> With meditation on the happy end. (604–606)

Michael's catalogue of knowledge inferior to this "happy end," asserting what Adam has already noticed—the superiority of his teachings to those of Raphael—is in effect a farewell in the name of higher providence to the soaring ambitions of the epic poet prior to his narration of the fall. We are warned in Book 7 that the song will be "narrower bound / Within the visible *Diurnal* Sphere" (21–22) in books to

come, and one mark of this compression is the difference between the symbolic home within the poet in Book 3—the vast abyss, ground of all—and the garden home of the final books. We miss the brimming urgency and hard-won joy of the poet who would see everything, thinking like a bird sings in harmonious numbers. The fall is fortunate in the private theodicy of the epic because a man seeks the renown of great prophets, creates in sorrow and in rapture, imagines wonders and believes them true, looks into the hell, the heaven, and the paradise within him, and completes the desire he prepared all his life to complete, countering the weights of necessity with a fulfilled wish. Milton gave far less to mankind in his public theodicy.

Mary Ann Radzinowicz, in one of the finest defenses of the last books, states that they "give to the mind of the responsive reader the last sweep down the vistas of time not into an unknown futurity when all shall somehow be well but straight into the heart of significant human action."[8] To be sure, the paradise within is only a sign of the happy end when cultivated in virtuous action. But although Michael tells Adam he must add faith to his knowledge of the time "when all shall somehow be well," add the virtues, and add love, the "significant human action" we regularly encounter in the Miltonic Bible is unyielding fortitude in the face of unyielding opposition, and the joys rendered are those of being right when others are wrong, knowing oneself righteous when being persecuted. The types of Abdiel, the Protestant angel who stood "unmov'd, / Unshak'n, unseduc't, unterrifi'd" (5.898–899), command this stage. There is no time in this history lesson for the prolonged exploration of a single hero that makes this negative fortitude come alive in *Paradise Regained.* Obstinacy in the service of truth is admittedly a great virtue; circumstances have forced many noble lives to take this shape; such persistence lies deep in the history of Judaism and Christianity. And the creation of the epic demanded this sort of obstinacy. We understand that Enoch, Noah, Abraham, Moses, and the Apostles are typological prefigurations of the creator of the epic. But while this relationship is one of strength, it is also peculiarly unhappy. The inner paradise of the final books, unlike the fluid consciousness of the poet, seems locked into conflict with external evil, and, as in any war, the opposed forces begin to look alike: Satan himself is unshaken, unterrified, relentless and obstinate in the service of evil. Reminiscent of the paralyzing reaction formations of *Comus*, virtue emerges as unceasing conflict with a wicked world. The cheer of the *felix culpa* lies less with the activity of being virtuous than "With meditation on the happy end."

In the promised end, the symbols and mysterious terms given to long-wandered man will, like time itself, "stand fixt" (12.555). God has foretold the eventual realization of the symbol of paradise when willing his throne to the Son:

Meanwhile
The World shall burn, and from her ashes spring
New Heav'n and Earth, where the just shall dwell
And after all thir tribulations long
See golden days, fruitful of golden deeds,
With Joy and Love triumphing, and fair Truth.
Then thou thy regal Sceptre shalt lay by,
For regal Sceptre then no more shall need,
God shall be All in All. (3.333–341)

After "tribulations long," God shall be all (entirely) in all (each one): the ultimate synecdoche. This is the closure in eternity of the hermeneutic circle. When whole is perfectly adjusted to part, the infinite perfectly unveiled to the finite, then each knows what everyone knows and there is nothing left unknown. God knows all of all of us, all of us know all of God: intuition being absolute, interpretation is obsolete. The arrival of the true "place of rest" is the end of otherness in the fullness of revelation, the demise of government in the ubiquity of righteousness, and the death of the symbol in the perfection of presence. Is there no happiness unique to the second paradise, no blessing for history itself?

Michael gives us little encouragement for transferring the "more" of the *felix culpa* from the literal realization of the promise to the time of our wandering. The happiness of the fall is developed in a decidedly Protestant manner. In Milton's Bible the actors seem hardened and joyless, wholly consumed by their prefigurational roles, as if time itself were the only real hero, pressing ever onward to the felicitous end where, at last, joy makes sense again. Providential time emerges as the hero of human history, and God is the Lord of time; the deity gains glory through the design of salvation, bestowing grace and good will on men. "Much more good thereof shall spring, / To God more glory, more good will to Men / From God, and over wrath grace shall abound." Adam does not rejoice in the good *we* make out of evil. Milton hoped for a hasty Apocalypse in his early pamphlets. When the soon-to-be-expected King proved a patient monarch, and the King unexpected at the end of the 1640s assumed his throne in 1660, Milton

learned a new patience in waiting for the end, but he did not cease to wait.[9] The typology of the three late poems aims at the Last Judgment unmistakably, and the day of restitution to the just crowns the theodicy of *Paradise Lost*. Understandable as it may be, the eager contemplation of Judgment fed by his domestic, political, and religious disappointments threatens the integral vision of the epic in several ways.

The poem becomes, first of all, otherworldly—a poem of the Christian letter. Milton has carefully vindicated God from the charge of allowing or necessitating our fault. This theodicial requirement drove the poem toward beginnings, since freedom—the condition that God will not revoke, even to save us from its consequence—comes to us in the beginning, the very nature of a distinct being. The first ten books of the epic are largely given over to an "alpha" mythology of origins, and these first things seem, in their spacious etiology, open to inexhaustible redescriptions of our worldly experience. When will we stop interpreting the myth of the fall, or Milton's myth of Satan, in our search for a comprehension of evil? For interpreters of mythical arguments, origins are generous: they enable us to reflect on our lives, illuminating what is. As a result, the symbols representing freedom and the loss of freedom in *Paradise Lost* are virtually nonsectarian. Almost all religions draw emotional authority from the celebration of createdness, and ethical motive from our indebtedness for the gift of being. Once the fault has freely appeared, however, the task of theodicy shifts to vindicating God from the complaint of mercilessness. This turn coincides with the decisive movement in the final books toward an "omega" mythology of Apocalypse, redemption, and blissful immortality. Discovering the proof of mercy in eternal life, Adam converts. At this point the interpreter stands in a new relationship to the poem. Adam is no longer mankind, but Christian man. The symbols look away from our situation in this life. With respect to the happy end, there appears to be no latitude for interpretation beyond the literal, no leeway for metaphor.

The myth of Apocalypse has psychological and political *functions*, but I would think that, if this myth has a *meaning* outside of faith's surrender to literality, it lies in charging us to adopt an ineradicable modesty in our efforts to fix the significance of past, present, or future. Until the end, when time stands fixed, no event is finished and no system absolute.[10] But Milton's is the Apocalypse of faith, set starkly against a representation of this life as sorrow and tribulation. The terms of the second education corner us: unless the reader can believe or achieve the illusion of believing that the sky will one day part,

and Christ, returned as judge, will reward and punish the human race, leading the just to immortal bliss, the kerygma of Milton's consolation is a dead letter. Even those for whom this corner is the heart of faith may have reason to feel cramped by Michael's lesson.

As Fowler notes, if we follow the pauses of Michael, his narrative divides into three sections—from Adam to the Flood, from the Flood to Christ, from Christ to the Second Advent, each period concluding with the intervention of God, first to destroy, then to redeem, and then at last to destroy *and* redeem.[11] Differing in their climactic interventions, the three ages share the same pattern in their human aspect. "Thus will this latter," Michael says of the second age, "as the former world, / Still tend from bad to worse" (12.105–106). And in the final age, our age,

> so shall the World go on,
> To good malignant, to bad men benign,
> Under her own weight groaning, till the day
> Appear of respiration to the just. (12.538–540)

These are mean lines. The weight under which the world groans is the bad fruit of Milton's reforming passions. "So shall the World go on," go on as it always has, driven by the law of bad to worse. In making the time from Apostles to Apocalypse into dead time, a repetitive time when symbols have calcified into invariant meanings and there is no triumph in virtue but patiently to await the end, Milton strikes his future readers with all the authority of his prophetic stance. For this period includes *all unlived time*, and those who read the poem now and hereafter will always be living in its dead time: Milton will allow for no future experience but his own, and only for the darkest side of that. Cursing the future is a commonplace of Renaissance literature; if Milton is to be scolded for this moment in *Paradise Lost*, Donne is apparently to be hanged for the *Anniversaries*. But the religious sublimity of this poem, lacking altogether the knowing hyperbole of Donne, invests the aggression in this topos with a firmer conviction. Hope rests, now and forever, on the happy end.

There are currents running counter to these in the second education. In the design of the last books, if not in the official history of Michael, Milton conveys a recognition of the creative ferment unique to historical time. Michael "ended" on the happy end. But Book 12 is framed by moments of suspension that restore to some extent the prestige of an *unfolding* symbol, crossing the "bad to worse" of human history with a capacity for surprise and renewal:

As one who in his journey bates at Noon,
Though bent on speed, so here the Arch-Angel paus'd
Betwixt the world destroy'd and world restor'd,
If *Adam* aught perhaps might interpose;
Then with transition sweet new Speech resumes. (12.1–5)

Written for the second edition of 1674, the year Milton died, these may well have been his final lines. In order to divide the outsized Book 10 of 1667 into two parts, he created a pause of symmetry shot through with the futural purpose of both storyteller and story told. The balance of "Noon," the hour of crisis in the epic, seems to long for motion, "bent on speed" toward restoration. This liminal moment "Betwixt" the Flood and the Covenant, "the world destroy'd and world restor'd," reiterates the pattern established in the opening five lines of the poem, where the pause between bliss destroyed and bliss restored is marked by the word "till." The new passage also foreshadows the conclusion of the poem, itself between a world hidden by fire and a "World . . . all before them"—a moment that must occur at noon, which is one reason why Michael is "bent on speed." The entirety of *Paradise Lost*, as Irene Samuel has noted, manifests the same design; its center lies between the damnation of the rebel angels at the end of Book 6 and the creation of a new world and new creatures in Book 7.[12] Milton punctuates heavily and repeatedly between just loss and grace-bestowing restoration, or (what is the same thing) between the two halves of his theodicy, and in this context the pause initiating the final book becomes a symbol for history itself, the middle time or "meanwhile" set between Eden lost and Eden regained. "If Adam aught might interpose": Adam and Eve are evicted into the pause or interstice of historical time in the closing lines of Book 12 (book of the hour of noon), wandering into the middle state of living symbols, neither first Eden nor third Eden. It is a pattern that, like an iamb, makes new energy final. The last book begins and ends with closures that are also transitions. To be continued . . . yet once *more*. These complementary passages temper the stark contemplation of the end by locating the contrary mood of the *felix culpa*, the regret and joy Adam expresses only when "time stands fixt," in the processes of time. They give to the vision of history something of the pleasure in the unfinished and oncoming so forcefully expressed in the first nine books of the epic. They evince an imaginative preference for the hermeneutic adventure of the life we wander through before the stasis of perfect meaning found at the symbol's death in the happy end.

This hinted emergence from devastation to the energy of the task,

replacing "bad to worse" with "bad to better," is clearest at the closure. The sudden flex of

> The World was all before them, where to choose
> Thir place of rest, and Providence thir guide;

suggests a post-Edenic aspiration that remains with us when we join the slow steps of a solitary couple wandering into time, space, and choice. Bearing the plenty of the world and the possibilities of man, the "all" of these lines rejuvenates the history closed and fixed by Michael. Whatever providence has determined, we are free creatures who will embody after our own fashion the divine scheme. The effect on us is not unlike that of the emptied book of creatures on the blind poet:

> and for the Book of knowledge fair
> Presented with a universal blanc
> Of Nature's works to me expung'd and ras'd.

"The World was all before them, where to choose" imaginatively presents Adam and Eve with an unmade Book of Scripture, a blank Bible intimating that we have the opportunity to *create*, within the limits of guiding providence, the meaning of the paradise within us. Milton is generous in the end. His epic started with the brisk settlement of hell through the arts of mining, architecture, government, martial training, oratory, poetry, philosophy, and theology. Here in the close, Adam and Eve begin to undertake the human analogue of that settlement. Michael has prophesied that the demonic forces released in the first books will weigh down on our history, like the damps and vapors of mortal physiology, driving us "from bad to worse." But the "World . . . all before them," a muffled echo of the sublime beforeness of the invocation to light, hints for an instant that we have space for innovation.

The freshened sense of this life, balancing the severity of our loss with a brief reminder of the poet's own "advent'rous Song," thus allowing the full emotional tonality of the epic to arise in a single passage, partially accounts for the success of the final lines. Heavens exist to be desired, yet it matters, theologically and psychologically, how they are desired.

The theology of the second education justifies God in another

world. Critics favorable toward the last books tend to stress the artistic difficulties inherent in a retelling of biblical history. But the problem Milton faced in this portion of the epic made demands on his character inseparable from those on his intellect, learning, and art. He could not avoid the internal tensions that result from grafting a theodicy onto the myth of the fall. The whole thrust of the myth is that we have chosen this world mistakenly, and by beginning his theodicy with the theme of freedom, Milton rendered our error with excruciating lucidity. There will be heavy penalties, however, unless the theodicy of mercy can in some manner contradict the theodicy of freedom, doubling back to discover values specific to fallen life. For if our situation in this world is endurable in the name of justice solely because of the blissful eschaton, Christianity becomes vulnerable to the sort of attack one finds in Nietzsche, who argued, also with excruciating lucidity, that the postulate of a transcendent world debases and devalues this world.[13] In its concern for reforming our institutions, its passion for liberty, and its spiritual materialism, Milton's Christianity eludes this critique triumphantly—but not in the final books of the epic. As Milton renegotiated the terms of our existence, demonstrating that men could suffer the agonies of this life and continue to worship their creator, I do not believe that he found in himself sufficient love of this life to underwrite for men who live in history the religious psychology of debt, of gratitude for the provision of being, rendered with unmistakable authority in the first two-thirds of the poem. Michael converts Adam to a revision of this psychology wherein the debt radiates from our end in another life, not our beginning in this life. But if the happiness of mortal existence lies primarily in the end, the third paradise literally conceived; if a man can exclaim "Goodness infinite!" only when contemplating the image of this end, which he prefers to the image of creation in Genesis; if we actually leave the poem feeling that our world is "To good malignant, to bad men benign," then Milton's great argument is hollow at the center.

A theodicy cannot succeed without undermining the God it seeks to justify unless we are able to be reconciled with what is because of, not wholly in spite of, what it is—this wandering and essentially homeless life where God is present through the (often unrecognized, often riddling) mediation of mysterious terms. There must be "more" in the condition of original sin than an occasion for God to outdo, beyond the confines of historical time, the initial act of Creation. The second paradise is the world we know. Theodicy arises from the threat of revulsion at this world, and insofar as the argument concludes without

transforming this attitude, it goes nowhere. Had the personal theodicy of the invocation to light concluded with "I shall see after I die and go to heaven" posterity would not have been able to distinguish artistic from personal weakness. All of "the ways of God to men" cannot be ways out of history. The fullness and, one might say, the justice of a theodicy should involve at the core an ontological endorsement, a peace with the given terms of our historical existence as such, that "redeems" the disobedient choice of Adam and Eve here and now. *The fall must be chosen again, and apart from any consideration of the afterlife, to prevent the belief in an afterlife from becoming another of the poisons that taint this life.*

This argument may be stated psychoanalytically. Aware as always of the abiding conservatism of human desire, with its many subterfuges for holding fast to its early objects, Freud recognized that the capacity for delay established by the reality principle does not really overcome the rule of wish:

> Actually the substitution of the reality principle for the pleasure principle implies no deposing of the pleasure principle, but only a safeguarding of it. A momentary pleasure, uncertain in its results, is given up, but only in order to gain along a new path an assured pleasure at a later time. But the endopsychic impression made by this substitution has been so powerful that it is reflected in a special religious myth. The doctrine of reward in the afterlife for the—voluntary or enforced—renunciation of earthly pleasures is nothing other than a mythical projection of this revolution in the mind. Following consistently along these lines, *religions* have been able to effect absolute renunciation of pleasure in this life by means of the promise of compensation in a future existence; but they have not by this means achieved a conquest of the pleasure principle. (*S.E.* XII, 223)

The reality principle can be nothing more than a wish with an expanded middle, delay being endured "for the sake of," not chosen in itself. Exactly this pattern dominates the second education: history is a long hiatus between paradises of happiness. The Miltonic God is not ferociously opposed to pleasure, but in the final books the poet has not generated a sense of mature joy won from talents gratefully received and fulfillingly spent vigorous enough to be sanctioned or opposed, which is why one misses, until the very end, the excitement of creative achievement conveyed in the invocations. The "more" progressively unveiled

through typology does not make history any the less a time of waiting, since it serves to perfect our image of the happy end. Christian man waits, albeit with an internal rather than an external law, for the fulfilled wish whose image cheers him. Psychoanalytically, the movement of figural time could be conceived as a persisting hunger for the return of paradisiacal immortality—the pressure of the pleasure principle demanding its illusion in order to disavow its loss.

Finding continuities between superego formation and religious symbolism, I have supplied an epigenesis for Milton's ambition, theology, mythopoesis, poetic inspiration, and elements of his style. I have maintained that his belated creation of an epic on the fall cannot be understood apart from certain harmonies between the religious and unconscious meanings of his blindness. In Chapter 3 I tried to comprehend on psychological grounds the authority of the Crucifixion as the seal of immortality. Throughout I have looked for an advancement, or in Hegel's sense an *aufhebung*—a repetition "sublimating" the repeated within a more inclusive synthesis—in which religion both expresses and frees the sediment of psychic history. The problem is before us again, and in its most acute form. What distinguishes the three paradises from the three moments of Freud's dreamer, whose future "has been moulded by his indestructible wish into a perfect likeness of the past"? In what way does Milton's presentation of the *felix culpa* pass beyond the oldest regulation of the psyche, the pleasure principle?

It is first necessary to establish what a positive answer to these questions would involve in psychoanalytic terms, and the obvious place to begin is with Freud's own attempt to go beyond the pleasure principle. In a famous passage from *Beyond the Pleasure Principle*, the author observes his grandson playing a game with a spool tied to a piece of string. When this child says "o-o-o-o," both grandfather and mother agree that he means *fort* ("gone"):

> What he did was to hold the reel by the string and very skillfully throw it over the edge of his curtained cot, so that it disappeared into it, at the same time uttering his expressive "o-o-o-o." He then pulled the reel out of the cot again by the string and hailed its reappearance with a joyful "*da*" [there]. This, then, was the complete game—disappearance and return. As a rule one only witnessed its first act, which was repeated untiringly as a game in itself, though there is no doubt that the greater pleasure was attached to the second act.
>
> The interpretation of the game then became obvious. It was

> related to the child's great cultural achievement—the instinctual renunciation (that is, the renunciation of instinctual satisfaction) which he had made in allowing his mother to go away without protesting. He compensated himself for this, as it were, by himself staging the disappearance and return of the objects within his reach. (*S.E.* XVIII, 15)

Commentators have seen in this domestic tableau the origins of phonetic and semantic difference, intentionality, work, ritual, and art.[14] Freud himself, arguing that this child has not gone beyond the pleasure principle, hesitates between two possible motives for the game. The boy acts under an "impulse to work over in the mind some overpowering experience so as to make oneself master of it" or else an impulse "to revenge himself on his mother for going away from him" (p. 16). Putting aside this uncertainty for the time being, one thing is clear enough: in the void of absence he creates a symbol. At this point everything depends on how one construes a symbol. The spool refers, always refers—to the mother? If so, the hermeneutic of psychoanalysis will be forever vigilant in tying old mothers to new spools. From a psychoanalytic viewpoint there would be no foreseeable alternative to Robert Rogers' definition of a metaphor:

> The transfer [*metaphorein*] creates the absence, the object-relational vacuum which the other meanings of the image rush to fill in. Figures of speech embody the figure of the lost object. Freud sees the schizophrenic's use of words as an attempt to regain the lost object. What he claims for schizophrenics holds for all of us. The restitution of lost objects occurs by virtue of the restitution of lost images the poet makes in metaphor.[15]

I take this to mean that in the semantic transfer from "my love" to "a rose" in the metaphor "my love is a rose" there is a small trauma of object loss analogous to the transfer from absent mother to present spool. But then I find my lost love amid the roses; the yield of pleasure generated by the metaphor is comparable to the restitution enjoyed by the happy child when his spool reappears over the edge of the cot. Language plays with presence and absence, and so far as psychoanalysis is concerned, we play this game to recover archaic objects. "What he claims for schizophrenics holds for all of us."

It does. But something more holds as well. Let us suppose that the child playing "*fort . . . da*" were Sigmund Freud. The game symbol-

izes his separation from his mother. Eventually, however, when this child is a man, the lost mother and the game restaging her loss will symbolize his concept of the pleasure principle. By possessing the idea of the pleasure principle Freud has gone beyond it, changing his psychic past from the symbolized into the symbol: he has preserved archaic meanings to gain access to another meaning, psychoanalysis, that our culture still labors to assimilate. What if the child playing this game were John Milton, who will one day recreate our lost paradise so that it might be lost again, then recovered again—appearing, disappearing, and reappearing as a symbol in the course of *Paradise Lost*?

The precondition for sublimation is that the bond between the symbol, whether paradise or a spool, and its referent in the past, whether first paradise or early mother, be loose enough to permit the advent of novel meaning. It is imperative that a work of mourning have occurred: Eden and mother must really be lost, or else their symbols, frozen in an original significance, will be employed defensively to deny this loss. This is the difference between a good symbol and an idol.

As his life would suggest and as some of his greatest poetry confirms, Milton understood the psychology of mourning uncommonly well, familiar in a variety of ways with the fact that everything in this fallen world is present in the negative mode of possible but canceled absence. He took care that man's entrance into the symbol of paradise not be idolatrous—a denial of his fall or an indefinite prolongation of his grief. Adam fell because he could not imagine that he would ever stop mourning for Eve:

> Should God create another *Eve*, and I
> Another Rib afford, yet loss of thee
> Would never from my heart; no no. (9.911–913)

Adam would have suffered precisely this fate with regard to the paradise he chose to lose instead of Eve had he been allowed to practice the piacular rites he designs for himself, making a shrine of Eden and conducting his children on regular pilgrimages:

> here I could frequent,
> With worship, place by place where he voutsaf'd
> Presence Divine, and to my Sons relate;
> On this Mount he appear'd, under this Tree
> Stood visible, among these Pines his voice
> I heard, here with him at this Fountain talk'd:

So many grateful Altars I would rear
Of grassy Turf, and pile up every Stone
Of lustre from the brook, in memory,
Or monument to Ages, and thereon
Offer sweet smelling Gums and Fruits and Flow'rs. (11.317–327)

By the end, however, there can be no doubt that a grieving has been provisionally concluded. The interior paradise borne into our history is an expiatory symbol, acknowledging our crime and our expulsion. "Some natural tears they dropp'd" at the last view of their native land. The education began with the penitential tears inspired by "Prevenient Grace descending" (11.3), but our final tears are "natural"—tears that from infancy signal the failure of wish. They "wip'd them soon," for the last books of the epic, like an expanded "Lycidas," depict mourning as a double movement toward, first, tears that actualize loss, purging the fond dreams and false surmises of denial, and second, a consolation allowing us to be free from our grief. "Tomorrow to fresh Woods, and Pastures new." As Adam and Eve turn from Eden to the "World . . . all before them," the meaning of the symbol of paradise turns with them. This symbol fixes, not merely the lost object symbolized, but the truth of its loss.

God teaches this lesson by profaning the ground of the first paradise during the Flood:

then shall this Mount
Of Paradise by might of Waves be mov'd
Out of this place, push'd by the horned flood,
With all his verdue spoil'd, and Trees adrift
Down the great River to the op'ning Gulf,
And there take root an Island salt and bare,
The haunt of Seals and Orcs, and Sea-mews' clang.
To teach thee that God attributes to place
No sanctity, if none be thither brought
By men who there frequent, or therein dwell. (11.829–838)

Milton has lovingly created paradise as the historical referent of all human dreams ("Not that fair field of Enna . . ."). Now, like a man awakening, he indicts the folly of such yearnings. Interpreted retrospectively, paradise as a place to be regained, a repository for desire, is a hopeless symbol—the illusion without a future. The journey of Eden "Down the great River to the op'ning Gulf" recalls the corpse of Lyci-

das visiting the monstrous world, and the head of Orpheus in the same poem, sent "Down the swift *Hebrus* to the *Lesbian* shore." The sanctity of the first paradise sprang from the nature of things. Nevermore: Adam and Eve must now work to create sanctity for an indifferent "place of rest." Authentic symbols begin in undenied catastrophe, making absense present not that desire may be bound to an indestructible wish, but that desire may be rerouted from an indestructible loss. The physical action at the end of *Paradise Lost*—turning back, weeping, then turning around to find another place in slow steps betraying the effort of detachment—represents, with tremendous force and brevity, the difficult acquisition of a good symbol. Gaining a space of indeterminacy in which "more" might flourish, the paradise within is being pried loose from its original referent.

Yet the emphatic literality of the third paradise appears to compromise the break between the first and the second. Is man provisionally able to end his mourning only because, the immortal bliss of the past referent having been transferred to the future referent, he has in effect lost nothing? To prevent the third paradise from becoming a defensive idol, our mortal state in the second paradise must be chosen without reference to the happy end—a psychological caution consistent with theological arguments against an otherworldly theodicy. The symbol of paradise should have a meaning that is neither past nor future, residing instead in the suspension between these fulfillments that Freud called "a new path." The last word in the poem is "way," and I believe the ultimate success of this great argument rests on the value of that word. It is again a question of the validity of meaning. To complete his task authentically, Milton must justify the ways of men to men.

Religiously or psychoanalytically, the fallen world is not our wish. I do not ask of Milton that he welcome this place of death with a lyrical expostulation to coffee and oranges. Nor am I claiming that Christian belief in an afterlife is defensive illusion; on the contrary, I am laying the groundwork for conceiving of a nondefensive belief in resurrection. I do ask of Milton, who aspires to answer the largest religious questions, that he locate something "more" in the delay of our historical existence, something after the loss and before the restitution that, in the extreme, would stand *even if the providential meaning of paradise, infinitely deferred, were to have no completion in eternal bliss.* The second paradise must have an integrity of its own. Otherwise, adapting Ricoeur's distinction in the epigraph to this chapter, we will be left with a false "resignation to Ananke"—this life tolerated because of a life to come, the Christian reality principle vitiated by the Christian pleasure

principle—rather than the spiritual *aufhebung* of that false submission in a genuine "love of Creation."

This safeguard against idolatry does indeed await us in the closure of the poem. We do not recognize its meaning in these terms, poetry being "more simple, sensuous and passionate" (*CP* II, 403) than logical or psychological discourse. We feel instead that the conclusion of the poem is appropriate to its ambition and achievement. I will come round to the nonsectarian completion of Milton's theodicy by exploring the symbol in which Satan resides.

The Uncanny Beauty of the End

His first book a translation of J. S. Mill, Freud had sufficient English to read Milton. We know that he admired this poet. When asked to name ten good books, Freud replied that he had not been asked to list his ten favorite books, "among which I should not have forgotten Milton's *Paradise Lost*" (*S.E.* XI, 245). If the founder of psychoanalysis had written about that favorite book, I am confident that a famous passage in Book 4 would have figured prominently. Satan has just landed in paradise. On Eden "his griev'd look he fixes sad" (28), then turns to the sun, blazing in the heavens in full noontime glory. Its beams remind him of his unfallen existence in heaven, for which he curses them, and with extraordinary acumen, indicts himself as the sole cause of his own fall:

> Nay curs'd be thou; since against his thy will
> Chose freely what it now so justly rues.
> Me miserable! which way shall I fly
> Infinite wrath, and infinite despair?
> Which way I fly is Hell; myself am Hell;
> And in the lowest deep a lower deep
> Still threat'ning to devour me opens wide,
> To which the Hell I suffer seems a Heav'n. (71–78)

Satan brings to life the engulfment of anxiety as unsparingly as Oedipus and Hamlet delineate the ambivalences of erotic desire. The mind, its own place particularly in anxiety, offers the flight reflex no traction; anxiety is the affect of a deserted ego fallen back upon itself, with no ground to land on, unable to flee the omnipresence of itself. This terrible solitude betrays the archaeology of the miserable or evil self in psychoanalysis. The precursor of the "Infinite wrath" of the punitive

superego is the engulfing mother, "Still threat'ning to devour me," and like the superego, she is not to be understood in Jungian fashion as an independent piece of mind-stuff given to us in a storehouse of mythy playmates. A relationship forged her. Insofar as the ego differentiates itself in times of absence and apparent abandonment, the knowledge of a separate self is correlated originally with the imago of the evil mother. At the bottom of solitude is the void first revealed in the failure and dispersal of the hallucinatory image of the nourishing mother. Defeat in a war, the loss of heaven, and the loss of the love of his creator have driven Satan down to the primordial disillusionment where unremitting attention to unremitting privation seems all that mind can be, and the starved becomes the devoured. At this depth there is no lowest. How to master such a trauma?

Satan knows loss purely. He, too, carries with him the symbol of his home, yet the "hot Hell that always in him burns" (9.467) has no providential sense beyond the perfection of his own defeat. He does not bear into exile a symbol of his first home, the time and place where he lived in the presence of God, but a symbol created by God to signify his eternal alienation in the not-God. What sense can he assign to this symbol? As Christ will at the end of *Paradise Regained*, God has seized from him the power of self-definition. He cannot mourn, living in the symbol as Belial and Mammon counsel in Book 2, for that would be to accept estrangement for what it is or, in other words, to be defined by God. So he wrests a motive from despair, extending the realm of hell to the newly created universe. There, like the child playing "*fort . . . da*," he will "work over in the mind some overpowering experience so as to make [himself] master of it." He will devote himself to the repetition of his hopelessness in a new and safer place:

So farewell Hope, and with Hope farewell Fear,
Farewell Remorse: all Good to me is lost;
Evil be thou my Good; by thee at least
Divided Empire with Heav'ns King I hold
By thee, and more than half perhaps will regain:
As Man ere long, and this new World shall know. (4.108–113)

Landing in Eden a second time, he hails his spool. There!

O Earth, how like to Heav'n, if not preferr'd
More justly, Seat worthier of God, as built
With second thoughts, reforming what was old!

For what God after better worse would build?
Terrestrial Heav'n, danc't round by other Heav'ns
That shine, yet bear thir bright officious Lamps,
Light above Light, for thee alone, as seems,
In thee concentring all thir precious beams
Of sacred influence: As God in Heav'n
Is Centre, yet extends to all, so thou
Centring receiv'st from all those Orbs. (9.99–109)

The ontological likeness between earth and heaven grounds a universal process of sublimation. Satan, possessed by the demonic energy of the repetition compulsion, ironically uses this very likeness to project onto our world the horror of his alienation. All joy is lost to the devil "Save what is in destroying"—destroying himself again and again in manipulating effigies of himself and the Godhead. For Satan as for fallen man, paradise means heaven, but the symbol he fashions is a manic idol bound absolutely to a trauma in the past.

Freud's grandson played either to master or to revenge his loss. Both these motives propel Satan's overdetermined game. His encomium to the earth leaves no doubt that his spool symbolizes lost objects: God and heaven. Yet the complexity of "*fort . . . da*" includes another and simultaneous substitution that Freud did not mention when reporting the game. Children play at being adults. Taking the active role in restaging a traumatic loss, the ego assumes the position of the frustrator. Like the child himself in the trauma as originally endured, the toy is thrown "over the edge of his curtained cot, so that it disappeared into it": putting his spool to bed, the child enacts the role of his mother. One can behold the double valence of the game in the psychological ploys that succeed with Eve. Playing himself, Satan creates her as the image of God ("sole Wonder," "Empress of this fair World," "Sovran of Creatures," "Queen of this Universe," "Goddess humane"), his fawning and licking restaging his own worship of the deity: "who more than thou," an angry Gabriel has charged, "Once fawn'd, and cring'd, and servilely ador'd / Heav'n's awful monarch" (4.958–960)? Simultaneously, playing God, he creates her as the image of himself—an envier of God who will produce sin and death, and for her transgression go to hell.

Proposed to the devils in Pandemonium, sworn before Chaos and Night, confirmed in soliloquy after soliloquy, the design must be executed. It is Satan's dark *Paradise Lost*, the long chosen scheme of one no longer sedulous to go to war except in symbols. Just before he ties

the string however, Satan is smitten by a power that he has not attributed to his spool. Eve has a meaning of her own:

Such Pleasure took the Serpent to behold
This Flow'ry Plat, the sweet recess of *Eve*
Thus early, thus alone; her Heav'nly form
Angelic, but more soft, and Feminine,
Her graceful Innocence, her every Air
Of gesture or least action overaw'd
His Malice, and with rapine sweet bereav'd
His fierceness of the fierce intent it brought:
That space the Evil one abstracted stood
From his own evil, and for the time remain'd
Stupidly good, of enmity disarm'd,
Of guile, of hate, of envy, of revenge;
But the hot Hell that always in him burns,
Though in mid Heav'n, soon ended his delight,
And tortures him now more, the more he sees
Of pleasure not for him ordain'd: then soon
Fierce hate he recollects, and all his thoughts
Of mischief, gratulating, thus excites.
  Thoughts, whither have ye led me, with what sweet
Compulsion thus transported to forget
What hither brought us, hate, not love, nor hope
Of Paradise for Hell, hope here to taste
Of pleasure, but all pleasure to destroy,
Save what is in destroying, other joy
To me is lost. Then let me not let pass
Occasion. (9.455–479)

The mind of Satan is preoccupied in the deepest sense. Eve can have no other significance than the one she has already been assigned. Turning against the "Sweet / Compulsion" of involuntary thoughts, the devil resumes his deadly game. All of Milton's late works contain passages about the ability of feminine beauty to distract, disarm, and deflate masculine purpose, and these passages are often used to convict him of some manner or degree of misogyny. But in *Paradise Lost*, as in *Paradise Regained*, these sentiments are voiced by demons. Why do we find the rejected transport of beauty here at the brink of our tragedy?

Immanuel Levinas has said that beauty, if approached from the field of implementation, is the excess of being over function.[16] The build-

ing is "more" than a shelter, the clothes "more" than protection. This surplus of ornamentation had been discerned in some of the earliest tools of man; today we live inundated by aesthetic excess, such that beauty itself has become a function and our economies supply glib symbols for a mechanized desire. In *Paradise Lost*, a poem suspicious of technology, the overflowing of the useful into the beautiful inheres in the order of nature. It stems from God, and it perplexes Adam and Eve.[17] Why so *many* stars? The "superfluous hand" (8.27) of the creator, himself the Son of a Father "Dark with excessive light," has bestowed excess on Eden as well, "Wild above Rule or Art, enormous bliss" (5.297). God has been "of his good / As liberal and free as infinite" (4.414–415)—"profuse" (4.243), "wanton" (4.629); Eden is "In narrow room Nature's *whole* wealth, *yea more* / A Heav'n on Earth" (4.207–208; my italics). Adam can never quite understand why his cup is overflowing, especially when superfluity falls on the floor and he must clean it up. To the caretaker of Eden's abundance, the blossoms "That lie bestrown unsightly and unsmooth, / Ask riddance, if we mean to tread with ease" (4.631–632), and the narrator himself observes that the sun dispenses "more warmth than *Adam* needs" (5.302).

Eve is the instance par excellence of beauty as the excess beyond function. Adam has requested a mate and a spiritual companion to ease his solitude. But the wife God creates for him is needlessly and incomprehensibly intricate, a nicer and subtler happiness than Adam believes himself to have desired. The experienced truth that her beauty disturbs reason cannot be separated from the fact that it defies reason:

here passion first I felt,
Commotion strange, in all enjoyments else
Superior and unmov'd, here only weak
Against the charm of Beauty's powerful glance.
Or Nature fail'd in mee, and left some part
Not proof enough such Object to sustain,
Or from my side subducting, took perhaps
More than enough; at least on her bestow'd
Too much of Ornament. (8.530–538; my italics)

In part because Eve enfolds and surpasses the beauty of Eden, Adam will choose to fall with her. Not simply uxorious, this willingness to submit to beauty distinguishes Adam from Satan, and in its highest

form represents the telling difference between human and demonic symbolism.

"Thoughts, whither have ye led me . . . " Satan will not meander off into the truth of beauty. Function, the bitter compulsion, rules him. There is psychic work pressing to be done, for which Eve is an instrument to be molded into the likenesses of the God he has lost and of himself, the being who has lost God. Her angelic beauty serves two functions in the production of these symbols: like the heavenly beauty of the earth, it makes Eve as a symbol appropriate to what she symbolizes, feeding Satan's illusion that he enjoys a second chance in his revolt against God; and it also identifies the potential gap between her earthly station and her heavenly worth into which the tempter will step, becoming the master of his symbol. But as even Satan has the opportunity to realize, symbols may lead our thoughts as well as follow from them, slaves to preordained functions. The play of Freud's grandson allayed a compelling need. At the same time, this game removed his catastrophe from the finality of psychic history to the as-if of imagination, where the spool given new meaning by a lost object might also give a new meaning *to* this loss—that it was, for example, a spatial absence the ego registered as emotional abandonment; that a differentiated ego may recover a "secondary omnipotence" in the process of creating and contemplating symbols. While they compensate children for their "great cultural achievement" in renouncing instinctual gratification, such games may also constitute another stage of that achievement. Strictly considered, a symbol transcending recompense escapes psychoanalysis as a hermeneutic, which is not to say that psychoanalysis is left behind in the advent of meaning. This meaning is known and authenticated by the residuum—the place of *rest*—in the aftermath of psychoanalytic interpretation. Addressing such instances, psychoanalysis becomes entwined in the various analytic discourses of the humanities. Idol-making Satan, however, pursues an entirely interpretable revenge. Insofar as the nature of man changes at the fall, subject thereafter to a play of forces that command him involuntarily and yet seem sufficiently intentional to be deemed sinful, we may say that Satan in this myth represents the permanent truth of Freud's "indestructible wish." He has no time for the dysfunctional instruction of the beautiful.

In Ricoeur's modern proverb, "The symbol gives rise to thought," beauty belongs to the giving rise.[18] It announces the possibility of meaning, inviting us to extricate ourselves from the functions and purposes of the world, and conceive the world anew: "That space the Evil

one abstracted stood / From his own evil." It is the rap of revelation asking for attention, but Satan, interminably evil, will not be distracted from his hard determination. He and his issue deteriorate into grotesques, and at our last sight of them, the devils are condemned annually to repeat in their own persons the fall of man, which is, after all, what the game really meant. Milton also plays a game with absence, loss, and symbol. The promises of his religion supply *Paradise Lost*, the "*fort . . . da*" of Christian man, with an otherworldly referent for the paradise within. Always the purveyor of excess, God finds a meaning in the fall beyond the ken of its instigator; mercy will abound over justice, and man will not be at home in hell. The providential meaning of the second paradise is the crushing of Satan—Satan's meaning, rebounding on himself. But there is a serpent to tempt us in the symbol of paradise, whispering that we need not mourn to make peace with God, since God is justified by the rescue of paradise lost in the happy end. Eve, who first lost paradise, guides our way toward the necessary expulsion of this serpent from our fallen home.

According to what Edward Phillips told John Aubrey, Milton lost Katherine Woodcock at about the time he began the epic in 1658 and married Elizabeth Minshull at about the time he finished in 1663. After Adam has seen the happy end, Michael directs him to the place where Eve has slept. She "Persisted, yet submiss, though last" (9.377) in the quarrel preceding their fall. Now she redeems herself, strong in her submission to Adam and to God, as she speaks the last words of man in paradise:

> In mee is no delay; with thee to go,
> Is to stay here; without thee here to stay,
> Is to go hence unwilling; thou to mee
> Art all things under Heav'n, all places thou,
> Who for my wilful crime art banisht hence.
> This further consolation yet secure
> I carry hence; though all by mee is lost,
> Such favor I unworthy am voutsaf't,
> By mee the Promis'd Seed shall all restore. (12.615–623)

Happily the epic returns to Eve. There is an excellent simplicity about her definition of the "further consolation." She states the essential fact of restoration; she feels "unworthy" to play the role of second Eve. This unworthiness seems, not a theological given that might be

already at war with aspirant pride, but the human truth of her situation, as if the "further consolation" were unnecessary for her. It is clear from the first consolation she articulates, for which her passion is reserved, that Eve has done her grieving apart from the happy end. To go with Adam is to stay here: love also inhabits the second paradise. Eve is the beauty at the threshold to something "more" in this symbol than past and future immortality.

Doctrine has had its say. Milton hands his great argument to poetry, and to us. In the beauty of the closure of the poem, our many experiences in reading *Paradise Lost* combine to enable an experience different from any of them—a unique catharsis in which we possess the symbol of a paradise within us while crushing its internal serpent of grief and illusion. The poet begins to sing tragic notes in Book 9. This genre is, in the preface to *Samson*, "the gravest, moralest, and most profitable of all other Poems: therefore said by *Aristotle* to be of power by raising pity and fear, or terror, to purge the mind of those and such like passions, that is to temper and reduce them to just measure with a kind of delight, stirr'd up by reading or seeing those passions well imitated." The catharsis of *Oedipus Rex*, Aristotle's model tragedy, resigns us to the fate of the hero. It arrives at a moment of detachment, when our identification with Oedipus, the source of our pity and fear, shifts to an identification with the cosmic order requiring his punishment and exile.[19] Milton creates small foreshadowings of this detachment throughout the epic, especially in the way his narrative voice comments, from the perspective of ultimate consequences, on the unfolding of the action. It is vital to his theodicy that the large design take precedence over the partiality of occasions. But the catharsis of *Paradise Lost* reverses the sequence of psychological positions in the classical and usually Elizabethan formula. Milton weans us from the immediacy of loss to the salvational *moira* of Books 11 and 12, summarized in the "Providence thir guide" of the final passage, then concludes in the last two lines with our complete identification, for the first time in the poem, with Adam and Eve. "In a process of artistic identification," Colie explains, "Adam's experience is brought abreast of that of Milton's readers." "In the last two books," Tayler concurs, "Milton completes the process hinted at in the ending of Book 10, where the narrative voice and the words of Adam and Eve precisely coincide: the reader, knowing the meaning of the 'mysterious terms,' watches as Adam's knowledge becomes progressively closer to his own." The Richardsons also proposed that this gradual convergence of the reader with the two human characters was

part of the secret of our satisfaction in the end: "This last Circumstance brings our Progenitors into the Condition in Which We Are, on *Even Ground* with us."[20] Whatever its problems as art or wisdom, the gulp of historical knowledge in the second education prepares us for a climactic intimacy with the human protagonists. The catharsis of the epic transpires in the lovely and unclassical contrivance of an identification with providential judgment that engenders, rather than thwarts, an identification with the figures judged.

Closures by their nature have a stillness about them, but Milton has artfully enhanced this effect. When critics talk about the dramatic irony purged at last in the final lines of the poem, they do not always stop to remind us what, in the experiencing, this irony has meant. For around 10,550 lines *Paradise Lost* has been constantly portentous. Omens and threats lurk everywhere. We first view paradise through the same eyes that opened in Book 1 on the first view of hell—the most noted of a legion of menaced beauties. Milton does not allow Adam and Eve to sleep peacefully in paradise without insisting that we realize, at this time of archetypal vulnerability, how precarious their rest is. The fall is foreknown, foretold, and forecast in detail upon detail, in the very language that describes paradise. After the fall the action is menaced by death, as when Adam ruminates mortality while outstretched on the cold ground, and by expulsion, as when Eve beholds "The Bird of *Jove*, stoopt from his aery tow'r," like the angel Michael, to drive "Two Birds of gayest plume before him" (11.185–186). The poet maintains in us the cold fire of a sharp anxiety that now and then, at his will, flares up into the center of our awareness. It is conflict again, whose management in art is Milton's singular talent: we wish the portents could be dispelled, and wish therefore to be unfallen; but recognizing the necessity of the past, we seek what pleasures there are in the vantage point of portentousness, feeling superior to our ignorant ancestors or appreciating the cunning architecture of fate.

The second education discharges much of this anxiety from our beholding of Adam and Eve. With it goes the gnawing intensity of the wish that things had been otherwise, and along with that, our feeling of superiority. There remains the portent of expulsion, which the angels hurriedly accomplish, obedient to divine schedule. There are tears. And then at last Adam and Eve and their spectators are "though sorrowing, yet in peace." Fate is still cunning, but now exile, labor, pain, death, and hope are shared by all the human beings involved in this drama. For everyone, things are as they are.

In this new and human peace Adam and Eve appear before us on "thir solitary way." No gods, no angels, no devils. Just the men and women in whom these supernatural beings live, ensnaring us in their complex and conflicting designs. What do we feel for each other? There is for each reader a private encounter with Adam and Eve at the end. When we identify with them, the content of this internal gesture is more than "now we are equally knowledgeable" or "now we are equally in history." These factors, Richardsons' "*Even Ground*," clear the way to profounder recognitions in our individual histories. For these are not our equals. An identification with Adam and Eve acknowledges them for what they are in this myth—parents, progenitors. Our old and difficult love awaits at the end of the poem. Mother, father, a reader who is their child: our identification at the end of the poem abbreviates oedipal resolution. As once we forfeited narcissistic illusion for vulnerable pride, so now, in order to participate in this catharsis, we must exchange the bliss of unfallen existence for the narrowed horizons and implacable sorrows of fallen existence. Milton defines his Godhead through the psychological anthropomorphism of an ideal oedipal submission; the obedience of ego to superego, enacting as psychic structure the marriage that created us, symbolizes the obedience of Son to Father. But the human analogues of these deities did not reign in serenity. The poet could receive artistic and religious strengths from this ideal only by placing himself in a sacred complex folded over on the crisis of a first and imperfect oedipal solution, and in the long course of becoming able to write, then actually writing his poem, the psychic survivals of his historical parents underwent manifold distortions: the superego was split, allowing him to recover oedipal rebellion in a conscience liberated from earthly authority; collecting a debt from the Father, he improved maternal with paternal omnipotence, enjoying with divine guarantee powers and securities originating in the figure of the earthly mother. In the end, however, the symbols of the sacred complex come down to earth, incarnate before us in the handed couple whom we are invited to join. The poem is making peace with its beginnings.

The uncanny beauty of the end comes to us and comes back to us, at once strange and familiar. We are invited to identify flesh, bone, and spirit with our mother and father. It is, again, an ideal peace. None of us have achieved an oedipal solution so sudden, final, and beautiful as this one. Somewhere in this triangle Milton has placed us in lies the reason why you went astray or were made to go astray, why you do not simply change, as a creature of pure reason might change, when

your behavior harms yourself and those you love, and why, to speak like Freud, you may have been destined to fall ill. If, however, we join them willingly and lovingly on their solitary way, we experience a healing sublimation of the drama by which, encountering terrors stronger than our wishes, we once entered the sphere of culture and became in our psyches sons or daughters. Encouraged by the design of the poem, this experience demands capacities that no artist can create in us, though some may bring them to light. There are viewers who are not moved when John Ford turns his camera on a dinner table, and readers for whom the identification offered in the last two lines of *Paradise Lost* will seem trite, sentimental, suspicious. But Milton has not compelled, and nor would I. This is where polemic ends. The grand style no longer agitates. Things are as they are. It is the quietest moment in the poem, and among the many attempts to produce a loving parental identification in art, conspicuous for its pitiless recognition of everything about us that might resist such an artistic effect. Within readers open to this experience, the passions reduced to "just measure" both own and defy the destiny of Oedipus. Milton's catharsis is a *pax christiana* for illnesses residual from the first conflicts of life.

There is more than our return to the private history of old loves. A child taking responsibility for itself in the structure of a family could not know what we know at the end of *Paradise Lost*. Thirty years before Milton brought children into the presence of their parents, reuniting a family close in structure to his own. The parents sat on thrones, and clandestine accusations lived in the pious language honoring them. Again he has brought children before their parents—but how fallen, and how much better the meeting as a result. Without lie or unvoiced murmur, victorious over what Ferenczi called the "confusion of tongues between adults and the child," this is a mature recognition prefigured by the oedipal solution.[21] Adam and Eve represent, not our "equals" in some neutral sense, but *our parents as our equals*. They are guilty parents, mourners themselves, victims of the same desires and afflictions we once thought to be ours alone. Milton lamented human discord unforgettably in Book 2:

> O shame to men! Devil with Devil damn'd
> Firm concord holds, men only disagree
> Of Creatures rational, though under hope
> Of heavenly Grace; and God proclaiming peace,
> Yet live in hatred, enmity, and strife

> Among themselves, and levy cruel wars,
> Wasting the Earth, each other to destroy. (496–502)

A shame to men, devils are one on the common ground of their despair and damnation. Might men find a better concord on the common ground of their fallen condition? And taking liberties with the language of this passage, might this better concord make the wish for "heavenly Grace" in the hereafter into a "hope," a wish that has internalized the possibility of its failure? The ontological reconciliation of *Paradise Lost*, completing its theodicy, occurs in the love between generations, able to convert our unchosen situation in history from a brutal throw or a meaningless burden into an opportunity for our mutual empathy. We know this marriage is the cause and consequence of the fall; Milton does not request an ignorant love, but one cleansed by our knowledge of the worst. That knowledge, echoed in our acknowledgment of the Adam and Eve in us, forges the bond. Affirming their union, we take our place—our place of rest, our ground in common: civilization—in the symbolic paradise they bear into history. As we perform this identification, we must accept "the Condition in Which We are," which is to forgive this mythical couple for the world they have bequeathed to us, and thus to choose the fall as the condition that unifies all men in mature concord.

No gods, no angels, no demons.[22] But in this solitude, humanity gathered to itself and manifested. The marriage of Adam and Eve relates in blood and destiny everyone who has ever lived and will ever live. The identifying reader stands, not only before mother and father in the privacy of his psyche but also before the entire family of man. Milton ends with the nonreligious moment that religion cannot afford to deny. This reunion of mankind with its progenitors supplies an existential counterpart to the reunion of creatures and creator in the bliss of the eschaton. Then God shall be all in all, entirely present to each one. But God the Father is excluded in a crucial way from the great gathering at the closure of the epic. The end is for fallen humankind: each of us stands related to the entirety of us, and man is all in all.

Everything began with a hateful turning away from the origins we rediscover in this moment. Each time an identifying reader finishes the poem, history stands perfectly balanced between a world destroyed and a world restored. For at that instant, the very last act in human time says *no* to the primal perversion of an angel who could not endure having been created:

Say first, for Heav'n hides nothing from thy view
Nor the deep Tract of Hell, say first what cause
Mov'd our Grand Parents in that happy State,
Favor'd of Heav'n so highly, to fall off
From thir Creator, and transgress his Will
For one restraint, Lords of the World besides?

If Satan be the cause, and we at the end of the poem the effect, then the creator of *Paradise Lost* has indeed brought good out of evil. Good symbols are a home. Milton has built us a home in the symbol of our unrest, transforming the trauma of original sin into an occasion for our renewed community.

NOTES

INDEX

NOTES

Introduction

1. Paul Ricoeur, *Freud and Philosophy: An Essay on Interpretation*, trans. Denis Savage (New Haven: Yale University Press, 1970), pp. 153–254 and passim.

2. Ibid., pp. 483–493. See also Kurt Eissler, "Psychopathology and Creativity," *American Imago*, 24 (1967), pp. 35–39.

3. Nelson Goodman, *Ways of Worldmaking* (Indianapolis: Hackett, 1978), p. 5. More than a few worthwhile things probably have been left undone out of fear of being charged with "reduction," which, if true, is a shame, for this is little more than a term of abuse appealing to intellectuals.

4. See *The Ego and the Id*, in James Strachey, trans., *Standard Edition of the Complete Psychological Works of Sigmund Freud*, 24 vols. (London: Hogarth, 1953–1964), XIX, 48, and *The Question of Lay Analysis*, XX, 223. All subsequent citations of Freud use this edition (*S.E.*) and will usually appear in the text.

5. I am paraphrasing, perhaps unfairly, the arguments to be found in Heinz Hartmann, *Psychoanalysis and Moral Values* (New York: International Universities Press, 1960), and Ernst Ticho, "The Development of Superego Autonomy," *Psychoanalytic Review*, 59 (1972), 217–233.

1. Beginning with a True Poem

1. Helen Darbishire, ed., *The Early Lives of Milton* (London: Constable, 1932), p. 1; hereafter this collection is cited in the text as *EL*. On this anecdote, see also William Riley Parker, *Milton: A Biography*, 2 vols. (Oxford: Clarendon Press, 1968), I, 4; II, 686–687.

2. See the somewhat whimsical account by Harry Rusche, "A Reading of John Milton's Horoscope," *Milton Quarterly*, 13 (1979), 6–11. Parker reproduces the flyleaf of the Milton Bible as the frontispiece to *Milton: A Biography*, vol. II. It may also be found in James Holly Hanford and James G. Taaffe, *A Milton Handbook*, 5th ed. (New York: Appleton-Century, 1970), plate 13. The various Bibles associated with Milton are discussed in Frank Patterson, et al., *The Works of John Milton* (New York: Columbia University Press, 1931–38), XVIII, 559–565.

3. Elegy VI, ll. 79–90, and *Paradise Lost*, 7.29–30. All quotations from

the poetry use the texts and translations of Merritt Y. Hughes, ed., *John Milton: Complete Poems and Major Prose* (New York: Odyssey, 1957).

4. See *Timber: or, Discoveries* in *Ben Jonson*, ed. C. H. Hereford and Percy and Evelyn Simpson (Oxford: Clarendon Press, 1932–1947), VIII, 625.

5. *The Works of George Herbert*, ed. F. E. Hutchinson (Oxford: Clarendon Press, 1941), p. 68.

6. Don M. Wolfe, et al., eds., *The Complete Prose Works of John Milton*, 8 vols. (New Haven: Yale University Press, 1953–1982), I, 890. Unless otherwise noted, all subsequent quotations from Milton's prose are from this edition and will be cited in the text as *CP*, followed by volume and page. Milton repeated the same sentiment in a letter to Henry De Brass: "he who would write worthily of worthy deeds ought to write with no less largeness of spirit and experience of the world than he who did them" (*CP* VII, 501); cf. *CP* VII, 497; IV, 305.

7. See the comments of Harold Bloom in *A Map of Misreading* (New York: Oxford University Press, 1975), pp. 127–128.

8. In *The Logical Epic* (London: Routledge, 1967), pp. 80–93, Dennis Burden notes that Milton has apportioned the argument of *Areopagitica* between his protagonists, doing his best to create a disagreement in which both parties are in the right. I think the precarious balance of the unfallen marriage at this point, a union tolerant of schism, is a domestic foreshadowing of political liberty as Milton understood it. Adam can no more require the presence of Eve than God can intervene to prevent the fall, or a state can legislate conscience.

9. Arnold Stein, *The Art of Presence: The Poet and Paradise Lost* (Berkeley: University of California Press, 1977), p. 4.

10. William Perkins, *Two Treatises* (Cambridge, 1597).

11. "The Temptation Motive in Milton" has been reprinted in *John Milton, Poet and Humanist: Essays by James Holly Hanford* (Cleveland: Press of Western Reserve University, 1966), pp. 244–263.

12. Hans Jonas, *The Phenomenon of Life: Toward a Philosophical Biology* (New York: Harper & Row, 1966), pp. 268–269.

13. One classic account of the soul in tribal cultures is Émile Durkheim's *The Elementary Forms of Religious Life*, trans. Joseph Swain (New York: Free Press, 1965), pp. 273–308.

2. The Root-Bound Lady

1. The element of familial ritual in *Comus* has been noticed by David Wilkinson, "The Escape from Pollution: A Comment on 'Comus,'" *Essays in Criticism*, X (1960), 34, and Barbara Breasted, "*Comus* and the Castlehaven Scandal," *Milton Studies* III, ed. James Simmonds (Pittsburgh: University of Pittsburgh Press, 1970), p. 211.

Breasted's argument, endorsed by Christopher Hill in *Milton and the*

English Revolution (New York: Viking, 1977), pp. 43–44, asks us to suppose that the Egerton family, wishing to dissociate themselves from the obscene behavior of a notorious relative, first commissioned a masque on the subject of their sexual virtue, then excised from the performance text passages explicitly touching on this theme. Is this not having it both ways? If the theme *was* commissioned, nothing prevents us from assuming that the Egertons (or Lawes) knew their man.

2. Hans Loewald, "On Internalization," *International Journal of Psychoanalysis*, 54 (1973), 16. I have discussed theories about the genesis of the superego in "Superego in Kierkegaard, Existence in Freud," in *Psychiatry and the Humanities*, vol. 5: *Kierkegaard's Truth: The Disclosure of the Self*, ed. Joseph Smith (New Haven: Yale University Press, 1981), pp. 119–165.

3. Paul Ricoeur, *The Symbolism of Evil*, trans. Emerson Buchanan (Boston: Beacon Press, 1967), esp. pp. 225–246. Owen Barfield develops a wide-ranging critique of idolatry from Levy-Bruhl's "collective representations (participations)" in *Saving the Appearances: A Study in Idolatry* (New York: Harcourt Brace, 1965): an idol is "a representation or image which is not experienced as such" (p. 110). Idolatry springs from an exclusively ontological interpretation of the "is" coupling two terms in a metaphor, which is why I suggest that the idol remains part of the internal dynamic of the symbol.

4. A. S. P. Woodhouse, *The Heavenly Muse: A Preface to Milton* (Toronto: University of Toronto Press, 1972), pp. 72–73.

5. E. M. W. Tillyard asserts the disunity of *Comus* in *Milton* (New York: Collier, 1967), pp. 58–66: "The final impression I get from *Comus* viewed as a whole is that when Milton wrote it he was not inspired by any compelling mood to give it unity" (p. 64). Parker found the plot "flimsy, devoid of suspense, and lacking in dramatic unity" in *Milton: A Biography*, I, 130. Edward Tayler, whose discussion of the temptation scene is close to my own, also feels that "Milton's concern with extreme forms of chastity threatens the artistic integrity of *Comus*"—*Milton's Poetry: Its Development in Time* (Pittsburgh: Duquesne University Press, 1979), p. 137.

6. Freud's insight was considerably advanced by the pioneering psycholinguistics of René Spitz, *No and Yes: On the Genesis of Human Communication* (New York: International Universities Press, 1956); Francesco Orlando explores some aesthetic implications of negation in *Toward a Freudian Theory of Literature,* trans. Charmaine Lee (Baltimore: Johns Hopkins Press, 1978).

7. See Enid Welsford, *The Court Masque* (Cambridge: Cambridge University Press, 1927), pp. 315–318, and for a sampling of criticism based on a Jonsonian view of the genre of the work, the well-known articles of Robert M. Adams ("Reading *Comus*"), Rosemond Tuve ("Image, Form, and Theme in *A Mask*"), and C. L. Barber ("*A Mask Presented at Ludlow Castle*: The Masque as a Masque"), all reprinted in John S. Diekhoff, ed., *A Maske at Ludlow: Essays on Milton's Comus* (Cleveland: Press of Case Western Reserve University, 1968).

8. See B. A. Wright's annotations for the masque in his *Shorter Poems of John Milton* (London: Macmillan, 1938), and John Arthos, *On "A Mask Presented at Ludlow's Castle"* (Ann Arbor: University of Michigan Press, 1954); "Milton, Ficino, and the *Charmides*," *Studies in the Renaissance*, VI (1959), 261–274.

Sears Jayne's "The Subject of Milton's Ludlow *Mask*" (included in Diekhoff, *A Maske at Ludlow*, pp. 165–187), proceeding at some distance from the language of the masque, finds a neoplatonic allegory of the ascent of the soul by assuming that mythological allusions in Milton have the same significance they possess in the philosophy of Ficino. One would think that the first test of the validity of such a reading is the coincidence between Ficino's philosophy and the philosophical passages in the masque itself. But despite his astrological mysticism of the bodily spirits, Ficino never to my knowledge proposes that the virtuous life will transform body into soul: the consistent goal of his philosophy is to escape from the confinements of the flesh. So while *Cosmus* has the look and feel of a neoplatonic cosmos, its central doctrine has no parallel in Ficino.

9. Ricoeur, *Symbolism of Evil*, pp. 279–305, 330–345. See also Leonard Eslick, "The Material Substrate in Plato," in Erson McMullin, ed., *The Concept of Matter in Greek and Medieval Philosophy* (Notre Dame: University of Notre Dame Press, 1963), pp. 39–54. The Orphic myth presumed to have given rise to Plato's philosophy is discussed by W. C. K. Guthrie, *The Greeks and Their Gods* (Boston: Beacon, 1954), pp. 305–353. Milton associates Circe with ignorance in Prolusion VII (*CP* I, 103).

10. *Caelica*, Sonnet C, in Geoffrey Bullough, ed., *Poems and Dramas*, 2 vols. (New York: Oxford University Press, 1945), vol. 1.

11. Durkheim, *Elementary Forms of Religious Life*, p. 53.

12. *The Collected Dialogues of Plato*, ed. Edith Hamilton and Huntington Cairns (New York: Pantheon, 1961), 108A-B (p. 89); my italics.

13. Northrop Frye, *The Secular Scriopture: A Study of the Structure of Romance* (Cambridge: Harvard University Press, 1976), pp. 81–86.

14. Woodhouse, *Heavenly Muse*, pp. 63–64.

15. Although she walks with "printless" feet, Milton's Sabrina has not, like Drayton's, dissolved into the stream—see my "The Heretical Milton: From Assumption to Mortalism," *English Literary Renaissance* (hereafter *ELR*), 5 (1975), 133, n. 20. This preservation of the integrity of a distinct and bounded body is consistent with psychoanalytic interpretations of virginity proposed later in this chapter.

16. Salvatore Lucia's *Wine and the Digestive System* (San Francisco: Fortune House, 1970) is an annotated bibliography containing classical and Renaissance items.

17. And not only in Galenism: see Allen Debus, *The Chemical Philosophy*, 2 vols. (New York: Science History Publications, 1977), II, 340–341, 353.

18. Roger Wilkenfeld, "The Seat at the Center: An Interpretation of

Comus," in *Critical Essays on Milton from ELH* (Baltimore: Johns Hopkins Press, 1969), pp. 123–150.

19. See J. H. Hanford, *John Milton, Englishman* (New York: Crown, 1949), p. 98. This biographer assumed that, when abroad, Milton "made it his special mission to preach 'the sage and serious doctrine of virginity' and conspicuously to exemplify it, sometimes . . . out of season" (p. 88).

20. Tillyard, *Milton,* pp. 318–326; Ernst Sirluck, "Milton's Idle Right Hand," *JEGP,* 60 (1961), 749–785. Disputing Tillyard, who found chastity as early as "At a Solemn Music," Sirluck claims that all the evidence for Milton's vow of chastity dates from 1637. But the two manuscripts of *Comus* are obviously interested in the subject, and John Imby has tried to reinstate Tillyard's opinion in his "Sensuality and Chastity in 'L'Allegro' and 'Il Penseroso,'" *JEGP,* 77 (1978), 504–529.

Divine Moly protects Milton from the girls in Elegy I.

21. Tillyard, *Milton*, p. 63, supposes the epigraph to mean that Milton was uneasy over the merit of *Comus.* In *The Life of John Milton*, 6 vols. (London: MacMillan, 1874–1880), I, 640–641, Masson argues that Milton was reticent about publication. I incline toward Masson's view, adding that Milton had probably decided not to pursue the sort of artistic career associated with this genre.

22. Parker, *Milton*, II, 975.

23. Lawrence Stone, *The Family, Sex and Marriage in England,* 1500–1800 New York: Harper & Row, 1977), pp. 93–119, 159–191, and passim. The relationship between the poet's family and the masquing family will convince some Miltonists, I hope, that Parker's insistence on the impersonality of *Comus* (*Milton: A Life*, I, 128–133), is at best a half truth. His unwillingness to confront the psychological power of this work, the longest and most revealing of all the early texts, is one of the reasons for the ultimate failure of his biography—his inability to supply us with a coherent, not to say arresting conception of his subject. The work of Saurat, Hanford, Tillyard, and Sirluck is treated as a wayward curiosity: "Modern criticism has grown increasingly fond of interpreting Milton's character development in terms of *Comus*, whereas the didactic element in the entertainment may not have been Milton's idea in the first place. Its elaboration was, of course; its elaboration may have been strongly influenced by parental criticism. . . ; but both expression and elaboration, we should remember, had passed a critical eye more accustomed to the masque tradition than our own" (pp. 128–129). Qualification by qualification, the issue of "Milton's character" is dropped in favor of the "masque tradition," which then launches a disastrously superficial discussion of the masque. I am not suggesting that Parker should have been a Freudian. But I do not see how it is possible to understand the first half of the life of a poet known to his classmates as "The Lady of Christ's"— probably an allusion, as Shawcross proposes in "Milton and Diodati: An Essay in Psychodynamic Meaning," *Milton Studies*, VII (1975), p. 137, to the Virgin

Mary—whose best poems during this period are all deeply involved in one way or another with virgins and virginity, without being direct and thoughtful about his sexual attitudes and their relation to his artistic achievement.

24. Tillyard, *Milton*, p. 59.

25. Erik Erikson, *Young Man Luther* (New York: Norton, 1958), p. 83.

26. Jon Harned, "Milton's Identity Crisis" (unpublished article).

27. Frank Kermode, "Milton's Crises," *The Listener* (19 Dec. 1968), p. 829.

28. Some of the poems in the 1645 volume—it is hard to believe unconsciously—were backdated by an author designing a new and more precocious youth for himself. See W. R. Parker, "Some Problems in the Chronology of Milton's Early Poems," *RES*, 11 (1935), 276–283, and Edward Le Comte, *Milton's Unchanging Mind* (Port Washington, N.Y.: Kennikat, 1973), pp. 5–69. The competent lines of a twenty-year-old might be the sign of prodigious genius if attributed to a seventeen-year-old. As an old man Milton would publish his academic exercises and his familiar letters, as if to give the illusion that his youth had been more conspicuous for literary industry than is in fact the case. The young Milton was stingy with the deposit of his talent, and his poems before the rebirth of his last years make a slim volume, no few pages of which are taken up with Latin juvenilia.

29. J. W. Flosdorf's "'Gums of Glutinous Heat': A Query," *Milton Quarterly*, 7 (1973), 4–5, inspired the opening of Edward Le Comte's *Milton and Sex* (New York: Columbia University Press, 1978), pp. 1–2.

30. Erikson outlines his epigenesis of identity in *Identity, Youth and Crisis* (New York: Norton, 1968), pp. 91–141.

31. Shawcross, "Milton and Diodati," pp. 141, 156–157.

32. Twice in *An Apology* Milton claims not to have indulged in an unnamed sin less serious than the whoring with which he is charged. His "niceness of nature" kept him "above those low descents of mind *beneath which* he must deject and plunge himself that can agree to saleable and unlawful prostitutions" (*CP* I, 890; my italics); again, "a certain reservedness of natural disposition" was enough, when supported by Greek philosophy, "to keep me in disdain of *far less* incontinences than this of the bordello" (892; my italics). What are these intermediate lusts? Drinking? The context suggests masturbation as well, the fraternal occupation of schoolboys in all ages. Comus himself realizes that beauty is "Unsavory in th'enjoyment of itself" (742), implying that masturbation is not a third alternative: it is either sexuality or virginity.

33. *Faerie Queene* I.x.xxx–xxxi. Charissa is wise enough to wean her brood, thrusting them away from mother and home.

There is still something peculiar about charity in *Paradise Lost*'s hymn to married love: "By thee adulterous lust was driv'n from men / Among the bestial herds to range, by thee / Founded in Reason, Loyal, Just, and

Pure, / Relations dear, and all the Charities / Of Father, Son, and Brother first were known" (4.753–757). No longer supplanted by unblemished chastity, charity is now withheld from the mother or her image, and if we assume that the "Father" has a "Son" and the "Son" a "Brother," withheld also from the cross-sex relationships of the family. By no accident, one would suppose, charity seems to exclude the only relationships that Milton as a father was able to generate.

There is some possibility that the author of *Comus* was being pedantic as well as symptomatic in replacing charity with chastity. James W. Earl, in an unpublished note on Augustinian allusions in the *Faerie Queene*, has shown that the inconography of charity as a mother can be traced to Book VIII of the *Confessions*, where continence, not charity, is "a fruitful mother of children"—trans. R. S. Pine-Coffin (Baltimore: Penguin, 1961), p. 176.

34. John Diekhoff, *Milton on Himself* (New York: Oxford University Press, 1939).

35. I have expanded on this point in the article on Kierkegaard and Freud cited in Note 2 to this chapter, drawing inspiration from Joseph Smith, "The Heideggerian and Psychoanalytic Concepts of Mood," *Journal of Existentialism*, 4 (1964), 101–111.

36. In the *Christian Doctrine*, Milton defines chastity as charity turned toward oneself (*CP* VI, 717, 719, 726–727).

37. Alluding to Edmund Wilson's *The Wound and the Bow*, Edward Tayler rightly suggests that castration anxiety underlies Milton's interest in wholeness of all kinds, including the monism of his late theology, in *Milton's Poetry: Its Development in Time* (Pittsburgh: Duquesne University Press, 1979), pp. 136–139, 146–147, 209–213.

38. W. B. C. Watkins, *An Anatomy of Milton's Verse* (Baton Rouge: Louisiana State University Press, 1955), p. 97.

39. The abstract logic of kinship systems can serve to demonstrate the derived or "secondary" character of sibling incest. Consider, for example, the position of a son in a totemic society with matrilineal descent. Because he shares a totem with his mother and his sister, the law of exogamy forbids him to fantasize intercourse with either of these females. The system also denies his capacity to be a progenitor; although he will become a husband, this son will not reproduce his image. But the children of his sister will bear his image, and in such cultures the brother is usually the major figure of authority in the life of his sister, living out the fantasy of an incestuous marriage to her. He thus replaces in his maturity the uncle with whom he shared his own mother, the man who occupied the slot "father." Within the psycho-logic of the kinship system, lateral incest has its impetus and authority in the cross-generational relationships. I think the importance of the avuncular role in Milton's life has features similar to these, especially when we consider that he could not, unlike his father, produce his "image" in a son. His desire to do so is evident in the fact that he gave to his dead boy the

name his own father had given to him, and in his famous disappointment with two of his daughters, likenesses of Mary Powell whom he eventually cast out of his home as if in delayed revenge for their mother's abandonment (*EL*, 77–78).

Because the taboo is directed against the son, Freud considered matrilineal descent the earliest kinship system (*S.E.* XIII, 5, n. 1). My own ruminations on the subject are in line with those of Robin Fox, "*Totem and Taboo* Reconsidered," in Edmund Leach, ed., *The Structural Study of Myth and Totemism* (London: Tavistock, 1967), pp. 161–178.

40. Robert M. Adams, "Reading *Comus*," in John Diekhoff, *A Maske at Ludlow*, p. 95. Easter eggs are sufficient to teach us that the Christian/pagan game must be played with care.

41. John Arthos, "The Realms of Being in the Epilogue of *Comus*," *MNL*, 76 (1961), 323.

42. Tillyard, "The Action of *Comus*," in Diekhoff, *A Maske at Ludlow*, p. 53. Edward Le Comte agrees: "Milton's Epilogue . . . points to the usual future of marital union and generation" (*Milton and Sex*, p. 17). Is our usual future beyond the stars?

43. *Faerie Queene* III.vi.xliv–lii.

44. God says that one day he will be a "husband" rather than a "master" in Hosea 2.16. The moral and exegetical principles underlying Protestant revulsion at the carnality of the devotional tradition surrounding such passages may be found in John Calvin, *On the Christian Faith*, ed. John McNeill (New York: Bobbs-Merrill, 1957), pp. 7–9, 10–11, 31–32, 79–80.

45. Angus Fletcher alerts us to this pun in *The Transcendental Masque: An Essay on Milton's Comus* (Ithica, N.Y.: Cornell University Press, 1971), p. 165.

46. Martin Heidegger, *Being and Time*, trans. John Macquarrie and Edward Robinson (New York: Harper & Row, 1962), pp. 179–181, 228–235, 279–311. Dasein, that moody creature, seems to have no erotic life: I am not aware of a single allusion to sexuality in all of Heidegger. My brief remarks here owe something to Maurice Merleau-Ponty, who explored the sexual being of the Husserlian "owned body" (*corps propre*) in *Phenomenology of Perception*, trans. Colin Smith (London: Routledge, 1962), pp. 154–173.

47. I am thinking of Joseph Summers' "Reply to Anthony Low's Address on *Samson Agonistes* (Le Moyne College, May 4, 1979)," *Milton Quarterly* 13 (1979), 102–106, and his talk entitled "The Crucifixion in Milton's English Poems" at the MLA session on Milton in 1979, abstracted in *Milton Quarterly*, 14 (1980), 72. A fuller statement of my own view that Milton subordinates the Christ of soteriology to the Christ of eschatology may be found in "The Heretical Milton," pp. 152–163.

48. "Milton's *Of Reformation* and the Dynamics of Controversy," in Michael Lieb and John Shawcross, eds., *Achievements of the Left Hand: Essays on the Prose of John Milton* (Amherst: University of Massachusetts Press,

1974), p. 72. Lieb goes on to supply a "rationale" for these tactics in rhetorical and political tradition.

49. Kermode, "Milton's Crises," p. 829.

50. Parker, *Milton*, I, 229.

51. See Edward Le Comte's *Milton's Unchanging Mind*, a development of themes first stated in his *Yet Once More: Verbal and Psychological Pattern in Milton* (New York: Columbia University Press, 1953).

52. Barbara Lewalski's *Milton's Brief Epic* (Providence: Brown University Press, 1966), which contains a great deal of information about typological patterns in Milton, should perhaps be given credit for anticipating the first concentrated full-length study of the theme, William Madsen's *From Shadowy Types to Truth* (New Haven: Yale University Press, 1968). Tayler cites the relevant materials in *Milton's Poetry*, pp. 227–229, n. 15.

53. Hanford, "The Youth of Milton," in *John Milton, Poet and Humanist*, pp. 73–74.

54. Stein, *The Art of Presence*, pp. 22–23.

55. I am drawing from the discussion of sexuality in Book XIV of *The City of God*, sections 17 (on the fallen penis) and 23 (on the penis in Eden).

Wolfgang Rudat's provocative "Milton, Freud, St. Augustine: *Paradise Lost* and the History of Human Sexuality," *Mosaic* XV / 2 (1982), 109–122, reached me in the late stages of preparing this book. Although I dispute his assertion that Milton followed Augustine's guidelines in imagining prelapsarian sexuality, his genital interpretation of the fall yields some alluring insights, especially with regard to the interlacing of penis envy and vagina envy in Adam and Eve's justifications for disobedience. Conducting psychoanalytic interpretation with reified equations yields some nonsense as well: if the fruit symbolizes the vagina, and Eve's fall is the masturbatory discovery of her genital insufficiency, must we not conclude that Adam's fall symbolizes the servitude of cunnilingus? Particulars aside, a sexualized disobedience is consistent with my emphasis on the dominance of oedipal trends in the epic. But my own discussion of the fall in Chapter 5 stresses what is there immediately in the text: the mortal taste of fatal food.

56. Kermode, "Milton's Crises," p. 829.

3. New Adam: Sacred History as Psychic History

1. I have also dealt with this subject in my "Superego in Kierkegaard," in Smith, ed., *Kierkegaard's Truth*, pp. 121–162.

2. I am thinking of the equation between faith and basic trust developed in *Young Man Luther*. Whereas Erikson stresses the smooth transition from one phase to another in epigenesis, I want initially to accent the rupture caused by the oedipal dilemma.

3. Jacques Lacan, *Écrits: A Selection*, trans. Alan Sheridan (New York: Norton, 1977). I have tried to isolate forms of Lacan's "mirror stage" specific

to Renaissance culture in "The Articulation of the Ego in the English Renaissance," in Joseph Smith, ed., *The Literary Freud: Mechanisms of Defense and the Poetic Will* (New Haven: Yale University Press, 1980), pp. 261–308.

4. Sören Kierkegaard, *Fear and Trembling and The Sickness Unto Death*, trans. Walter Lowrie (Princeton: Princeton University Press, 1954), p. 147.

5. Lacan, *Écrits*, pp. 61–77. See also Anika Lemaire, *Jacques Lacan*, trans. David Macey (London: Routledge, 1977), pp. 59, 61–64, 91–92.

The naming of the Name-of-the-Father has justly angered feminists, but the name has nothing to do with the concept of an inherent relationship between oedipal resolution and placement in culture. I preserve it because it seems so appropriate to the culture and the poet I am studying.

6. Consult, besides Freud's "Remembering, Repeating and Working-Through" (*S.E.* XII, 145–156), Hans Loewald's "Some Considerations on Repetition and Repetition Compulsion" and "Perspectives on Memory" in *Papers on Psychoanalysis* (New Haven: Yale University Press, 1980), pp. 87–101, 148–173.

7. Kerrigan, "Superego in Kierkegaard," pp. 125–131, 156–162.

8. Barbara K. Lewalski, *Donne's Anniversaries and the Poetry of Praise: The Creation of a Symbolic Mode* (Princeton: Princeton University Press, 1973), pp. 92–107.

9. Kerrigan, "Articulation of the Ego," pp. 284–285.

10. Pope, *Imitations of Horace*, Ep. II, i, ll. 99–102. Coleridge agreed; see Joseph Wittreich, *The Romantics on Milton* (Cleveland: Press of Case Western Reserve University, 1970), p. 240.

11. Tillyard, *Milton*, p. 259.

12. Merritt Y. Hughes, *Ten Perspectives on Milton* (New Haven: Yale University Press, 1965), p. 62.

13. Walter MacKellar, *A Variorum Commentary on the Poems of John Milton*, vol. 4 (New York: Columbia University Press, 1975), p. 87.

14. For elaboration see my "The Heretical Milton," 149–151.

15. Philip Gallagher is particularly forceful on this point in "'Real or Allegoric': The Ontology of Sin and Death in *Paradise Lost*," *ELR*, 6 (1976), 317–333, and "*Paradise Lost* and the Greek Theogony," *ELR*, 9 (1979), 121–148.

16. A. S. P. Woodhouse, "Theme and Pattern in *Paradise Regained*," *UTQ*, 25 (1955–56), 181.

17. Almost everyone finds a semantic complexity in Christ's final reply; my own reading is similar to that of Stanley E. Fish in "Inaction and Silence: The Reader in *Paradise Regained*," in Joseph Wittreich, ed., *Calm of Mind* (Cleveland: Press of Case Western Reserve University, 1971), p. 43. There is something akin to a Hegelian *aufhebung* (simultaneous cancellation and preservation) in the great monosyllabic line recording both the speech and the act of Christ.

18. E. H. Gombrich, *Symbolic Images: Studies in the Art of the Renaissance*

(2nd ed.; London: Phaidon, 1978), pp. 159–160, 168–170, and passim.

19. Henry Reynolds, *Mythomystes*, in Edward Tayler, ed., *Literary Criticism of Seventeenth-Century England* (New York: Knopf, 1967), p. 238.

20. See Canto II of Thomas Evans' unpaginated *Oedipus* (London, 1615).

21. William Kerrigan, *The Prophetic Milton* (Charlottesville: University Press of Virginia, 1974), pp. 183–186.

22. Tayler, *Milton's Poetry*, pp. 172–174, 176–177.

23. In H. J. Todd, ed., *The Poetical Works of John Milton*, 6 vols. (London, 1826), IV, 298. Joan Webber called Satan the "shadow" of Christ in *Milton and His Epic Tradition* (Seattle: University of Washington Press, 1979), pp. 178, 192.

24. *Theogony*, trans. Norman O. Brown (Indianapolis: Bobbs-Merrill, 153), pp. 59–62.

25. Certain of my observations about the poem have been anticipated in J. B. Broadbent's "The Private Mythology of Paradise Regained," in Wittreich, *Calm of Mind*, pp. 77–92. Although sympathetic with efforts to connect Milton and Freud, my delight over this iconoclastic essay is marred by what Freud called "wild psychoanalysis" and what might just as well be called, for lack of another term, "wild anthropology."

26. Roy Wagner, "The Talk of Koriki: A Daribi Contact Cult," *Social Research*, 46 (1979), 160.

27. M.-D. Chenu, *Nature, Man, and Society in the Twelfth Century*, trans. Jerome Taylor and Lester Little (Chicago: University of Chicago Press, 1968), pp. 119–121. What follows is also indebted to Dan Sperber's *Rethinking Symbolism*, trans. Alice Morton (Cambridge: Cambridge University Press, 1974).

28. Augustine, *On Christian Doctrine*, trans. D. W. Robertson, Jr. (New York: Bobbs-Merrill, 1958), pp. 87–88.

29. *CP* VI, 136.

30. Chenu, *Nature, Man, and Society*, pp. 123–138.

31. Paul Ricoeur, *The Rule of Metaphor: Multi-Disciplinary Studies of the Creation of Meaning in Language*, trans. Robert Czerny (Toronto: University of Toronto Press, 1977). A shorter version of the argument may be found in Ricoeur's *Interpretation Theory: Discourse and the Surplus of Meaning* (Fort Worth: Texas Christian University Press, 1976), pp. 45–70.

32. E. D. Hirsch, Jr., *The Aims of Interpretation* (Chicago: University of Chicago Press, 1976), pp. 1–16.

33. Lewalski, *Milton's Brief Epic*, pp. 227–241, 318–319.

34. See Hans-Georg Gadamer, *Truth and Method* (New York: Seabury, 1975), pp. 331–332.

35. Stanley Fish, *Surprised by Sin* (New York: St. Martin's, 1967). I am about to propose an interpretation of the symbolic God in *Paradise Lost* similar to Denis Saurat's in *Milton, Man and Thinker*, pp. 130–133. Tillyard seemed to endorse this idea in *Milton*, p. 193; see also Alastair Fowler's sug-

gestive remark on 5.117 in Fowler and John Carey, eds., *The Poems of John Milton* (London: Longmans, 1968), and E. H. Visiak, *The Portent of Milton* (New York: Humanities, 1968), p. 134.

36. Ricoeur, *Symbolism of Evil*, p. 243.

37. Ibid., pp. 171–210, 232–260.

38. Albert Cirillo, "Noon-Midnight and the Temporal Structure of *Paradise Lost*," *ELH*, 29 (1962), 372–395; see Fowler's notes to 3.555f., 10.668f., and pp. 446–450 in the edition cited in Note 35. I discuss the passage on Lucifer in *The Prophetic Milton*, pp. 229–231.

39. Woodhouse, *Heavenly Muse*, p. 322.

40. *S.E.* IV, 261; VII, 195; IX, 135. Géza Róheim, emphasizing the confusion about the number of legs, associates the riddle with the primal scene (discussed in the next chapter) and notes that riddles are often found at the gateway to sexuality in myth and folklore; see *The Riddle of the Sphinx*, trans. R. Money-Kyrle (New York: Harper, 1974), pp. 1–9. Lévi-Strauss argues for a structural affinity between incest and riddling in *The Scope of Anthropology*, trans. Sherry Paul and Robert Paul (London: Jonathan Cape, 1967), pp. 35–39.

41. Watkins, *An Anatomy of Milton's Verse*, p. 116.

42. On the distinction between need and desire see Jean Leplanche, *Life and Death in Psychoanalysis*, trans. Jeffrey Mehlman (Baltimore: Johns Hopkins University Press, 1976), pp. 8–24, 74–75; Jean Leplanche and J.-B. Pontalis, *The Language of Psycho-Analysis*, trans. Donald Nicholson-Smith (New York: Norton, 1973), pp. 481–483. Although I have not used these terms with technical precision, I have kept the spirit of the distinction.

43. See Dunster's note in Todd, *Poetical Works*, pp. 105–106.

44. See Lacan, "The Signification of the Phallus," in *Écrits*, pp. 281–291.

45. Tillyard, *Milton*, p. 239.

46. John Smith, "The Excellency and Nobleness of True Religion," in C. A. Patrides, ed., *The Cambridge Platonists* (Cambridge, Mass.: Harvard University Press, 1970), p. 161.

47. John Smith, "The True Way of Attaining to Divine Knowledge," in ibid., p. 134.

48. Like Tayler (*Milton's Poetry*, p. 261) and others, I find a pronounced emphasis on will and intuition in the last two poems; Northrop Frye has stated the matter succinctly in *The Return of Eden* (Toronto: University of Toronto Press, 1973), pp. 96–98. Lewalski (*Milton's Brief Epic*) has taught all of us to hear the echoes of Job in the poem, and it is astonishing to realize that Milton the justifier turned for inspiration to the book that offers the most powerful refutation we have of the theodicial project of *Paradise Lost*. It is significant that while Christ scorns the thought of Socrates, he has praise for the man (3.96–99). Life, not doctrine, is central to the latest Milton.

49. Saurat, *Milton, Man and Thinker*, p. 237.

50. G. F. Waller, "Transition in Renaissance Ideas of Time and the Place

of Giordano Bruno," *Neophilologus*, 55 (1971), 5.

51. Giordano Bruno psychoanalyzed the widespread contempt for matter in his culture as a case of misogyny in his *Cause, Principle, and Unity*, trans. Jack Lindsay (New York: International Publications, 1962), pp. 59–60, 74–75, and esp. 117–121.

52. My "Heretical Milton," pp. 144–149, contains a fuller discussion.

53. Hans Loewald, *Psychoanalysis and the History of the Individual* (New Haven: Yale University Press, 1978), pp. 55–77. Loewald might reply, and perhaps correctly, that the concept of eternity is not yet the idea of an afterlife. I still am not convinced that this concept derives from an intuition about the unconscious.

54. Arthur Barker is interesting on the design of the volume and its title page in "Calm Regained Through Passion Spent," in Balachandra Rajan, ed., *The Prison and the Pinnacle* (Toronto: University of Toronto Press, 1973), pp. 12–20.

4. The Way to Strength from Weakness

1. Jonathan Richardson, *Explanatory Notes and Remarks on Milton's Paradise Lost* (London: Knapton, 1734), p. cxix.

2. See Maurice Kelley, *This Great Argument: A Study of Milton's De Doctrina Christiana as a Gloss upon Paradise Lost* (Princeton: Princeton University Press, 1941), pp. 8–24.

3. Phillips told Aubrey that the epic was begun "about 2 yeares before the King came in, and finished about 3 yeares after the King's restoration" (*EL*, 13). Phillips confirmed the second date in his life of Milton, adding that "he finisht his noble Poem" after marrying Elizabeth Minshull and moving "to a House in the Artillery-walk" (*EL*, 75); but aside from placing it sometime after Milton composed his last pamphlet against More (published in 1655), he did not assign a precise date to the beginning of the labor. The case for 1658 is one seal short of being airtight. Scholars have, nonetheless, pushed the beginning back as far as 1655–56, even 1652 and earlier.

The simplest hypothesis is that Aubrey correctly transcribed "about 2 yeares before the King came in" from the testimony of one who must have known. It is worth remembering that Aubrey also recorded from Phillips that the epic was written in "4 or 5 yeares" (*EL*, 13): anyone assuming that Aubrey mistook Phillips must assume that he made the same error twice. Rough calculations suggest that, working about half the year and producing around 100 lines a week, the epic would have taken precisely four or five years. If Phillips in his own text did not confirm 1658, he said nothing contradicting 1658, which is the date accepted in this book.

4. Anticipating what has come to be known as the hermeneutic circle, Hugh of St. Victor distinguished the "order of knowledge" from the "order of time" in reading the Bible—*Didascalion*, trans. Jerome Taylor (New York:

Columbia University Press, 1961), p. 145. Frances Yates includes a chapter on Ramism in *The Art of Memory* (Chicago: University of Chicago Press, 1966), pp. 231–242.

5. The doctrine of climate has been ably studied by Z. S. Fink, "Milton and the Theory of Climatic Influence," *MLQ*, 2 (1941), 67–80.

6. Richardson, *Explanatory Notes*, pp. cxiii–cxiv.

7. John Donne's "A Nocturnall upon S. Lucies Day," l. 18.

8. Parker, *Milton*, I, 479.

9. Dilthey's classic paper on "The Rise of Hermeneutics" has been translated by Fredric Jameson in *New Literary History*, 3 (1972), 229–244. The most extreme development of the idea is doubtless the philosophy of Heidegger, for whom the hermeneutic circle, underlying perceiving and thinking, is the ontological structure of our Being.

10. Jackson Cope, *The Metaphoric Structure of Paradise Lost* (Baltimore: Johns Hopkins University Press, 1962), p. 120.

11. This is the view of creativity and loss implied by analysts so otherwise diverse as Kurt Eissler, *Talent and Genius* (New York: Quadrangle, 1971), pp. 248–320, and Heinz Kohut, *The Restoration of the Self* (New York: International Universities Press, 1977), pp. 1–62.

12. See Ernst Cassirer, *The Individual and the Cosmos in Renaissance Philosophy*, trans. Mario Domandi (Philadelphia: University of Pennsylvania Press, 1963), pp. 7–72.

13. From Meditation III in *Descartes: Selections*, ed. Ralph Eaton (New York: Scribner's, 1927), p. 125.

14. As David Morris has shown in *The Religious Sublime: Christian Poetry and Critical Tradition in 18th-Century England* (Lexington: University of Kentucky Press, 1972), the new interest in Longinus issued in a literary commentary focused on the specific "beauties" of a work, cutting against the old emphasis on the entireties of genre, plot, and decorum.

15. Marjorie Nicolson, *The Breaking of the Circle*, rev. ed. (New York: Columbia University Press, 1960), p. 186.

16. Thomas Browne, *Religio Medici* II.11 in *Religio Medici and Other Works*, ed. L.C. Martin (Oxford: Oxford University Press, 1964), p. 70.

17. "Upon Appleton House," Stanza LXXI.

18. This plate is reproduced, with commentary, in S. K. Heninger, *Touches of Sweet Harmony: Pythagorean Cosmology and Renaissance Poetics* (San Marino, Calif.: Huntington Library, 1974), p. 191. See also the chapter on "Pyramids and Monochords" in Joscelyn Godwin, *Robert Fludd: Hermetic Philosopher and Surveyor of Two Worlds* (Boulder, Col.: Shambala, 1979), pp. 42–53.

19. *Timaeus* 50–56; the diagrams and commentary of Desmond Lee are helpful in *Timaeus and Critias* (London: Penguin, 1965), pp. 68–77.

20. Richardson, *Explanatory Notes*, p. 88.

21. *Enneads*, I.6.1. in *The Essential Plotinus*, trans. Elmer O'Brien (New

York: New American Library, 1964), p. 35. Later we read: "fire glows with a beauty beyond all other bodies, for fire holds the rank of Idea in their regard" (p. 37).

22. See the discussions in Alexandre Koyré, *From the Closed to the Infinite Universe* (Baltimore: Johns Hopkins Press, 1957), pp. 46–47, and Paul-Henri Michel, *the Cosmology of Giordano Bruno* (Ithaca, N.Y.: Cornell University Press, 1973), pp. 67–68, 160–165.

23. Rosalie Colie, *Paradoxia Epidemica: The Renaissance Tradition of Paradox* (Princeton: Princeton University Press, 1966), pp. 176–177. Isabel MacCaffrey's views, stressing the nature of imagination, are closer to my own in "The Theme of *Paradise Lost*, Book III," in Thomas Kranidas, ed., *New Essays on Paradise Lost* (Berkeley: University of California Press, 1969), pp. 80–85.

24. Jonas, *Phenomenon of Life*, p. 145. "I have nothing to do but to look, and the object is not affected by that: and once there is light, the object has only to be there to be visible, and I am not affected by that: and yet it is apprehended in its self-containment from out of my own self-containment, it is present to me without drawing me into its presence" (p. 146). This is consistent with the hierarchy of the senses familiar in Renaissance thought (in Ficino, for example), where sight is the noblest sense because at the greatest distance from its object.

25. Arthur Schopenhauer, *The World as Will and Representation* III.39, trans. E. F. Payne, 2 vols. (New York: Diver, 1969), I, 206.

26. Ibid., p. 205.

27. Marsilio Ficino, *De Sole* XII, in *Renaissance Philosophy*, ed. and trans. Arturo Fallico and Herman Shapiro, 2 vols. (New York: Random House, 1967), I, 136.

28. David C. Lindberg, *John Pecham and the Science of Optics* (Madison: University of Wisconsin Press, 1970), pp. 40–41.

29. Nicholas H. Steneck, *Science and Creation in the Middle Ages* (Notre Dame, Ind.: University of Notre Dame Press, 1976), pp. 47–53; the most influential text in this regard was Aristotle, *De Anima* 2.7.13–19, where light is defined as the form of a luminous body.

30. For a spirited modern defense of beauty as a Platonic *methexis* see Hans-Georg Gadamer, *Truth and Method* (New York: Seabury, 1975), pp. 431–447.

31. Ludwig Wittgenstein, *Tractatus Logico-Philosophicus*, trans. D. F. Pears and B. F. McGuinness (London: Routledge, 1963); the theory of the "Picture" is sketched from 2.1 to 3.1.

32. Jacques Derrida, "White Mythology: Metaphor in the Text of Philosophy," *New Literary History*, 6 (1974), 5–74. See also Ronald Bruzina, "Heidegger on the Metaphor and Philosophy," in Michael Murray, ed., *Heidegger and Modern Philosophy* (New Haven: Yale University Press, 1978), pp. 184–200.

33. Heidegger, *Being and Time*, pp. 24–28, 256–273; also Heidegger's essay "The Origin of the Work of Art," in *Poetry, Language, Thought*, trans. Albert Hofstadter (New York: Harper and Row, 1971), pp. 57–78.

34. Browne, *Religio Medici* I.32 (Martin, p. 31).

35. Kelley, *This Great Argument*, p. 94. Of the immense literature on Milton and the metaphysics of light, the following items are particularly helpful: Don Cameron Allen, *Harmonious Vision: Studies in Milton's Poetry*, enlarged ed. (Baltimore: Johns Hopkins Press, 1970), pp. 122–142; Hughes, "Milton and the Symbol of Light," in *Ten Perspectives*, pp. 63–103; Walter Clyde Curry, *Milton's Ontology, Cosmology, and Physics* (Lexington: University of Kentucky Press, 1957), pp. 114–143, 189–204; A. B. Chambers, "Wisdom at One Entrance Quite Shut Out," *PQ*, 42 (1963), 114–119. I regret that Michael Lieb's treatment of this subject in *Poetics of the Holy: A Reading of Paradise Lost* (Chapel Hill: University of North Carolina Press, 1981), pp. 185–210, 212–245, reached me after this chapter was completed, for the materials he has gathered allow one to appreciate more fully than before the tension between philosophical and biblical traditions in the opening lines of the invocation.

36. William Hunter, "The Meaning of 'Holy Light' in *Paradise Lost*," *MLN*, 74 (1959), 589–592. The essay appears, expanded, under the title "Milton's Muse" in Hunter, C. A. Patrides, J. H. Adamson, *Bright Essence: Studies in Milton's Theology* (Salt Lake City: University of Utah Press, 1971), pp. 149–158.

37. C. S. Lewis, *A Preface to Paradise Lost* (Oxford: Oxford University Press, 1969), pp. 82–93; C. A. Patrides, "The Godhead in *Paradise Lost*: Dogma or Drama," *JEGP*, 64 (1965), 29–34, which also appeared, expanded, in *Bright Essence*, pp. 71–80.

Although it has antecedents, Milton's Godhead is an original religious idea, and *Paradise Lost* is the most profound presentation of this idea. Anyone who in legalistic fashion comes to the poem looking for language that an orthodox Trinitarian could assent to, will find it; part of Milton's argument is that the biblical language for the Godhead, if properly understood, proves *his* idea. But this reader, by taking the language in the traditional way, misses altogether the fresh union of dogma and drama in Milton's meaning. Troubling tradition, a heresy such as Milton's must be understood from the bottom up, all its ramifications explored. Its comprehension invites, in effect, a rethinking of Christianity.

38. A recent expositor of this position is Naseeb Shoheen, "Milton's Muse and the *De Doctrina*," *Milton Quarterly*, 8 (1974), 72–76; see also Karl Winegardner, "No Hasty Conclusions," *Milton Quarterly*, 12 (1978), 101–107.

39. Nicholas J. Perella, *The Kiss Sacred and Profane* (Berkeley: University of California Press, 1969), pp. 53–54, 253–259.

40. Victor Turner's brilliant "Betwixt and Between: The Liminal Period in *Rites de Passage*" appears in *The Forest of Symbols* (Ithaca, N.Y.: Cornell

University Press, 1967), pp. 93–111. For his subsequent development of this concept, see Victor Turner, *The Ritual Process: Structure and Anti-Structure* (Chicago: Aldine, 1969), pp. 94–130; *Dramas, Fields, and Metaphors* (Ithaca, N.Y.: Cornell University Press, 1974), pp. 231–288; and with Edith Turner, *Image and Pilgrimage in Christian Culture* (New York: Columbia University Press, 1977).

Liminality, as Turner repeatedly emphasizes, is not a state. Its ultimate model is time—the indeterminate present, forever between the "no longer" of the remembered past and the "not yet" of the anticipated future. Milton, poet of time's motion, refused to hypostasize or "stop" the portion of the traditional Godhead devoted to this principle.

41. Heiko A. Oberman, "Some Notes on the Theology of Nominalism with Attention to its Relation to the Renaissance," *Harvard Theological Review*, 53 (1960), 47–76, explicates this distinction, which is also related to early Renaissance humanism by Charles Trinkaus, "*In Our Image and Likeness*," 2 vols. (London: Constable, 1970), I, 59–102.

42. For a general discussion of the antipathy between Greek and Christian thought, consult W. J. Verdenius, "Plato and Christianity," *Ratio*, 5 (1963), 15–32. Milton's God cannot, however, will a contradiction (*CP* VI, 146). Otherwise he is "utterly free," and not to be described, therefore, as an "Actus Purus" or actuality; for this reason he includes, as the potentiality that uncircumscribes his being, the force of creative motion normally attributed to the Holy Spirit.

43. *Descartes: Selections*, p. 121.

44. See the selections conveniently grouped together in A. H. Armstrong, *Plotinus* (New York: Collier, 1962), pp. 65–83; also his *An Introduction to Ancient Philosophy* (London: Methuen, 1977), pp. 186–187: in the first moment, the Nous "is radiated as an unformed potentiality" without content; this answers to the light addressed at the beginning of the invocation, not to be hypostasized as Holy Spirit.

45. Stella Revard, "The Dramatic Function of the Son in *Paradise Lost*: A Commentary on Milton's 'Trinitarianism,' " *JEGP*, 66 (1967), 45–58. Hugh MacCallum has followed in her footsteps in " 'Most Perfect Hero': The Role of the Son in Milton's Theodicy," in *Paradise Lost: A Tercentenary Tribute*, ed. Balachandra Rajan (Toronto: University of Toronto Press, 1969), pp. 79–105.

46. Bloom, *A Map of Misreading*, p. 138.

47. See John Swan, *Speculum Mundi* (Cambridge, 1635), p. 301. This commonplace also appears in Andrew Willet, *Hexapla in Genesin* (London, 1632), p. 31.

48. Watkins, *An Anatomy of Milton's Verse*, pp. 31–41.

49. Edmund Bergler, *The Basic Neurosis: Oral Aggression and Psychic Masochism* (New York: Grune and Stratton, 1949), p. 189. See also David Allen, *The Fear of Looking or Scopophilic-Exhibitionistic Conflicts* (Char-

lottesville: University Press of Virginia, 1974).

50. Jean-François Lyotard, "The Psychoanalytic Approach," a contribution to Mikel Dufrenne, *Main Trends in Aesthetics and the Sciences of Art* (New York: Holmes and Meier, 1979), p. 146. Dufrenne himself presents a nicely judged critique of this view of art, psychoanalysis, and their relationship on pp. 213–222.

51. William Hunter, "Milton's Materialistic Life Principle," *JEGP*, 45 (1946), 68–76; Michael Lieb, "Milton and the Metaphysics of Form," *SP*, 71 (1974), 206–224.

52. See Laplanche and Pontalis, *Language of Psycho-Analysis*, pp. 111–114, where credit is given to Lacan for calling attention to this notion. In recent years the most ambitious papers on the primal scene have been those of Henry Edelheit—"Mythopoiesis and the Primal Scene," *Psychoanalytic Study of the Child*, 5 (1972), 213–222, and "Crucifixion Fantasies and their Relation to the Primal Scene," *International Journal of Psycho-analysis*, 55 (1974), 193–99. The fullest discussion in Freud is the Wolf Man case, "From the History of an Infantile Neurosis" (*S.E.* XVII). See also A. Esman, "The Primal Scene: A Review and a Reconsideration," *Psychoanalytic Study of the Child*, 28 (New York: International Universities Press, 1973), 49–81; Wayne A. Myers, "Split Self-Representation and the Primal Scene," *Psychoanalytic Quarterly*, 42 (1973), 525–538, and "Clinical Consequences of Primal Scene Exposure," *Psychoanalytic Quarterly*, 47 (1979), 1–26; and Harold Blum, "On the Concept and Consequences of the Primal Scene," *Psychoanalytic Quarterly*, 48 (1979), 27–47.

53. Tzvetan Todorov, "On Linguistic Symbolism," *New Literary History*, 6 (1974), 134.

54. Laplanche and Pontalis, "Fantasty and the Origins of Sexuality," *International Journal of Psycho-analysis*, 49 (1968), 1–18.

55. Thomas Weiskel, *The Romantic Sublime: Studies in the Structure and Psychology of Transcendence* (Baltimore: Johns Hopkins University Press, 1976). See also Neil Hertz, "The Notion of Blockage in the Literature of the Sublime," in Geoffrey Hartman, ed., *Psychoanalysis and the Question of the Text* (Baltimore: Johns Hopkins University Press, 1978), pp. 62–85.

56. Heinz Kohut, *The Analysis of the Self* (New York: International Universities Press, 1971), pp. 1–34.

57. Fish, *Surprised by Sin*, esp. pp. 38–56; Tayler, *Milton's Poetry*, pp. 66–68.

58. Marcel Mauss, *The Gift: Forms and Functions of Exchange in Archaic Societies*, trans. Ian Cunnison (New York: Norton, 1967), pp. 37–41.

59. "Performative Utterances" in J. L. Austin, *Philosophical Papers*, ed. J. O. Urmson and G. J. Warnock (Oxford: Oxford University Press, 1961), pp. 233–252.

60. See Heidegger's "'Poetically Man Dwells,'" in *Poetry, Language, Thought*, pp. 213–229.

61. Note to 3.26f. in *Paradise Lost, Books III–IV*, ed. Lois Potter and John

Broadbent (Cambridge: Cambridge University Press, 1976), p. 49.

62. Kerrigan, *The Prophetic Milton*, p. 132.

63. *The Interpretation of Dreams* in *S.E.* V, 576.

64. Parker, *Milton*, I, 155–157.

65. I refer to the indispensable Ernst Robert Curtius, *European Literature and the Latin Middle Ages*, trans. Willard Trask (New York: Harper, 1953), pp. 302–348.

66. Reproduced in Godwin, *Robert Fludd*, p. 24. I claim no direct allusion. But in Fludd's blank picture of the nothingness that became everything I believe that I behold, imaginatively, Milton's blindness.

67. *Religio Medici*, I.16 (Martin, p. 15). For Milton, "the universall and public manuscript, that lies expans'd unto the eyes of all" was truly, as the man says, *sous rature*.

68. See Bertram Lewin's "Sleep, the Mouth, and the Dream Screen," *Psychoanalytic Quarterly*, 15 (1946), 419–433, and the historical study by Wilfred Abse, "The Dream Screen: Phenomenon and Noumenon," *Psychoanalytic Quarterly*, 46 (1977), 256–286.

69. One of the earliest and best expositions of this theme is by Joseph Summers, *The Muse's Method: An Introduction to Paradise Lost* (Cambridge, Mass.: Harvard University Press, 1970), pp. 176–185.

70. Patrides believes, with considerable textual evidence to support him, that "when we encounter the Godhead in action beyond the confines of heaven, the distinction between the two Persons is arrested abruptly" (*Bright Essence*, p. 74); put another way, however, what Milton portrays is Christ manifesting the Father, which is consistent with his subordinationalism.

71. "The Origin and Structure of the Superego," in Henrik Ruitenbeck, ed., *Heirs to Freud* (New York: Grove Press, 1966), p. 35. The next quotation appears on p. 38. See also Peter Blos, "The Genealogy of the Ego Ideal," in *The Psychoanalytic Study of the Child*, 29 (1974), 43–88, for a theory about the origins of the ego-ideal in the negative oedipus complex.

5. "One First Matter All": Spirit as Energy

1. Ricardo Quinones, in *The Renaissance Discovery of Time* (Cambridge, Mass.: Harvard University Press, 1972), pp. 444–493, and Le Comte, *Milton's Unchanging Mind*, pp. 5–69, have written well about Milton and fame.

2. On Milton and the Hartlib group, consult Charles Webster, ed., *Samuel Hartlib and the Advancement of Learning* (Cambridge: Cambridge University Press, 1970), pp. 41–43.

3. *Renaissance Letters*, ed. Robert J. Clements and Lorna Levant (New York: New York University Press, 1976), p. 180.

4. *Mataeotechnia Medicinae Praxeus: The Vanity of the Craft of Physick* (London, 1651), sig. a4. The chemical philosophy in Puritan thought has been explored by P. M. Rattansi, "Paracelsus and the Puritan Revolution," *Ambix*,

11 (1963), 24–32; Allen Debus, ed., *Science and Education in the Seventeenth Century: The Webster-Ward Debate* (London: Macdonald, 1970); Audrey B. Davis, *Circulation Physiology and Medical Chemistry in England, 1650–1680* (Lawrence, Kans.: Coronado, 1973); Charles Webster, *The Great Instauration: Science, Medicine, and Reform, 1626–1660* (New York: Holmes & Meier, 1975), pp. 32–99, 122–129, 246–323, 384–402; J. R. Jacob, *Robert Boyle and the English Revolution: A Study in Social and Intellectual Change* (New York: Burt Franklin, 1977).

5. Though traditionally attributed to Hartlib, *A Description of the Famous Kingdome of Macaria* is almost certainly by Gabriel Plattes, as Charles Webster has shown in "The Authorship and Significance of *Macaria*," *Past and Present*, 56 (1972), 34–48.

6. James Holly Hanford, "Dr. Paget's Library," *Bulletin of the Medical Library Association*, 33 (1945), 91–99; Christopher Hill, *Milton and the English Revolution*, pp. 492–495. An auction catalogue of Paget's library, *Bibliotheca . . . Nathanis Paget*, was printed in London in 1681. Oldenburg, active in the founding of the Royal Society, corresponded with Boyle and other members of the Harlib group as well as with Milton (Jacob, *Robert Boyle*, pp. 124–125, 153–158); the Palatine exile Theodore Haak participated in the mysterious "London group" of the 1640s, championed the program of Comenius, and translated half of *Paradise Lost* into Dutch.

7. Quoted from Hughes, ed., *Complete Poetry and Major Prose*, p. 625.

8. Kester Svendsen, *Milton and Science* (Cambridge, Mass.: Harvard University Press, 1956), pp. 209–210; see also his "Milton and Medical Lore," *Bulletin of the History of Medicine* (*BHM* hereafter), 13 (1943), 158–184.

9. As will become evident, to be conversant in the medicine of the period was to know something about the occult arts. Christopher Hill has recently suggested that "there is a book to be written" on Milton and Fludd (*Milton and the English Revolution*, p. 6), a subject broached some years ago by Saurat (*Milton, Man and Thinker*, pp. 301–309). See also William Hunter, "Milton and Thrice Great Hermes," *JEGP*, 45 (1946), 327–336.

10. Sister Mary Irma Corcoran, *Milton's Paradise Lost with Reference to the Hexameral Background* (Washington: Catholic University of America Press, 1945), p. 125; John Weemse, *The Portraiture of the Image of God in Man* (London, 1639), pp. 31–35.

11. Most of the arguments summarized and interpreted here may be found in Book I, Chapters VII and XIII, of the *Christian Doctrine* (*CP* VI, 299–325, 399–414). Of the many commentaries, I have been enlightened by these particularly: Marjorie Nicolson, "The Spirit World of Milton and More," *SP*, 22 (1925), 433–452; William Hunter, "Milton's Materialistic Life Principle," *JEPG*, 45 (1946), 68–76; A. S. P. Woodhouse, "Notes on Milton's Views on the Creation: The Initial Phases," *PQ*, 28 (1949), 211–236; Walter Clyde Curry, *Milton's Ontology, Cosmogony and Physics* (Lexington: University of Kentucky Press, 1957); Harry F. Robins, *If This Be Heresy: A Study of Milton*

and Origin, Illinois Studies in Language and Literature, No. 51 (Urbana: University of Illinois Press, 1963).

12. Bruno, *Cause, Principle, and Unity*, p. 123. Bruno attributes this belief to Plotinus, who did indeed argue that the Nous contains ideas of individuals. For a full exposition of this point, to which I am indebted, see John Peter Rumrich, "Matter of Glory: Motivation in Paradise Lost" (unpub. diss., University of Virginia, 1981), pp. 69–84.

13. *Two Treatises Concerning Eie-Sight* (Oxford, 1616), p. 14.

14. *A Treatise*, Chapter 1, Section 9. The point about vaporous digestion is repeated with greater emphasis in Sect. 1 of Chapter 6. My favorable assessment of Banister derives from Arnold Sorsby, "Richard Banister and English Ophthalmology," in *Science, Medicine, and History: Essays . . . in Honour of Charles Singer*, ed. E. Ashworth Underwood, 2 vols. (London: Oxford University Press, 1953), II, 42–55. Here and elsewhere in this book the conception of Milton's blindness depends on the Renaissance idea of *gutta serena*, not on the much-debated question of what he actually suffered from. For that subject consult Arnold Sorsby, "On the Nature of Milton's Blindness," *British Journal of Ophthalmology*, 14 (1930), 339–354; W. H. Wilmer, "The Blindness of Milton," *JEGP*, 32 (1933), 301–315; Eleanor G. Brown, *Milton's Blindness* (New York: Columbia University Press, 1934); and William Hunter, "Some Speculations on the Nature of Milton's Blindness," *Journal of the History of Medicine*, 17 (1962), 333–341. The weight of opinion favors glaucoma.

15. Browne, *Religio Medici and Other Works*, p. 3; *The Vanity of Dogmatizing* (London: Henry Eversden, 1661), p. 248. For the general reputation of physicians, see Herbert Silvette, *The Doctor on the Stage: Medicine and Medical Men in Seventeenth-Century England* (Knoxville: University of Tennessee Press, 1967).

16. I will argue against the interpretation of this passage proposed by James Holly Hanford in "John Milton Forswears Physic," *Bulletin of the Medical Library Association*, 32 (1944), 23–34.

17. Paget, besides being about the age Diodati would have been, had a library that reveals a keen interest in poetry. See Note 6 to this chapter.

18. French, *Life Records*, V, 220. The information comes from the deposition of the maid during the contesting of Milton's will.

19. J. V. Cunningham, *Tradition and Poetic Structure* (Denver: Alan Swallow, 1960), p. 17.

20. T. S. Eliot, *Selected Essays* (New York: Harcourt Brace, 1960), p. 17; Christopher Ricks, *Milton's Grand Style* (Oxford: Oxford University Press, 1963), p. 111.

21. *The Letters of Marsilio Ficino*, Vol. 1, trans. School of Economic Science (London: Clowes & Sons, 1975), p. 128. Mather's work has been edited by Gordon Jones (Charlottesville: University Press of Virginia, 1972).

22. For the Spenserian parallel, see James Nohrnberg, *An Analogy of the*

Faerie Queene (Princeton: Princeton University Press, 1976), pp. 68–76, 238–251, and passim. As usual, Milton exceeds the earlier poet in the clarity of his design, even when the design at issue is partially unconscious.

23. A. J. A. Waldock, *Paradise Lost and Its Critics* (Cambridge: Cambridge University Press, 1961), pp. 109, 114.

24. C. S. Lewis, *A Preface to Paradise Lost* p. 113; Joseph Summers, *The Muse's Method*, p. 93. The sort of fantasy that Milton released in his portrayal of angelic sexuality would not have been alien to Montaigne, who remarked in "Of Experience" that it would be nice if we could beget children "voluptuously with our fingers and heels." *The Complete Essays of Montaigne*, trans. Donald Frame (Stanford: Stanford University Press, 1965), p. 855.

25. On the applicability of "science fiction" to the study of older literature, see Kent Kraft, "Incorporating Divinity: Platonic Science Fiction in the Middle Ages," in *Bridges to Science Fiction*, ed. George Slusser, George Guffey, and Mark Rose (Carbondale: Southern Illinois University Press, 1980), pp. 22–40.

26. Nicolson, "The Spirit World of Milton and More" (cited in Note 11 to this chapter) and "Milton and the *Conjectura Cabbalistica*," *SP*, 6 (1927), 1–18; Lewis, *A Preface*, pp. 108–115.

27. For an expansion of this argument, see my "The Articulation of the Ego in the English Renaissance," pp. 290–303.

28. The same skepticism can be found at 3.598–605, where once again the point is not that alchemy is nonsense, but rather that nature has already accomplished the alchemical wonders—in this case, the creation of the philosopher's stone.

29. Thomas Vaughan, *Lumen de Lumine: or, A New Magicall Light (London:* 1651), sig. C4.

30. Robert West, *Milton and the Angels* (Athens: University of Georgia Press, 1955), p. 183.

31. The following discussion, like all the remarks about medical history in this book, is greatly indebted to Walter Pagel, especially *Paracelsus: An Introduction to Philosophical Medicine in the Era of the Renaissance* (Basel: S. Karger, 1967), pp. 277–281.

32. Quoted by Robert P. Multhauf in "Van Helmont's Reformation of the Galenic Doctrine of Digestion," *BHM*, 29 (1955), 155.

33. Quoted from James Winney, ed., *The Frame of Order* (London: Allen & Unwin, 1957), p. 55.

34. "Alchemical and Paracelsian Medicine," in Charles Webster, ed., *Health, Medicine and Mortality in the Sixteenth Century* (Cambridge: Cambridge University Press, 1979), p. 323. F. N. L. Poynter considers the rash of alchemical translation during the 1650s in "Nicholas Culpepper and the Paracelsians," in Allen Debus, ed., *Science, Medicine, and Society in the Renaissance: Essays in Honor of Walter Pagel*, 2 vols. (New York: Neale Watson, 1972), I, 201–220.

35. *Lectures on the Whole Anatomy*, trans. C. D. O'Malley, F. N. L. Poynter, K. F. Russell (Berkeley: University of California Press, 1961), p. 215.

36. Walter Pagel has provided us with an impressive chronicle of Van Helmont's advances. See *The Religious and Philosophical Aspects of Van Helmont's Science and Medicine*, Supplements to the Bulletin of the History of Medicine, No. 2 (Baltimore: Johns Hopkins Press, 1944); "J. B. Van Helmont's Reformation of the Galenic Doctrine of Digestion—and Paracelsus," *BHM*, 29 (1955), 563–568; "Van Helmont's Ideas on Gastric Digestion and the Gastric Acid," *BHM*, 30 (1956), 524–536; "The Wild Spirit (Gas) of Van Helmont and Paracelsus," *Ambix*, 10 (1962), 1–13.

37. Along with Webster, Rattansi, and Davis in Note 4 to this chapter, see Rattansi's "The Helmontian-Galenist Controversy in Restoration England," *Ambix*, 12 (1964), 1–23.

38. Webster, *Great Instauration*, p. 139.

39. *Metaphysics* A.982a, trans. Hippocrates Apostle (Bloomington: Indiana University Press, 1966), p. 15.

40. *Docta Ignorantia*, I, Prologue (19–23).

41. Pico della Mirandola, *On the Dignity of Man, On Being and the One, Heptaplus*, trans. Charles Wallis, Paul Miller, Douglas Carmichael (New York: Bobbs-Merrill, 1965), p. 3. The Milton of Prolusion VII and "Il Penseroso," in his energetic pursuit of universal knowledge, could almost be imitating Pico.

42. Ibid., p. 7.

43. Ricks, *Milton's Grand Style*, p. 17.

44. *The Analogy of the Faerie Queene* (cited in Note 22 to this chapter). My ideas about the enfolded sublime are meant to counter strictures against the long poem offered by Yvor Winters in *The Function of Criticism* (Denver: Alan Swallow, 1957), pp. 40–47.

45. William Madsen, *From Shadowy Types to Truth*, pp. 88–89.

46. For more detail, see Kerrigan, "The Fearful Accommodations of John Donne," 40–50.

47. I refer to Ryle's attack on the Cartesian "orthodoxy" in *The Concept of Mind* (New York: Barnes & Noble, 1949), pp. 11–24. For the adversary relationship of Cudworth to both Descartes and Hobbes, see J. A. Passmore, *Ralph Cudworth: An Interpretation* (Cambridge: Cambridge University Press, 1962), pp. 63–133, and Meyrick Carré, "Ralph Cudworth," *Philosophical Quarterly*, 3 (1953), 342–51.

48. Pagel, "Medieval and Renaissance Contributions to Knowledge of the Brain and Its Functions," in *The History and Philosophy of Knowledge of the Brain and Its Functions* (Springfield, Ill.: Charles Thomas, 1958), pp. 95–114; Charles Singer, "Some Galenic Sources and Animal Sources of Vesalius," *Journal of the History of Medicine*, 1 (1946), 6–24.

49. Pagel, "Medieval and Renaissance Contributions," pp. 100–101.

50. Ibid., p. 112.

51. Michael Lieb, *The Dialectics of Creation* (Amherst: University of Massachusetts Press, 1970).

52. Watkins, *An Anatomy of Milton's Verse*, pp. 15–16.

53. Ricks, *Milton's Grand Style*, p. 76.

54. *Galen on the Passions and Errors of the Soul*, trans. Paul W. Harkins, ed. Walter Riese (Columbus: Ohio State University Press, 1963), pp. 14–16; Charles Singer notes the difficulties of this doctrine for Christian medicine in his edition of *Vesalius on the Human Brain* (London: Oxford University Press, 1972), p. xvii. The best general history of Galenism is Owsei Temkin's *Galenism: Rise and Decline of a Medical Philosophy* (Ithaca, N.Y.: Cornell University Press, 1973).

55. The distinction and the metaphor may be found in Nemesius, *The Nature of Man*, trans. George Wither (London, 1636), pp. 87–89, 126–127, 181–183, 197–198, and also in Daniel Widdowes, *Natural Philosophy* (London, 1631), p. 59.

56. Pierre Charron, *Of Wisdome*, trans. Samson Lennard (London, n.d.), p. 29; Weemse offered the same argument in *The Portraiture of the Image of God*, pp. 51–52.

57. For a brief summary of "mana" see Evans-Pritchard, *Theories of Primitive Religion*, pp. 59–77, 110.

58. Ricoeur, *Freud and Philosophy*, pp. 65–157. A less enthusiastic view of energics has been often expressed within the Freudian tradition—for example, by Roy Schafer, "Heinz Hartmann's Contributions to Psychoanalysis," *International Journal of Psycho-analysis*, 51 (1970), 425–446. Dislike of energics is almost universal in the existential tradition; see Gerald N. Izenberg, *The Existentialist Critique of Freud: The Crisis of Autonomy* (Princeton: Princeton University Press, 1976), pp. 13–69.

59. The pneumatology of Descartes has been studied by Walter Riese in "Descartes's Ideas of Brain Function," in *The History and Philosophy of Knowledge of the Brain*, pp. 115–134. For Bacon, Telesio, and Donio, see Lynn Thorndyke, "The Attitude of Francis Bacon and Descartes Towards Magic and Occult Science," in *Science, Medicine, and History*, ed. Underwood, I, 451–454, and D. P. Walker, "Francis Bacon and Spiritus," in *Science, Medicine and Society in the Renaissance*, ed. Debus, II, 121–130. The second edition of the *Principia* (1713) speculates about a "most subtle spirit which pervades and lies hid in all gross bodies"—*Principia Mathematica*, ed. F. Cajori (Berkeley: University of California Press, 1947), p. 547.

60. D. P. Walker, *Spiritual and Demonic Magic from Ficino to Campanella* (London: Warburg Institute, 1958), pp. 213–236. See also Robert Klein, *Form and Meaning: Writings on Renaissance and Modern Art*, trans. Madeline Jay and Leon Wieseltier (New York: Viking, 1979), pp. 62–85.

61. Alexander Ross, *Medicus Medicatus* (London, 1645), pp. 17–19. This book is largely an attack on the Cartesian pneumatology of Kenelm Digby's

Two Treatises (Paris, 1644). In his *Philosophical Touch-Stone: Or Observations Upon Sir Kenelm Digby's Discourses* (London, 1645), Ross also scored Digby for confusing the several meanings of "spirit" (pp. 67–68). Weemse issued a similar caution in *Portraiture of the Image of God*, pp. 52–57, 107.

62. Lawrence Babb, *The Moral Cosmos of Paradise Lost* (East Lansing: Michigan State University Press, 1970), p. 41. See also Svendsen, *Milton and Science*, p. 181.

63. Kerrigan, "The Heretical Milton," pp. 125–66.

64. Fish, *Surprised by Sin*, pp. 241–254. Granting the arbitrariness of the commandment, Michael McCanles argues that reason could have prevented the fall: it could have known rationally not to try to rationalize the commandment. See *Dialectical Criticism and Renaissance Literature* (Berkeley: University of California Press, 1975), p. 127. It seems to me that what McCanles calls rationalism is exactly what Fish calls voluntarism. Somehow, of course, man had to fall. I think we come closer to the tenor of Milton's attitude toward the commandment when we consider how it is presented to a reader of the poem, which is consistent with Fish's assumption that the choice presented to Adam and Eve is the choice variously presented to the reader. But in my view the reader, who is not a few days old and not caught up in the sweep of great events, but a person able to spend as much time as he wishes in figuring out the coherence of a poem, is in a better position than Adam or Eve to behold the ultimate good sense of this commandment within Milton's design.

65. Masson, *Life*, VI, 535.

66. Quoted from James Thorpe, ed., *Milton Criticism: Selections from Four Centuries* (New York: Octagon, 1966), pp. 81–82.

6. The Place of Rest

1. Louis Martz, *Poet of Exile: A Study of Milton's Poetry* (New Haven: Yale University Press, 1981), pp. 79–94. For Milton's unwavering loyalty to the liberty of conscience, see the recent treatment of the late pamphlets by Robert Thomas Fallon, "Milton in the Anarchy, 1659–1660: A Question of Consistency," *SEL*, 21 (1981), 123–146.

2. French, *Life Records*, V, 425.

3. Tayler, *Milton's Poetry*, p. 87. See also Arthur O. Lovejoy's classic "Milton and the Paradox of the Fortunate Fall," *ELH*, 4 (1937), 161–179, whose full implications, Tayler suggests, "remain even now to be assessed" (p. 242), and Peter F. Fisher, "Milton's Theodicy," *JHI*, 17 (1956), 28–53.

4. M. H. Abrams, *Natural Supernaturalism: Tradition and Revolution in Romantic Literature* (New York: Norton, 1971), pp. 32–70, 141–236.

5. The lecture on Novalis is quoted in Arne Naess, *Four Modern Philosophers*, trans. Alastair Hannay (Chicago: University of Chicago Press, 1968), p. 174.

6. Richardson, *Explanatory Notes and Remarks*, pp. clxii–clxiii.

7. Lewis, *A Preface*, p. 129.

8. Mary Ann Radzinowicz, "'Man as a Probationer of Immortality': *Paradise Lost* XI–XII," in Patrides, ed., *Approaches to Paradise Lost*, p. 34.

9. Milton's enthusiasm for the Apocalypse has been the object of many studies, among them Michael Fixler, *Milton and the Kingdoms of God* (Evanston, Ill.: Northwestern University Press, 1964); Austin C. Dobbins, *Milton and the Book of Revelation: The Heavenly Cycle* (University: University of Alabama Press, 1975); and Joseph Wittreich, *Visionary Poetics: Milton's Tradition and His Legacy* (San Marino: Huntington Library, 1979), pp. 3–78.

I do not know all of the reasons for the apocalyptic emphasis in the final books. The possibility of dark motives—that he wished no great and God-favored man to be happier than himself; that he looked forward to some special commendation at the Last Judgment; that he wanted to savor the damnation of his opponents; that the theology of the happy end is a way of obligating God to supply this end, since without it, God would be unjust—has not escaped me.

10. Ricoeur has amplified this point in *History and Truth*, trans. Charles Kelbley (Evanston: Northwestern University Press, 1965), pp. 52–56, 73–77, 81–97.

11. Fowler, note to 12.466–467 in *Poems of John Milton*, pp. 1049–1050.

12. Irene Samuel, "*Paradise Lost* as Mimesis," in Patrides, ed., *Approaches to Paradise Lost*, p. 23.

13. For references and discussion, see Walter Kaufmann, *Nietzsche: Philosopher, Psychologist, Antichrist* (Princeton: Princeton University Press, 1974), pp. 347–350.

14. See, for example, Alphonse de Waelhens, "Sur l'inconscient et la pensee philosophique," *Journees de Bonneval sur l'inconscient* (1960), pp. 16–21 (summarized by Ricoeur in *Freud and Philosophy*, p. 385), and *Schizophrenia*, trans. Wilfried Ver Eecke (Pittsburgh: Duquesne University Press, 1972), pp. 118–119; Jacques Derrida, "Coming into One's Own," in Hartman, ed., *Psychoanalysis and the Question of the Text*, pp. 114–148. Ricoeur suggests that much of the fascination of this game lies in hinting that the death drive may play a nonpathological, integrative role in psychic development (*Freud and Philosophy*, pp. 285–286, 314).

15. Robert Rogers, *Metaphor: A Psychoanalytic View* (Berkeley: University of California Press, 1978), pp. 111–112.

16. Immanuel Levinas, *Totality and Infinity: An Essay on Exteriority*, trans. Alphonso Lingis (Pittsburgh: Duquesne University Press, 1969), pp. 130–140.

17. The question of *Comus* ("Wherefore did Nature pour her bounties forth / With such a full and unwithdrawing hand?") still rattles in the head of the Adam of *Paradise Lost*. Geoffrey Hartman points to Adam's vexation over the abundance of Eden in his "Adam on the Grass with Balsamum," in

Beyond Formalism: Literary Essays, 1958–1970 (New Haven: Yale University Press, 1970), pp. 124–150.

18. See Ricoeur, *The Symbolism of Evil*, pp. 347–357.

19. This is substantially the position of Simon O. Lesser in *Fiction and the Unconscious* (Chicago: University of Chicago Press, 1957), pp. 248–251.

20. Colie, *Paradoxia Epidemica*, pp. 198–199; Tayler, *Milton's Poetry*, p. 89; Richardson, *Explanatory Notes*, p. 535.

21. Sandor Ferenczi, *Final Contributions to the Problems and Methods of Psycho-Analysis*, ed. Michael Balint (New York: Basic Books, 1955), pp. 156–167.

22. After Michael's lesson in typology, which has urged us to look ahead, it may seem perverse to deny Christ entrance into the family of man gathered imaginatively about our Grand Parents at the end of the poem. Christ will also pursue a "solitary way" into the wilderness. But the power of the moment derives from our resonant confrontation with what every child of this marriage has, but Christ does not: two fallen parents. Inviting us to forgive these guilty makers, the poem concludes within the oedipal rather than the sacred complex.

INDEX

Multiple Intelligences Can Help Us Learn More about Batteries

- **VERBAL/LINGUISTIC**
  Develop a booklet entitled "Everything You Wanted to Know about Batteries but Were Afraid to Ask."

- **LOGICAL/MATHEMATICAL**
  Create a chart to compare and contrast a dry battery with a wet battery.

- **VISUAL/SPATIAL**
  Demonstrate in a drawing how batteries are used in one of the following categories: home entertainment, health care, law enforcement, communications, and travel.

- **BODY/KINESTHETIC**
  Role play a variety of battery-operated toys such as robots, trains, and race cars, and see if others can guess the toy you are imitating.

- **MUSICAL/RHYTHMIC**
  Imagine that you are the world's most successful battery salesperson. Develop a creative sales promotion campaign that includes music, jingles, slogans, and lyrical phrases.

- **INTERPERSONAL**
  Work with a partner to prepare a display that shows how batteries can sometimes mean the difference between life and death.

- **INTRAPERSONAL**
  Think back over your preschool years and determine whether your favorite toys were battery-operated toys or not. How do battery-operated toys keep children from using their imaginations and creativity?

Research Topics

1 ● Find out how a car battery works.

2 ● Find out why batteries have different labels such as A, B, and C.

3 ● Find out the relationship of batteries to electricity.

4 ● Find out who invented the battery.

5 ● Find out where batteries are located in a variety of retail settings such as hardware stores, drugstores, department stores, and discount stores.

6 ● Find out how batteries are manufactured.

7 ● Find out how people should dispose of dead batteries.

8 ● Find out why batteries can be poisonous to the environment.

9 ● Find out how the life of a battery can be determined.

10 ● Find out some possible uses for a used battery.

Creative Writing Topics

1 ● Write a short story using one of the following titles:
- **The Blue-Ribbon Battery**
- **The Battery That Refused to Work**
- **The Day Batteries Grew on Trees**
- **The Feud between the Dry Cell Battery and the Wet Cell Battery**

2 ● Select a "battery of words" to describe each of the following events or situations:
- **A baseball or football game**
- **A circus or carnival**
- **A parade or pageant**
- **A movie or book**
- **A video arcade or record store**
- **A park or beach**

3 ● Write a creative response to each of the following questions:
- **Would you rather be a battery in a hearing aid or a pacemaker?**
- **Would you rather be a battery in a camera or a radio?**
- **Would you rather be a battery in a flashlight or a searchlight?**
- **Would you rather be a battery in a toy train or toy boat?**
- **Would you rather be a battery in an automatic razor or an automatic toothbrush?**

4 ● What would a battery say to each of the following objects?
- **A stalled car**
- **A burned-out flashlight**
- **A toy that is not battery-operated**
- **An electric can opener**
- **A house without lights**

Assessment Product and Performance Options

PRODUCTS

1 ● Create a battery tongue twister.

2 ● Compose a tribute to the "Almighty Battery."

3 ● Write a letter of apology to a battery that died because you forgot to turn something off.

4 ● Design a greeting card that uses a battery pun in its message.

5 ● Prepare an argument to limit the use of batteries in children's toys.

6 ● Prepare a critique of a resource book that teaches the reader about batteries.

7 ● Compose a letter to the editor warning citizens about the environmental dangers of the improper disposal of dead batteries.

8 ● Describe a battery and its function to an alien from outer space.

9 ● Write a letter to a battery manufacturer and inquire about the quality of his or her product.

10 ● Compile a set of trivia facts about the battery.

Assessment Product and Performance Options

Page 2

PERFORMANCES

1 ● Invent a puppet resembling a battery, and use it to teach others about the multiple uses of batteries in today's world.

2 ● Draw a series of diagrams showing the different parts of a battery and use these to give a mini-speech about the battery.

3 ● Deliver a brief campaign speech that urges eliminating the use of batteries in children's toys.

4 ● Prepare and give a sales pitch to promote one brand of battery.

5 ● Write and read aloud an original story about a battery that saved a life.

6 ● Make predictions about the future uses of batteries and discuss them with a small group of peers.

7 ● Give an oral explanation of the differences between dry and wet cell batteries.

8 ● Conduct an inventory of battery uses in your household and verbally summarize your findings for the rest of the class.

9 ● Lead a class discussion on the many ways batteries have improved our lifestyles.

10 ● Make a list of true and untrue statements about batteries, and use these to stage a mock television quiz show.

Battery Challenges

Page 1

Write a complete sentence to explain what you would do in each of the following situations.

1 ● Your mother tries to start the family car in order to take you to school, but the battery is dead.

____

____

____

2 ● You are preparing to conduct a science experiment involving the use of batteries, but you have forgotten how to wire the batteries correctly.

____

____

____

3 ● You have bought a battery-operated toy for a younger brother or sister and the directions for installing the batteries are missing.

____

____

____

Name ______________________________

Battery Challenges Page 2

Write a complete sentence to explain what you would do in each of the following situations.

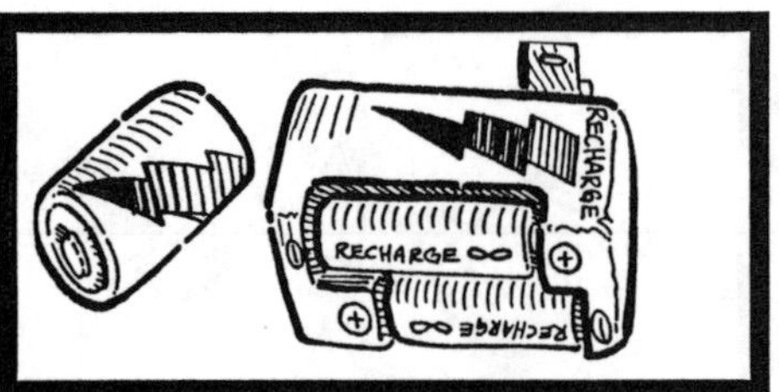

4 ● You, a battery salesperson, have been asked to write a statement explaining why your product is the best on the market.

____

____

____

5 ● You are on the way to the beach and discover that you do not have batteries for your transistor radio.

____

____

____

6 ● You are ready to leave for a camping trip and can't find the batteries for your flashlight.

____

____

____

EXPLORING

Bicycles

Exploring Bicycles à la Bloom

Page 1

● KNOWLEDGE

Study the bicycle diagram below. Find at least one representation in the diagram of each of the six simple machines listed below. List the examples on the lines below each simple machine.

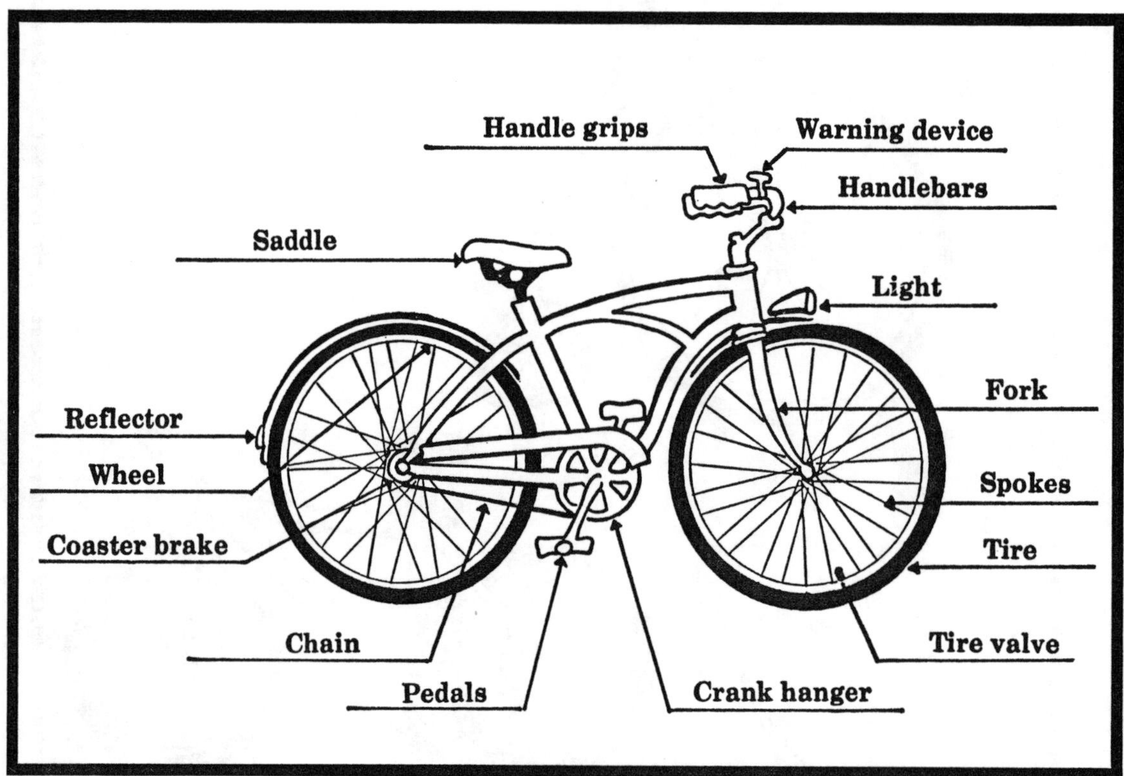

SIX SIMPLE MACHINES

1. Wheel and Axle ______ ______

2. Screw ______ ______

3. Lever ______ ______

4. Pulley ______ ______

5. Wedge ______ ______

6. Inclined Plane ______ ______

Name ____________________

Exploring Bicycles à la Bloom
Page 2

• COMPREHENSION

Explain how a bicycle illustrates each of the following scientific principles: application of force, friction acting against motion, acceleration, velocity, and transfer of energy.

• APPLICATION

Compile a list of safety rules that you think could prevent bicycle injuries.

• ANALYSIS

Make an inference: how are bikes related to health? Prepare a three-minute speech to share your ideas with the class.

• SYNTHESIS

Invent a new way to prevent bicycle theft. Consider such things as locks, warning devices, identification procedures, laws, or self-help programs.

• EVALUATION

Do you think that a bicycle rider should be required to pass an operator's test? Summarize your ideas in a one-page position paper.

Answers for KNOWLEDGE Activity:

1. wheel, crank hanger
2. tire valve, handle grips
3. warning device, pedals
4. crank hanger, chain
5. saddle, light
6. handlebars, fork

Multiple Intelligences Can Help Us Learn More about Work, Energy, Motion, and Machines

- **VERBAL/LINGUISTIC**

  In your own words, define **energy** and explain the difference between **potential** and **kinetic** energy.

- **LOGICAL/MATHEMATICAL**

  Identify **Newton's Laws of Motion** and give a real-life example to illustrate each one.

- **VISUAL/SPATIAL**

  Draw a series of illustrations to show how **work,** in the scientific sense, is done by each of the six simple machines.

- **BODY/KINESTHETIC**

  Work with a group of peers to act out each of the following concepts as part of your work with simple machines:

  - WORK is done when . . .
  - ENERGY is . . .
  - FORCE is . . .
  - POWER is . . .
  - MOTION is . . .

- **MUSICAL/RHYTHMIC**

  Make up a series of sounds to represent each of the six types of simple machines. Combine the sounds to represent the concepts of **work, energy,** and **motion.**

- **INTERPERSONAL**

  Working with a partner, use a set of tinker toys to construct a simple model of one of the following types of machines: freight elevator, windmill, lawnmower, harvester, crane, or space shuttle.

- **INTRAPERSONAL**

  Decide on the five most important simple machines that you use every day because they make your life easier, safer, more comfortable, or more enjoyable. Be able to defend your choices by establishing a set of criteria against which you are going to select from alternatives.

Using Williams' Taxonomy to Study Simple Machines

• FLUENCY

List as many different examples of simple machines as you can think of that are found in the home.

• FLEXIBILITY

Classify each of these items from your list according to one of the six types of simple machines that it best represents: **lever, inclined plane, wedge, screw, wheel,** and **pulley.**

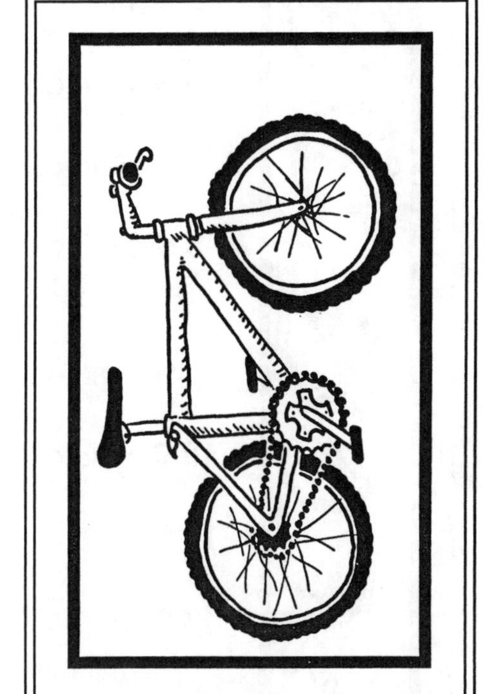

• ORIGINALITY

Design a rainbow-making machine, a homework machine, a meal-making machine, or a baby-sitting machine that incorporates the use of all six simple machines.

• ELABORATION

Use the knowledge acquired during your study to reflect on this statement by inventor Thomas A. Edison: "Genius is one percent inspiration and ninety-nine percent perspiration."

• RISK TAKING

Determine which simple machine is most like you and give reasons for your choice.

• COMPLEXITY

Explain how using simple machines has helped humans harness forces of nature.

• CURIOSITY

Decide which topics and questions you would want to discuss with each of the following inventors: Lewis E. Waterman, Leo H. Baekeland, and Elisha Otis.

• IMAGINATION

Imagine you have invented a time machine and can go back in history to relive the invention of an important lifesaving machine. Tell where you would go and what you would want to do.

Research Topics

1. Find out about friction.
2. Find out about first-class, second-class, and third-class levers.
3. Find out about fixed and movable pulleys.
4. Find out about the many sources of energy.
5. Find out about Sir Isaac Newton.
6. Find out about work and force.
7. Find out about gasoline engines and diesel engines.
8. Find out how people did their work before they had machines.
9. Find out about machine tools.
10. Find out about flying machines and space machines.
11. Find out about floating machines.
12. Find out about building machines and road-making machines.
13. Find out about farming and dairy machines.
14. Find out about rescue machines and fire-fighting machines.
15. Find out about power sources.
16. Find out about inventors and their work.

Creative Writing Topics

1 ● Write a short story using one of the following titles:

- **The Lifesaving Machine**
- **Finding the World's Largest Inclined Plane**
- **A Machine That Raced against Time**
- **The Bully Pulley**
- **Reinventing the Wheel**
- **The Wedge That Came between Us**

2 ● Create a series of six couplets, devoting one to each of the six simple machines.

3 ● Compose a tall tale about an inventor who could design any type of machine in the world. Be sure to stretch the truth. Write your tall tale on a long piece of adding machine tape that has to be unraveled in order to be read.

4 ● Pretend that you are the manager of a complaint department in either a machine shop, an appliance store, or an automobile plant. Make a list of complaints you might receive from your customers and a possible resolution for each one.

5 ● Compose an invitation to the National Convention for Junior Inventors. Tell the guests when and where it is located, what to expect to see or do, and what to wear or bring.

6 ● Think up ten uses for a broken radio, blender, toaster, or clock.

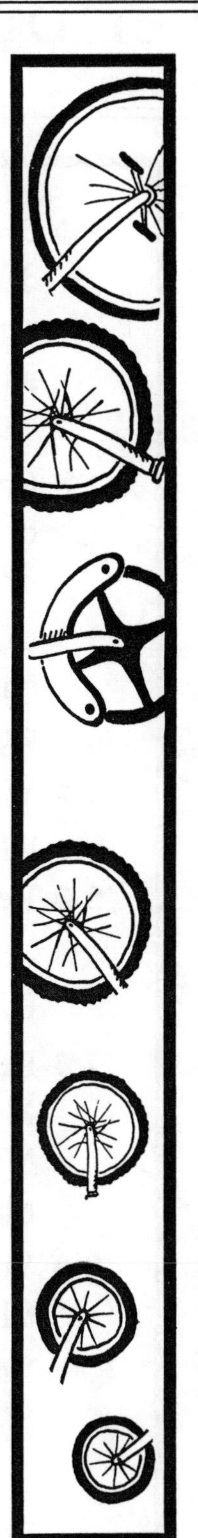

Assessment Product and Performance Options

Page 1

PRODUCTS

1 ● Create a simple magazine for Future Inventors of America that is designed to encourage students to attempt the invention of something. Prepare a Table of Contents for items that might be of interest to the readers. Complete one of the articles listed in your Table of Contents.

2 ● Conduct a survey of students and adults in your school to obtain their ideas on which kinds of inventions our world needs most. Compile your results in chart or graph form.

3 ● Create a Rube Goldberg cartoon that shows the invention of an outrageously unnecessary object or machine such as a window-closer, a dog-walker, or a back-scratcher.

4 ● Write an essay that compares the mechanics of writing with the mechanics of auto repair.

5 ● Research a famous inventor and write his or her biography.

6 ● Compile a set of questions about inventions for a Trivial Pursuit® game.

7 ● Design a pamphlet that discusses one of these topics: Flying Machines, Floating Machines, Fighting Machines, Rescue Machines, Building Machines, Farming Machines, or Home Machines.

8 ● Visit a place that repairs some type of machine such as televisions, appliances, automobiles, clocks, bicycles, or vacuum cleaners. Maintain a learning journal to record both your questions and your observations.

Assessment Product and Performance Options

PERFORMANCES

1 ● Prepare a lab demonstration to show the function of several simple machines and the scientific principle(s) behind each one.

2 ● Prepare a set of Jeopardy® cards with only answers on them and use these to lead a Jeopardy game of terms and concepts related to your study of simple machines.

3 ● Deliver an oral explanation of how the concepts of work and force are used in physics.

4 ● Locate magazine pictures of machines and use these to give a short speech entitled "The Machine Age in Which We Live."

5 ● Prepare an oral report on the topic of "friction" and use props to help clarify or emphasize your major points.

6 ● Use simple materials such as rubber bands, spools, string, or wire to show an audience how to construct a type of machine.

7 ● Organize an Inventor's Fair for your classroom where students can display simple student-designed inventions with moving parts.

8 ● Prepare and deliver a television commercial to sell a favorite brand of machine that is used in your home. Be sure you compare it to other brands and explain why your choice is a better one.

9 ● Prepare a meal for your family or friends that requires the use of as few tools or machines as possible in its preparation. Plan to describe your cooking experience to a group of peers.

10 ● With a group of friends, select five to ten complex machines to act out a charade, and see if everyone can guess which machines you have chosen to portray.

The Sport of Cycling

Pretend you are a cyclist who has been asked to perform the following tasks. Write your responses on the lines below the questions.

Task One

You have been asked to plan a bike-a-thon for your school. What route would you suggest for the bike-a-thon?

____

____

____

____

Task Two

You have been asked to plan a six-week training program for junior cyclists. What would be included in the program?

____

____

____

____

Task Three

You have been asked to plan a window display for a new bicycle shop that caters to middle-graders. What would the display look like?

____

____

____

____

Name ____________________

EXPLORING

Bionic People, Robots, and Artificial Intelligence

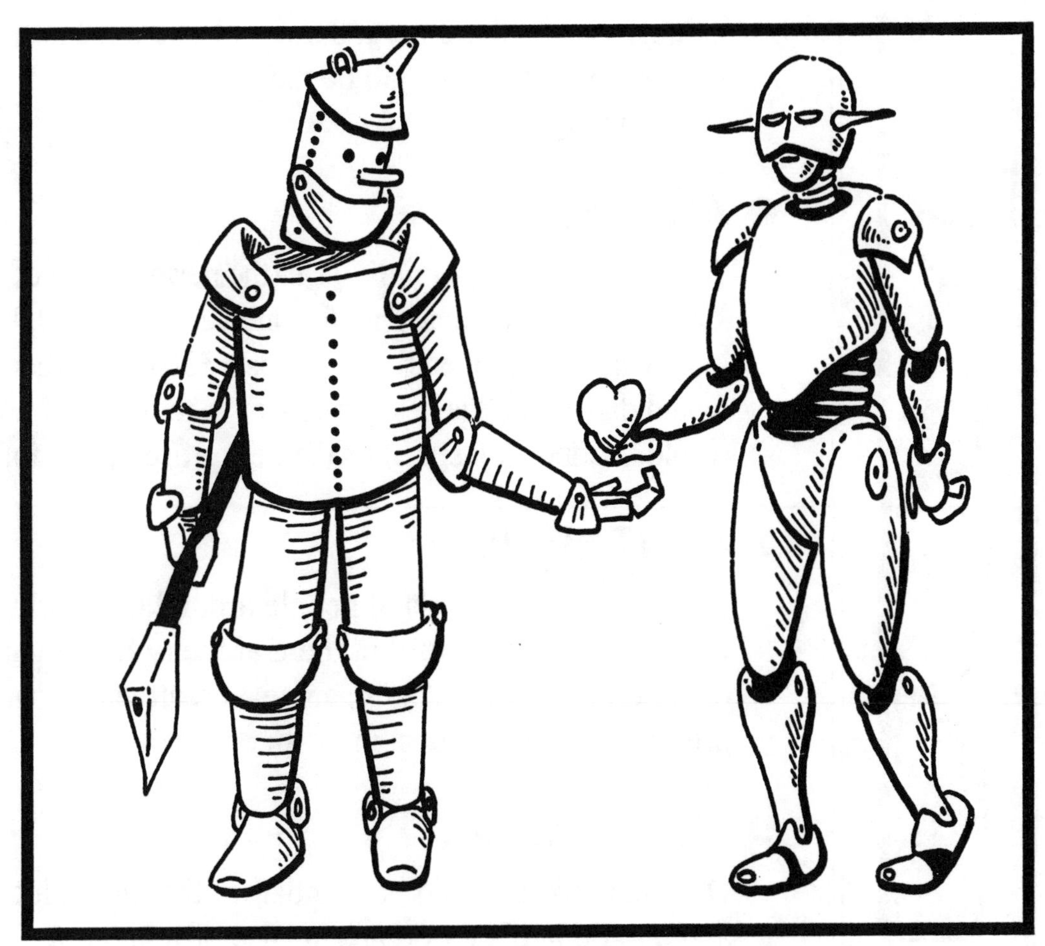

Exploring Bionic People à la Bloom

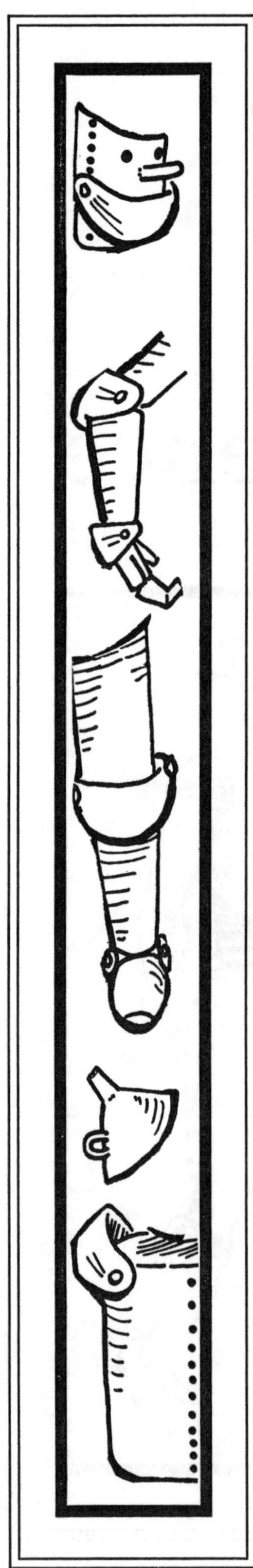

• KNOWLEDGE

Write the dictionary definition of each of the following words: **transplant, organ,** and **donor.**

• COMPREHENSION

Give at least three examples of a situation in which an organ transplant might be the only solution to a serious medical problem.

• APPLICATION

Suppose that you are a transplant candidate waiting for a kidney. Predict the criteria that would be used to decide whether or not you would be the recipient of the next available organ.

• ANALYSIS

To maintain a person on **dialysis** (the performance of kidney function by machine) for one year costs $30,000. A kidney transplant operation also costs $30,000. When would dialysis be the best procedure to follow, and when would a kidney transplant be the best procedure to follow?

• SYNTHESIS

Hundreds of thousands of people around the world are waiting for transplants, but there are very few organs or organ donors available. Organize a creative campaign to promote the idea of becoming an organ donor.

• EVALUATION

Do you think it should be mandatory that individuals agree to donate their organs in case of premature death? Be able to support your opinion with at least three good arguments.

Multiple Intelligences Can Help Us Learn More about Artificial Intelligence

Page 1

• VERBAL/LINGUISTIC

Write a good paragraph that discusses the new science that is forming around the idea of creating machines that can mimic human intelligence (often referred to as artificial intelligence).

• LOGICAL/MATHEMATICAL

Use the Yellow Pages® of your local telephone directory to identify a company or university in your area that is working on robotics and/or artificial intelligence. Write to the organization and ask that information about the organization and its work be sent to you.

• VISUAL/SPATIAL

A computer is an important part of a robot because it serves as the robot's "brain," or source of intelligence. Create a chart or Venn diagram that compares a human with both a computer and a robot. Show how these three things are alike and how they are different.

• BODY/KINESTHETIC

You will need to work with a partner to do this activity. One of you will act as the robot, and the other will be its programmer. Select a function that you would like to have the robot perform. Use some or all of the terms below to issue commands for completing the function. The robot should obey the command. Then switch roles and choose another function and set of commands to perform.

Lift	Lower	Right	Left
Move	Carry	Up	Down
Turn	Walk	Forward	Backward

Multiple Intelligences Can Help Us Learn More about Artificial Intelligence

Page 2

- **MUSICAL/RHYTHMIC**

  A speech "recognizer" in a robot or computer works in much the same way as the human eardrum and brain. A microphone picks up sound waves, which are simply vibrations in the air, and changes them into electrical waves that a computer can recognize. The computer measures the amplitude (height of the sound waves) and the frequency (number of waves per second). Every human word has its own unique pattern of frequencies and amplitudes. Practice saying several words with varying frequency and amplitude waves. Draw each word according to the letter sounds it makes.

- **INTERPERSONAL**

  Work with a group of peers to produce a panel discussion about whether we should put a high priority on the continued development of computers, robots, and artificial intelligence tools that can think, learn, feel, and act like human beings.

- **INTRAPERSONAL**

  Take an inventory of the items in your home and make a list of objects which contain a computer. Write a paragraph explaining how computers have enriched or complicated your life as a human being.

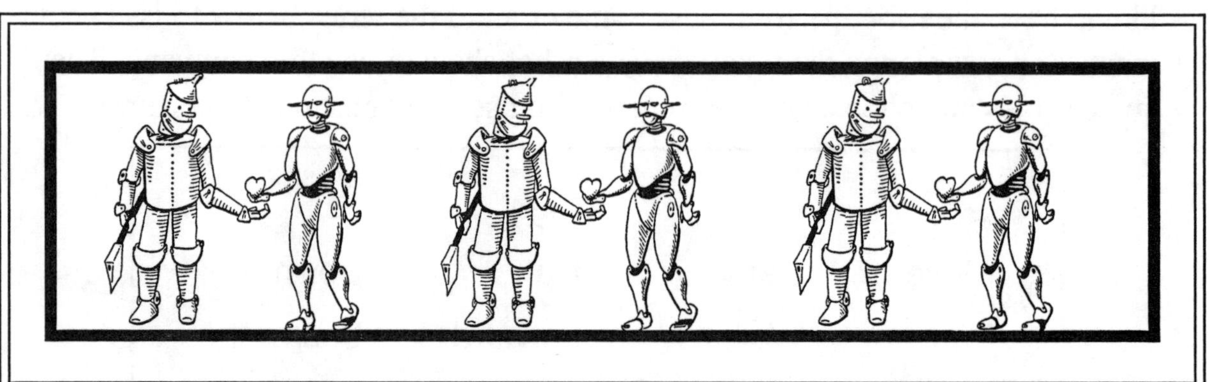

Using Williams' Taxonomy to Study Artificial Intelligence

● FLUENCY

Make a list of criteria for what makes a job unpleasant. Then make a list of as many unpleasant jobs or careers as you can that could be performed by robots.

● FLEXIBILITY

Write down as many situations as you can recall in industry, space, or the movies in which robots have played an important part.

● ORIGINALITY

Draw a picture of a personal robot designed to perform a simple task for you. What shape would be best for your robot? Would your robot move best with wheels, tracks, or legs?

● ELABORATION

Discuss the implications of automation on the job market for many workers.

● RISK TAKING

Complete each of these starter statements:

- If I could have a personal robot that I could program to do anything, it would . . .
- With the time I could save by having my robot work for me, I would . . .
- I would not like my robot to . . . because . . .

● COMPLEXITY

Explain how one can determine whether or not a particular machine is a robot.

● CURIOSITY

If the use of robots increases in the future, there will be a need for mechanical doctors. Ponder the following ideas and write down your thoughts:

- What qualifications would a robot doctor need?
- Where would a robot doctor get its training?
- What robot ailments are likely to need treatment?
- What robot ailments would be most difficult and/or most expensive to treat?

● IMAGINATION

Pretend you are a robot specialist in one of the following areas and design a business card that tells about your services:

- robot repair
- robot manufacturing
- robot customer service

Research Topics

1 ● Find out how widespread the use of organ transplants is in your community or state.

2 ● Find out about the procedures for donating organs in your community or state.

3 ● Find out how organ transplants prolong the lives of individuals who receive them.

4 ● Find out how values and ethics play a part in the organ transplant decision-making process.

5 ● Find out which types of organ transplants are most common, most difficult to perform, most likely to be successful, or most difficult to supply.

6 ● Find out about the varying shapes, sizes, and abilities found in personal and commercial robots.

7 ● Find out how scientists have come up with ways to test the intelligence of animals and which animals are considered to be the most intelligent.

8 ● Find out how robots see, hear, and feel.

9 ● Find out how an electronic sensor works.

10 ● Find out more about amplitude and frequency of sound waves.

11 ● Find out how the form and appearance of a robot's body depends on the job it does.

12 ● Find out about the following types of robots: jointed-arm robots, spherical or polar robots, XYZ robots, spine robots, and cylindrical robots.

13 ● Find out how a robot is taught or programmed to do a specific job.

14 ● Find out more about the science of artificial intelligence.

Creative Writing Topics

1● Pretend you are a bionic boy or girl living in the year 2050. Write an autobiography telling about your bionic body and your life. Tell about your place of manufacture, your damaged or worn-out body parts, your new organ parts, and the unusual facts/activities/hobbies/interests you experience that are related to your mechanical body parts.

2● Write an essay describing your feelings, concerns, and expectations surrounding the extensive use of organ transplant banks.

3● The Tin Man in *The Wizard of Oz* was a "robot" who was searching for a heart. Write your own version of this story in which a robot is searching for another human characteristic.

4● Pretend that you are the writer and producer of a national television network. You have been asked to write and produce a one-hour science fiction production with a cast of three robots and two human beings. The setting is New York City in the year 2020. Develop a plot and outline the script. Include the names of all of the characters as well as specifications for costumes, scenery, and special sound effects.

5● Pretend you are a scientist who specializes in the field of artificial intelligence. Compose a letter explaining why you need funding to continue your important research. Describe several of your key projects that would be threatened if funding was not available.

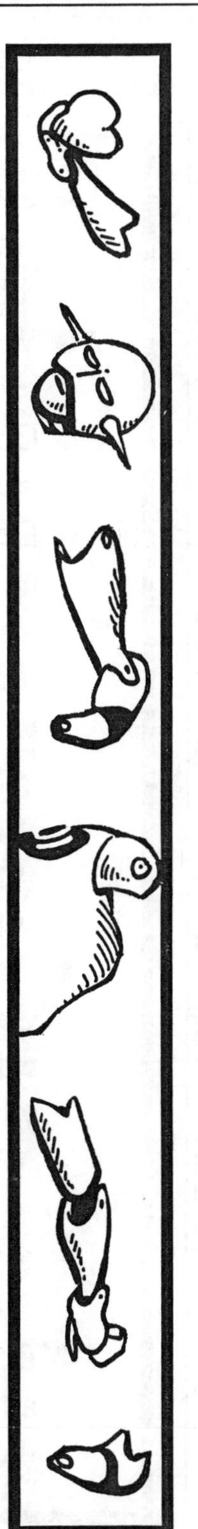

Assessment Product and Performance Options

Page 1

PRODUCTS

1 ● Plan a shopping center for the 21st century. Draw a picture and write a set of descriptions of the shopping center showing how robots will provide most of the required services.

2 ● Write a factual report which demonstrates how an organ transplant can be referred to as "the gift of life" or "an extraordinary miracle."

3 ● Compare what is involved in a cornea transplant with what is involved in a liver transplant. Show your findings in outline form.

4 ● Design a donor card for organ donors. What information would you need or want to know about a donor?

5 ● You have been hired to improve the operation of a human organ transplant bank. Write a description of the improvements you plan to make in each of the following areas:

- procedures for organ donors
- procedures for organ recipients
- procedures for organ storage, identification, and transportation
- procedures for marketing organ transplant services

6 ● Plan a project about artificial intelligence for a science fair. Combine creative flair with scientific facts to create an outstanding exhibit. Use the set of questions below as a guide in structuring your project.

- From what sources will you get your facts and information?
- What type of project will you prepare?
- What kind of graphics and illustrations will you use?
- How will you evaluate your project?

7 ● Design a display that would provide information about robots, artificial intelligence, or organ transplants. Include pictures, illustrations, graphs, diagrams, and/or drawings.

Assessment Product and Performance Options

Page 2

PERFORMANCES

1 ● Prepare and deliver a short transparency talk on the parts of a computer and what they do or on the phenomenon of artificial intelligence and how it mimics human intelligence.

2 ● Create a timeline that shows the major developments in computers and robotics over the past several years. Use this timeline as a visual prop for delivering a short talk on the subject.

3 ● Compare what is involved in a cornea transplant with what is involved in a heart transplant. Show your findings in outline form.

4 ● Pretend you are a sales representative for a computer company that manufactures speech recognition units or electronic seeing or hearing sensors. Create and perform a lively commercial to sell your product.

5 ● Create a tape recording of different sounds from the natural or human environments that represent varying amplitudes and frequencies. Play it for your class and conduct a class discussion about the sounds represented.

6 ● Locate a speaking toy in a retail toy outlet and give a demonstration for your class using this object to explain the basics of a speech synthesizer and how it functions.

7 ● Compose and read aloud a short story that describes a robot that had to be protected from the humans who owned or created it.

8 ● Pretend that you are to interview a potential organ donor. Design interview questions that demonstrate how one goes about becoming an organ donor. Work with a friend to perform a mock interview, using your questions and his or her answers as the basic script.

Dot-to-Dot Robot

Connect the dots from 1 to 91 to find a most unusual robot. Add details of your own to make the robot uniquely yours.

On another sheet of paper, write a program that would allow the robot to perform a service that would make your life easier or more fun.

ROBOT'S NAME: ______________________________

Name ______________________________

EXPLORING

Careers in Science

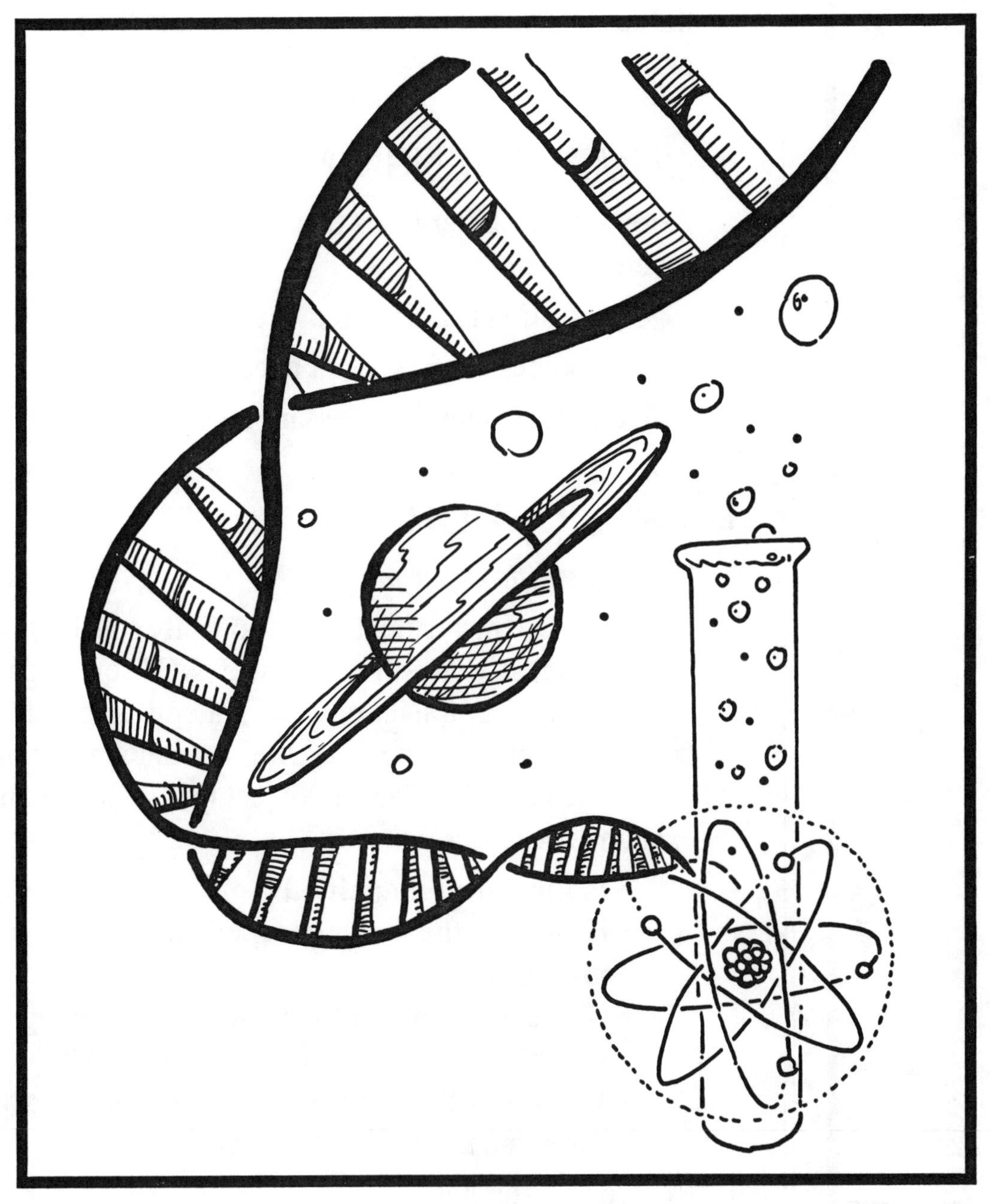

Exploring Science Careers à la Bloom

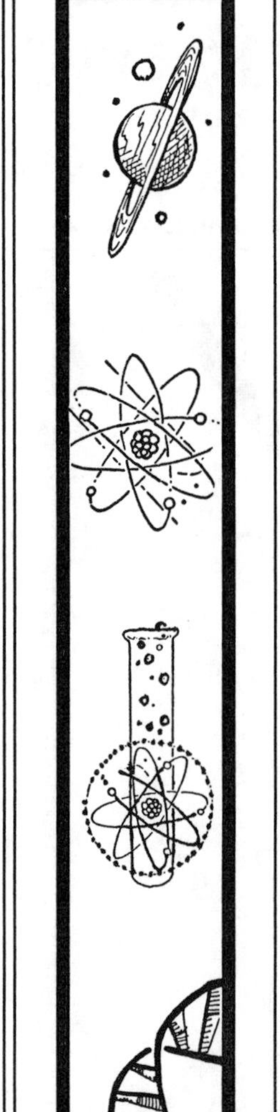

● KNOWLEDGE

Prepare for a career in science. Write a brief outline of the scientific method and explain the method to a classmate.

● COMPREHENSION

Extend these starter ideas that are related to the scientific method:

Observation is a first step because . . .

A hypothesis is important because . . .

Experimentation is necessary because . . .

● APPLICATION

Try a scientific experiment:

Water can do interesting things, optically speaking. For this experiment you will need one 3" x 3" square of aluminum foil, a straight pin, and water. Make a small hole in the center of the foil. Drop a small amount of water into the hole. Place an object under the foil and examine the object.

- **Is the object magnified?**
- **How does the water compare to the lens in a microscope?**
- **If you use a larger hole and a larger drop of water, will the magnification increase?**
- **Will other liquids such as vinegar or oil give the same magnification results?**

Record your findings.

Exploring Science Careers à la Bloom

Page 2

• ANALYSIS

Write an analysis of your conduction of the experiment on the previous page. How did the conduction of this experiment follow the scientific method? Which parts of the experiment were related to which parts of the scientific method? Did you enjoy the experiment, or did you find it uninteresting? Would you like to use the scientific method to find out something in a field that may interest you more?

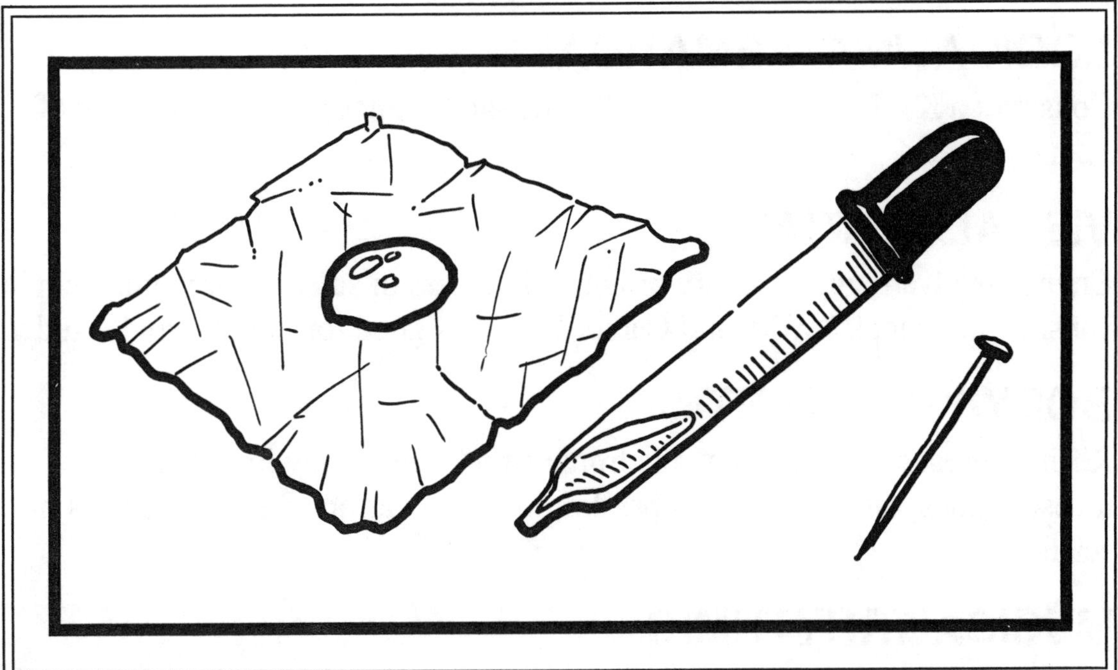

• SYNTHESIS

Prepare a short dialogue between a scientist who enjoys his or her job and a student who so far has not shown any interest in a career in science.

• EVALUATION

Determine if/how the activities on these two pages can help you decide if you should pursue a science career.

Multiple Intelligences Can Help Us Learn More about Careers in Science

• VERBAL/LINGUISTIC

Write a short essay discussing the following issues related to a career in science:

- **What type of person do you think is most likely to pursue a career in science?**
- **What characteristics do you think most scientists are likely to have?**
- **What are some of the more interesting science careers to you and why?**

• LOGICAL/MATHEMATICAL

Construct a KWL (Know, Want to Know, and Learned) chart for a study of science careers.

• VISUAL/SPATIAL

Create an advertising brochure promoting careers in science for young people to consider at a mock "Work and Career Fair" designed for middle-grade students.

• BODY/KINESTHETIC

Reenact great scenes or moments from history concerning important investigations, discoveries, and breakthroughs in various fields or careers of science.

• MUSICAL/RHYTHMIC

Turn a biography of a famous scientist into a musical play or skit.

• INTERPERSONAL

Organize a team competition for your class by discussing various science career options and then writing team position papers on the best science career opportunities predicted for the future.

• INTRAPERSONAL

Write a short journal entry on the following topic: "If I could qualify for any science-related career today, I would want to be . . . because . . ."

Using Williams' Taxonomy to Study Science Careers

- **FLUENCY**
  List as many careers related to science as you can in two minutes.
- **FLEXIBILITY**
  Come up with a unique classification system for the science-related careers you listed in the Fluency activity.
- **ORIGINALITY**
  Create a slogan to promote student interest in science careers.
- **ELABORATION**
  Expand on this statement by Stephen Jay Gould: "Science is an integral part of culture . . ."
- **RISK TAKING**
  Tell how you really feel about the possibility of pursuing a career in science. Is there any science-related career that appeals to you?
- **COMPLEXITY**
  How could a person balance his or her own needs for a personally satisfying career with America's need for more top-notch scientists?
- **CURIOSITY**
  What specific questions would you like to ask an older scientist about how the field of science and careers in science have changed since the early days of his or her career?
- **IMAGINATION**
  Imagine a world without a standard set of measurements such as the metric or English systems. How would this affect scientists and society?

Research Topics

1 ● Find out which resources are available in your school media center for studying careers in science.

2 ● Find out about the lives of famous scientists and what these lives have in common.

3 ● Find out about the Nobel Prize for Science and who has received it in recent years.

4 ● Find out about the education and training requirements for various science careers and the monetary rewards for each one.

5 ● Find out why far more boys than girls tend to choose science careers after graduation from high school.

6 ● Find out about careers that require the combination of a knowledge of science with some expertise in the humanities. (Example: the writing of science fiction).

7 ● Find out about a scientific/technological achievement that has caused serious problems as well as provided some benefits.

8 ● Find out which areas of science overlap.

9 ● Find out about the daily life of a scientist who works in a field in which you are particularly interested.

10 ● Find out about how some scientists and nonscientists are taking steps to help ensure that knowledge gained from scientific research is put to the best uses.

11 ● Find out about the many possible careers in the field of medicine.

Creative Writing Topics

1 ● Create a short story using one of the following titles as a springboard:

- **The Mad Scientist**
- **Meet Mr. Wizard**
- **Microscope Mania**
- **The Secret Science Experiment**

2 ● Write a diary entry as if you were a marine biologist who has been alone on a Pacific island for two weeks.

3 ● Write an essay that begins with this topic sentence: "One shouldn't be afraid of looking into the possibility of a career in science."

4 ● Create a coloring book for young children that will help to familiarize them with various careers in science.

5 ● Write a short story about a student who comes from a family of artists, and who is considering a career in science.

6 ● Write a character sketch of a scientist you know or have read about.

7 ● Write a short essay that tells why cooperation and communication among scientists have become increasingly important over the years.

Assessment Product and Performance Options

Page 1

PRODUCTS

1 ● Create a simple test to determine how much people know or do not know about careers in science. Administer your test to five to ten people and graph the results.

2 ● Maintain a series of journal entries on "Questions I have that might be answered if I pursue a career in science."

3 ● Compile a notebook of newspaper and magazine articles related to careers in the field of science.

4 ● Create a series of word problems whose solutions depend on a knowledge of careers in science. Provide an answer key.

5 ● Research a famous scientist and write a Fact Book about him or her, recording one fact on each page of the booklet. Find a magazine picture to illustrate it. Add a cover, title page, and dedication page to your booklet.

6 ● Create a diagram that would visually demonstrate the steps that would be helpful to take when pursuing a career in science.

7 ● Make a booklet that shows various ways a science career can be related to the scientist's personal beliefs (such as a chemist who uses his or her knowledge in the pursuit of a project that will increase environmental health and safety).

Assessment Product and Performance Options

Page 2

PERFORMANCES

1 ● Stage an imaginary talk or interview with a famous scientist of the past.

2 ● Describe the "before and after" of important scientific discoveries in a series of skits or talks.

3 ● Plan and deliver a presentation on one of the following topics:

- **Careers in Chemistry**
- **Careers in Astronomy**
- **Careers in Biology**
- **Careers in Physics**
- **Careers in Medicine**
- **Careers in the Social Sciences**
- **Careers in Paleontology**
- **Careers in Psychology**
- **Careers in Meteorology**
- **Careers in Genetics**

4 ● Design, implement, and evaluate a one-week "Promote Careers in Science" project for your school or community.

5 ● Use mind-mapping as a tool for teaching a mini-lesson about careers in science.

What Is Left to Explore?

There are many areas of science in which there are still many questions to be answered. Below are listed a few of the fields in which new ideas and discoveries are being made every day. Do some research to find out some of the exciting new discoveries that are being made or pursued in each of these fields. Write something about a new discovery in each field in the appropriate box.

Genetic Engineering

Astronomy

Physics

Medicine

Name ______________________________

EXPLORING

Color, Light, and Optical Illusions

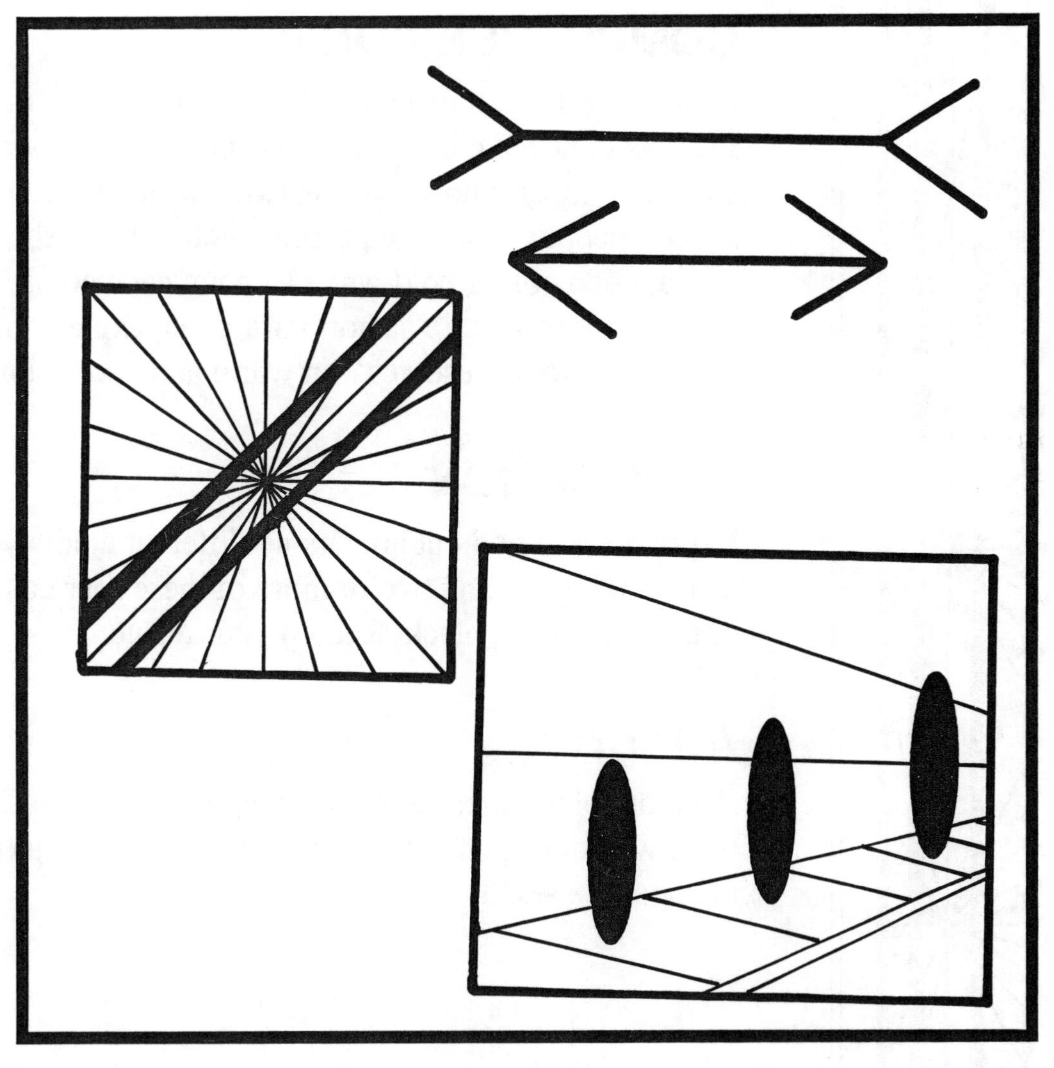

Exploring Color à la Bloom Page 1

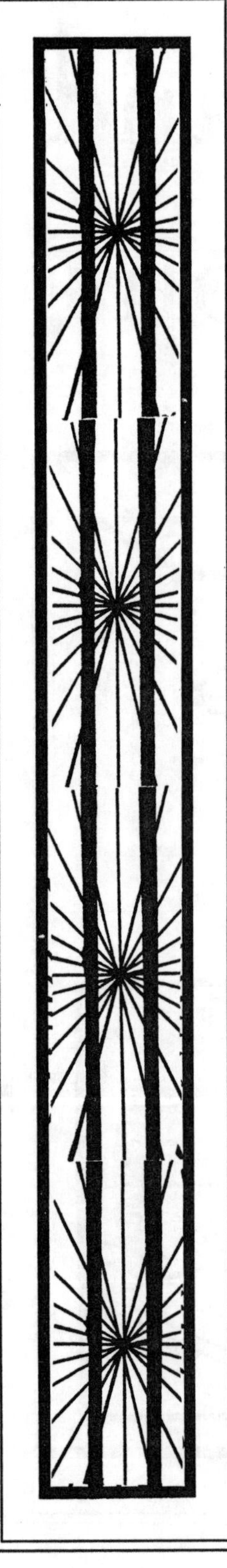

• KNOWLEDGE

Make a list of the basic colors. Write other color words that are shades or hues of each basic color.

Example: GREEN
- mint
- lime
- chartreuse

• COMPREHENSION

People say that colors can make them feel warm or cool. Pick six or seven colors and use each one to color a square of paper. Show the squares, one at a time, to several people. Ask each person whether he or she feels warm, hot, cool, or cold when viewing each color. Graph your results. In a short paragraph, explain the findings of your "research" on warm and cool colors.

• APPLICATION

Experiment to find out how many different colors you can create by mixing two or more of these four colors: blue, red, yellow, black. Record your results.

• ANALYSIS

Paint or color a picture that is completely yellow. Decide what would happen if everything were yellow. List all of the possible results.

Exploring Color à la Bloom Page 2

• SYNTHESIS

Using this chart of color meanings, design a coat of arms for your family.

TRADITIONAL COLOR MEANINGS

yellow or gold	=	honor, loyalty	red	=	bravery
brown	=	flexibility	blue	=	sincerity
black	=	grief, sorrow	purple	=	royalty
white	=	faith, purity	green	=	hope

• EVALUATION

Artists use color to "speak" and poets use color to "paint" visual word pictures. Scientists view color as something that is partially in the object, partially in the light, and partially in the eye. Determine how three people in different professions can view color very differently. Support your responses with examples.

Multiple Intelligences Can Help Us Learn More about Light

- **VERBAL/LINGUISTIC**

  In your own words, explain the nature of light.

- **LOGICAL/MATHEMATICAL**

  Compare and contrast a light wave and a sound wave. How are they alike and how are they different?

- **VISUAL/SPATIAL**

  Lights come in many kinds and sizes. Draw a mural showing the multiple sources of light in a wide variety of settings.

- **BODY/KINESTHETIC**

  Design a science experiment to demonstrate how light is reflected and how light can be bent or refracted.

- **MUSICAL/RHYTHMIC**

  Experiment with the effect of vibration on a metal plate. Relate these sound vibrations to your study of light.

- **INTERPERSONAL**

  Work with a partner to explore the concept of the shadow. Prepare a short demonstration to show the meaning of "umbra" and "penumbra."

- **INTRAPERSONAL**

  Use a camera (or locate photographs from home) to prepare a photo essay about you and your family.

Using Williams' Taxonomy to Study Optical Illusions

- **FLUENCY**

  Locate a book on optical illusions and write down as many examples of these illusions as you can find, eliminating any duplicates that you see.

- **FLEXIBILITY**

  Make a list of any optical illusions that you have personally experienced in your life.

- **ORIGINALITY**

  Write a story about a main character in a crisis who is constantly fooled by a series of optical illusions he or she encounters during an attempt to survive.

- **ELABORATION**

  Defend this statement with specific examples: "Magicians often rely upon illusions or tricks of sight to fool their audiences."

- **RISK TAKING**

  Use a mirror to examine your image up close. Write a paragraph describing what you see or would like to see in the mirror.

- **COMPLEXITY**

  Write an explanation of how a rainbow is a type of optical illusion.

- **CURIOSITY**

  Examine a number of paintings by famous artists to determine how they used "tricks of light" in their work. Record your ideas.

- **IMAGINATION**

  Visualize a world without light, with darkness 24 hours a day. How would life be different in such a world?

Research Topics

1 ● Find out how light travels.

2 ● Find out how mirrors and lenses work.

3 ● Find out how we see with the eye.

4 ● Find out about invisible light.

5 ● Find out why objects have different colors.

6 ● Find out how light energy can be changed.

7 ● Find out about laser light and laser light shows.

8 ● Find out how light is reflected and refracted.

9 ● Find out how rainbows are formed.

10 ● Find out about artificial light.

11 ● Find out how a camera works.

12 ● Find out how photography and printing make use of light and shadow.

13 ● Find out how shadows are formed.

14 ● Find out the sources of light.

15 ● Find out about the Electromagnetic Spectrum.

16 ● Find out how optical illusions are used to create elaborate sets for movie productions and/or in advertising of products on television.

17 ● Find out why optical illusions can be fascinating to examine and study.

18 ● Find out what role "light" and "color" play in the creation of several optical illusions.

Creative Writing Topics

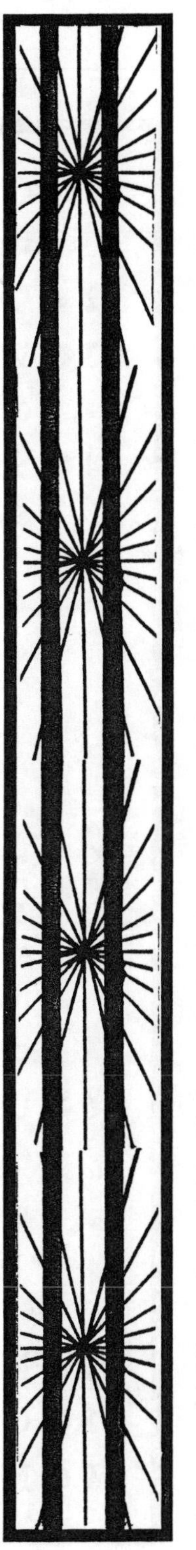

1 ● Read *Hailstones and Halibut Bones* by Eve Merriam and create a series of color poems on your own.

2 ● Write an original story using one of the following titles:

- **The Day Everything Turned Red**
- **The Shadow That Refused to Go Away**
- **The Magic Mirror**
- **The Light Trick**

3 ● Write an essay concerning color, light, and optical illusions that begins with this topic sentence: "Beauty is in the eye of the beholder."

4 ● Create an anti-coloring book for students of middle grades.

5 ● Make up a "Recipe for Orange" or "Purple" or "Peach."

6 ● Use a set of ink blots as a basis for a cartoon or comic strip.

7 ● Respond to this question: "If you could eat colors, how would each one taste?"

8 ● Do a character sketch of someone called "The Color Broker."

9 ● Compose a personal essay that illustrates an experience you have had or know about that reflect one of these expressions:

- **He/she was the light of my life.**
- **It was a lighthouse community.**
- **My friend helped me to see a light at the end of the tunnel when I was very upset.**
- **It was a time when I needed someone to light the way.**

10 ● Write a one-page report on the magic of optical illusions.

Assessment Product and Performance Options

Page 1

PRODUCTS

1 ● Construct a **periscope,** a **kaleidoscope,** or a **color wheel,** and write down the directions for its construction.

2 ● Draw a diagram of the eye and explain how it (the eye) works.

3 ● Write a report on how several occupations require a knowledge of color.

4 ● Show you have an eye for color by completing each of these color tasks:

- **Be a fashion designer and create a color-coordinated outfit for yourself.**
- **Be a dietician and draw an appealing, nutritious meal on a plate.**
- **Be an artist and design a geometric, modern masterpiece for a frame.**

5 ● Decide on the best color to represent each of the following situations and tell why you chose the colors you did: anger, fear, love, happiness, patriotism, friendship, illness, autumn, courage, and envy.

6 ● Use each of the following "colorful" expressions in a sentence. Color or paint a picture to show what each of these expressions means:

- **white as a sheet**
- **green thumb**
- **colorful personality**
- **tickled pink**
- **show one's true colors**

7 ● Make a booklet of optical illusions.

8 ● Construct and demonstrate a **stroboscope.**

9 ● Make a **camera obscura** and use it in your classroom.

Assessment Product and Performance Options

Page 2

PERFORMANCES

1 ● Plan and perform a shadow show on a wall at school.

2 ● Plan and perform a magic show that depends on optical illusions for its tricks.

3 ● Plan and deliver a presentation on one of the following topics:
- **use of light or color in art**
- **use of light or color in photography**
- **use of light or color in fashion design**
- **use of light or color in interior design**
- **use of light or color in advertising**

4 ● Give a "chalk talk" on how a sundial uses a shadow made by the sun to tell the time.

5 ● Use mirrors to give a demonstration of "mirror writing."

6 ● Prepare and deliver a mini-lesson on **convex, concave,** and **convex/concave** lenses.

7 ● Prepare and administer a simple vision or optical illusion test for your classmates.

8 ● Give a set of verbal explanations of light waves and sound waves.

9 ● Give a short speech on the impact of color on our daily lives.

10 ● Create and perform a drama on the nature of light.

Make a Set of 3-D Glasses

1. **Draw a pair of glasses on heavy paper or cardboard (see the illustration). Cut out the glasses.**
2. **Paste a piece of red cellophane on the glasses to make the right lens. Paste a piece of green cellophane on the glasses to make the left lens.**
3. **Draw a picture using a red pencil. On a separate sheet of paper, draw a picture using a green pencil.**
4. **Put on the glasses. Shut your left eye and look at the red drawing through the red cellophane lens. Then shut your right eye and look at the green drawing through the green cellophane lens. Record what you see in each instance. Give an explanation if you can.**

Looking through the Red Lens:

What happens? ____________________

Why does it happen? ____________________

Looking through the Green Lens:

What happens? ____________________

Why does it happen? ____________________

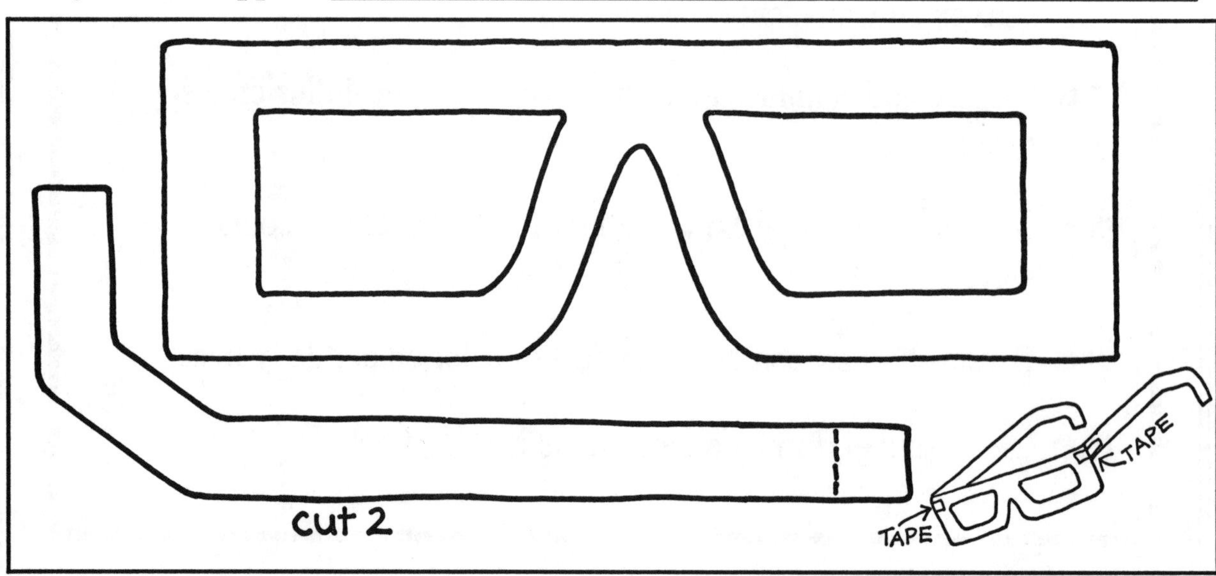

Name ____________________

EXPLORING

Dinosaurs, Predators and Prey, and Endangered Species

Exploring Dinosaurs à la Bloom

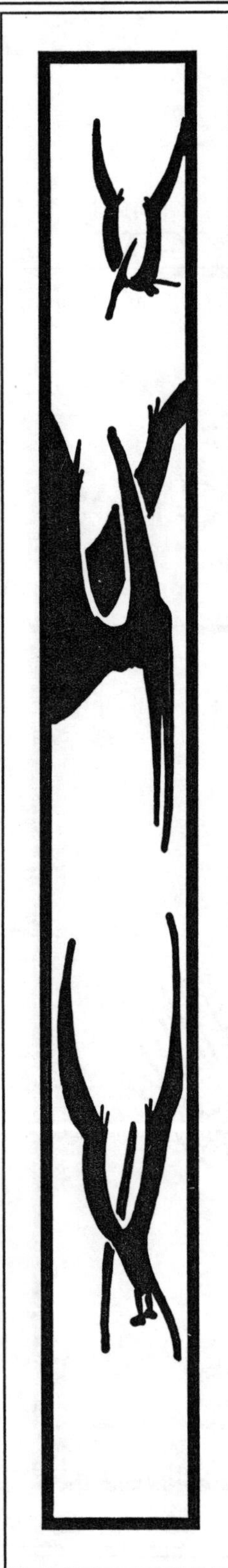

• KNOWLEDGE

Define each of the following words:

prehistoric	**fossil**
paleontologist	**fossil fuel**
herbivorous	**carnivorous**

• COMPREHENSION

Explain the adaptive or defensive features and structural characteristics of dinosaurs.

• APPLICATION

Construct a series of "Who Am I?" cards by describing several different species of dinosaurs on 3" x 5" index cards.

• ANALYSIS

Devise a theory to explain why dinosaurs became extinct. Predict how things would be different today if the dinosaurs had lived.

• SYNTHESIS

Pretend that you are a paleontologist and have just uncovered the bones of an unknown species of dinosaur. Write a journal entry to describe when and where the bones were found, what you named the dinosaur, what the dinosaur probably looked like, and approximately how big the dinosaur was. Draw a diagram of the dinosaur.

• EVALUATION

Set up a list of criteria to use in assessing at least five books about dinosaurs to determine which book is the best reference for your class. Write a paragraph summary of your conclusions.

Multiple Intelligences Can Help Us Learn More about Predators and Prey

Page 1

• VERBAL/LINGUISTIC

Discuss the concepts of predator and prey in a substantial paragraph.

• LOGICAL/MATHEMATICAL

Predators catch prey by using a variety of adaptations. Construct a chart showing different predators that exemplify each of the following adaptations: superb eyesight, muscular power, strong jaws, silence, speed, agility, persistence, patience, and camouflage.

• VISUAL/SPATIAL

Select an animal that uses camouflage as part of its protection such as a lizard, butterfly, or grasshopper. Draw a picture to show the animal in an environment in which it might hide. Next, think of a predator that would prey on the animal you have drawn and add it to the picture.

• BODY/KINESTHETIC

Design and play an original game called "PREDATOR or PREY."

Multiple Intelligences Can Help Us Learn More about Predators and Prey

Page 2

• MUSICAL/RHYTHMIC

Create a musical collage of different compositions or musical scores that seem to portray a predator and prey relationship.

OR

Listen to the famous musical composition "Peter and the Wolf" and describe the instruments that are used to represent the predator and the instruments that are used to represent the prey. Explain why you think the composer made these choices.

• INTERPERSONAL

Working with a partner, pick a common predator and one of its prey. Use a reference book to locate five to ten facts that describe the predator and five to ten facts that describe the prey. Sketch the basic shape of each creature on a sheet of paper. Then write all or some of your factual statements around each animal's shape.

• INTRAPERSONAL

Create a personal superhero/heroine who is named after a well-known predator that you admire. List this person's special powers and characteristics. Design a costume that "fits" the name you have chosen.

Using Williams' Taxonomy to Study Endangered Species

• FLUENCY

List as many endangered species of animal life as you can.

• FLEXIBILITY

Write down many examples of how human beings have made changes in the environment which have endangered several animal species.

• ORIGINALITY

Animals provide people with products to meet their needs and wants. Discover the unusual or unique products that are provided by at least ten different animals.

• ELABORATION

Explain how each of the following groups have worked to protect the world's endangered species: **Audubon Society, Sierra Club, National Geographic Society, park and forest rangers, National Wildlife Federation.**

• RISK TAKING

Pretend you are on a city council that must decide if a community baseball field/stadium should be built on a meadow that serves as a bird and wildlife sanctuary. Prepare a logical argument for or against the field/stadium that might be offered by each of the following individuals. Make a decision.

- **wildlife biologist**
- **local bird watcher**
- **baseball player**
- **sports fan**
- **building contractor**
- **local tax collector**

• COMPLEXITY

Choose an endangered species of interest to you. Do some research to find out why the organism is endangered and how the organism might be saved. Outline a plan of action that could be used to prevent extinction.

• CURIOSITY

It has been said that people "can be shaped by their environments." Is this good or bad? Give some examples and/or arguments for others to ponder or wonder about in answering this question.

• IMAGINATION

Imagine you had the power to save one animal species from extinction. Write about the one you would choose to save and why.

Research Topics

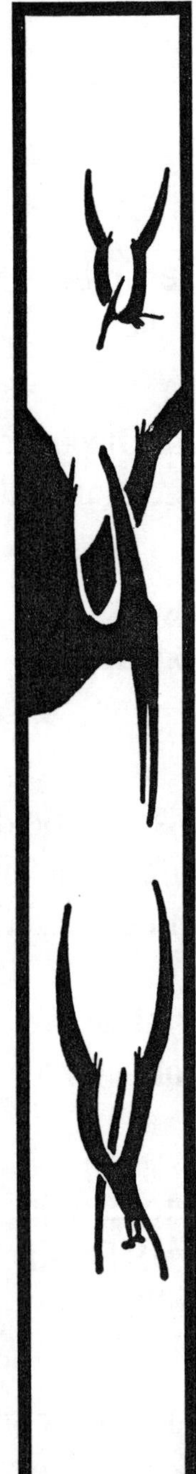

1 ● Find out about the work of a paleontologist.

2 ● Find out about plant-eating dinosaurs and meat-eating dinosaurs.

3 ● Find out about the plant and animal life of prehistoric times.

4 ● Find out about the various sizes, habitats, and food sources of ten different dinosaurs.

5 ● Find out about the tools, equipment, and resource materials that are important to a scientist studying prehistoric life.

6 ● Find out about some of the archeological digs that have uncovered dinosaur remains.

7 ● Find out about predator-and-prey relationships in the ocean, in the desert, in the rain forest, and in the mountains.

8 ● Find out about food chains and their implications for predator-and-prey relationships.

9 ● Find out about adaptations and characteristics that make various animals effective predators or susceptible prey.

10 ● Find out how these predators trap prey: octopus, sea anemone, sea urchin, triggerfish, and angler fish. Then find out how the following prey protect themselves: barracuda, grouper, jellyfish, lantern fish, and lobsters.

11 ● Find out how animal protection organizations are funded and organized to do their important work.

12 ● Find out the steps that are being taken to protect various animals that are in danger of becoming extinct.

13 ● Find out which animals have become extinct over the years.

14 ● Find out which animals are on the endangered species list and how they came to be in their precarious situation.

15 ● Find out what you can do to protect the animals in your community, state, or country.

Creative Writing Topics

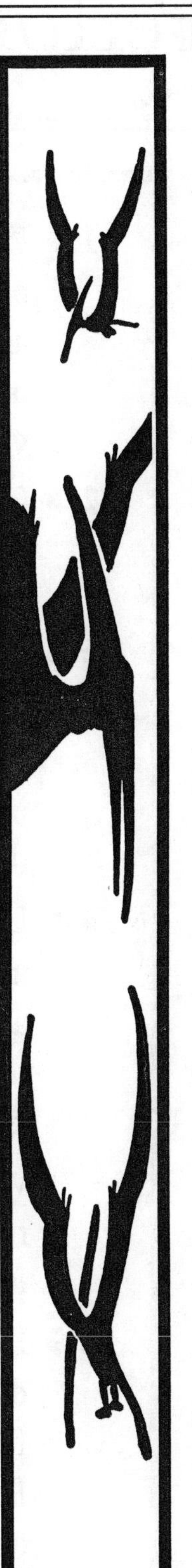

1 ● Create a series of name poems to tell key facts about an endangered species. Consider the orangutan, armadillo, sloth, manatee, ferret, gazelle, panda, ocelot.

2 ● Write an original fairy tale, fable, or legend about one of the following endangered species: snow leopard, whooping crane, spider monkey, white rhinoceros, Ceylon elephant, or Andean condor.

3 ● Create a play about a pet dinosaur that came to live with you and your family.

4 ● Select a predator-and-prey relationship that interests you. Construct paper bag puppets to represent these animals and put on a puppet show entitled "The Day We Became Friends."

5 ● Write a "public service" announcement to make others aware of several of the groups and organizations that help protect the world's endangered species.

Assessment Product and Performance Options

Page 1

PRODUCTS

1 ● Use reference materials to research one animal in danger of becoming extinct. Write a flip book report that describes the animal's natural habitat, facts about the animal, steps that are now being taken to protect the animal, and some questions that you have about the protection of that species of animal.

2 ● Collect magazine and newspaper articles related to endangered species. Use the articles to show causes and effects of human carelessness and indifference.

3 ● Write a letter to one of the organizations devoted to the preservation of animals in danger of extinction. Express your specific concern for an endangered species and ask for more information about the animal as well as suggestions for ways that you can help.

4 ● Design a badge that kids could wear to show that they care about the world's endangered species. Write a short pamphlet about the badge color/design/symbols and tell what you think these various elements represent.

5 ● Summarize what you have learned about prehistoric life by writing a feature article suitable for publication in your local newspaper. Be sure to use accurate facts and to organize your information so that the article is easy to read.

6 ● Design a survey on the different theories that have been developed to explain the disappearance of dinosaurs from the face of the earth. Administer this survey to a number of your friends to determine which theory makes most sense to them and why. Share your findings in chart or graph form.

Assessment Product and Performance Options

Page 2

PERFORMANCES

1 ● Compose a series of role plays showing various conflicts that might arise between a group of environmentalists and a group of community planners with different interests.

2 ● Organize a mock archeological dig for your classmates.

3 ● Plan an "endangered species" rally for your school to inform others about the many ways that human behavior has contributed to the near-extinction of various animal species.

4 ● Prepare a series of television commercials to encourage the public to join at least one community, state, or national organization devoted to the care of the plant and animal environments throughout our world today.

5 ● Create a set of transparencies showing various ways that animals protect themselves from predators and/or ways that animals entrap their prey. Use the set of transparencies as a prop that will help you give an oral report on the topic.

6 ● Prepare a set of fact and/or flash cards about various endangered species from around the world. Use these with a group of friends to prepare for a test, a panel, or a quiz show.

7 ● Prepare a dramatic reading of an excerpt from a fiction or nonfiction book that depicts the terror of the dinosaur age.

8 ● Mime a series of predator-and-prey relationships, and have your friends guess what situations you have chosen to portray.

9 ● Write a series of questions you would like to ask a park or forest ranger. Use them to conduct a personal or telephone interview.

Dinosaurs Can Be Puzzling

Solve this puzzle to find the "state" of dinosaurs today. To find the hidden word, carefully read each sentence below. If the statement is true, color the numbered spaces as directed. If the statement is false, color nothing.

1. **If the scientific name of a person who studies plants and animals of the past is paleontologist, color the #1 spaces.**
2. **If *Pteranodon* was a winged reptile that soared through the air, color the #2 spaces.**
3. **If *Stegosaurus* was fast and slender, color the #3 spaces.**
4. **If *Protoceratops* is known as the "egg-laying dinosaur," color the #4 spaces.**
5. **If *Tyrannosaurus* was a huge, fierce, meat-eating creature, color the #5 spaces.**
6. **If *Plesiosaurus* lived in caves, color the #6 spaces.**
7. **If some dinosaurs are still living in jungles of the Amazon, color the #7 spaces.**
8. **If *Anatosaurus* is known as the "duck-billed dinosaur," color the #8 spaces.**
9. **If *Diplodocus* was a small, meat-eating creature, color the #9 spaces.**

6	3	9	7	3	6	9	7	6/3	9	7	6	3	9	6	3	7	6	9	3	9	3	6/3	9	7	3	9
7	6	7	3	9	3	6	3/6	7/6	3	6	7	6	3	9	6	9	6	7	6	7	6	6	3	6/3	6	/3
3	9	7	6	7	6	7	7/6	9	7	3	6	7	9	7	9	3	3	9	7	3/6	9	7	6	7	9	7
7	2	5/8	6	1	3	5	9	8/5	4	8	9	1	3	4	2\3	9	8	7	3/5	4	8	6	2	1/4	8	9
6	4	3	7	6\8	7	2/6	3	6	2	3	9	4	6	1	6\5	4\7	5	9	8	6	3	7	6	8	3	7
9	8	5	9	3	1	9	6	7	8	7	3	8	9	5	3	9\2	1	6	1	7	6	3	9	5	6	9
6	2	3	6	3/4	9	5\7	3	6	4	9	7	2	6	8	7	9	4	3	2	3	7	9	6	8	3	7
3	4\5	1	3	2	7	4	9	7	2/1	6	3	1	7	5	6	7	2	7	6\4	5	1	7	3	1	7	6
7	6\3	6	7	3	9	6	7	9	7	6	7	3	6	3	9	3	7	6	7	9	3/7	6	7/9	6	7\9	3
6	7\9	6\7	9	7	6	3	7	3	9/3	7	6\3	7	7	9	6	7	3	7	9	7	9	3	6	9	9	7
3	6	7	3	9	7	7	6	9	7	7	3	6	9	6/3	9	7	6	7	6	9	3	7	3	6	3	6

Name ______________________________

EXPLORING

Insects

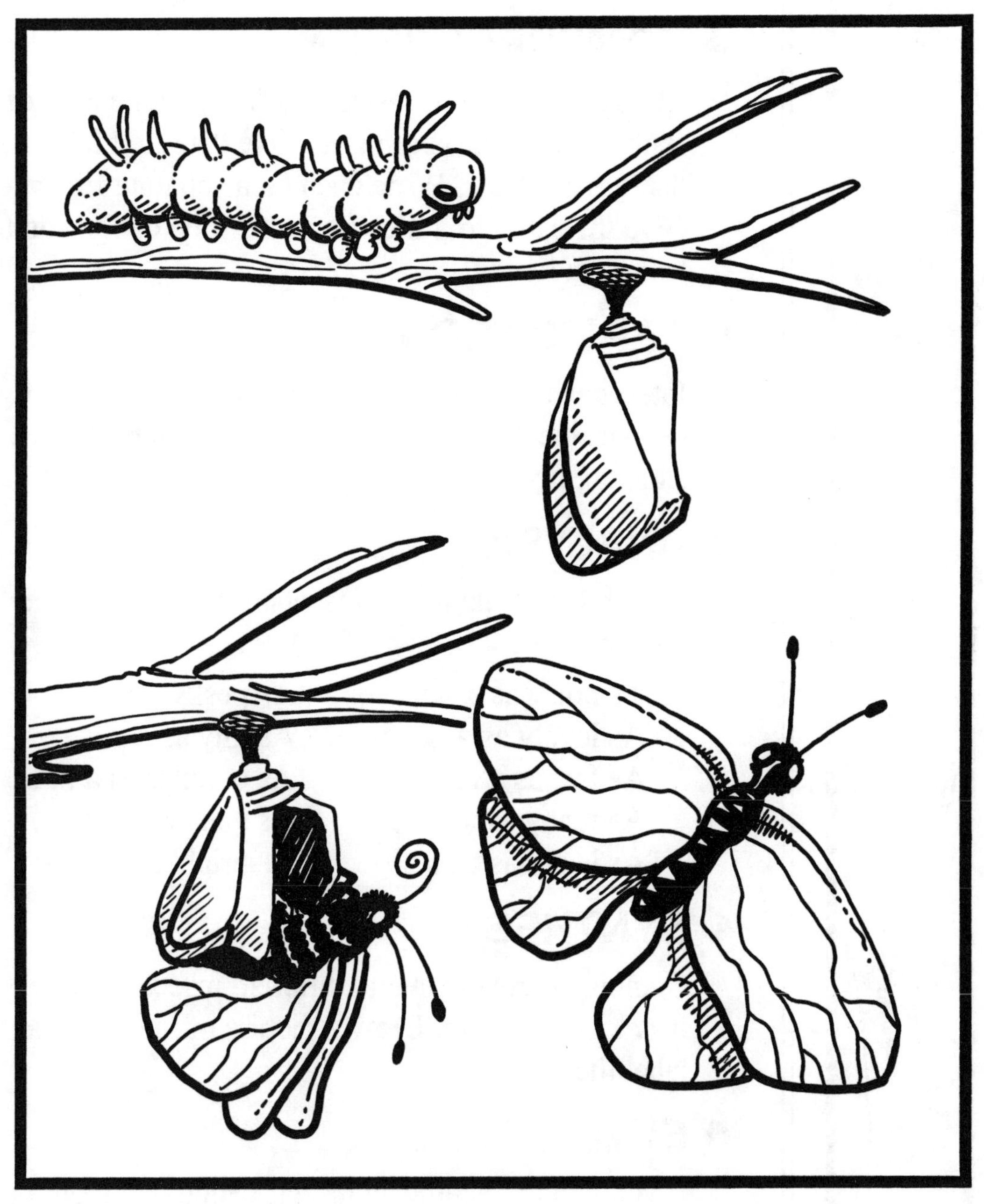

Exploring Insects à la Bloom

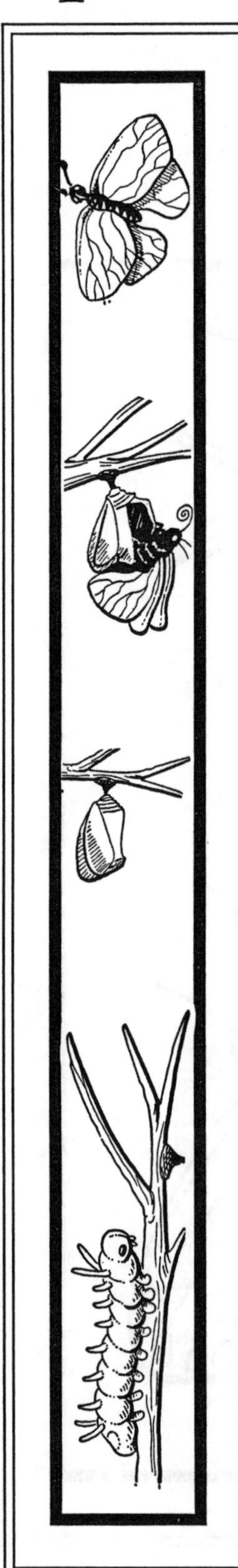

• KNOWLEDGE

Brainstorm a list of insects, at least one insect for each letter of the alphabet. Group your list of bugs as helpful or harmful insects.

• COMPREHENSION

Choose an insect to research. Make a collection of ten statements about your insect, five that are facts and five that are not facts. Write each on a separate file card. Give the cards to a friend and see if he or she can tell fact from fiction.

• APPLICATION

Interview ten people to find out which insect each dislikes the most. Graph the results and draw conclusions from your findings.

• ANALYSIS

Search for the names of insects to finish these similes or comparisons.

As noisy as a . . .
As lovely as a . . .
As fast as a . . .
As funny as a . . .
As leggy as a . . .
As pesky as a . . .
As tiny as a . . .
As popular as a . . .
As unusual as a . . .
As bright as a . . .

• SYNTHESIS

Combine parts of several different insects to create a new kind of insect. Draw a picture of the insect and label the parts.

• EVALUATION

If you had to become an insect, which one would you be? Support your choice with at least five reasons.

Multiple Intelligences Can Help Us Learn More about Bugs Page 1

Note: **Although many insects are often called bugs, true bugs are only a small group, whose scientific name is "Hemiptera." Some examples of bugs are aphids, pond skaters, bedbugs, water boatmen, and shield bugs. Bugs all have sharp mouth-parts designed to pierce plants and skin and suck up sap or blood into their mouths. Finally, bugs are like beetles, except that only half of their wing-cases are hard.**

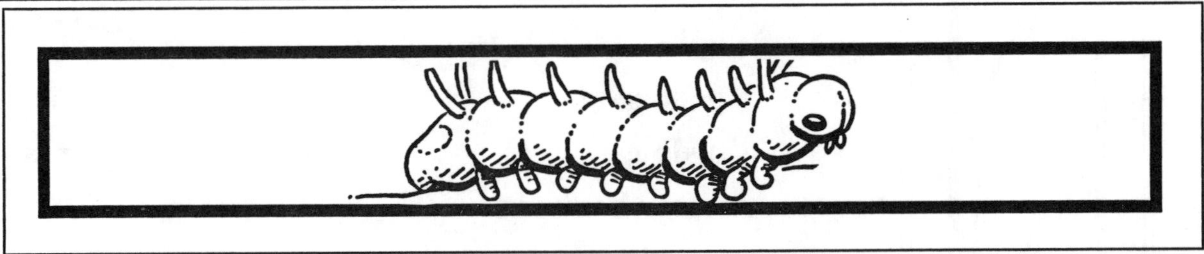

• VERBAL/LINGUISTIC

Pretend you are a bug found in your backyard. Write your life story, making sure your ideas are as scientific as possible.

- ***When were you born?***
- ***What is your family history?***
- ***What work do you do?***
- ***Who are your friends?***
- ***What special skills or talents do you have?***
- ***Where do you live?***
- ***What are your hobbies?***
- ***Where do you like to travel?***
- ***What do you look like?***

• LOGICAL/MATHEMATICAL

Do some research to find out at least five ways that humans protect themselves from harmful bugs.

• VISUAL/SPATIAL

Invent a new kind of bug spray. Be sure to tell what the bug spray is for and what it will do. Think of a clever name for your repellent and draw a picture to show what your repellent does.

Multiple Intelligences Can Help Us Learn More about Bugs Page 2

● BODY/KINESTHETIC

How would you draw each of the bugs listed below if you did not know what each bug was but you did know what the words in their names meant? Be creative in your drawings. Then try imitating the movements of each insect you have drawn.

- **aphid**
- **pond skater**
- **bedbug**
- **water boatman**
- **shield bug**

● MUSICAL/RHYTHMIC

Learn Morse Code and use it to communicate the names of common bugs.

● INTERPERSONAL

Work with a friend to draw the habitat or natural environment of each of these insects: **grasshopper, mole cricket, termite,** and **dragonfly.** Use reference materials in the library to find information about each insect.

● INTRAPERSONAL

Which bug has the strongest effect—positive, negative, or both—on you and your family? Defend your answer in a detailed paragraph.

Using Williams' Taxonomy to Study Creepy Crawlies

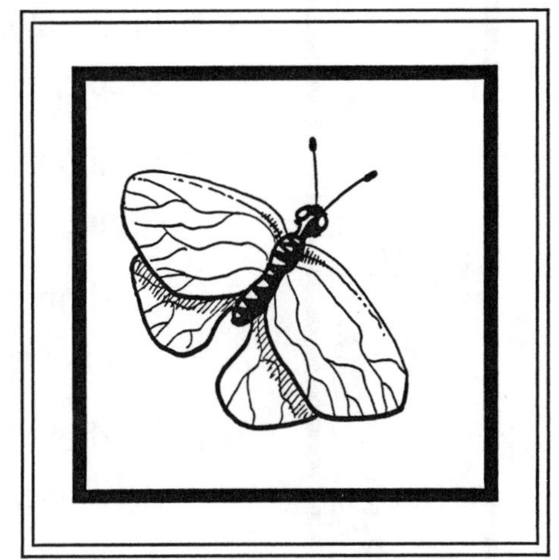

• FLUENCY

List as many facts about the insect world as you can.

• FLEXIBILITY

Write down the names of all the body parts of a typical insect and tell what is unique about each part.

• ORIGINALITY

Pretend you are a scientist who has just discovered a new species of insect. Draw it, describe it, and distinguish it from other "critters."

• ELABORATION

Determine whether or not you agree with the following statement and tell why: "Spiders are a type of insect."

• RISK TAKING

Define **metamorphosis** as it relates to the insect world, and then explain what type of metamorphosis *you* would like to undergo at the present time.

• COMPLEXITY

Decide why the ocean is the only place on Earth where insects are not found.

• CURIOSITY

Write down a list of questions you would like to ask a scientist about the social lives of bees, ants, hornets, and termites.

• IMAGINATION

Imagine life as a flea. Tell about the animal and the family with which it lives.

Research Topics

1 ● Find out why the insect group is one of the most successful groups of animals.

2 ● Find out how insects play an important part in the world of nature.

3 ● Find out about insects that undergo complete or partial metamorphoses.

4 ● Find out how various insects protect themselves from their enemies.

5 ● Find out about **arachnids** and **arthropods.**

6 ● Find out about social insects.

7 ● Find out about the techniques used in collecting insects.

8 ● Find out about moths and butterflies.

9 ● Find out about some relatives of insects such as **spiders, centipedes,** and **millipedes.**

10 ● Find out about different insect homes and habitats.

11 ● Find out how insects touch, feel, taste, talk, see, and hear.

12 ● Find out about insect eggs.

13 ● Find out about beetles and ladybugs.

14 ● Find out about prehistoric insects.

15 ● Find out how human beings attempt to control insects.

16 ● Find out how bugs differ from insects (from a scientific point of view).

Creative Writing Topics

1● Compose a short story using one of the following titles:

- **The Garden Pest**
- **Molting Magic**
- **Flashing Lights in the Night**
- **The Anthill Hotel**
- **Bees, Buzzes, and Me!**
- **Beetles Are Like That!**

2● Construct a series of riddles about different insects for others to answer. Read this example first:

> **I am a type of insect with chewing mouth parts, two pairs of wings (front wings narrow and leathery and hind wings broad and membranous), medium to long antennae, and incomplete metamorphosis. What type of insect might I be?**

Answers: **crickets, grasshoppers, locusts, mantids,** and **cockroaches.**

2● Keep a diary for one of the social insects, telling about your life as a member of a special group and living as a member in a special community.

4● Create a series of rhymes about different types of insects. Read through this sample first:

> **A glowworm looks bright**
> **as it glows in the night.**
> **It's a strange kind of beetle**
> **whose light can cause fright.**

5● Invent an "Insectabet" whose basic shape reflects one letter of the alphabet. Draw your critter and describe it in detail.

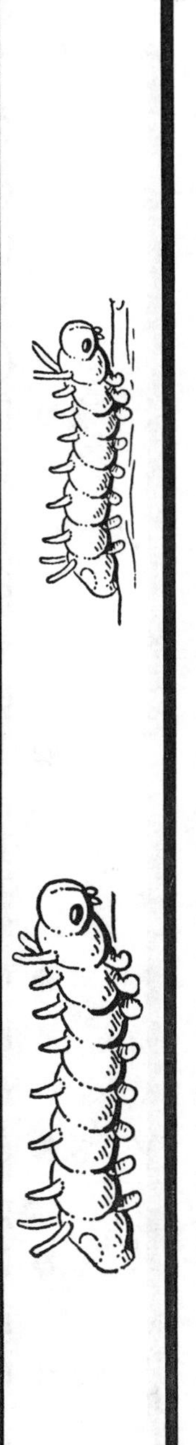

Assessment Product and Performance Options

Page 1

PRODUCTS

1 ● Construct a chart that compares and contrasts at least ten different insects. Consider the legs (how many, what size, if they are jointed), the body (number of parts, outer covering, size, length), antennae (length, shape, number), and wings (shape, size, number).

2 ● Before your study of insects, draw an insect from memory without looking at a picture of one. Include all of the body parts you can think of. Next, make a list of all the insects you can think of. After your study of insects, draw an insect from memory and make a list of all the insects you can think of. How do your drawings and lists compare with one another?

3 ● Take a walk around your school and neighborhood and look for habitats of many different insects. Draw each habitat you see and write down where you found it. Be sure to look under stones, in the grass, and under leaves.

4 ● Make an insect collection of your own using specimens you have caught or insects you have constructed from simple materials. Label each insect and briefly tell something about it.

5 ● Construct a labeled diagram of several insects and their body parts.

6 ● Make a project cube which identifies ways insects are helpful and/or harmful to man.

7 ● Create a glossary of important insect terms and concepts. Make a set of flash cards from your glossary.

8 ● Design a flip book to illustrate the movement and activities of a favorite insect.

Assessment Product and Performance Options

Page 2

PERFORMANCES

1 ● Write a script for a play and perform it on one of the following topics: a day in the life of a **wasp queen,** a **soldier ant,** a **drone bee,** or a **king termite.**

2 ● Create a series of dances to show the movements and habitats of several different insects.

3 ● Invent a science fiction television series with insects as the central characters. Describe the characters, setting, and plot for the pilot program and act it out for an audience.

4 ● Create a set of action scenes depicting the behavior within a hornet's nest, an ant colony, a beehive, or a termite's nest.

5 ● Mime the life stages of one or more insects going through a complete metamorphosis.

6 ● Assume the role of a chief construction engineer for a **termitarium.** Give your orders to your crews of workers.

7 ● Perform a creative and interpretive "dance of the insects" to a piece of appropriate music.

8 ● "Become" various insects (act out their activities).

9 ● Demonstrate how to catch and mount an insect for display.

10 ● Design a "live mural" of various insects in their surroundings.

Name an Insect Who . . .

Write the name of an insect to go with each of the characteristics below:

1. An insect who is a music-maker ______________________
2. An insect with a powerful jaw for chewing ______________________
3. An insect with a coiled tube for sipping liquids ______________________
4. An insect that is helpful to humans ______________________
5. An insect who is a "social" insect ______________________
6. An insect who has only two wings ______________________
7. An insect that lays its eggs in or on the water ______________________
8. An insect that molts ______________________
9. An insect without a *pupa* stage ______________________
10. An insect that is active all year, even when winters are cold ______________________

Name ______________________

EXPLORING

Oceans

Exploring Oceans à la Bloom

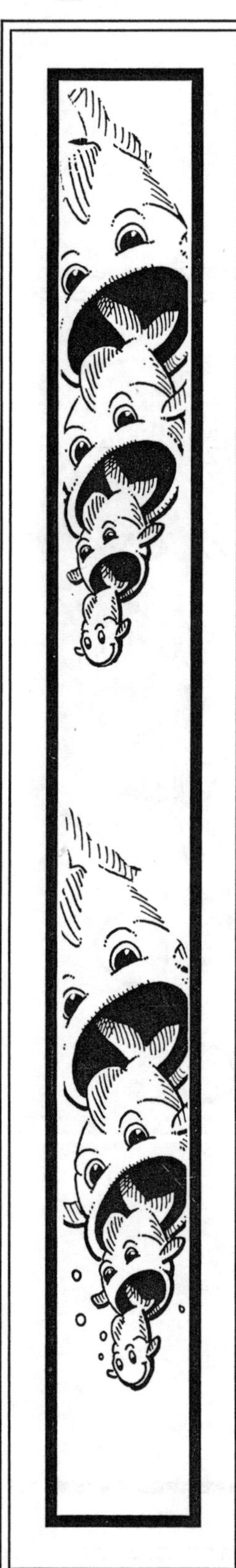

• KNOWLEDGE

List the major oceans and seas of the world and write down an important fact about each one.

• COMPREHENSION

Use each of the following ocean-related terms in a sentence to convey the meaning of the term: **continental slope, currents, tides, ocean floor, coast, waves, swells,** and **salinity.**

• APPLICATION

Construct a series of word problems with subjects relating to the ocean. For example, a family harvests 500 pounds of seaweed a week. Thirty percent is sold to a processing plant to be made into medicine and 40 percent is sold to a local market. How many pounds does the family have left to eat?

• ANALYSIS

Compare/contrast an ocean with a lake, a sea, and a river.

• SYNTHESIS

Design an underwater resort of the future. Create a travel brochure to describe it in detail.

• EVALUATION

You are about to take a long ocean voyage in a handcrafted sailboat. Since space is limited, you will be able to take only five of the following items: blanket, oar, first-aid kit, life preserver, anchor, compass, hunting knife, fresh water/food supply, and one other object of your choice. Which items will you take with you ? Why?

Multiple Intelligences Can Help Us Learn More about the Mysteries of the Ocean

- **VERBAL/LINGUISTIC**

  Write a series of paragraphs to describe the phenomena of waves, currents, and tides.

- **LOGICAL/MATHEMATICAL**

  Do some research to find information about the tools of oceanography and underwater exploration such as the **periscope** or the **bathysphere.** Record your findings in chart form.

- **VISUAL/SPATIAL**

  Draw a map of the ocean floor that includes the following: **continental shelf, estuary, abyssal plain, canyon, continental slope, rift valley, guyot, sea mount, trench,** and **island.**

- **BODY/KINESTHETIC**

  Paint a mural of either a coral reef and all of the marine life associated with it or marine life at the different levels of the sea.

- **MUSICAL/RHYTHMIC**

  Compose an ocean dance to show the actions and movements of different **mollusks, crustaceans, echinoderms, anemones,** and **coelenterates.**

- **INTERPERSONAL**

  Work with a partner to compare exploration of the seas and oceans with exploration of outer space. Determine which of these two options is most likely to provide us with better resources in the years ahead.

- **INTRAPERSONAL**

  Write a short essay explaining which "mystery of the ocean" is of most interest to you and why.

Using Williams' Taxonomy to Study Marine Life

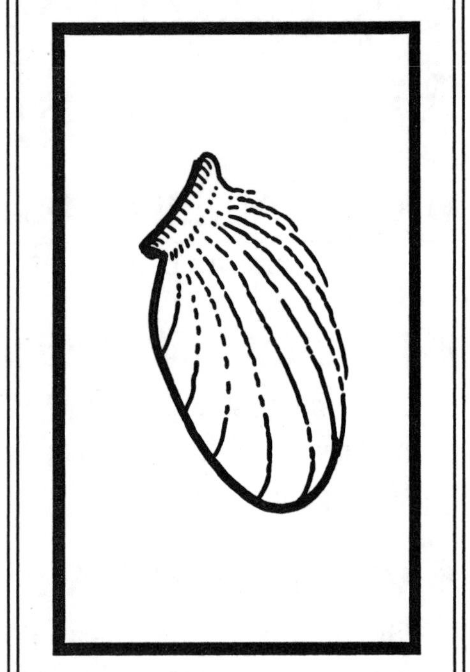

- **FLUENCY**

  List all of the different types of marine life one might find living in the sea.

- **FLEXIBILITY**

  Group your list of marine life items in some way and explain your classification scheme.

- **ORIGINALITY**

  Do some research to find information about the most unusual or unique type of mammal that makes its home in the sea.

- **ELABORATION**

  Defend or negate this statement:
  "Seaweed and kelp are important products of the sea."

- **RISK TAKING**

  Determine which of the following kinds of fish is most like you and why: **blue ribbon eel, clown triggerfish, eyed electric ray, lionfish, mudskipper, powder blue surgeonfish,** or **viperfish.**

- **COMPLEXITY**

  Explain how oil spills, sewage, pesticides, and dumping of toxic wastes contribute to the demise of ocean life.

- **CURIOSITY**

  If you could talk with a deep-sea diver, what would you would want to ask about the marine life he or she has seen in underwater explorations?

- **IMAGINATION**

  Imagine what life would be like if all of the continents were connected instead of surrounded by the ocean or the seas. How would things be different?

Research Topics

1 ● Find out about underwater exploration tools and challenges.

2 ● Find out about **squids, stingrays,** and **octopi.**

3 ● Find out about ocean giants such as the **blue whale, whale shark,** or **Pacific leatherback turtle.**

4 ● Find out about poisonous sea creatures such as the **Portuguese man-of-war,** the **blue-ringed octopus,** and the **stonefish.**

5 ● Find out about the topography of the sea.

6 ● Find out about what keeps the oceans from freezing.

7 ● Find out about measuring ocean depths and speed.

8 ● Find out about the unsolved mystery of "Mary Celeste."

9 ● Find out about icebergs.

10 ● Find out about important nautical terms.

11 ● Find out about the sinking of the Titanic.

12 ● Find out about endangered species of the sea.

13 ● Find out about careers in oceanography.

14 ● Find out about metals and oil obtained from the sea.

15 ● Find out about lighthouses, flag messages, radar, and sonar as safety measures for ships traveling the high seas.

16 ● Find out about famous ports and harbors of the world.

17 ● Find out about the plant life in the ocean.

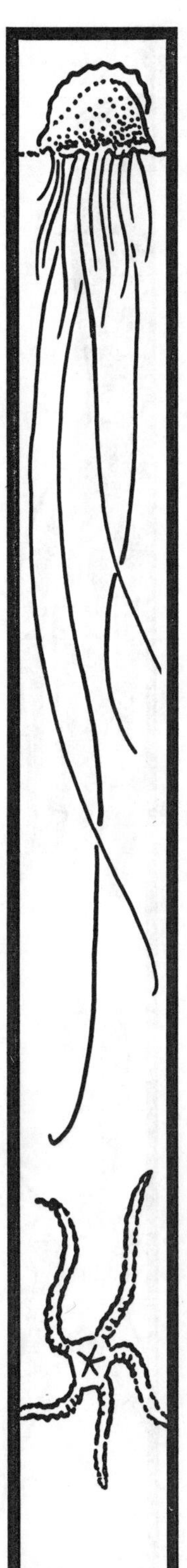

Creative Writing Topics

1 ● Write a short story that has one of the following titles:

- **Finding the Sunken Treasure**
- **My Adventures with the Frogmen**
- **Scuba Diving Can Be Dangerous**
- **The Day the Octopus Stole My Camera**
- **It's a Whale of a Story**
- **Shark Hunting**

2 ● Write a series of diary entries that might have been recorded by someone who was on the Titanic in the Atlantic Ocean on April 15, 1912.

3 ● Personify an animal of the sea and write about your adventure.

4 ● Collect a series of poems about the sea and arrange them in a poetry collage. Add your own original poem to the collage.

5 ● Compose a simple essay telling what you like about the sea.

6 ● Write a tall tale about a fish you have caught or a fish that got away.

Assessment Product and Performance Options

PRODUCTS

1 ● Write an ABC Report about the ocean and begin each idea and page with a different letter of the alphabet.

2 ● Create a series of Top Ten lists around a sea theme, including such items as:

- **Ten Things You Should Know about Deep-sea Diving**
- **Ten Famous Sea Voyages from History**
- **Ten Interesting Fish to Study**
- **Ten Secrets of the Ocean**
- **Ten Ways the Sea Helps People**
- **Ten Things to Remember about Waves, Tides, and Currents**

3 ● Do some research to find out more about pirates and develop an audiotape of stories about them.

4 ● Pretend you are a newspaper reporter and write a true feature or news story about a helicopter sea rescue, an oil spill, or the finding of sunken treasure.

5 ● Locate information about building a dike and construct a series of pictures or diagrams to share what you know.

6 ● Use the *Guinness Book of World Records* to locate some interesting records of the sea and its animal life. Create a mural to illustrate these facts.

7 ● Collect shells and prepare a hobby display of the shell pieces in your collection.

8 ● Construct a timeline that shows when life first appeared in the seas and when life began on land.

9 ● Design a postage stamp to commemorate a person, place, or thing related to the sea. Write a brief commentary on it.

Assessment Product and Performance Options

Page 2

PERFORMANCES

1 ● Read a book about the sea such as *Kon-Tiki, Jaws, Robinson Crusoe, Moby Dick, Voyage to the Bottom of the Sea,* or *20,000 Leagues Under the Sea,* and give a book talk about it.

2 ● Prepare and deliver a skit about sea scavengers who feed on dead and decaying plants and animals.

3 ● Give a chalk talk explaining a food chain that has a killer whale at the top of the chain.

4 ● Prepare a chart comparing seals and sea cows. Make a transparency of it and present a mini-lesson to the rest of the class.

5 ● Write and illustrate a children's picture book about a special mammal of the sea. Organize a story hour for younger children and read it aloud to them.

6 ● Prepare an experiment and demonstration for the class on one of the following topics:

- **Archimedes' Principal of Buoyancy**
- **How Salt Adds Buoyancy to Water**
- **Effects of Temperature on Water**
- **Movements of the Earth and Its Influences on Tides**
- **How Depth Affects Water Pressure**

7 ● Design a floating city and give a "travelogue" about it.

8 ● Role play a mock interview with a deep-sea diver and discuss his or her experiences with salvaging cargoes and treasures of sunken ships.

9 ● Demonstrate the use of a periscope.

Protection Please

Select one marine animal that is near extinction. Use reference books to find information about the animal. Organize the information on the form below. Then design a poster (page 88) that encourages protection of the animal.

ANIMAL: ______________________________

Habitat

History

Behavior

Other

Reference Materials Used (include title, author, copyright date, and page numbers):

____

____

____

Name ______________________

Name ______________________________

EXPLORING

Planets

Exploring Planets à la Bloom

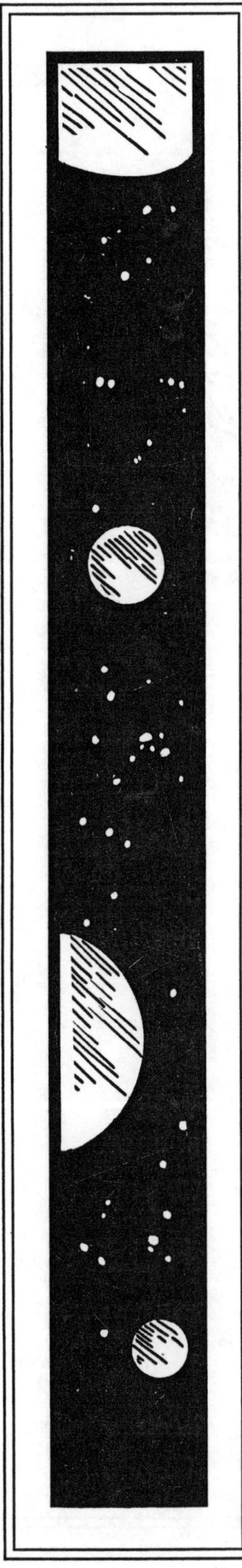

● KNOWLEDGE

Define each of the following terms associated with our study of the Sun, the Moon, and the planets: **solar system, orbit, asteroids, comets, meteors, craters, gravity, light year, moon, sunspots,** and **phases of the Moon and Earth.** Add any other words you think are important. Use these words and their definitions to make a set of flash cards.

● COMPREHENSION

Briefly describe the major characteristics of each of the nine planets.

● APPLICATION

Construct a model of our solar system using a variety of differently textured materials to represent the planets.

● ANALYSIS

Many planets are named after gods and goddesses from mythology. Explain why you think some of these names were chosen.

● SYNTHESIS

Construct an original mnemonic device to help you remember the names of the planets in order of their distance from the sun, beginning with Mercury and ending with Pluto.

● EVALUATION

Many people believe there is life on other planets which is more advanced than the life on Earth. Describe how you feel about this idea and try to justify your thoughts with important facts and figures.

Multiple Intelligences Can Help Us Learn More about Outer Space

- **VERBAL/LINGUISTIC**

  Read stories, poetry, and articles about outer space. Compile an annotated bibliography of your readings.

- **LOGICAL/MATHEMATICAL**

  Develop a set of recommendations to Congress encouraging more (or less) exploration of space. Be sure to justify your ideas with logic and facts.

- **VISUAL/SPATIAL**

  Design a poster, mural, or scroll that shows in some interesting way everything you have learned about space.

- **BODY/KINESTHETIC**

  Reenact great moments in the history of space exploration.

- **MUSICAL/RHYTHMIC**

  Make up a creative and interpretive dance to show different dimensions of life in outer space.

- **INTERPERSONAL**

  Working with a group of peers, and using *The Magic School Bus* format with Ms. Fizzle, write your own version of a classroom trip through outer space!

- **INTRAPERSONAL**

  Create a résumé that would convince the people at NASA that you should be the first student in outer space. What subjects, hobbies, work experience, school activities, and personal attributes would make you a good candidate?

Using Williams' Taxonomy to Study the Stars

Page 1

- **FLUENCY**
  Name as many different constellations as you can think of in two minutes.
- **FLEXIBILITY**
  Draw diagrams of the constellations that make up the zodiac.
- **ORIGINALITY**
  Make up a new constellation and describe it in detail.
- **ELABORATION**
  Create a drawing to illustrate at least ten different facts about the stars.
- **RISK TAKING**
  Discuss which constellation is most like you, and tell why.

Using Williams' Taxonomy to Study the Stars

Page 2

● COMPLEXITY

Develop a theory to determine the possible origin of the stars. You may want to do some research first.

● CURIOSITY

If a telescope could listen and talk, what questions would you like to ask it? Write the questions in a mock conversation, using "balloons" to capture the dialogue between you and the telescope.

● IMAGINATION

Visualize yourself as a star and write your biography, tracing your life from beginning to end.

Research Topics

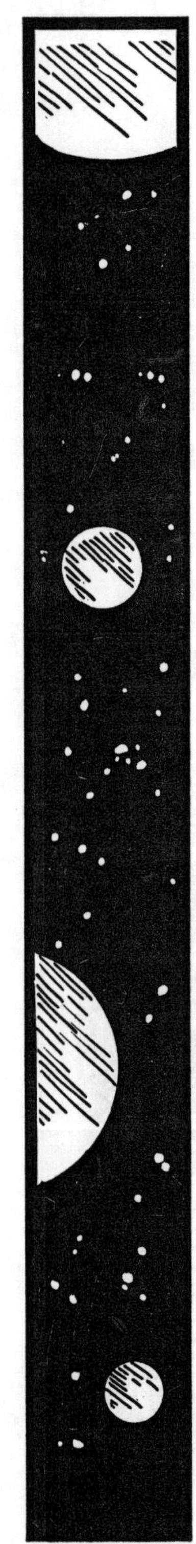

1 ● Find out about what makes night and day and the seasons.

2 ● Find out about the moon's gravity.

3 ● Find out about Halley's Comet.

4 ● Find out about Galileo Galilee.

5 ● Find out about the training of astronauts.

6 ● Find out about the history of rocketry.

7 ● Find out about the zodiac and how it has changed since ancient times.

8 ● Find out about the cause of an eclipse.

9 ● Find out about the work of an astronomer.

10 ● Find out about space labs and space shuttles.

11 ● Find out about the dangers and problems of space travel.

12 ● Find out about the power of the sun.

Creative Writing Topics

1 ● Write a short story that has one of the following titles:

- **Mission Alpha**
- **Mysteries of Space**
- **The Star That Got Lost**
- **The Planet That Hates Its Name**
- **Sky Pictures**
- **Hyperspace!**
- **It's Not a Bird, It's Not a Plane, It's a UFO!**
- **How Jupiter Really Got Its Rings**
- **3 - 2 - 1 - LIFTOFF!**
- **My Ride in a Skymobile**
- **Moon Rescue**

2 ● Use various poetry forms to write about the Sun, the Moon, or the stars. Consider **haiku, tanka, diamante, free verse,** and **concrete poems**.

3 ● Use personification to write a series of adventures or episodes about Rod-Nee Robot and his work in the space program.

4 ● Draw a set of comic strips concerning the topic of The Big Bang Theory or The Black Hole Concept.

Assessment Product and Performance Options

Page 1

PRODUCTS

1 ● Write an ABC Report about a topic related to your study of the solar system and outer space. Use each letter of the alphabet to describe an important concept in your topic.

2 ● Write a script for a television panel show entitled "The Lives and Times of America's Astronauts." Consider such heroes as Alan Shepard, Jr., Virgil Grisson, John Glenn, Jr., Scott Carpenter, Walter Schirra, Jr., Gordon Cooper, John Young, James McDivitt, Frank Borman, and James Lovell, Jr., to interview.

3 ● Create a chart giving important facts about the planets.

4 ● Illustrate several different constellations by creating a mini-planetarium with a light box, flashlight, or diorama.

5 ● Write a rebus story around a space-related concept.

6 ● Construct a simple working model of a rocket and rocket launcher.

7 ● Invent a new space toy. Create an advertisement to promote it.

8 ● Design a booklet of riddles based on your study of space and the solar system.

9 ● Make a mobile of instruments used to observe and measure the stars and other celestial phenomena.

10 ● Compile a scrapbook of articles, jokes, cartoons, and pictures about the space program.

Assessment Product and Performance Options

Page 2

PERFORMANCES

1 ● You are an alien who has just returned from Earth to your home planet. Create a series of "Earthling Reports" similar to the one suggested here and deliver them before a live audience to get their reactions.

CAN YOU TELL WHAT IS BEING DESCRIBED?

- **Earthlings are looking up at the sky with their eyes closed. (A group of sun bathers at the beach).**
- **Earthlings are dressed identically and fighting with each other. (Players at a football game).**
- **Earthlings have red faces as some shiny, colorful things grow from their mouths. (Blowing up a balloon).**

2 ● Plan a mini-lesson to teach a group of students about the concepts of **revolution** and **rotation.**

3 ● Role play the situation of a guide who conducts a group of students through a planetarium.

4 ● Give a "space talk" about your favorite astronaut or space flight.

5 ● Imagine yourself as a famous astronomer who has just discovered a new solar system in the universe. Describe your findings to a group of excited newspaper reporters.

6 ● Write a story entitled "A Star Is Born," and share it with a group of students in a story-telling session.

7 ● Plan and perform a science demonstration to teach others about the concept of **gravity.**

What Will the Future Hold?

As recently as fifty years ago most people did not believe that humans would walk on the moon in this century. Write a story describing how you think space exploration will develop in the next fifty years. Be as imaginative as possible, but include factual details and historical data to make your story interesting and believable.

Name ______________________

EXPLORING

Plants

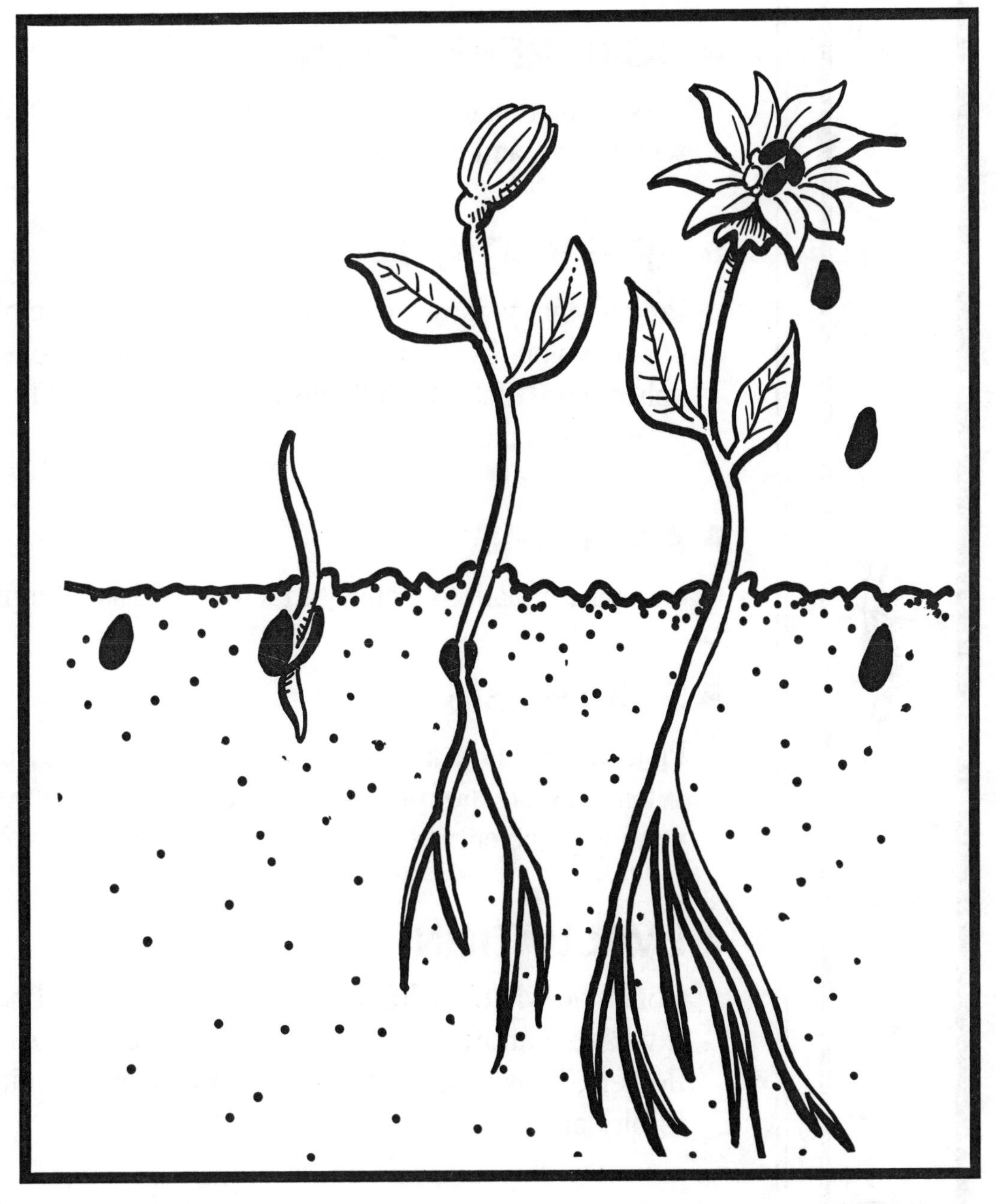

Exploring Plants à la Bloom

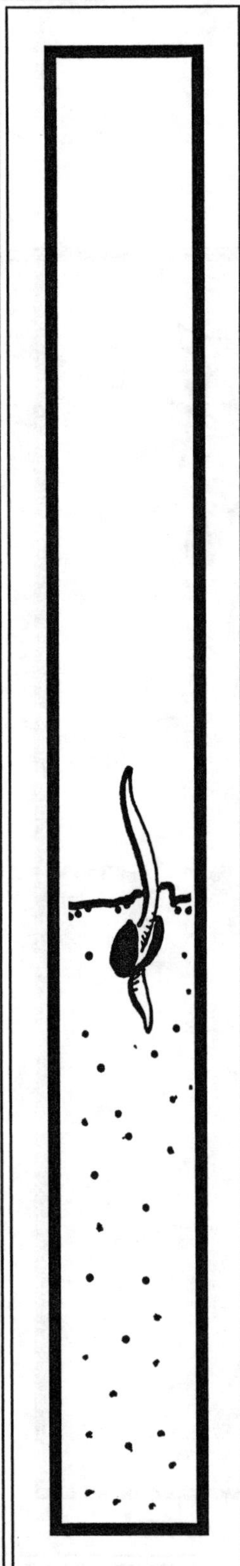

• KNOWLEDGE

Draw several illustrations that show how to grow a particular type of plant. Label your illustrations.

• COMPREHENSION

Make a dictionary to define common terms associated with plants, and give examples of each term.

• APPLICATION

Pretend that you own a valuable species of house plant. One day you discover the plant is wilting and looks very unhealthy. Think of ten things that you might do to try to save your plant.

• ANALYSIS

Compare a plant to a child, a factory, and a community.

• SYNTHESIS

Plan and organize a special plant show. Who may enter? What categories of plants will you display? How will you judge the entries? What prizes will you award?

• EVALUATION

Conclude what our life and our economy would be like if we were allowed to grow only edible plants. All flowers, house plants, and decorative plants would be forbidden.

Multiple Intelligences Can Help Us Learn More about Trees

• VERBAL/LINGUISTIC

Write a short essay on this topic: "Trees Are Important to Us Because . . . "

• LOGICAL/MATHEMATICAL

Make a list of as many products as you can think of that are made from trees. Decide which of these products are necessities and which are luxuries.

• VISUAL/SPATIAL

Draw a diagram to show the role played by trees in the oxygen cycle.

• BODY/KINESTHETIC

Examine a tree in detail. Record your observations on its size, leaf color, bark color, leaf design, branch arrangement, environment, distinguishing characteristics, and living inhabitants.

• MUSICAL/RHYTHMIC

Read aloud several poems about trees with a background of appropriate music. Analyze these poems to discover what poets find so appealing about trees. Describe your findings while the background music is still playing.

• INTERPERSONAL

Work with a partner to select a hypothetical tree and one of its probable locations—a palm tree in Florida, an evergreen in Maine, or a redwood in California. Make inferences about the events the tree might witness in its lifetime, such as historical happenings, national catastrophes, ecological events, human progress, and everyday occurrences.

• INTRAPERSONAL

Determine which tree is most important to you and humankind. Support your choice with personal feelings, opinions, observations, and facts.

Using Williams' Taxonomy to Study Flowers

- **FLUENCY**

  Write down the names of as many different flowers as you can think of in a reasonable period of time.

- **FLEXIBILITY**

  List the many real-life situations or occasions when these flowers might be used or displayed.

- **ORIGINALITY**

  Invent a new flowering plant and describe it in detail. Give reasons for your choice.

- **ELABORATION**

  Many flowers have unusual names such **black-eyed Susan, bleeding heart,** and **buttercup.** Think of several other plants with unusual names and illustrate each one in a creative way.

- **RISK TAKING**

  Decide which flower would best "describe" your personality. Support your choice with examples.

- **COMPLEXITY**

  Comment on the observation that "weeds are just flowers that grow in the wrong place."

- **CURIOSITY**

  If you could interview a florist or horticulturist, what would you like to know? Make a list of questions.

- **IMAGINATION**

  Visualize yourself as a flower in one of the most beautiful gardens in the world. Describe your life, including the kind of attention you get.

Research Topics

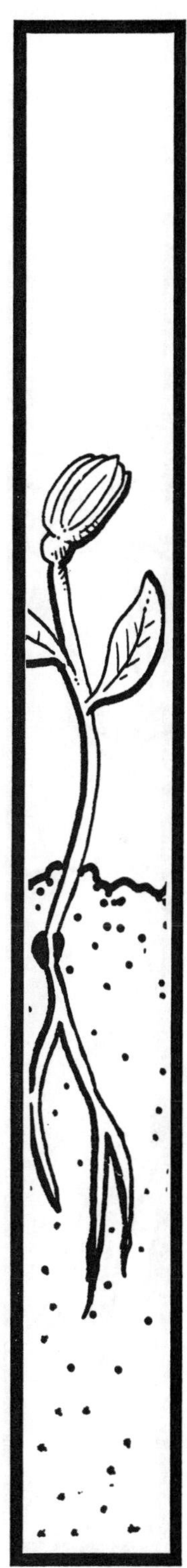

1 ● Find out about **monocots** and **dicots.**

2 ● Find out about the germination of seeds.

3 ● Find out about the parts and functions of a tree or plant.

4 ● Find out about **photosynthesis.**

5 ● Find out about **pollination** and **fertilization** of flowers.

6 ● Find out how seeds travel.

7 ● Find out about parts of plants that provide us with food.

8 ● Find out about the life cycle of a seed plant.

9 ● Find out about the proper care or nourishment of plants.

10 ● Find out about **fungi** and **molds.**

11 ● Find out about **annuals, biennials,** and **perennials.**

12 ● Find out about the anatomy of a plant cell.

13 ● Find out about **cone-bearing plants** and **spore-boring plants.**

14 ● Find out about **growth rings** on trees.

15 ● Find out about the enemies and diseases of plants and trees.

16 ● Find out about famous ports and harbors of the world.

17 ● Find out about medicine trees.

Creative Writing Topics

1 ● Use one of the following titles to write a creative story:

- **The Travel Tales of Susie Seed**
- **The Flower That Failed to Bloom**
- **The Lost Weed**
- **Ode to a Tree**
- **The Secrets of My Green Thumb**
- **The Time Mother Nature Fooled Us All**

2 ● Write an editorial from the point of view of a tree that is about to be cut down in order to build a housing development.

3 ● Pretend that you are a tree of the tropics. Consider a coconut palm tree in the South Pacific, a banana tree in the African jungle, a rubber tree in the East Indies, or a coffee tree in South America. Write a story about your life in those surroundings. Discuss your problems, adventures, dangers, and lifestyles that might occur in that environment.

4 ● Create a series of tongue twisters about popular flowers. Example: Daisies dare to dance with dignity during dark days and dawdle with daffodils on December days.

5 ● Create a haiku booklet about flowers, plants, and trees in nature.

6 ● Create a "Greek myth" that tells why leaves change color or why flowers drop their petals or why weeds grow wild.

Assessment Product and Performance Options

Page 1

PRODUCTS

1● Develop a glossary of terms related to the world of plants and trees.

2● Create a menu for a special garden party at which you serve only edible plant foods such as berries, roots, leaves, plants, fruits, bulbs, and mushrooms.

3● Write a report on the **sassafras** tree, which has been called the "everything tree." Then design your own "everything tree" that provides for all your needs.

4● Design the ideal garden of plants and flowers for your home or school. Draw it and describe it in detail.

5● Use the Yellow Pages® of the telephone directory and list the many occupations that you find which are related to the world of plants, flowers, and trees.

6● Design a set of postcards or greeting cards that use plants in their design and flowery words in their messages.

7● Create a series of wallpaper, wrapping paper, or fabric designs using different flowers or plants.

8● Construct a terrarium environment using a glass or plastic container and collected materials from a wooded area near your home or school.

9● Grow a plant and maintain a growth journal.

10● Measure the circumference of various trees using a piece of string. Then convert these measurements to various English and metric measures.

Assessment Product and Performance Options

Page 2

PERFORMANCES

1. Organize a nature walk for your school or community to teach others about the plants and trees in your area.
2. Collect a variety of leaves and flowers and press these specimens between waxed paper to preserve them. Mount these on cardboard and use them as a basis for an informal "show and tell" session for your classmates.
3. Create a photo essay of flowers, plants, and trees in your area and prepare an audiotape to accompany your album.
4. Design a game to tell others about your study of plants and trees. Teach others to play it and learn from it.
5. Create a dance or drama that focuses on the different concepts and behaviors of varied plant life.
6. Give a speech to persuade others that a vegetarian diet is a good way to eat.
7. Locate and memorize the Robert Louis Stevenson poem "I Think That I Shall Never See a Poem as Lovely as a Tree," and use it as a basis for creating a choral reading.
8. Construct pine cone puppets or critters and use these to discuss the differences between pollen cones and seed cones.
9. Write a one-act play entitled "Flower Power" and perform it.
10. Create a field guide for local plants and use it to give a lecturette to other classes in the school.

The Spice of Life

Page 1

Poetry, prose, art, music, history, ocean journeys, world exploration, business empires, and diets have been influenced over the ages by the quest for new and exotic herbs and spices. Spices have always been prized for their wonderful smells, tastes, and "extraordinary" qualities.

Some spices are derived from the bark of trees, some from plant leaves or fruit, and some from bulbs. Use reference materials to find information about the spices below. Check the appropriate box to show the plant source of each spice. Then draw a small picture of the herb or spice in the box.

Cinnamon

☐ Bulb ☐ Leaves

☐ Bark ☐ Fruit

Cloves

☐ Bulb ☐ Leaves

☐ Bark ☐ Fruit

Dill

☐ Bulb ☐ Leaves

☐ Bark ☐ Fruit

Garlic

☐ Bulb ☐ Leaves

☐ Bark ☐ Fruit

Name ______________________________

The Spice of Life

Page 2

ॐॐॐॐॐॐॐॐॐॐॐॐॐॐॐॐॐॐ

Ginger

☐ Bulb ☐ Leaves

☐ Bark ☐ Fruit

Mint

☐ Bulb ☐ Leaves

☐ Bark ☐ Fruit

Nutmeg

☐ Bulb ☐ Leaves

☐ Bark ☐ Fruit

Parsley

☐ Bulb ☐ Leaves

☐ Bark ☐ Fruit

Red Pepper

☐ Bulb ☐ Leaves

☐ Bark ☐ Fruit

EXPLORING

Rocks and Minerals

Exploring Rocks and Minerals à la Bloom

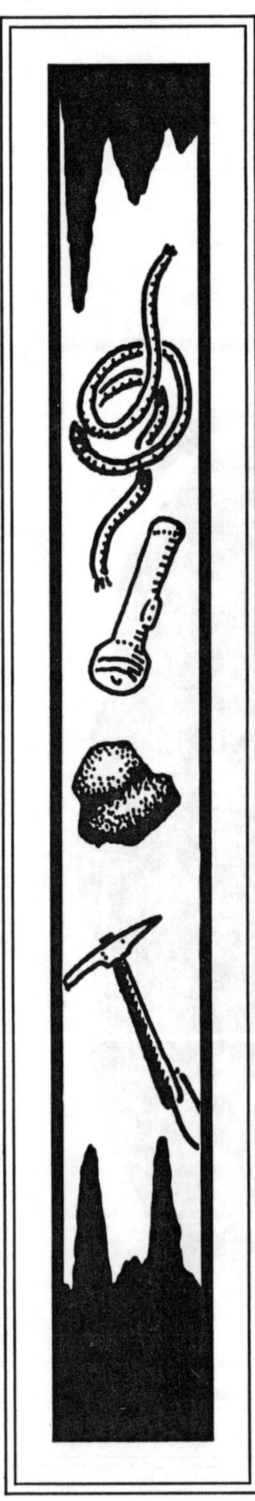

• KNOWLEDGE

Collect ten rocks and identify where each was found.

• COMPREHENSION

Explain how the following types of rocks were formed: **igneous, metamorphic,** and **sedimentary.**

• APPLICATION

Arrange the ten rocks from the Knowledge activity in a display to share with your class. Label each rock and write a brief description of the rock's characteristics and origin.

• ANALYSIS

Compare and contrast the importance of a rock as seen through the eyes of a geologist, a landscape designer, and a construction worker.

• SYNTHESIS

Write an obituary for a rock. Include where and how the rock was "born," the important events of the rock's "life," and how the rock "met its end."

• EVALUATION

Develop a plan that you would use if you were going on a rock and mineral expedition. Include what you would take, where you would go, the tasks you would plan to accomplish, and the problems you might encounter. Establish a set of criteria to use in determining if the expedition was a success or not.

Multiple Intelligences Can Help Us Learn More about Rocks and Minerals

- **VERBAL/LINGUISTIC**

  Write a paragraph describing the following elements of the earth's surface: **crust, mantle,** and **core**.

- **LOGICAL/MATHEMATICAL**

  Construct a chart that compares and contrasts rocks, minerals, and fossils as they relate to the planet Earth.

- **VISUAL/SPATIAL**

  Draw a diagram showing the various layers of the soil.

- **BODY/KINESTHETIC**

  Act out the various processes of weathering and erosion.

- **MUSICAL/RHYTHMIC**

  Use appropriate sounds and music to enhance a report on how ancient people made up stories (called myths and legends) to explain the happenings in their environments.

- **INTERPERSONAL**

  Organize a panel discussion to talk about the "before and after" of key geological events.

- **INTRAPERSONAL**

  Maintain a geology journal or learning log of special questions, observations, feelings, opinions, and facts that you have accumulated throughout your study of planet Earth.

Using Williams' Taxonomy to Study the Earth's Surface Page 1

• FLUENCY

List as many ways as you can think of that people have changed Earth's surface.

• FLEXIBILITY

Divide your items from the list above into two categories: **Ways People Have Improved Earth's Surface** and **Ways People Have Damaged Earth's Surface.**

• ORIGINALITY

Write down the most unusual thing that someone has done in your community to change Earth's surface for better or for worse.

• ELABORATION

Do some research to determine why each of the following locations is considered to be a "geological wonder of the world": **San Andreas Fault, Petrified Forest, Crater Lake, Mammoth Cave, Glacier National Park.**

Using Williams' Taxonomy to Study the Earth's Surface Page 2

• RISK TAKING

Complete the following statement and give details to support your response: "If I could change Earth's surface and the environment in which I live now, I would change . . . "

• COMPLEXITY

"In order for a cultural environment to grow, it must assimilate part of the natural environment." Decide whether you agree or disagree with this statement and justify your response.

• CURIOSITY

Survey the surface of Earth around you to discover something that looks ugly to you and something that looks beautiful to you.

• IMAGINATION

Imagine you could be an environmentalist with magical powers. Describe the things you would want to do.

Research Topics

1 ● Find out about growing crystals.

2 ● Find out about the **porosity** of soils.

3 ● Find out about **stalactites** and **stalagmites.**

4 ● Find out about **glaciers** and **the Ice Age.**

5 ● Find out about the identification of rocks and minerals.

6 ● Find out about prehistoric plants and animals.

7 ● Find out about how Earth was formed.

8 ● Find out about **geysers.**

9 ● Find out how wind and rain carve rocks.

10 ● Find out about the **scale-of-hardness** of minerals.

11 ● Find out about precious gems and stones.

12 ● Find out about the composition of the earth.

13 ● Find out about metals and ores.

14 ● Find out about the multiple uses of rocks.

15 ● Find out about careers in geology.

Creative Writing Topics

1 ● Write a short story using one of the following titles:

- **Traveling the Rocky Road**
- **The Mysterious Vibrations in the Earth**
- **A Hidden Cave**
- **Discovering the Mystery of the Missing Fossil**
- **The Secret Gem**
- **The Return of the Dinosaurs**

2 ● Compose a series of "What Am I?" riddles about different rocks and minerals.

3 ● Write an original "myth" or "legend" that explains one of Earth's many secrets.

4 ● Role-play a set of scenarios to show methods of soil conservation.

5 ● Compose a series of haiku poems about the geological wonders of the earth.

Assessment Product and Performance Options

Page 1

PRODUCTS

1. Develop a field guide for identification of common rocks and minerals.
2. Construct a timeline to show the various stages in the evolution of Earth.
3. Create a museum display about erosion and weathering.
4. Compile a scrapbook of pictures showing various ways that man has altered Earth's surface over time.
5. Design a bulletin board that shows how the world is changing shape.
6. Research and prepare a wordfind puzzle of Earth words.
7. Construct a mobile showing various products that "come from the ground."
8. Draw a series of illustrations to tell the story of **caves.**
9. Grow a crystal garden and maintain a log of your activities.
10. Maintain a Rock Chart that shows the different kinds of sedimentary, igneous, and metamorphic rocks found in the earth and important information about each one.

Assessment Product and Performance Options

Page 2

PERFORMANCES

1 ● Perform a series of identification tests on a collection of unknown rocks and minerals. Consider the acid test, the streak test, the hardness test, and the crystal shape test.

2 ● Plan and conduct an experiment to show the process of weathering or erosion.

3 ● Organize a field trip to a rock quarry and share your plans with a group of students.

4 ● Give a talk about rock collecting as a hobby.

5 ● Conduct a flash card drill exercise on important terms and concepts learned in your study of earth science.

6 ● Organize a debate to discuss the pros and cons of mining activities on our environment today.

7 ● Write and act out a script depicting the eruption of a volcano.

8 ● Make up an imaginary conversation between the three classifications of rocks.

9 ● Prepare an oral explanation of how glaciers are formed.

10 ● Stage a geology spelling and pronunciation bee.

Household Survey

Draw pictures below to show three rocks or minerals in and around your home. Write the use or purpose of each one below the picture.

Make an X in the corner of the box of the most expensive one.

Make an 0 in the corner of the box of the most useful one.

Make a % in the corner of the box of the one you use most often.

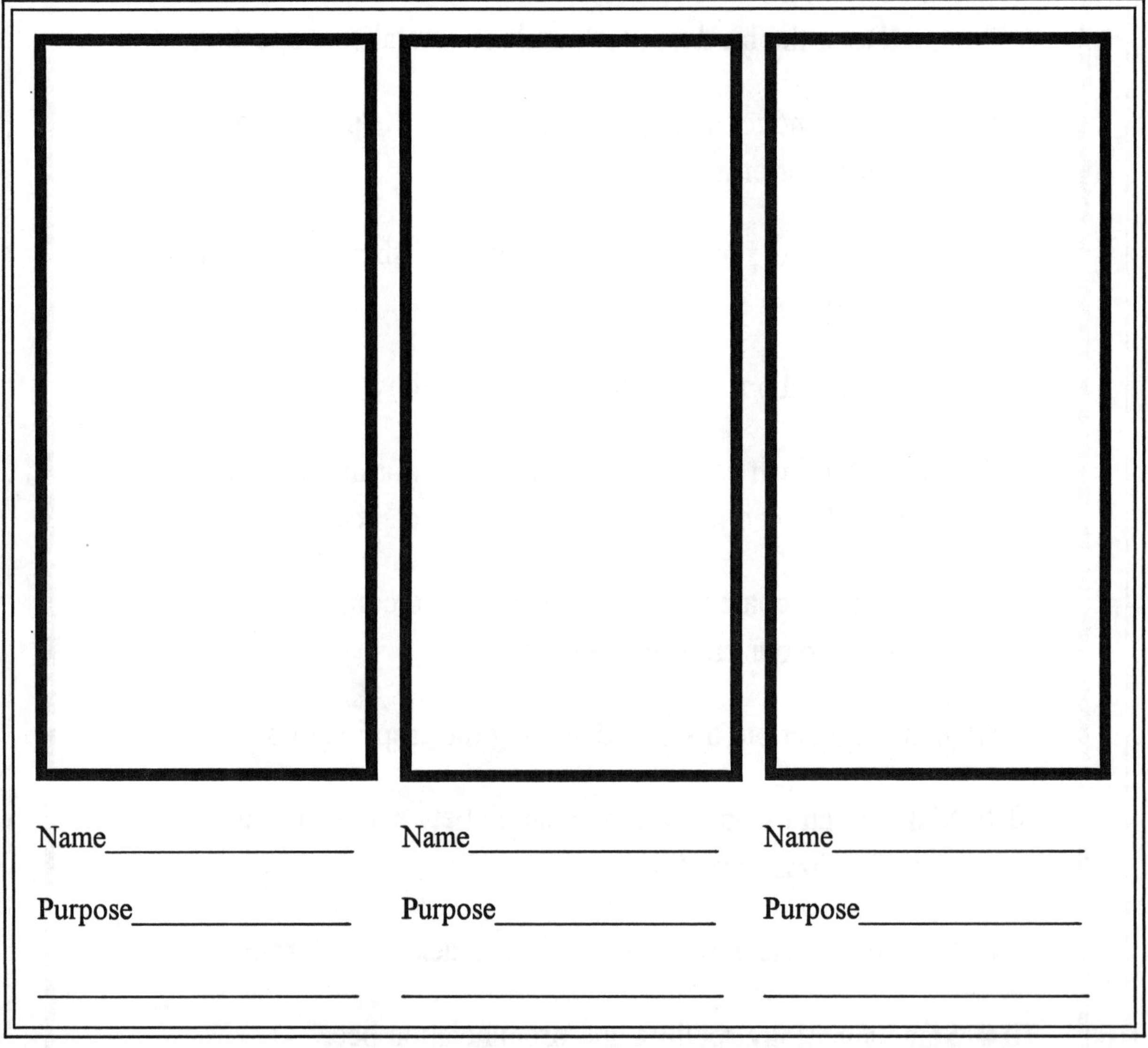

On a separate sheet of paper, design a new household product to be made from rock. Tell the kind of rock from which it will be made, what its use will be, and the approximate cost of the product.

Name ______________________________

EXPLORING

The Weather

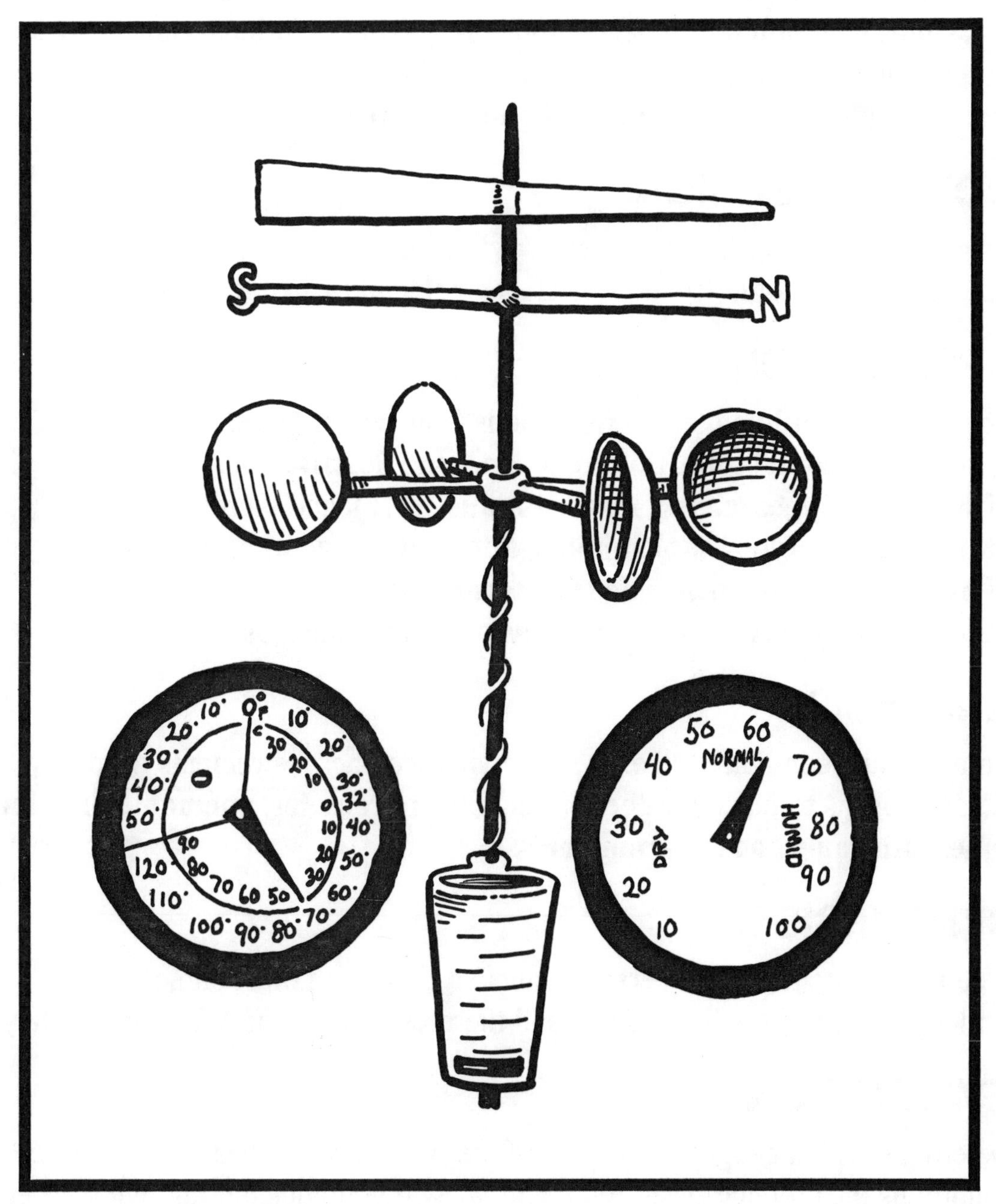

Exploring the Weather à la Bloom

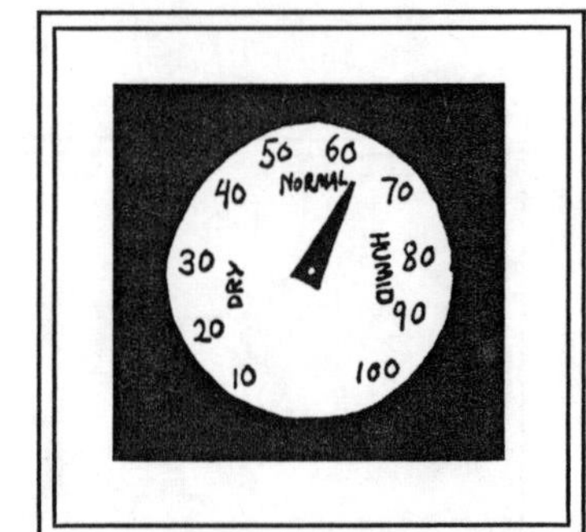

• KNOWLEDGE

Compile a glossary of the following weather-related terms and their definitions: **fog, snow, dew point, hail, wind, lightning, front, condensation, convection, current, evaporation, forecast, frost, humidity, relative humidity, meteorologist, precipitation, wind chill, weather,** and **climate.**

• COMPREHENSION

Using your own words, explain each of the following important weather concepts: **Hydrologic Cycle, Beaufort Scale,** and **Coriolis Effect.**

• APPLICATION

Construct four different cloud formations from construction paper, magic markers, and cotton balls. Be sure to label and describe each one:

- High Cloud Types: **cirrus, cirrocumulus (rare),** and **cirrostratus**
- Middle Cloud Types: **altocumulus, altostratus,** and **nimbostratus**
- Low Cloud Types: **stratus** and **stratocumulus**
- Clouds Through All Levels: **cumulus** and **cumulonimbus**

• ANALYSIS

Compare and contrast each of the following weather instruments used by meteorologists to make weather predictions: **barometer, anemometer, wind vane, rain gauge,** and **hygrometer.**

• SYNTHESIS

Pretend you are a weatherperson in ancient times and must come up with explanations for the falling of hail and the presence of fog. What would they be?

• EVALUATION

Determine which geographic region of the world has the best weather or climate conditions on a regular basis. Which areas will you consider and what criteria will you use? Be able to defend your decision.

Multiple Intelligences Can Help Us Learn More about Weather Phenomena

Page 1

● VERBAL/LINGUISTIC

Throughout history, we have collected weather sayings and beliefs that are based either on superstition or on science. A list of these is given below. Select any five of them and write a paragraph about each one, speculating on how or why it might be true.

WEATHER LORE

1. Red sky in the morning, sailors take warning; red sky at night, sailors' delight.
2. Long foretold, long last; short warning, soon past.
3. Birds suddenly sit on a wire, pack more, or roost before a storm.
4. If the moon shows a silver shield, be not afraid to reap your field, but if she rises haloed round, soon we'll tread on rainy ground.
5. Aching corns and other pains can forecast bad weather.
6. A groundhog leaves his den on February 2nd. If he sees his shadow, it means that cold weather will continue for six more weeks.
7. Dew on grass at night or in early morning is a sign of fair weather.
8. A veering wind, fair weather; a backing wind, foul weather.
9. When the wind backs and the barometer falls, then be on your guard against gales and squalls.
10. Smells are stronger before rain.
11. Duck hunters know that ducks fly higher in good weather than in bad.
12. When distant land across salt water seems closer, it usually means rain the next day.
13. Ants traveling in straight lines means rain is coming; ants scattering means clear weather.
14. Crickets act like a thermometer, chirping faster as temperature rises. Adding 37 to the number of chirps in 15 seconds will give you the temperature in Fahrenheit degrees.
15. The wider the middle brown band on the woolly bear caterpillar, the milder the winter will be.

. . . Weather Phenomena Page 2

● LOGICAL/MATHEMATICAL

Use the Weather Symbol Chart below and the Weather Observation Chart on page 123 to keep track of the weather for the next week. In addition to writing actual measurements in some of the spaces on the Weather Observation Chart, you may draw symbols or abbreviations.

WEATHER SYMBOLS

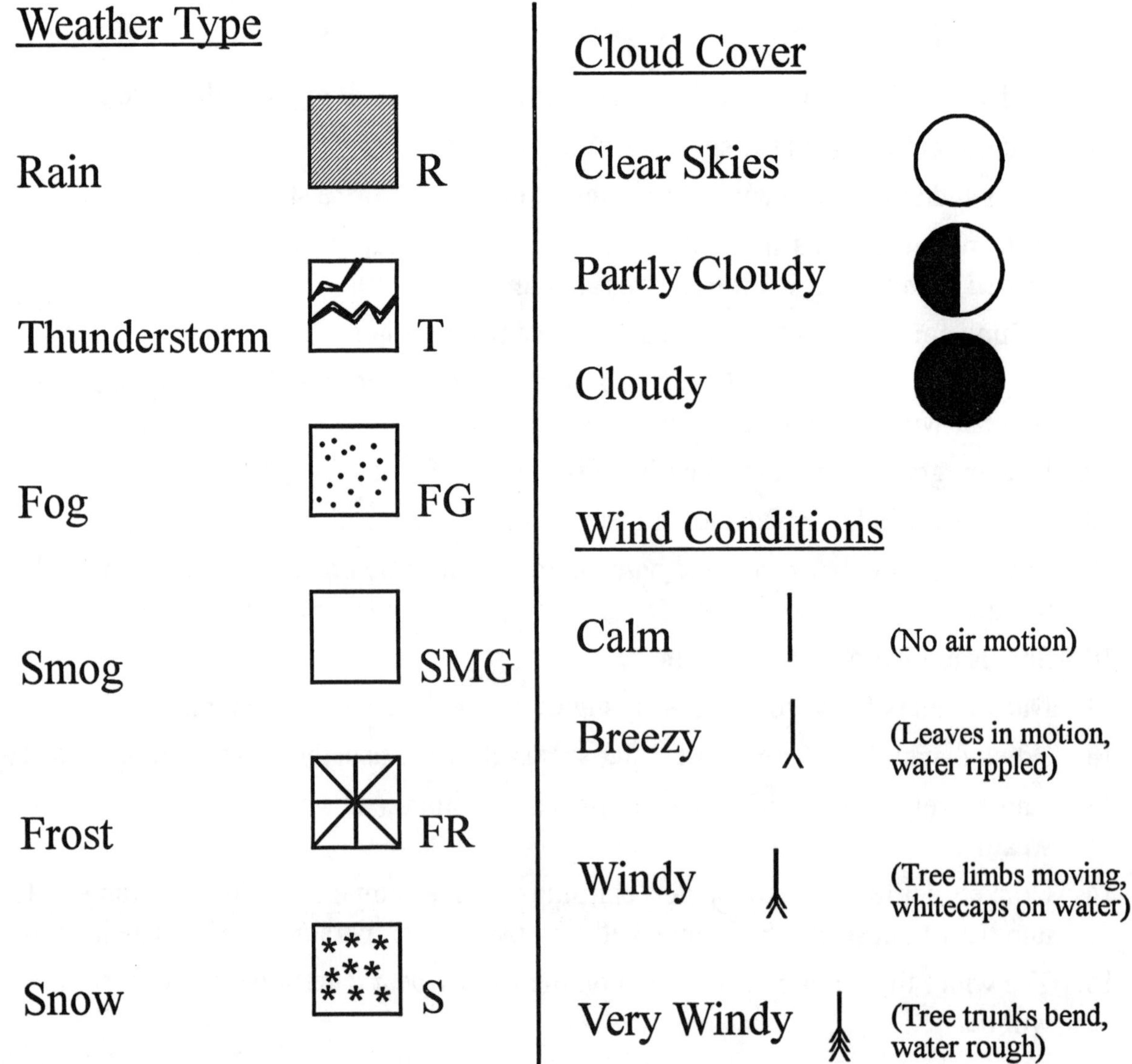

. . . Weather Phenomena

Page 3

WEATHER OBSERVATION CHART

Date	Time	Pressure	Temperature	Humidity	Cloud Cover	Weather Type	Wind Conditions	Wind Chill Temperature	Forecast

Name ______________________________

. . . Weather Phenomena Page 4

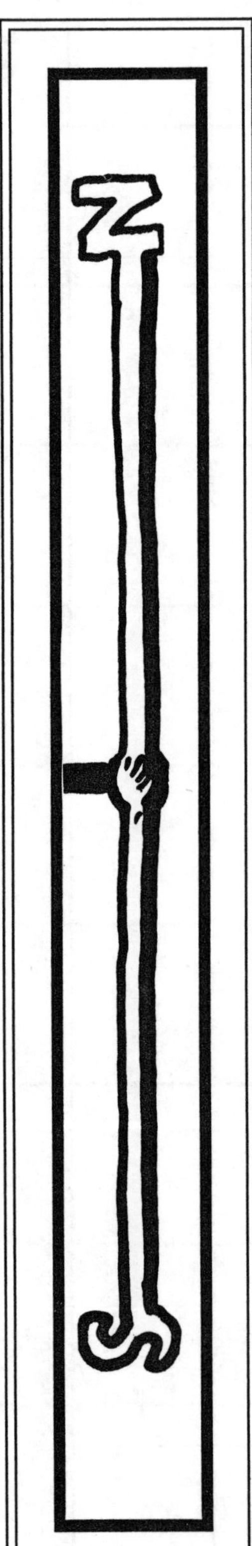

• VISUAL/SPATIAL

Make a diagram of a lightning rod. Write a brief description of how it works. Next, construct a chart of plants and trees that either attract lightning or repel lightning. Consider size, location, and why some attract lightning and why some repel it.

• BODY/KINESTHETIC

Use a prism to create a rainbow and describe what you see. Make a wish on the rainbow and write down the wish.

• MUSICAL/RHYTHMIC

Pretend you are a professional rainmaker of ancient times. Create a rain dance, speech, chant, and script for a rainmaking ceremony.

• INTERPERSONAL

Work with a partner to develop a short presentation on the interpretation and use of the **Beaufort Wind Scale** for estimating the speed of wind.

• INTRAPERSONAL

Keep a weather journal for a week and describe how the weather is often responsible for physical and emotional changes in yourself.

Using Williams' Taxonomy to Study Weather Disasters

• FLUENCY

List as many different weather disasters as you can think of that have occurred over the last few years. Consider **storms, hurricanes, tornadoes, earthquakes, droughts, blizzards, volcanoes,** and **floods.**

• FLEXIBILITY

Write down other types of disasters that are often results of such weather conditions as those listed above.

• ORIGINALITY

Compose a short story about an unpleasant weather condition that you or someone else you know once encountered.

• ELABORATION

Respond to this question: "How is a storm like a battle?"

• RISK TAKING

Explain how you are most like a tornado, hurricane, drought, earthquake, storm, blizzard, volcano, or flood.

• COMPLEXITY

Determine how some people have actually benefited from weather disasters.

• CURIOSITY

Ponder the question of how weather has influenced folktales and folklore over the years.

• IMAGINATION

Visualize a greeting card that you might send to your favorite meteorologist after a weather disaster, and then draw it.

Research Topics

1 ● Find out what causes earthquakes and what causes volcanoes.

2 ● Find out what causes storms and blizzards.

3 ● Find out what causes tornadoes and hurricanes.

4 ● Find out what causes floods and what causes droughts.

5 ● Find out how rainbows are formed.

6 ● Find out what special training is required to become a meteorologist.

7 ● Find out what weather terms are especially important to airline pilots.

8 ● Find out the multiple causes of evaporation and condensation.

9 ● Find out how clouds are formed.

10 ● Find out about sea and land breezes.

11 ● Find out how fog is formed.

12 ● Find out how to track a weather system.

13 ● Find out how to use various weather instruments.

Creative Writing Topics

1 ● Write a short story that has one of the following titles:

- **The Day of the Blizzard**
- **A Drought to Remember**
- **Beaufort's Bonanza**
- **Big Raindrops and Puddle Play**
- **Magical Weather Friends**
- **Weather Wisdom**

2 ● Draw a picture to show what a storm looks like when it rolls in, and compose a poem about it.

3 ● Plan and perform a short skit based on a weather legend.

4 ● Personify a rainbow.

5 ● Compose a myth or fable that has a "windy" individual as its main character.

6 ● Create your own myth about thunder.

Assessment Product and Performance Options

PRODUCTS

1 ● Write a report describing how each of the following plants and animals can and do help predict weather through their reactions to certain conditions: ants, squirrels, flies, groundhogs, sunflowers, cats, horses, sheep, frogs, deer and elk, and birds.

2 ● Construct a model to demonstrate the **Greenhouse Effect.**

3 ● Make a graph of the temperatures in your neighborhood for a week.

4 ● Summarize the types of information one can find about weather in the *Farmer's Almanac.*

5 ● Combine five important facts about air masses or fronts into one statement, and use that statement as the springboard for a learning poster on the topic.

6 ● Design and construct a home weather station.

7 ● Make a wordfind or crossword puzzle of the terms and instruments used by meteorologists.

8 ● Design an interior lounge for an airport using weather or clouds as the theme.

PERFORMANCES

1 ● Create an original dance that depicts the feelings and movements of a volcanic eruption and perform it for the class.

2 ● Assume the role of a meteorologist for a mock television news broadcast.

3 ● Prepare an oral report on how to prepare for a weather disaster such as a tornado, hurricane, blizzard, or earthquake.

4 ● Invent a hypothetical weather disturbance and use a variety of media to share it orally with the rest of the class.

5 ● Assume the role of a Roman god from mythology and explain how you control the elements of the weather.

6 ● Dramatize one or more of the popular stories, myths, legends, or fables that are associated with the weather.

Record a Tornado

Use reference materials to research the development, appearance, and effects of a tornado. Use the information to make a pictoral record of a tornado and its consequences.

●1●	●2●
●3●	●4●

Name ______________________

EXPLORING

Things That Fly

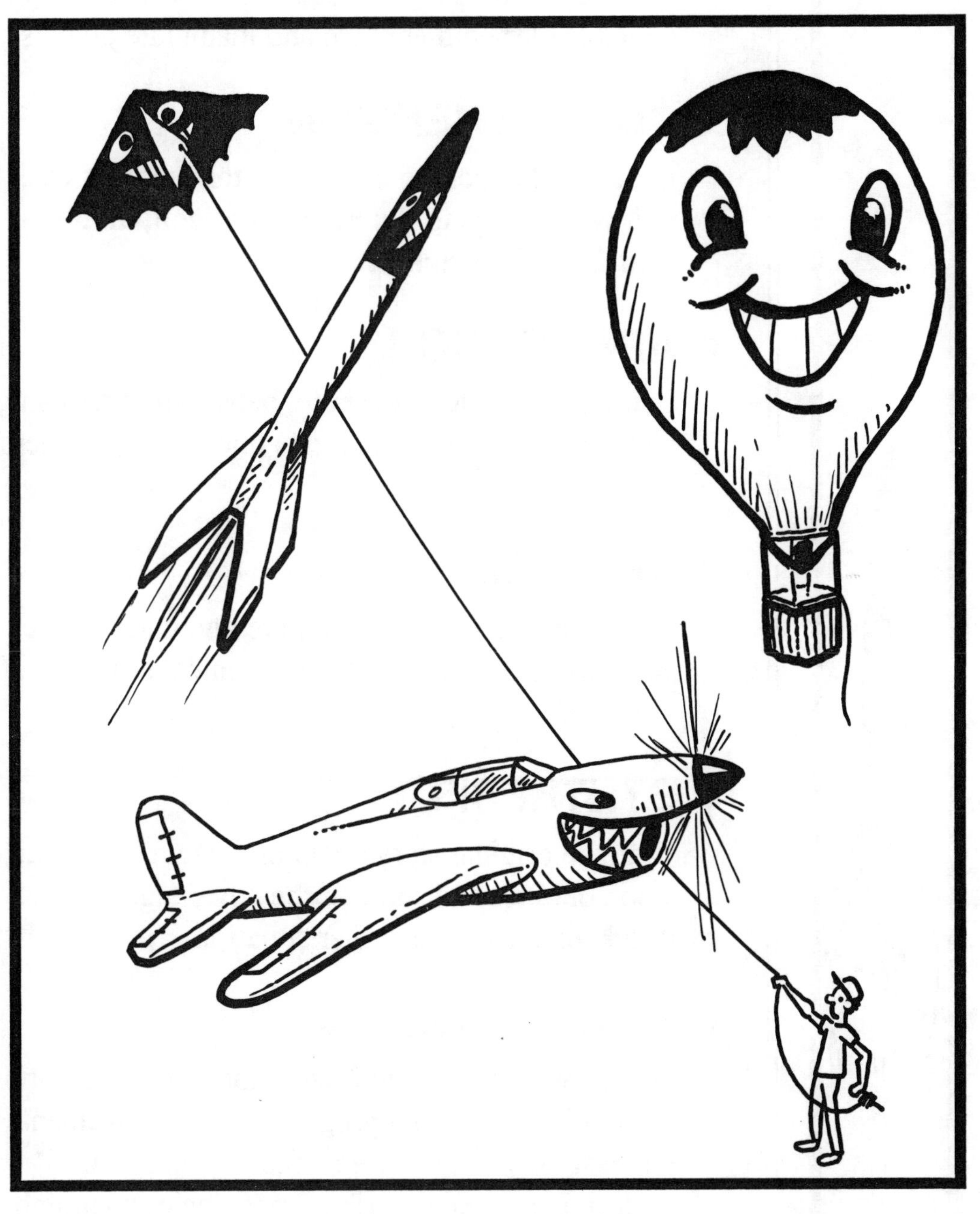

Exploring Things That Fly à la Bloom

• KNOWLEDGE

Name as many things as you can think of that fly in the air. Consider both animate and inanimate objects.

• COMPREHENSION

Describe the method or process that enables each of the following things to fly: **hot air balloon, airplane, space shuttle,** and **parachute.**

• APPLICATION

Use a simple kite, parachute, balsa-wood airplane, or balloon design to demonstrate one or more principles of flight.

• ANALYSIS

Use a Venn diagram to compare and contrast the anatomy of a bird with the mechanical structure of an airplane.

• SYNTHESIS

Write an original story about an imaginary character who could fly and some of the advantages and disadvantages of his or her situation.

• EVALUATION

Decide which of the following forms of transportation has made the greatest progress and improvements over the past 50 years in overall design, function, performance, and contribution to society: automobile, train, airplane, or boat.

Multiple Intelligences Can Help Us Learn More about Airplanes

Page 1

● VERBAL/LINGUISTIC

Write an explanation of **aerodynamics** that classmates can understand.

● LOGICAL/MATHEMATICAL

Determine the important problems humans had to solve before they could fly.

● VISUAL/SPATIAL

Create a sensory display of things related to the world of airplanes or aviation. Try to include objects, pictures, or artifacts representative of all five senses.

● BODY/KINESTHETIC

Create a game of Jeopardy®, Trivial Pursuit®, or Pictionary® based on terms and concepts related to your study of airplanes and aviation.

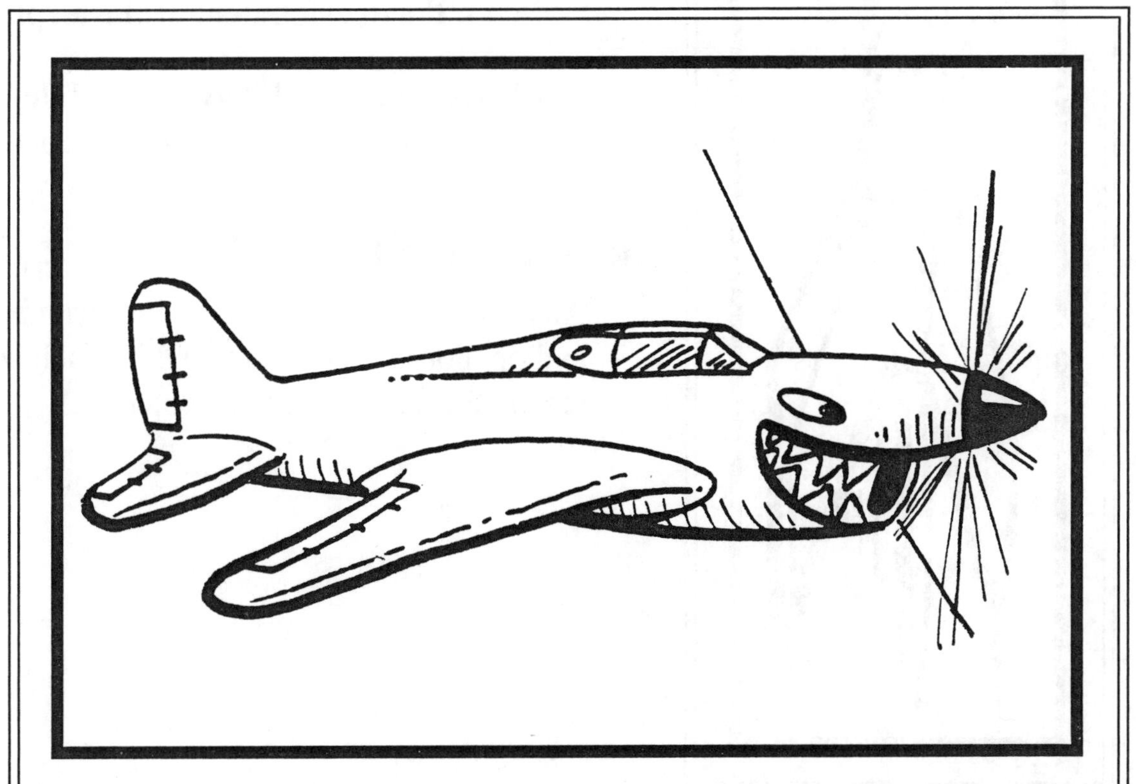

Multiple Intelligences Can Help Us Learn More about Airplanes

Page 2

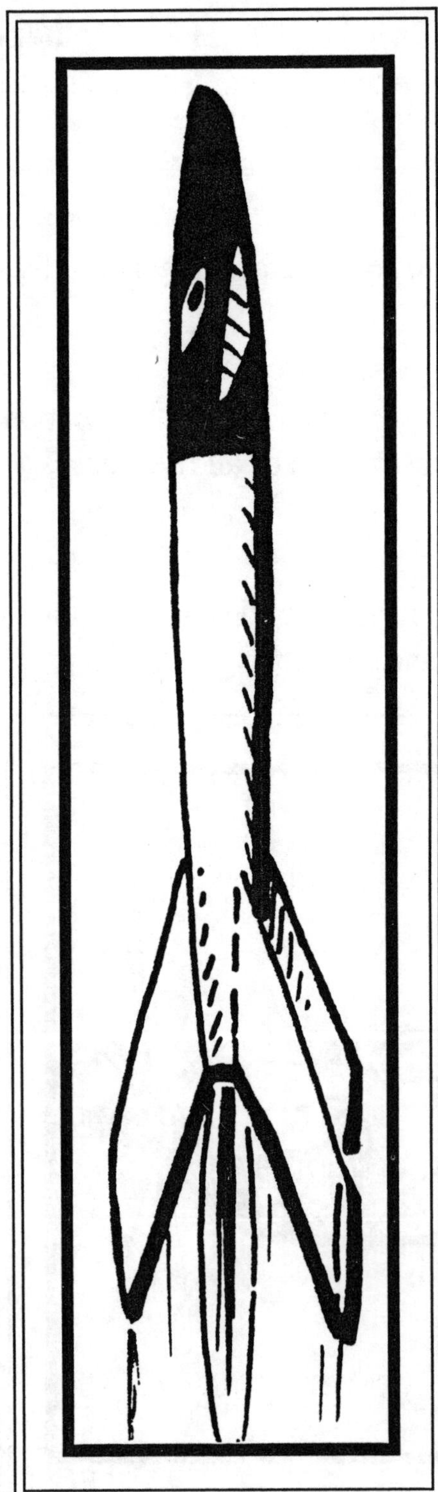

- **MUSICAL/RHYTHMIC**

Compile a collection of songs about aviation or flying.

- **INTERPERSONAL**

Organize a discussion about one of these topics:

- **Ways our lives would be changed if there were no way to fly.**
- **The importance of airplanes to man.**
- **Ways our lifestyles have changed with the development of the airplane.**
- **Inventions which have helped the airplane to improve and progress.**

- **INTRAPERSONAL**

Answer this question using your own ideas and opinions: "What do you think is the real reason the airplane was invented?"

Using Williams' Tax Study the Airport

- **FLUENCY**

  List as many words as you can think of that are associated with the operation of a large, commercial airport.

- **FLEXIBILITY**

  Categorize your list of words in some meaningful and interesting way.

- **ORIGINALITY**

  Think up a new and unusual service provided by an airport that would improve customer service or relations. How about a video arcade that has pilot simulator games?

- **ELABORATION**

  Expand on this quotation about flying: "Airplanes and airports have made the world smaller."

- **RISK TAKING**

  List all the reasons you would—or would not—make a good pilot.

- **COMPLEXITY**

  Determine the qualities of an efficient and user-friendly airport.

- **CURIOSITY**

  Write down a set of questions that you would like to ask an air controller in the world's busiest airport.

- **IMAGINATION**

  Visualize what it would be like to be a doctor who treats patients who have the disease of **acrophobia.** Create a prescription for its cure.

Research Topics

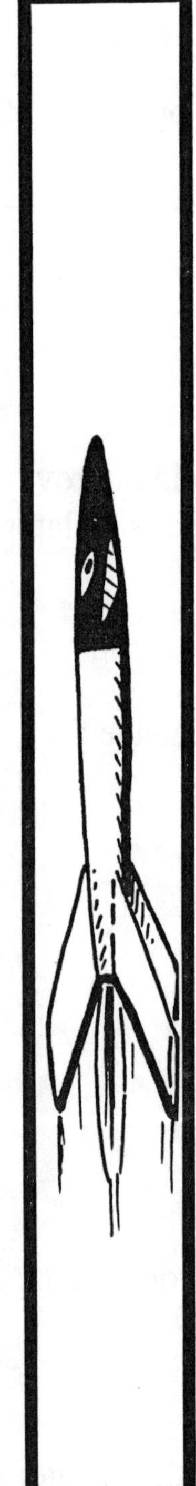

1 ● Find out about an air disaster such as that of the Hindenburg.

2 ● Find out about a famous aviator such as Amelia Earhart.

3 ● Find out about an interesting vocation associated with airplanes such as fire fighting or hurricane scouting.

4 ● Find out about sports associated with things that fly such as hang gliding or stunt flying.

5 ● Find out about the work of a bush pilot or a fighter pilot.

6 ● Find out about **sonic booms** and **sound barriers.**

7 ● Find out about flight regulations governing airplanes.

8 ● Find out about the history of flying.

9 ● Find out about **aeronautics.**

10 ● Find out about UFOs.

Creative Writing Topics

1 ● Pretend you are one of the following individuals, and write a series of diary entries telling about your exciting adventures:

- **An astronaut on a space flight**
- **A pilot of a seaplane on a rescue mission**
- **A stunt pilot on a barnstorming adventure**
- **A pilot of a flying saucer from another planet**
- **A flying ace from World War II**
- **A weather pilot scouting the eye of a hurricane**

2 ● Write a short story that has one of the following titles:

- **The Airplane Flight I Want to Forget (or Remember)**
- **The Day I Took My First (or Last) Flying Lesson**
- **My Most Exciting (or Disappointing) Flying Experience**
- **The Time I Was a Substitute Pilot (or Flight Attendant)**

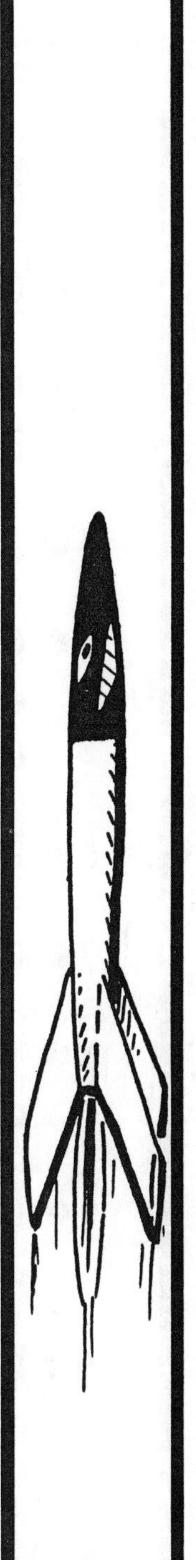

Assessment Product and Performance Options

Page 1

PRODUCTS

1 ● Design a series of comic strips in which the main character is a talking aircraft.

2 ● Create a collage of pictures of airplanes or flying objects.

3 ● Construct an Airplane Hall of Fame book about contributions of individuals to the world of aviation.

4 ● Assemble a model airplane and prepare a report to go with it.

5 ● Conduct a survey of students in your class to determine the extent of their experiences and/or perceptions of air travel. Compile your results.

6 ● Prepare a critique of a book you have read about aviation.

7 ● Design an airplane of the future.

8 ● Tape record a series of mock conversations among air controllers facing a potential air disaster.

9 ● Draw a picture of what your school, neighborhood, or community might look like to a person in an airplane.

10 ● Construct a series of aviation mazes for others to complete.

Assessment Product and Performance Options

Page 2

PERFORMANCES

1 ● In a game of charades, act out a series of words associated with aviation.

2 ● Read a book about airplanes to a younger student, and prepare a set of follow-up discussion questions.

3 ● Prepare a five-minute speech or monologue about a famous person you have learned about in your study of aviation.

4 ● Plan and conduct a Paper Airplane Flying Contest. Give prizes for airplane entries that can fly the farthest, fly the highest, fly the longest, are smallest in size, are largest in size, are most unusual in design or material, and are most decorative.

5 ● Write and perform a skit or play about the adventures of the Red Baron.

6 ● Prepare and present a demonstration to explain one or more of the principles of flight.

7 ● Experiment with several types of paper airplanes to combine the designs of several in order to create a new and better one.

8 ● Design a television commercial for one of the airlines.

9 ● Locate and conduct a series of experiments to demonstrate the properties of air.

Airplane Drama

Write a brief play, from a pilot's point of view, about taking an airplane trip. Include information about parts of the airplane, principles of flight, and the roles of different workers on the ground crew.

Outline your play below.

Characters
Setting (Time and Place)
Problem or Conflict
Special Words, Terms, or Principles to Be Included in Play

Name ______________________________

Topics for Student Reports

ASTRONOMY

The Earth-Moon System
The Solar System
Stars and Galaxies
The Sun

BIOLOGY

Basic Units of Life/Jobs of Cells
Breathing/Respiration
Circulation
Moving the Body (bones, muscles)
The Nervous System
Reproduction/Heredity

BOTANY

How Plants Reproduce
Photosynthesis

CHEMISTRY

Chemical Reactions
Organic Chemistry
The Periodic Table of Elements

ECOLOGY

Acid Rain
Changing Ecosystems
Environmental Issues and Conservation
Global Warming/Greenhouse Effect
Pollution

GEOLOGY

Common Rocks
Earthquakes
Elements and Minerals
Energy Resources
Forces inside Earth
The Rock Cycle
Soil
Volcanoes

METEOROLOGY

Air
Atmospheric Pressure
Weather and Climate

OCEANOGRAPHY

Ocean Floor/Shore Zone 
Water

PALEONTOLOGY

Dinosaurs

PHYSICS

Electricity and Magnetism
Electromagnetic Waves
Gases, Atoms, and Molecules
Heat and Temperature
Light
Machines
Motion
Moving Water
Sound
Structure of the Atom
Thermal Energy
Waves
Work and Energy

ZOOLOGY

Animal Communication
Animal Life
How Animals Reproduce
Interactions in the Living World
Invertebrates
Vertebrates

Annotated Bibliography

This annotated bibliography of Incentive Publications titles was selected to provide additional help in reinforcing science concepts and strengthening thinking skills.

Breeden, Terri. *Cooperative Learning Companion.* Nashville, TN: Incentive Publications, 1992. (Grades 5–8)
A winning collection of creative teaching aids, including reproducible charts, forms, and posters, along with comprehensive instructions for setting up a smoothly running and effective cooperative classroom environment.

Breeden, Terri and Emalie Egan. *Strategies and Activities to Raise Achievement.* Nashville, TN: Incentive Publications, 1995. (Grades 4–8)
Comprehensive manual contains high-interest activities and esteem-building exercises that will motivate students to become more effective test-takers and lifelong learners.

Forte, Imogene and Sandra Schurr. *The Cooperative Learning Guide and Planning Pak for Middle Grades.* Nashville, TN: Incentive Publications, 1992. (Grades 5–8)
A collection of high-interest thematic units, thematic thinking skills projects, and thematic poster projects. Includes reference skills sharpeners and much more.

—. *The Definitive Middle School Guide: A Handbook for Success.* Nashville, TN: Incentive Publications, 1993. (Grades 5–8)
This comprehensive, research-based manual provides the perfect overview for educators and administrators who are determined to establish a school environment that stimulates and motivates the middle grade student in the learning process.

—. *Interdisciplinary Units and Projects for Thematic Instruction for Middle Grade Success.* Nashville, TN: Incentive Publications, 1994. (Grades 5–8)
A jumbo-sized collection of thematic-based interdisciplinary activities and assignments that was created to spark interest, encourage communication, and promote problem solving as well as decision making.

—. *Making Portfolios, Products, and Performances Meaningful and Manageable for Students and Teachers.* Nashville, TN: Incentive Publications, 1995. (Grades 4–8)
Filled with valuable information and specific suggestions for incorporating authentic assessment techniques that help students enjoy a more active role in the evaluation process. Includes a convenient pull-out Graphic Organizer with creative ideas for integrating content instruction and appraising student understanding.

—. *Middle Grades Advisee/Advisor Program.* Nashville, TN: Incentive Publications, 1991. (Grades 5–8)
A comprehensive program dedicated to meeting the needs and confronting the challenges of today's young adolescent students. A flexible, manageable curriculum, available at three different levels, that contains both a teacher's guide and over 300 reproducible activities on essential topics.

—. *Tools, Treasures, and Measures for Middle Grade Success.* Nashville, TN: Incentive Publications, 1994. (Grades 5–8)
This practical resource offers a wide assortment of teaching essentials, from ready-to-use lesson plans and student assignments to valuable lists and assessment tools.

Frender, Gloria. *Learning to Learn.* Nashville, TN: Incentive Publications, 1990. (All grades)
This comprehensive reference book is filled with creative ideas, practical suggestions, and "hands on" materials to help students acquire the organizational, study, test-taking, and problem-solving skills they need to become lifelong effective learners.

—. *Teaching for Learning Success: Practical Strategies and Materials for Everyday Use.* Nashville, TN: Incentive Publications, 1994. (All grades)
This ready-to-use resource has the materials needed to successfully implement cooperative learning techniques, organize and manage the classroom environment, adapt teaching to suit varied learning styles, and promote the home-school connection.

Graham, Leland and Darriel Ledbetter. *How to Write a Great Research Paper.* Nashville, TN: Incentive Publications, 1994. (Grades 5–8)
Simplify the research process with these mini-lessons. Help students choose and narrow topics, locate appropriate information from a variety of sources, take notes, organize an outline, develop a rough draft, document sources, as well as write, revise, and evaluate their final papers.

Opie, Brenda, Lori Jackson, and Douglas McAvinn. *Masterminds Decimals, Percentages, Metric System, and Consumer Math.* Nashville, TN: Incentive Publications, 1995. (Grades 4–8)
These high-interest student activities, designed for today's middle grades science programs, are based on measurement instruction using metric units and decimal values.

Science YELLOW PAGES for Students and Teachers. Nashville, TN: Incentive Publications, 1988. (Grades 2–8)
A virtual "treasure chest" containing valuable facts, lists, charts, and definitions related to science laws and principles, formulas, experiments, investigations, and more.

Spivack, Doris and Geri Blond. *Inventions and Extensions.* Nashville, TN: Incentive Publications, 1991. (Grades 3–7)
A unique resource containing activities based on information about famous inventors and their inventions; designed to expand kids' natural curiosity about how things work, and to stimulate use of critical thinking and problem-solving skills.

Index

R

S

T

U

W